Animal Tissue Techniques

Animal Tissue Techniques

FOURTH EDITION

Gretchen L. Humason

OAK RIDGE ASSOCIATED UNIVERSITIES

W. H. FREEMAN AND COMPANY
San Francisco

Sponsoring Editor: Arthur C. Bartlett
Project Editor: Pearl C. Vapnek
Copyeditor: Susan Weisberg
Designer: Robert Ishi
Production Coordinator: Fran Mitchell
Compositor: Bi-Comp, Inc.
Printer and Binder: The Maple-Vail Book Manufacturing Group

Library of Congress Cataloging in Publication Data

Humason, Gretchen L
    Animal tissue techniques.

    Bibliography: p.
    Includes index.
    1. Histology—Technique. 2. Stains and staining
(Microscopy) 3. Histochemistry—Technique. I. Title.
QM556.H85   1979   578'.9   78-17459
ISBN 0-7167-0299-1

9 8 7 6 5 4 3 2 1

Contents

Part III HISTOCHEMISTRY AND
         MISCELLANEOUS SPECIAL PROCEDURES

Preface

This book of basic and standard histological procedures (and some specialized techniques) was designed to meet the diverse needs of premedical students, medical technicians, zoology majors, and research assistants. Most histological reactions follow a logical and specific sequence, and I have attempted to include simplified discussions of the basic methods that are applicable both to normal and to pathological conditions in zoology and medicine.

It is not intended that this text should be a complete reference book on histology; the experienced worker knows of numerous such tomes, as well as journals that specialize in histology and related topics. However, special methods of wide usage and exceptional merit are included, particularly those that are not overly complicated or unpredictable. It is hoped that technicians, once familiar with the material covered here, will watch the literature for modifications and improvements of standard techniques; in this way, with this book as a foundation, their work can be kept up to date and, perhaps, simplified.

Methods for fixation are fairly well established, with only occasional variations. The section on fixation presented herein is as modern as I can make it, and it includes a brief description of the chemicals employed. Old staining techniques continue to be perfected and new ones developed; I have tried to include the best of these and, for the sake of the student, to adapt them to the standard three-hour laboratory period and to the kinds of equipment most widely available. Some special methods that are more time-consuming have been included for special projects and research. They have been simplified wherever possible to serve as introductory techniques for the student who plans to proceed to more complicated techniques later.

Some instructors may not agree with the way I have organized the text, but to me it is logical. Thus fixation is treated first because it is usually the first process in tissue preparation; this is followed by embedding in some kind of medium, sectioning on a microtome, mounting sections on slides, and, finally, staining them with the help of a microscope. A logical arrangement of staining methods is hard to come by, so I have followed my own inclinations: Some sections are organized by related tissues, others by related methods. The latter was considered desirable for such processes as silver impregnation, metachromasia, and the use of Schiff reagent. The final chapters include such specialized techniques as histochemistry, chromosome preparation, autoradiography, and invertebrate mounts. Wherever possible, I have referred to my own experience with these methods to help students succeed with their first efforts, and I have included modifications that might appeal to other adventurous technicians.

This book is in four parts. Part I covers those basic procedures and general considerations with which every tissue technician should be familiar. Part II provides detailed information about specific staining methods for most tissues. An instructor might choose a few favorite methods from this section to round out a course, while the professional technician will find here most of the specific methods required on the job. Part III deals with special procedures, those that are special in the sense that they are not common in most laboratories, although they may be very important in some. Although the discussion of some of these procedures is brief, references have been cited extensively for the benefit of those who might wish to refer to more thorough discussions. Part IV is devoted largely to laboratory aids and the preparation of solutions—useful information in any laboratory. In this edition some methods are removed and new ones added, others are modified. The arrangement of chapters has been altered by moving freezing, nitrocellulose, and other techniques to the Special Procedures section. This leaves the first two parts devoted almost exclusively to the paraffin method and the use of the microscope to examine the finished slides.

In the fourth edition, *Animal Tissue Techniques* has been extensively revised and updated. Many of the changes have been to improve its usefulness for graduate and undergraduate teaching. The typography has been altered and the design improved with an eye to making the book more readable and, hence, more useful to students and technicians alike. The list of references has been carefully emended to cover recent important publications in the field. Information on all suppliers cited in the text is consolidated for easier reference in one list beginning on p. 574.

Many of us have not regarded with proper respect the potentially dangerous materials that are used in the laboratory. Accidents do happen, so a section is included on pp. 572–574 concerning the hazards of laboratory materials. This should be called to the attention of all students and technicians.

To have included everything necessary to satisfy everyone and still to have kept the price of the book within the means of the average student would have been impossible. Some topics, necessarily, have been treated only in passing. The electron microscope, for example, is much too specialized for students in beginning technique classes, and an entire book could be devoted to instructing students in its operation alone. The topic of photomicrography is equally complex. Methods for preparing plastic whole mounts have not been included; excellent leaflets on the subject are published by the companies that supply the materials necessary for their preparation. Good color photographs are helpful, but they are also, unfortunately, expensive—even a few of them can add appreciably to the cost of a book. In my teaching, I have used a demonstration set of slides to help my students recognize proper staining. The set started with a few of my own slides, and it was gradually enlarged by additions from the students in my classes. The students were happy to contribute examples of their best work, and the collection eventually increased to several hundred excellent slides. Other instructors might consider building a study collection of slides in the same way.

I have derived invaluable personal satisfaction from my association with students. I am grateful to them for helping me to develop my tolerance and patience—two qualities that are essential in my profession. I am grateful to them, too, for what they have helped me to learn, for there is no surer way to master a subject than to teach it to others. One former student in particular should receive credit for her encouragement and for prodding me toward writing this book—Marlies Natzler of the University of California at Los Angeles.

Grateful acknowledgments are also due to Marvin Linke, Jeanne Simmons, and Leta Burleson, the three artists who contributed to the four editions of this book; to Julie Langham, for help with photography; to Nellie M. Bilstad, for valuable suggestions; to the Cytogenetics Division of Oak Ridge Associated Universities, for information about late developments in chromosome preparation; to the Zoology Department of the University of California at Los Angeles, for the lessons I learned there as a student, a departmental technician, and a lecturer; and to Dr. C. C. Lushbaugh, for his continued encouragement.

September 1978   *Gretchen L. Humason*

Part I
BASIC PROCEDURES

Chapter 1
Fixation

As soon as a tissue ceases to be alive, its cells start to change. Multiplying bacteria begin to destroy them, and the process of autolysis (self-digestion) by contained enzymes begins to dissolve them. The activity of these enzymes is reversed from that in live cells; instead of synthesizing amino acids into proteins, they begin to split proteins into amino acids. These amino acids diffuse out of the cells; as a result cell proteins are no longer coagulable by chemical reagents. These cell changes are called postmortem conditions and must be prevented if tissue is to be examined in the laboratory.

The prevention of postmortem conditions is the primary objective of tissue preparation, but it is also necessary to treat tissue to differentiate the solid phase of the protoplasm from the aqueous phase, to change cell parts into materials that will remain insoluble during subsequent treatment, and to protect cells from distortion and shrinkage when subjected to such fluids as alcohol and hot paraffin. Other important objectives of tissue preparation are to improve the staining potential of tissue parts and to alter their refractive indices for better visibility.

The procedure used to meet these requirements is called fixation, and the fluids used are called fixatives or fixing solutions. Any fixative should:

1. Penetrate rapidly to prevent postmortem changes.
2. Coagulate cell contents into insoluble substances.
3. Protect tissue against shrinkage and distortion during dehydration, embedding, and sectioning.
4. Allow cell parts to be made selectively and clearly visible.

Some fixatives may have a mordanting effect on the tissue—that is, they combine insolubly with it—and enhance the attachment of dyes and proteins to each other.

Tissues should be placed in fixatives as soon as possible after death. If delay is unavoidable, they should be put in a refrigerator, thus reducing autolysis and putrefaction to a minimum until the fixative can be applied.

Because a single chemical seldom has all the qualities of a good fixative, a fixing solution is rarely composed of only one chemical—familiar exceptions are formalin and glutaraldehyde. Most reliable fixatives contain one or more coagulant chemicals and one or more noncoagulant chemicals. Coagulants change the spongework of proteins into meshes through which paraffin can easily pass, thus forming a tissue of the proper consistency for sectioning. They also strengthen the protein linkages against breaking down during later procedures. Used alone, however, coagulants may form too coarse a network for the best cytological detail or may induce the formation of artificial structures (artifacts). Noncoagulants produce fewer artifacts, but if used alone they give the tissue a poor consistency for embedding.

Thus the most efficient fixing fluids are combinations of protein coagulants and protein noncoagulants. It should be emphasized, however, that no fixing solution is ideal: all chemical agents cause some chemical change in the protein structure of the cells. The choice of fixative will depend on the type of investigation to be undertaken, and an effort should be made to discover what adverse effect the fixing agent may have on the cellular components under study.

Because they contain ingredients that act upon each other, many mixtures are most efficient when made up fresh. The individual ingredients can usually be made up into stock solutions which can then be mixed together immediately before use. Among the frequently used chemicals are formaldehyde, glutaraldehyde, ethyl alcohol, acetic acid, picric acid, potassium dichromate, mercuric chloride, chromic acid, and osmium tetroxide (osmic acid). Since every chemical has its own set of advantages and disadvantages, each component should, whenever possible, compensate for a defect in some other component. For example, in the widely used fixative, Bouin solution: (1) Formaldehyde fixes the cytoplasm, but in such a manner that it retards paraffin penetration. It fixes chromatin poorly and makes cytoplasm basophilic. (2) Picric acid coagulates cytoplasm so that it admits paraffin, leaves the tissue soft, fixes chromatin, and makes the cytoplasm acidophilic. Its disadvantages are that it shrinks the tissue and that it makes chromatin acidophilic. (3) Acetic acid compensates for the defects of both formaldehyde and picric acid.

Another example: Birge and Tibbitts (1961), by adding 0.7% sodium chloride to two fixing solutions (formalin and Bouin), reduced the amount of shrinkage caused in nuclei and cytoplasm.

When the future use of a tissue is in doubt or if it is to be stored for an indefinite time, formalin is usually the conservative choice for a fixative; it permits secondary fixation (postfixation) and will not harden excessively. If the primary objective of tissue preparation is the compilation of a simple anatomical study of cell components, then routine fixatives can be used: formalin, Gomori, Susa, Zenker, Helly, or Bouin. Special fixatives for cell inclusions are Carnoy, Flemming, Champy, Helly, Schaudinn, Regaud, and others. For histochemistry the researcher is limited to aldehydes, acetone, or ethyl alcohol (p. 392).

Most fixing solutions are named after the person originating them, e.g., Zenker and Bouin. If the same person originated more than one combination of chemicals, additional means of designating them have been used: Flemming weak and strong solutions; Allen B3, B15 series; etc. A few fixatives have been named arbitrarily: The name Susa was coined by Heidenhain from the first two letters of the words *su*blimate and *sä*ure.

CHEMICALS COMMONLY USED IN FIXATIVES

Acetic Acid

Acetic acid (CH_3COOH) is one of the oldest fixatives on record: in the eighteenth century vinegar (4–10% acetic acid content) was used to preserve hydras. In modern techniques it is rarely used alone but is an important component of many fixing solutions because of its efficient fixing action on the nucleus and its rapid penetration. It fixes the nucleoproteins, but not the proteins of the cytoplasm. It does not harden the tissue; actually it prevents some of the hardening that, without it, might be induced by subsequent alcohol treatment. In some techniques, however, acetic acid must be avoided because it dissolves out certain cell inclusions, such as Golgi and mitochondria, and some metals, such as calcium. Many lipids are miscible with acetic acid or are soluble in it. It neither fixes nor destroys carbohydrates. Pure concentrated acetic acid is called glacial acetic acid because it is solid at temperatures below 17°C.

Acids in general cause swelling in tissues—in collagen in particular—by breaking down some of the cross linkages between protein molecules and by releasing lyophilic radicles that associate with water molecules. An acid's swelling action can be a desirable property because it counteracts some of the shrinkage caused by most fixing chemicals. To curtail swelling after fixation with acetic or trichloracetic acid solutions, tissues should be transferred to an alcoholic washing solution rather than to water.

Acetone

Acetone (CH_3COCH_3) is used only for tissue enzymes, such as phosphatases and lipases. It is used cold and penetrates slowly. Only small pieces of tissue are fixed in this chemical.

Chromium Trioxide (Chromic Acid)

Crystalline chromium trioxide (CrO_3) is called chromic acid when it is added to water, usually in a 0.5% amount. Chromic acid is a valuable fixative but is rarely used alone. It penetrates slowly, hardens moderately, causes some shrinkage, forms vacuoles in the cytoplasm, and often leaves the nuclei in abnormal shapes. It is a fine coagulant of nucleoproteins and increases the stainability of the nuclei. It oxidizes polysaccharides and converts them into aldehydes—an action forming the basis of the Bauer histochemical test for glycogen and other polysaccharides. To fix water-soluble polysaccharides, it is better to use acetic acid and then posttreat them with chromic acid.

Chromic acid can also be used to partially oxidize fats to make them insoluble in lipid solvents. The oxidizing action may go too far, however, and potassium dichromate, which acts in a similar fashion, is safer for this purpose and is therefore more commonly used.

Excess chromic acid must be washed out, because it may later be reduced (undesirably, for our purposes) to green chromic oxide (Cr_2O_2). Because formalin and alcohol are reducing agents, they must not be mixed with chromic acid until immediately before use.

Alcohols

Alcohol cannot be used as a fixative for lipids because it makes them soluble. It does not fix carbohydrates, but neither does it extract mucins, glycogen, iron, and calcium. Alcohol is seldom used alone, although it is occasionally used for fixing enzymes.

Ethyl alcohol (ethanol, C_2H_5OH) hardens tissue but causes serious shrinkage. It is a strong cytoplasmic coagulant but does not fix chromatin. When alcohol is used, nucleic acid is transformed into a soluble precipitate and is lost in subsequent solutions and during staining.

Methyl alcohol (methanol, CH_3OH) is also used as a fixative, principally for hematologic tissues.

Aldehydes

Formaldehyde (HCHO) is a gas sold as a solution (approximately 40%), in water, in which form it is known as formalin. The use of the term formol is incorrect, since a terminal -ol designates an alcohol or phenol. Unless the author of a technique specifies the dilution of formalin in terms of actual formaldehyde content, dilutions must be made from the commercial product; e.g., a 10% solution would be 10 volumes of concentrated formalin (40% formaldehyde-saturated water) to 90 volumes of water.

Formaldehyde, if left standing for long periods, may either polymerize to form paraformaldehyde or oxidize into formic acid. A white precipitate in a stock solution indicates polymerization; the solution has been weakened. Cares (1945) suggests the following correction: Shake the solution to suspend the sediment; pour into Mason jars and seal them tightly; autoclave at 15 lbs. for 30 minutes; and cool. This should produce a clear solution. More dilute solutions (such as 10%) tend to oxidize more readily than do stock solutions. Marble chips placed in the jars will keep the solution neutralized, or the solution may be buffered (p. 15).

Formalin, when reacting with proteins, appears to form links between adjacent protein chains. The final effect therefore depends on the number of chemical reactions taking place between the formalin and the reactive groups of the protein molecules. Although some of these protein linkages will be broken down by washing in water, excess formalin must be removed; enough formalin-bound proteins will remain to react with future reagents when they are applied. Histologists frequently do not realize that the quantity of irreversibly bound protein drops as the pH of the solution rises above pH 10. For this reason, formalin reacts most efficiently as a buffered solution around the neutral point of pH 7.5–8.0. Never use it unbuffered on tissues for histological study.

Formalin has a moderate speed of penetration, but its action is slow. Although formalin preserves the cells adequately, it may not protect them completely unless it is given a long time to harden the tissue: Shrinkage may take place if dehydration, clearing, and infiltration are started before the hardening action is complete.

Formalin is a good fixative for lipids; it does not dissolve lipoids or fats. It does not fix soluble carbohydrates, and it does dissolve some glycogen and urea.

So-called formalin pigment may appear in tissues rich in blood. This pigment is formed when hematein of the hemoglobin escapes from red blood corpuscles before or at death and reacts with the formalin. Its formation may be prevented by a short period of fixation in formalin followed by a prolonged soaking in 5% mercuric chloride. Once formed, the pigment can be removed in a solution of 1% potassium hydroxide in 80% alcohol or in picric acid dissolved in alcohol (Baker 1958). A 30 minute treatment in either of these solutions is recommended (p. 260).

Glutaraldehyde and paraformaldehyde have special uses in histochemistry and electron microscopy (see p. 386). See Ericsson and Biberfeld (1967) for studies on aldehyde fixation.

Mercuric Chloride (Corrosive Sublimate)

Mercuric chloride ($HgCl_2$) is usually applied as a saturated aqueous solution (approximately 7%) and is acidic in action, owing to the release of H^+ and Cl^- ions in water. It is a powerful protein precipitant and forms intermolecular mercury links between SH, carboxyl, and amino groups. It penetrates reasonably well, but not as rapidly as acetic acid. It shrinks tissue less than the other protein coagulants, hardens it moderately, and distorts the cells less than other fixatives. It is excellent for mucin.

One disadvantage of mercuric chloride is that it deposits in the tissue a precipitate of uncertain chemical composition. This precipitate is in part crystalline (needle-shaped)—perhaps mercurous chloride—and in part metallic mercury (amorphous, irregular lumps) (Fig. 1–1). It can be removed by iodine, the following reaction probably taking place, $2HgCl + I_2 = HgCl_2 + HgI_2$, and any metallic mercury being converted into mercuric iodide. Mercuric iodide is soluble in alcohol, but the brown color of iodine may remain in the tissue. This can be removed by prolonged soaking in 70% alcohol, or more quickly by treatment with sodium thiosulfate, 5% aqueous. A further disadvantage is that mercuric chloride crystals inhibit freezing, making it difficult to prepare good frozen sections.

Most stains react brilliantly on tissue fixed in mercuric chloride. Chromatin stains strongly with basic stains and lakes; cytoplasmic structures react equally well with acidic or basic stains.

Osmium Tetroxide (Osmic Acid)

In solution, usually 1% aqueous, osmium tetroxide (OsO_4) takes up a molecule of water and becomes H_2OsO_5, erroneously but commonly called osmic acid, as it is in this book. (Osmic acid would be H_2OsO_4.) The substance is chemically neutral, is not an acid, and cannot be isolated. Baker (1958) suggested that it might be named hydrogen per-persomate.

In solution, ionization of osmic acid is so minute that the pH is almost

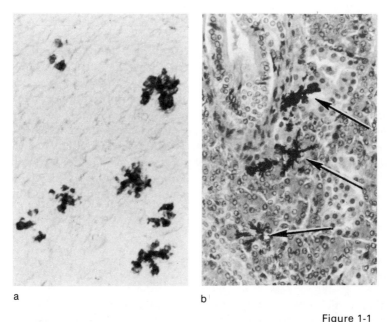

a   b

Figure 1-1
Mercuric chloride crystals: (a) in paraffin sections as seen under
microscope before deparaffinization; (b) in stained and cleared sections.

exactly that of the distilled water used in making the solution. The solution penetrates poorly and leaves tissue soft, the tissue later becoming so
friable in paraffin that it sections badly. Osmic acid preserves the cytoplasm and nuclei, but, although it increases the stainability of chromatin in
basic dyes, it reduces the stainability of the cytoplasm. Fats and other
lipids reduce osmium tetroxide or combine with it to form a dark compound. Thus, fat sites become black and insoluble in absolute alcohol,
cedarwood oil, and paraffin, but they remain soluble in xylene, toluene,
and benzene, and if left more than 5 minutes in xylene or toluene become
colorless. Osmic acid is not a fixative for carbohydrates.

When fixation is complete, excess osmic acid must be washed out of the
tissue or it will reduce to an insoluble precipitate during treatment in
alcohol. Since osmic acid is also easily reduced by light and heat, it must
be stored in a cool, dark place.

Osmic acid is a valuable chemical for electron microscopy, and many
tissues are postfixed in this solution. Good preservation can be achieved
with the controlled pH of a buffer solution, which can be a veronal,
symcollidine, or phosphate buffer (considered a physiological solution)
(Glauert 1965).

Picric Acid

Picric acid [$C_6H_2(NO_2)_3OH$] is usually used in a saturated solution, 0.9–1.2% aqueous. It is an excellent protein coagulant, forming protein picrates that have a strong affinity for acid dyes. However, it penetrates slowly, causes extreme shrinkage, and offers no protection against subsequent shrinkage. The shrinkage has been found to be close to 50% of the original tissue volume by the time the tissue has undergone paraffin infiltration. In spite of its acidity, it does not cause swelling. It does not dissolve lipids, nor does it fix carbohydrates, but it is recommended as a fixative for glycogen. It is a desirable constituent of many fixatives because it does not harden tissue, but it cannot be used alone because of the shrinkage it causes. Acetic acid is frequently used with it to counteract this undesirable characteristic.

Potassium Dichromate

Potassium dichromate ($K_2Cr_2O_7$), often incorrectly called potassium bichromate, is a noncoagulant of proteins; it makes them more basic in action but dissolves nucleoproteins. Chromosomes are therefore poorly shown, if at all. It fixes the cytoplasm without forming gross precipitates. Potassium dichromate leaves tissues soft and in poor condition for paraffin sectioning. One valuable use, however, is in preparations for fixation of mitochondria; it makes the lipid components insoluble in lipid solvents. After fixation, tissues may be soaked in 3% aqueous potassium dichromate solution to insure that lipids will be well preserved.

Potassium dichromate can be mixed with mercuric chloride, picric acid, and osmic acid, but it reacts with formalin and must not be mixed with it until immediately before use. If acetic acid is added, mitochondria will be lost, but chromosomes will be more clearly shown than they are when the chromate is used alone. It is usually necessary to wash out the dichromate with water in order to prevent the formation of an insoluble precipitate of Cr_2O_3; this will form if the tissue is placed directly in alcohol.

Trichloracetic Acid

Trichloracetic acid (CCl_3COOH) is never used alone and is similar to acetic acid in action. It swells tissues and has a slight decalcifying property. As was noted under acetic acid, washing should be done in alcoholic solutions instead of in water.

Indifferent Salts

Baker (1958) applied this term to a group of chemicals (sodium sulfate, sodium chloride, and others) whose action is not clearly understood. Zenker and Helly solutions often have sodium sulfate added. Sometimes sodium chloride is added to formalin or mercuric chloride fixatives, particularly if marine forms are to be fixed or if it is necessary to hold shrinkage to a minimum.

For detailed discussions of fixation, see Baker (1945, 1958); Birge and Tibbitts (1961); Pearse (1968); and Wolman (1955). See p. 386 for fixatives used in histochemistry.

MACERATION

Some tissues are extremely dense and cannot be manipulated while fresh. When working with these it may be desirable to separate the individual fibers of a muscle or nerve, and this is simplified by maceration. Some macerating fluids are not fixatives; these must usually be followed by some form of fixation. Other maceration fluids rely on the differential effect of weak fixatives on interstitial substances to separate the main fibers of a tissue. For example, if a fixative contains a connective tissue solvent, the solution will produce dissociation along with fixation. If a small piece of Bouin-fixed muscle is shaken in water or in 70% alcohol, the block of tissue will separate into individual fibers. Sublimate–acetic acid does not perform quite as well, but the acetic acid dissolves some connective tissue.

The following are the most commonly used macerating fluids (Hale 1958; McClung 1939, 1950).

1. 30% alcohol: 24 hours or longer (up to 4 days).
2. Formalin, 1 part in 10% salt (NaCl) solution, 100 parts: 24 hours or longer.
3. 1% sodium chloride: 24 hours or longer.
4. Chromic acid, 0.2% aqueous: 24 hours.
5. Nitric acid, 20% aqueous: 24 hours. Particularly recommended for smooth muscle of bladder.
6. Boric acid, saturated solution in saline (sea water for marine forms), plus 2 drops Lugol iodine solution (p. 544) each 25 ml: 2–3 days.
7. Potassium hydroxide, 33% aqueous. Good for isolation of smooth and striated muscle. After $1-1\frac{1}{2}$ hours, tease tissue apart with needles.

8. Maceration by enzymes. Good for connective tissue, reticulum (Galigher 1934). Place frozen sections in:

pancreatin siccum	5.0 g
sodium bicarbonate	10.0 g
distilled water............................	100.0 ml

Wash thoroughly and stain.
9. 0.01% osmic acid dissociates and fixes muscle fibers in a few days. 1% acetic acid added to it causes quicker dissociation and almost as good fixation. Use only small pieces of tissue.

FIXING THE TISSUE

First consideration in the choice of fixative should be the purpose to be served in preparing the tissue for future use. Is a routine all-purpose fixative adequate, or must some special part of a cell be preserved? An aqueous fixing fluid will dissolve glycogen, and an alcoholic one will remove lipids. Thought should be given to the rate of penetration of the fluid and the density of the tissue to be fixed; obviously an extremely dense tissue will not fix well in a fixative that penetrates slowly and poorly. If fixatives of poor penetration are used, the pieces of tissue must be as small as possible. In any case, pieces should never be any larger than is absolutely necessary—the smaller the bulk, the better the fixation.

The hardening effect of fixatives should be considered. A fixative that produces excessive hardening may lead to difficulties with liver and muscle, and a fixative with the desired attributes but with less hardening effect should be used. If there is any doubt concerning future needs for a tissue, fix it in formalin; this can be followed by postfixation treatments.

Use a large volume of fixing fluid—at least ten times the volume of the tissue if possible. Remove the tissues from the animal and place them in the fixative as rapidly as is feasible, thereby keeping postmortem changes at a minimum. In most cases, do not attempt to fix an entire organ; it will be too large to allow rapid and complete penetration of the fixative, particularly if it is covered by a tough membrane. An ideal piece would be 1 to 2 cm in diameter; place a larger piece in the fixative for 15 to 30 minutes, then trim it to smaller size and return it to the fixative. This sometimes is necessary when tissue is very soft or easily crushed. Trim the piece with a new razor blade or sharp scalpel. This will cause less damage than the squeezing action of scissors. Do not crush or tear the tissue while removing it; such material is worthless. Never allow tissue to become dry before placing it in the solution; keep it moist with normal saline. Wash off accumulated blood with normal saline, as blood retards the penetration of a fixative. After placing the tissue in the solution, shake the container gently several times to make certain that the fluid has reached all surfaces and that pieces are not sticking to the bottom or sides. A chunk of glass wool placed in the container will aid in keeping the tissue free of the bottom.

Thin pieces of tissue that show muscular contraction or that may turn inside out (tissues of the gastrointestinal tract are particularly likely to do this) may be placed on thick filter paper, outside wall against the paper, and dropped in the fixative. Tiny, easily lost specimens—biopsies, bone marrow, and so on—may be wrapped in lens paper or coffee filter paper.

The length of time required for complete fixation depends on the rate of penetration and action of the fixative. Most coagulant fixatives produce complete fixation as fast as they can penetrate the tissue. But some fixatives, such as formalin, exhibit progressive improvement in fixing action after the tissue is completely penetrated. Prolonged action in this case improves the condition of the tissue and is rarely harmful. Occasionally some type of postfixation treatment is advisable (p. 28).

WASHING THE TISSUE

After fixation is accomplished, the excess of many fixatives must be washed out of the tissue to prevent interference with subsequent processes. Often washing is done with running water; sometimes the tissue may be carried directly to alcohol, 50% or higher. Some technicians maintain, for instance, that if tissue prepared with Bouin fixative is washed in water, it loses some of its soluble picrates and should therefore be transferred from the fixative directly to 50% alcohol. When freezing is planned, formalin-fixed tissue may be washed briefly in water, but if necessary it can be taken from the fixing solution directly to the freezing microtome. Alcohol inhibits freezing, and if it is used it must be thoroughly washed out before any attempt to freeze the tissue.

When tissue has been fixed with a mercuric chloride solution, additional treatment is necessary. After washing in 50% alcohol, transfer the tissue to 70% alcohol containing enough iodine-alcohol (saturated solution of iodine in 70% alcohol) to give the solution a deep brown color. Leave the tissue in this solution from 5 to 8 hours, but no longer. If during this time the solution loses color, add some more iodine-alcohol. The iodine removes some, but probably not all, of the excess mercuric chloride precipitate in the tissue. Later, when staining the sections, iodine-alcohol or Lugol solution (p. 544) must be included in the staining series to eliminate the remaining crystals that would otherwise persist as dark clumps and needles in the finished slides.

After washing, the tissue may be stored for several weeks or months in 70% alcohol, but it is always safer to dehydrate and embed it as soon as possible. Storage in alcohol for long periods of time (a year or longer) tends to reduce the stainability of tissues, as do long immersion in chromate and decalcifying solutions, or the presence of traces of acid or iodine.

Do not use corked bottles for fixing or for alcoholic solutions; extractives from corks can be injurious to tissues. Do not use metal caps on bottles of fixatives containing mercuric chloride.

FIXATIVES AND THEIR USES

Routine Fixatives and Fixing Procedures for General Microanatomy

Bouin Fixative

Fix for at least 24 hours—several weeks cause no damage, but long periods (months) results in poor nuclear staining.

picric acid, saturated aqueous	75.0 ml
formalin, concentrated	25.0 ml
glacial acetic acid	5.0 ml

Because of acetic acid content, do not use for cytoplasmic inclusions. Large vacuoles often form in tissues. Wash in 50% alcohol. The yellow color must disappear before sections are stained. It is usually removed in the alcohol series, but if not, treat slides in 70% alcohol plus a few drops of saturated lithium carbonate until the yellow disappears.

Bouin-Duboscq (Alcoholic Bouin) Fixative (Pantin 1946)

Fix for 1–3 days.

80% ethyl alcohol	150.0 ml
formalin, concentrated	60.0 ml
glacial acetic acid	15.0 ml
picric acid crystals	1.0 gm

Prepare just before using. This is better than Bouin for tissues difficult to penetrate. Go directly to 95% alcohol.

Buffered Formalin

May remain in fixative indefinitely; action is progressive.

10% formalin	1000.0 ml
sodium acid phosphate ($NaH_2PO_4 \cdot H_2O$)	4.0 g
anhydrous disodium phosphate (Na_2HPO_4)	6.5 g

Wash in water.

Formalin (Baker 1944)

May remain in fixative indefinitely; action is progressive.

calcium chloride	1.0 g
cadmium chloride	1.0 g
formalin, concentrated	10.0 ml
distilled water...........................	100.0 ml

Wash in water.

Formalin (Baker 1958)

May remain in fixative indefinitely; action is progressive.

formalin, concentrated	10.0 ml
calcium chloride (anhydrous), 10% aqueous solution (10 gm/100 water)	10.0 ml
distilled water...........................	80.0 ml

Wash in water.

Glutaraldehyde, see p. 387.

Gomori 1–2–3 Fixative[1]

May remain in fixative 2–3 weeks.

formalin, concentrated	1 part
mercuric chloride, saturated aqueous	2 parts
distilled water...........................	3 parts

---

[1] Personal communication, C. C. Lushbaugh.

Excellent quick penetration, good for general cell morphology. Wash tissue 6–8 hours. Posttreat for mercuric chloride. The tissue will be hard and require soaking (p. 61), but stains well. I prefer Susa or Stieve if acetic acid can be used.

Helly Fixative (Zenker formol)

Fix for 6–24 hours. If tissue seems to harden excessively, follow a maximum of 18 hours in fixative with 12–24 hours in 3% potassium dichromate.

potassium dichromate......................	2.5 g
mercuric chloride	4.0–5.0 g
sodium sulfate (may be omitted; see	
Indifferent Salts, p. 11)...................	1.0 g
distilled water.............................	100.0 ml
formalin, concentrated (add just before using)	5.0 ml

Formalin reduces dichromate and should not be left in stock solutions. The above stock, without formalin, can be used for Zenker fixative, p. 18. Excellent for bone marrow, blood-forming organs, and intercalated discs. Wash tissue in running water overnight; posttreat for mercuric chloride.

Maximow modification (Romeis 1948)

Add 10% of formalin instead of 5%, and, for some techniques, 10% of osmic acid (fix in a dark container).

Hollande Bouin Fixative (Romeis 1948)

Fix for 8 hours to 3 days.

copper acetate	2.5 g
picric acid crystals	4.0 g
formalin, concentrated	10.0 ml
distilled water.............................	100.0 ml
glacial acetic acid	1.5 ml

Dissolve copper acetate in water without heat; add picric acid slowly while stirring. When dissolved, filter and add formalin and acetic acid. Keeps indefinitely. Wash for several hours in 2 or 3 changes of distilled water. Hartz (1947) recommends this for fixation of calcified areas as in lymph nodes or fat necroses. It is a good general fixative.

Orth Fixative (Galigher 1934; Gatenby and Beams 1950)

Fix for 12 hours at room temperature or 3 hours at 37°C.

formalin, concentrated	10.0 ml
potassium dichromate	2.5 g
sodium sulfate	1.0 g
distilled water	100.0 ml

Mix fresh. Wash in running water overnight. A good routine fixative, also for glycogen and fat.

Lillie variation (1965)

formalin, concentrated	10.0 ml
2.5% potassium dichromate	100.0 ml

Paraformaldehyde, see p. 388.

Stieve Fixative (Romeis 1948)

Fix for 24 hours.

mercuric chloride, saturated aqueous	76.0 ml
formalin, concentrated	20.0 ml
glacial acetic acid	4.0 ml

Similar in effect to Susa, but simpler to prepare. Penetrates rapidly, good for large pieces. Time not critical. Go directly to 70% alcohol. Posttreat for mercuric chloride.

Susa Fixative (Romeis 1948)

Fix for 24 hours.

mercuric chloride saturated in 0.6% NaCl	50.0 ml
trichloracetic acid	2.0 g
glacial acetic acid	4.0 ml
formalin, concentrated	20.0 ml
distilled water	30.0 ml

Good substitute for Zenker if potassium dichromate is not required; Susa hardens tissue less. Rapid penetration. Go directly to 70% alcohol. Posttreat for mercuric chloride.

Zenker Fixative

Fix for 4–24 hours. If tissue seems to harden excessively, follow a maximum of 18 hours in Zenker with 12–24 hours in 3% potassium dichromate. Fixation time must be controlled. The nuclei shrink during the first couple of hours and then begin to swell to normal size after the first 3–4 hours. Treatment beyond 8–10 hours reduces the quality of nuclear staining and the sections should be treated with sodium bicarbonate (p. 565).

potassium dichromate	2.5 g
mercuric chloride	4.0–5.0 g
sodium sulfate (may be omitted, see	
Indifferent Salts, p. 11)	1.0 g
distilled water	100.0 ml
glacial acetic acid	5.0 ml

Excellent general fixative, fairly rapid penetration. Wash in running water overnight. Posttreat for mercuric chloride. Zenker may be substituted for Helly if acetic acid is not added to the stock solution.

Fixatives for Cell Inclusions and Special Techniques

Altmann Fixative (Gatenby and Beams 1950)

Fix for 12 hours.

5% potassium dichromate	10.0 ml
2% osmic acid	10.0 ml

Good for fat and mitochondria. Wash in running water overnight.

Altmann Fixative (Gatenby and Beams 1950)

Fix for 2 hours; do not fix longer or material will swell.

chromium potassium sulfate (chrome alum) ...	3.0 g
formalin, concentrated	30.0 ml
glacial acetic acid	2.0 ml
distilled water	238.0 ml

Good for yolk-rich material and insect larvae. Does not harden and gives good cytological detail. Wash in 70% alcohol or water.

Carnoy Fixative (Gatenby and Beams 1950)

Fix for 3–6 hours.

Formula A

acetic acid	20.0 ml
absolute ethyl alcohol	60.0 ml

Formula B (The True Carnoy)

glacial acetic acid	10.0 ml
absolute ethyl alcohol	60.0 ml
chloroform	30.0 ml

Chloroform is said to make action more rapid. Important fixative for glycogen and Nissl substance, but dissolves most other cytoplasmic elements. Wash 2–3 hours in absolute alcohol to remove chloroform, particularly if embedding in nitrocellulose.

Methacarn (Puchtler et al. 1970)

absolute methyl alcohol	60.0 ml
chloroform	30.0 ml
glacial acetic acid	10.0 ml

For embedding, transfer through two changes of methyl alcohol (2–3 hours and 4 hours); two changes of methyl benzoate (3 hours and 1 hour); methyl benzoate–xylene or toluene (1 hour); xylene or toluene; infiltrate and embed. Do not wash tissues in water or saline before fixation and do not contaminate with water during processing; shrinkage and alterations of tissue structure will result. Do not overexpose to alcohol; it causes shrinkage and hardening. Store over a weekend in methacarn or methyl benzoate. Methacarn is recommended over Carnoy since it causes less hardening.

Carnoy-Lebrun Fixative (Lillie 1965)

Fix for 3–6 hours.

absolute ethyl alcohol	15.0 ml
glacial acetic acid	15.0 ml
chloroform	15.0 ml
mercuric chloride crystals	4.0 g

This is a very fine fixative, but it does not keep; mix fresh. Penetrates rapidly. Wash well in alcohol to remove chloroform.

Champy Fixative (Gatenby and Beams 1950)

Fix for 12 hours.

potassium dichromate, 3% aqueous (3 g/100 ml water)	7.0 ml
chromic acid, 1% aqueous (1 g/100 ml water).	7.0 ml
osmic acid, 2% aqueous (2 g/100 ml water)	4.0 ml

Prepare immediately before use. Good for cytological detail, mitochondria, lipids, etc. Penetrates poorly; use only for small pieces. Wash in running water overnight.

Dafano Fixative (Romeis 1948)

Fix for 12–24 hours.

cobalt nitrate	1.0 g
formalin, concentrated	15.0 ml
distilled water.............................	100.0 ml

Wash in water. Good for Golgi apparatus.

Flemming Fixatives

Fix for 12–24 hours.

Strong solution

chromic acid, 1% aqueous (1 g/100 ml water) ..	15.0 ml
osmic acid, 2% aqueous (2 g/100 ml water)	4.0 ml
glacial acetic acid	1.0 ml

Mix just before using. Acts slowly: use small pieces with no blood on outside. Wash in running water for 24 hours. High acetic acid content makes it poor for cytoplasm fixation.

Modification for cytoplasmic study

chromic acid, 1% aqueous (1 g/100 ml water) ..	15.0 ml
osmic acid, 2% aqueous (2 g/100 ml water)	4.0 ml
glacial acetic acid	2–3 drops

Weak solution

chromic acid, 1% aqueous (1 g/100 ml water) ..	25.0 ml
osmic acid, 2% aqueous (2 g/100 ml water)	5.0 ml
acetic acid, 1% aqueous (1 ml/99 ml water) ...	10.0 ml
distilled water............................	60.0 ml

Lewitsky saline modification (Baker 1958)

Omit acetic acid and replace the water with 0.75% sodium chloride.

Flemming solutions are good for mitotic figures, but are not suitable for general work because they penetrate poorly, harden excessively, blacken material, and interfere with the action of many stains. The weak solution is a fine fixative for minute, delicate objects. For dense tissues use the strong solution. Iron hematoxylin is excellent following these fixatives.

Formol-Alcohol (Tellyesniczky) (Lillie 1965)

Fix for 1–24 hours.

70% ethyl alcohol	100.0 ml
formalin, concentrated	5.0 ml
glacial acetic acid	5.0 ml

Good for insects and crustaceans. Widely used by botanists. Transfer to 85% alcohol.

Gendre Fluid (Lillie 1965)

Fix 1–4 hours at 25°C.

95% ethyl alcohol saturated with picric acid ...	80.0 ml
formalin, concentrated	15.0 ml
glacial acetic acid	5.0 ml

Good for glycogen. Wash in several changes of 80% alcohol.

Gilson Fixative (Gatenby and Beams 1950)

Fix for 24 hours; may be left several days.

nitric acid, concentrated	15.0 ml
glacial acetic acid	4.0 ml
mercuric chloride crystals	20.0 g
60% ethyl alcohol	100.0 ml
distilled water.............................	880.0 ml

Good for invertebrate material. Does not give a good histological picture; shrinks cytoplasm badly. Wash in 50% alcohol. Posttreat for mercuric chloride. Good for beginners, easy to work with.

Johnson Fixative (Gatenby and Beams 1950)

Fix for 12 hours.

potassium dichromate, 2.5% aqueous (2.5 g/100 ml water)	70.0 ml
osmic acid, 2% aqueous (2 g/100 ml water)	10.0 ml
platinum chloride, 1% aqueous (1 g/100 ml water)	15.0 ml
glacial acetic acid	5.0 ml

Shrinks spongy protoplasm less than Flemming. Wash in water.

Hermann modification (Gray 1954)

Fix 12–16 hours.

platinum chloride, 1% aqueous (1 g/100 ml water)	15.0 ml
glacial acetic acid	1.0 ml
osmic acid, 2% aqueous (2 g/100 ml water)	2.0–4.0 ml

Better for protoplasm than chromic mixtures. Good staining of nuclei, but not of plasma. Some shrinkage of chromatin. Without acetic acid it is good for mitochondria.

Kolmer Fixative

Fix for 24 hours.

potassium dichromate, 5% aqueous (5 g/100 ml water)	20.0 ml
10% formalin	20.0 ml
glacial acetic acid	5.0 ml
trichloracetic acid, 50% aqueous (50 g/100 ml water)	5.0 ml
uranyl acetate, 10% aqueous (10 g/100 ml water)	5.0 ml

Good for entire eye (Walls 1938) or nerve tissue, due to presence of uranium salts. Wash in running water.

Lavdowsky Fixative (Gray 1954)

Fix for 12–24 hours.

distilled water...............................	80.0 ml
95% ethyl alcohol	10.0 ml
chromic acid, 2% aqueous (2 g/100 ml water)..	10.0 ml
glacial acetic acid	0.5 ml

Good for glycogen (Swigart et al. 1960). Transfer to 80% alcohol.

Navashin Fixative (Randolph 1935)

Fix for 24 hours.

Solution A

chromic acid...............................	1.0 g
glacial acetic acid	10.0 ml
distilled water...............................	90.0 ml

Solution B

formalin, concentrated	40.0 ml
distilled water...............................	60.0 ml

Mix equal parts of A and B just before using. At end of 6 hours change to a new solution for another 18 hours. Useful for preserving cellular detail in plant materials; as good as Flemming on root tips and less erratic. Transfer to 75% alcohol.

Perenyi Fixative (Galigher 1934)

Fix for 12–24 hours.

chromic acid, 1% aqueous (1 g/100 ml water) ..	15.0 ml
nitric acid, 10% aqueous (10 ml/90 ml water) ..	40.0 ml
95% ethyl alcohol	30.0 ml
distilled water..............................	15.0 ml

Good for eyes. When fixing the entire eye always make a small hole near the ciliary body so the fluids of both chambers can exchange with the fixing fluid. For the best fixing results, inject a little of the fixative into the chambers. Decalcifies small deposits of calcium; good fixative for calcified arteries and glands. Trichromes stain poorly; hematoxylin is satisfactory. Wash in 50% or 70% alcohol.

Regaud Fixative (Kopsch) (Romeis 1948)

Fix for 4–24 hours.

potassium dichromate, 3% aqueous (3 g/100 ml water)	40.0 ml
formalin, concentrated	10.0 ml

Mix immediately before use. Recommended for mitochondria and cytoplasmic granules. Tends to harden. Follow fixation by chromating several days in 3% potassium dichromate, which should be renewed every 24 hours. Wash in running water overnight.

Rossman Fixative (Lillie 1965)

Fix for 12–24 hours.

absolute alcohol, saturated with picric acid	90.0 ml
formalin, concentrated	10.0 ml

Good for glycogen. Wash in 95% alcohol.

Sanfelice Fixative (Baker 1958)

Fix for 4–6 hours.

chromic acid, 1% aqueous (1 g/100 ml water) ..	80.0 ml
formalin, concentrated	40.0 ml
glacial acetic acid	5.0 ml

Mix immediately before use. Good for chromosomes and mitotic spindles. Fix small pieces. Produces less final shrinkage than others of this type. Wash in running water 6–12 hours.

Schaudinn Fixative (Kessel 1925)

Fix smears for 10–20 minutes at 40°C.

mercuric chloride, saturated aqueous	66.0 ml
95% ethyl alcohol	33.0 ml
glacial acetic acid	5.0–10.0 ml

Recommended for protozoan fixation, and for smears on slides or in bulk. Not for tissue; produces excessive shrinkage. Transfer directly to 50% or 70% alcohol. Posttreat for mercuric chloride.

Sinha Fixative

Specimens may remain in fixative 4–6 days.

picric acid, saturated in 90% alcohol	75.0 ml
formalin, concentrated	25.0 ml
nitric acid, concentrated	8.0 ml

Sinha (1953) adds that 5% mercuric chloride may be included; he probably means 5 g per 100.0 ml of above solution. Recommended for insects; softens hard parts with no damage to internal structure. Transfer directly to 95% alcohol.

Smith Fixative (Galigher 1934)

Fix for 24 hours.

potassium dichromate	5.0 g
formalin, concentrated	10.0 ml
distilled water	87.5 ml
glacial acetic acid	2.5 ml

Mix immediately before use. Good for yolk-rich material (Laufer 1949). Wash in running water overnight.

The pH of some fixatives changes after mixing and again after tissue is added. This may be a factor worth considering when stainability or silver impregnation is important (Freeman et al. 1955).

Fixation by Perfusion

Perfusion (forceful flooding of tissue) is advantageous only for tissue that requires rapid fixation but is not readily accessible for rapid removal. A prime example is the central nervous system. Many organs are not adequately fixed by this method because the perfusion fluid may be carried away from the cells rather than to them.

Special equipment necessary for perfusion includes a glass cannula that fits the specific aorta to be used, and rubber tubing to connect the cannula to the perfusion bottle.

When the animal is dead or under deep anesthesia, cut the large vessels in the neck and drain out as much blood as possible. Expose the pericardium by cutting the costal cartilages and elevating the sternum. Cut the pericardium and reflect it back to expose the large arteries. Free part of the aorta from the surrounding tissue and place a moistened ligature behind it. Make a small slit directed posteriorly in the wall of the aorta and insert the moistened cannula. Bring the two ends of the ligature together and tie the cannula firmly in place. Cut open the right atrium to permit escape of blood and other fluids.

Precede fixation with an injection of a small amount of saline (50–100 ml) to wash out residual blood before it attaches to the vessel walls. Fill just the rubber tubing leading from the perfusion bottle to the cannula. (Separate the saline from the fixative with a clamp near the attachment of the rubber tubing to the bottle.) If a formalin-dichromate fixative is being used, substitute 2.5% potassium dichromate for normal saline. Fill the perfusion bottle with fixative (500–1000 ml depending on size of animal) warmed to body temperature.

When ready to start perfusion, hold the bottle at table level and open the clamp on the rubber tubing. Gradually raise the bottle to a height of 4 to 5 feet, at which point enough pressure will be exerted to force out most of the blood. After 5 minutes, open the abdomen and examine the organ to be perfused. If the surface vessels are still filled with blood and the organ has not begun to take on the color of the perfusing solution, it is possible that the perfusion has failed. However, stubborn cases may require 10 to 30 minutes to perfuse. Whenever the blood color is absent, perfusion is complete, but it should be continued long enough to keep shrinkage to a minimum.

Observe these suggestions: (1) The cannula used should be as large as possible to permit as rapid a flow as possible. This aids in washing out blood ahead of the fixative. (2) If the head alone is to be fixed, clamp off the thoracic duct, and if the brain and spinal cord are to be fixed, clamp off intestinal vessels. The fixative is then directed toward brain and spinal cord. (3) When the perfusion bottle is being filled, allow some of the fluid to flow through the rubber tubing and cannula to release air bubbles. This can also be done with the saline. Air bubbles will block the perfusion. (4) Do not allow the injection pressure to exceed the blood pressure; artifacts will result.

If liver is to be perfused—particularly if by glutaraldehyde, which penetrates slowly—perfusion should be done through the portal vein followed by perfusion through the hepatic vein (Fahimi 1967).

If only a small piece of an organ is to be fixed, a modified and easier perfusion is usually adequate. Inject the organ with a hypodermic syringe of fixative, and immediately after injection cut out a small piece of tissue close to the injection site and immerse in the same type of fixative.

Lillie (1954b) lists two disadvantages of perfusion: (1) The blood content of the vessels is lost, and (2) perfusion is not possible if postmortem clotting is present. But he does favor perfusion as the outstanding method for brain fixation, saying that immersion of the whole brain without perfusion "can only be condemned." He suggests the following as the preferred method of fixation for topographic study, if whole brain perfusion is not possible.

1. Cut a single transverse section anterior to oculomotor roots and interior margin of anterior colliculi, separating the cerebrum from midbrain and hindbrain.
2. Make a series of transverse sections through the brain stem and cerebellum (5–10 mm intervals), leaving part of meninges uncut to keep slices in position.
3. Separate two cerebral hemispheres by a sagittal section. On the sagittal surface identify points through which sections can be cut to

agree with standard frontal sections. Make cuts perpendicular to sagittal surface. Cut rest of brain at 10 mm intervals.

4. Fix in a large quantity of solution.

Less shrinkage occurs in the brain after perfusion than after immersion without perfusion. Lodin et al. (1967) recommend formalin-saline (1 : 1) as the best fixative, ethanol (not tertiary butyl alcohol) for dehydration, and embedding in celloidin or epoxy (or frozen cut sections). Paraffin embedding results in more shrinkage than the above methods; if it is used, try vacuum to reduce the shrinkage. Mounting on slides with 80% ethanol, rather than with water, will recover a good portion of the shrinkage in the sections.

For additional references on perfusion see Bensley and Bensley (1938), Emmel and Cowdry (1964); and Lillie (1965). See also Koenig et al. (1945) and Eayrs (1950).

POSTFIXATION TREATMENTS

Chromatization

Chromatization improves preservation and staining, particularly of mitochondria and the myelin of nervous tissue.

Place the fixed tissue in a 2.5–3% aqueous potassium dichromate solution (2.5–3 g/100 ml water): overnight for small gross specimens (1–2 cm); 2–3 days for larger ones; 1–2 hours for sections on slides before staining (may be left overnight).

Wash thoroughly in running water: overnight for large gross tissues; 15–30 minutes for slides.

Deformalization

The removal of bound formalin is frequently necessary in silver impregnation methods, such as the Ramón y Cajal and del Río-Hortega methods, and the Feulgen technique.

Lhotka and Ferreira Method (1950)

1. Wash tissue blocks in distilled water: 15 minutes.
2. Transfer through 2 changes of chloral hydrate (a drug, may be unobtainable), 20% aqueous (20 g/100 ml water): 24 hours each.

3. Wash in distilled water: 15 minutes.
4. Proceed to any method of staining. With silver stains, time of impregnation needs to be lengthened: Ramón y Cajal to 2 weeks; del Río-Hortega to 24 hours.

Krajian and Gradwohl Method, on Slides (1952)

1. Place in ammonia water (40 drops ammonia in 100.0 ml water): 1 hour.
2. Wash in running water: 1 hour.
3. Fix in special fixative if necessary: 1 hour.
4. Wash in running water: 1 hour.
5. Proceed to stain.

Removal of Mercuric Chloride Crystals

Iodine Method, see p. 13.

Gonzalez Method (1959*a*)

Gonzalez suggests the following if it is necessary to avoid the mordant action of iodine.

1. After fixation, wash briefly or immerse directly in Cellosolve (ethylene–glycol–monoethyl ether): 24–48 hours, 3 changes.
2. Follow with toluene: 1–2 hours.
3. Infiltrate and embed.

DECALCIFICATION

Calcium deposits may be so heavily concentrated in the tissue that they interfere with sectioning and result in torn sections and nicks on the knife edge. If deposits are sparse, overnight soaking of blocked tissue in water will soften the deposits sufficiently for sectioning (p. 61). Heavy deposits may be removed by any of several methods, but do not leave tissues in any of the fluids longer than necessary.

If any doubt arises about the completion of decalcification, check for calcium by the following method.

To 5 ml of the solution containing the tissue, add 1 ml of 5% sodium or ammonium oxalate. Allow to stand 5 minutes. If a precipitate forms, decalcification is not complete. A clear solution indicates it is complete. Sticking needles in the tissue to check hardness is a sloppy technique that can damage the cells.

An excellent decalcifying fluid, RDO,[2] can be purchased by the gallon. After using RDO for several years, I recommend it as superior to other solutions. Its rapidity of action is remarkable and the quality of staining and histological detail following its use is excellent. Old bones cut down to 1 cm in thickness, if possible, require a 6 hour treatment; small and young pieces, only 1½–2 hours. Teeth will require overnight and up to 18–24 hours. Do not overdecalcify; this detracts from the staining quality. Decalcifying may be followed by a brief washing in water, but this is not necessary. Fixation and decalcification may be combined in a mixture of 1 part undiluted formalin with 9 parts RDO.

Acid Reagents

Following the use of an acid for decalcification, prevent swelling in the tissue and impaired staining reactions by transferring into 70% alcohol: 3–4 hours or overnight. If acidity will interfere with staining, treat with 2% aqueous lithium carbonate solution, or 5% sodium sulfate: 6–12 hours, then into 70% alcohol. Chromate treatment (p. 28) also improves staining.

Because calcium ions are soluble at pH 4.5, buffer solutions may be used. They are slower than other methods, but cause no perceptible tissue damage.

Formic acid A

formic acid	5.0–25.0 ml
formalin, concentrated	5.0 ml
distilled water to make	100.0 ml

With 5 ml formic acid content, 2–5 days. If increased to 25 ml, less time is required, but there is some loss of cellular detail.

---

[2] DuPage Kinetic Laboratories.

Formic acid B

> formic acid, 50% aqueous (50 ml/50 ml
> water)............................... 50.0 ml
> sodium citrate, 15% aqueous
> (15 g/100 ml water) .... 50.0 ml

Kristensen Fluid (1948), *pH 2.2*

> 8 *N* formic acid (see p. 542) .... 50.0 ml
> 1 *N* sodium formate .... 50.0 ml

Treat for 24 hours. Wash in running water for 24 hours. Highly recommended.

Chelating Agents for Decalcification

Chelating agents offer the advantage of maintaining good fixation and sharp staining potential during decalcification. They are organic compounds that have the power of binding certain elements, such as calcium and iron. A commercial preparation of the disodium salt of ethylene diamine tetracetic acid (EDTA)—Versene (Dow Chemical Co.) is the most commonly used agent. The method does have two disadvantages: The tissue tends to harden, and decalcification is slow.

Hilleman and Lee (1953) recommend 200 ml of a 5.5% solution of Versene in 10% formalin, for pieces 40 × 10 × 10 mm. Decalcification may require up to 3 weeks. Renew the solution at the end of each week. Transfer directly to 70% alcohol.

Vacek and Plackova (1959) report that a 0.5 *M* solution of EDTA at pH 8.2–8.5 yields better results in silver methods than does decalcification with acids.

Schajowicz and Cabrini (1955) adjust the solution to pH 7.0 with NaOH and HCl. Hematoxylin and eosin stain as usual, but glycogen is lost, and alkaline phosphatase has to be reactivated after the use of chelating agents.

Decalcification Combined with Fixation (See RDO, p. 30)

Lillie Fluid (1965)

This is used for 1–2 days.

picric acid, 1–2% aqueous (1–2 g/100 ml water)	85.0 ml
formalin, concentrated	10.0 ml
formic acid, 90–95% aqueous (90–95 ml/10–5 ml water)	5.0 ml

To extract some of the yellow: 2–3 days in 70–80% alcohol.

Lillie Alternate Fluid (1965)

Add 5% of 90% formic acid to Zenker fixative.

Perenyi Fluid

This is also a fixative, p. 24. Good for small deposits. Little hardening effect. Excellent cytological detail. Good for calcified arteries, and glands such as thyroid: 12–24 hours. Wash in 50–70% alcohol.

Schmidt Fluid (1956)

Schmidt uses the following for 24–48 hours, pH 7–9:

4% formalin (4 ml/96 ml water) plus 1 g sodium acetate	100.0 ml
disodium versenate (EDTA)	10.0 g

No washing necessary. Transfer tissue directly into 70% alcohol, dehydrate and embed as usual.

Decalcification Combined with Dehydration

Jenkin Fluid (Culling 1957)

absolute ethyl alcohol	73.0 ml
distilled water	10.0 ml

chloroform	10.0 ml
acetic acid	3.0 ml
hydrochloric acid, concentrated	4.0 ml

The swelling action of the acid is counteracted by the inclusion of alcohol. Large amounts of solution should be used, 40–50 times bulk of tissue. After decalcification, transfer to absolute alcohol for several changes, then clear and embed.

OTHER METHODS OF TISSUE PREPARATION

The major portion of this book is devoted to sectioning methods for preparing tissue for staining because of the complexity and quantity of such methods. However, the student should recognize that there are other means of examining tissues.

Exceedingly thin membranes can be examined directly by mounting them in glycerol or other aqueous media. Considerable detail can be observed with reduced light or under the phase microscope. Sometimes a bit of stain can be added to sharpen or differentiate certain elements. More permanent preparations can be secured by fixation.

"Touch" preparations (impression films) are made by pressing the cut surface of fresh tissue against a dry slide. Cells adhere to the surface and can be examined unstained, or the slide can be immediately immersed in a fixative and then stained.

Smears are one of the commonest devices for simple slide preparation, such as blood and bone marrow (p. 219), Papanicolaou (p. 461), fecal (p. 517), and chromosomes (squash preparations, a modification of smears, p. 474).

Dried smears and touch preparations can develop artifacts by structural distortion of the cells. Rehydration destroys some of the cellular detail. More satisfactory results follow a 3 minute treatment in a mixture of glycerin and water, equal parts. Follow by washing in running water, fix, and stain.

Mikat and Mikat (1973) flatten cells for better visibility of their contents. 5 mm cubes of tissue are placed on formalin-moistened filter paper and centrifuged 16 hours at 2200 rev/minute. The nuclear detail is improved with the spreading of the chromosomes.

"Cell blocks"—concentrated clusters of individual cells or grouped cells—are described in detail on p. 44.

Chapter 2
Dehydration:
Preparation for Embedding

Tissues fixed in aqueous solutions maintain a high water content, which can hinder later processing. Except in special cases (freezing method, water-soluble waxes, and special cell contents), the tissue must be dehydrated (water removed) before some steps in processing can be successful.

During fixing and washing, tissues lack the ideal consistency for sectioning, or cutting thin slices of a few microns in thickness. They may be soft, or they may contain hollow spaces (lumen) and be easily deformed by sectioning. If the cells are pierced by a knife, their fluid content may be released, causing them to collapse. To preclude these problems, it is necessary to replace the fluids in the tissue by a medium that hardens to a firm, easily sectioned material. The cells are filled intracellularly (impregnated or infiltrated) and enclosed extracellularly (embedded) with the medium and are thereby protected during handling. Universally used media for this purpose are paraffin and nitrocellulose and variations of these. Other media, less frequently used, are gelatin, water-soluble and ester waxes, and plastics.

Various conditions determine the choice of medium. Paraffin is suitable for most histological and embryological purposes when thin sections (1–15μ) are required. Nitrocellulose and plastics can also be used, but serial sections are made more easily with paraffin-impregnated tissue, and paraffin preparation requires much less time. Impregnating with nitrocellulose has distinct advantages when it is desirable to avoid heat and when a tissue becomes hard too readily or is too large for the paraffin technique. With nitrocellulose shrinkage is kept at a minimum, whereas with paraffin it can amount to as much as 20%. Gelatin can be used for extremely friable tissue in the freezing technique, and water-soluble waxes are used when alcohols, hydrocarbons, and the like must be avoided. Ester waxes are recommended for hard and smooth tissues and when it is necessary to avoid hardening agents, such as hydrocarbons.

Before embedding the tissue in paraffin, nitrocellulose, plastics, or ester waxes, all water must be removed from it. This dehydration is usually achieved by immersing the tissue in a series of solutions of ethyl alcohol (ethanol) in water, with gradually increasing percentages of alcohol. Changing through solutions of 30%, 50%, 70%, 80%, 95%, and absolute alcohol is said to reduce some of the shrinkage occurring in the tissue. If time does not permit such a series, the 30% and 80% steps, and even the 50% one, may be eliminated without great damage to the tissue. The time required for each step depends on the size of the object—30 minutes to 2 hours, maybe 3 hours for large dense pieces (greater than 1 cm). A second change of absolute alcohol should be included to insure complete removal of water; a total of 2–3 hours in both changes should be ample, even for large pieces. Too long an exposure to either 95% or absolute alcohol tends to harden the material, making it difficult to section.

There are other agents which are just as successful dehydrants as ethyl alcohol. The ideal dehydrating fluid would be one that would mix in any proportion with water, ethyl alcohol, xylene, or paraffin. Two such solutions are dioxane (p. 38) and tertiary butyl alcohol (butanol). Absolute butyl alcohol is miscible with paraffin, and after infiltrating tissues with a warm (50°C) butyl alcohol–paraffin mixture, infiltration with pure paraffin can follow. If isobutyl alcohol is used, there may be some difficulty with impregnations, probably because of the limited miscibility of this alcohol with paraffin and water. The butyl alcohols have the added disadvantage of a disagreeable odor. With both tertiary and isobutyl alcohol there is more shrinkage than with other alcohols. Isopropyl (99%) alcohol (isopropanol) is an excellent substitute for ethyl alcohol and is sufficiently waterfree for use as absolute alcohol. Actually isopropyl alcohol produces less shrinkage and hardening than ethyl alcohol, and is free from Internal Revenue restrictions. There are two disadvantages to the use of isopropyl alcohol for dehydrating purposes: (1) It cannot be used on tissues that are to be embedded in nitrocellulose, since nitrocellulose is practically insoluble in it; and (2) most dyes are not soluble in it so, with a few exceptions, it cannot be used during staining procedures. Methanol can be used. Solox,[1] ethanol denatured with methanol, is an inexpensive substitute for pure ethanol or isopropanol.

For the preparation of dilutions of ethyl alcohol, it is customary to use 95% alcohol and dilute it with distilled water in the following manner: If a 70% solution is required, measure 70 parts of 95% alcohol and add 25 parts of distilled water to make 95 parts of 70% dilution. In other words,

---

[1] Industrial Chemicals, Inc.

into a 100 ml graduated cylinder pour 95% alcohol to the 70 ml mark, and then add distilled water up to the 95 ml mark.

Absolute alcohol is not exactly 100%, but may contain as much as 1 or 2% water. If the water content is no higher than this, the absolute alcohol is considered 100% for practical purposes in microtechnique. If it is necessary to make certain that the water content is no more than 2%, add a few ml of the alcohol to a few ml of toluene or xylene. If turbidity persists, there is more than 2% water present. But if the mixture remains clear, the alcohol is satisfactory as an absolute grade.

Dilutions of isopropyl alcohol (99%) can be handled as a 100% solution; that is, for 70% use 30 ml of water to 70 ml of alcohol, etc.

If distilled water is not provided in the laboratory building, a Barnstead or Corning still[2] can provide a sufficient amount of water for the average microtechnique laboratory. The stills are available in sizes that produce $\frac{1}{2}$ to 10 gallons of water per hour. A special model will produce 30 gallons per hour. Deionized water[3] is a satisfactory substitute for distilled water in some solutions.

See Chapter 3, p. 38, for solutions that both dehydrate and "clear" tissues.

Special Treatment for Small, Colorless Tissues

A small, colorless tissue often seems to disappear into the opaqueness of the paraffin. An easy answer to this problem is to add some eosin to the last change of 95% alcohol; the tissue can then be seen more readily and oriented with greater facility. (However, eosin cannot be added if iso-propyl alcohol is being used, since eosin is not soluble in isopropyl.) The eosin will not interfere with future staining; it is lost in the hydration series following deparaffinization.

---

[2] Barnstead Still and Sterilizer Co.; Corning Glass Works.
[3] Produced by a Barnstead or Corning demineralizer, with cartridges containing ion-exchange resin.

Chapter 3
Clearing, Infiltrating, and Embedding: Paraffin Method

CLEARING

In most techniques that require dehydration and infiltration or impregnation, an intermediary step is necessary between the two. Because the alcohol used for dehydration will not dissolve or mix with molten paraffin (an exception is tertiary butyl alcohol), the tissue must be immersed in some fluid miscible with both alcohol and paraffin before infiltration can take place.

Clearing may seem a strange name for this intermediary step, but it describes a special property of the reagents that are used. They remove or *clear* opacity from dehydrated tissues, making them transparent. Blocks of tissue appear to deepen in color; they seem almost crystalline, never milky.

Three hydrocarbons—benzene, toluene, and xylene—are reagents commonly used for clearing. However, if the tissue contains much cartilage or is fibrous or muscular, it is wise to avoid these because of their tendency to harden such tissue. Xylene, formerly one of the most widely used reagents, has the highest tendency of the three to harden such tissue and therefore is not recommended.

An additional problem with these three reagents is their rapid rate of evaporation; unless they are handled quickly, they evaporate out of the tissue and leave air pockets that will not infiltrate with paraffin. D. Tully (M.I.T. Education Research Center) suggests correctly that the rate of evaporation of these reagents is more directly correlated to their vapor pressures than to their boiling points. (Boiling points are: xylene, 142°C; benzene, 80°C; and toluene, 111°C.) Xylene has a vapor pressure of only 5 mm Hg but, as noted above, hardens tissue excessively. Benzene, with a vapor pressure of 80 mm Hg, has the highest rate of evaporation of the three, although it hardens least. Toluene, then, of the three reagents, is probably the safest; it has a vapor pressure of 25 mm Hg, thus eliminating some of the hazards of evaporation, and it does not harden as excessively as xylene.

Caution: See Laboratory Safety, p. 572.

Chloroform is used in many laboratories but has outstanding disadvantages. It desiccates some tissues, connective tissue in particular, and has a boiling point of 61°C and a vapor pressure of 160 mm Hg, making it highly volatile and difficult to use. Aniline can be used with good results but is difficult to remove during infiltration. A mixture of equal parts of methyl salicylate (oil of wintergreen) and aniline followed by pure methyl salicylate, and then methyl salicylate–paraffin offers quicker and surer results. Methyl salicylate can also be used alone. Other oils, such as bergamot, clove, creosote, terpinol, can be used. Amyl acetate and Cellosolve (ethylene–glycol–monoethyl ether) do not harden excessively; however, Cellosolve is highly volatile.

Oils, including those above, can be used for clearing if tissue hardening is a serious problem. Cedarwood oil is well known and is relatively safe for the beginner, but it has disadvantages. Overnight immersion is usually required to insure complete replacement of the alcohol in the tissue. Also, as with all oils, all traces must be removed before infiltration, which is sometimes difficult since the oil may have a boiling point in the 200s and be slow moving out of the tissue. Oil will move out of the tissue more rapidly with the addition of an equal part of toluene. Furthermore, cedarwood oil and the absolute alcohol which must be used with it are both expensive.

During the use of xylene, toluene, or benzene for clearing, if the solution becomes turbid, it means that water is present and the tissue is not completely dehydrated and may shrink. The only remedy is to return the tissue to absolute alcohol to eliminate the water and then to place it in a fresh clearing solution. Embedded tissue containing water can shrink as much as 50% and cause difficulties in sectioning and in mounting sections.

DEHYDRATION AND CLEARING COMBINATIONS

Dioxane Method

Any procedure that shortens the preparation time of tissues to be embedded has merit and finds favor among technicians. Graupner and Weissberger in 1931 proposed the use of dioxane (diethyl dioxide), which dehydrates and clears tissue in a minimum of steps. Dioxane is miscible with water, alcohol, hydrocarbons, and paraffin. It seems to eliminate some shrinkage and hardening, and is a relatively inexpensive method because fewer solutions are required. (Dioxane is far more expensive than alcohol, however.)

Some technicians warn of disadvantages to the use of dioxane, indicating that dioxane is cumulatively toxic and leads to conditions that can become acute. It should not be used by anyone with liver or kidney trouble. Perhaps it is well to be cautious with dioxane: Use it only in a well-ventilated room; avoid inhalation and unnecessary contact with hands; and keep dioxane containers tightly closed at all times.

Another disadvantage of dioxane is that it frequently contains water. Stowell (1941) condemned it, saying that some brands contained about 10% water and other impurities; he reported shrinkage as high as 35% with dioxane. If water is present, tissue shrinkage will appear during infiltration and may be as high as 50%. Dioxane solution is cloudy with 1% water content. Many technicians place anhydrous calcium chloride ($CaCl_2$) in the bottom of the dioxane container during dehydration to aid in removing all the water. However, Mossman (1937) observes that dioxane reacts with calcium chloride and swells as though water were present; he suggests the use of calcium oxide (CaO), but feels that the use of either chemical is unnecessary.

If a fixative containing potassium dichromate is used, tissues must be washed thoroughly before using dioxane or the dichromate will crystallize.

Miller Schedule for Dioxane (1937)

1. Move tissue directly from fixative into 3 parts dioxane, 1 part distilled water: $\frac{1}{2}$–1 hour.
2. Dioxane: 1 hour.
3. Fresh dioxane: 2 hours.
4. Paraffin, 1 part, to 1 part dioxane (in paraffin oven): 2–4 hours.
5. Paraffin: overnight.
6. Embed.

Tetrahydrofuran (THF) for Dehydration

Haust (1958, 1959) recommends tetrahydrofuran (THF) for dehydration for several reasons. It mixes with water in all proportions; it also mixes with paraffin depending on the temperature, becoming increasingly miscible as the temperature rises. It is miscible with nearly all solvents and can be used as a solvent for mounting media. Most dyes are not soluble in THF, but iodine, mercuric chloride, acetic acid, and picric acid are soluble in it. THF has a low boiling point of 65°C, so it evaporates rapidly and must be kept in a tightly closed container at all times. It has a tendency to form peroxides, but storing it in amber bottles lessens this problem. All mixtures of THF should be stored at 4°C (Salthouse 1958). It can be irritable to the eyes and, if used in a technicon, that instrument should be kept in an exhaust-ventilated room.

Haust Method

Following fixation, proceed to step 1.

1. 1 part THF to 1 part water: 2 hours.
2. THF, 3 changes: 2 hours each.
3. 1 part THF to 1 part paraffin, 53–54°C: 2 hours.
4. Paraffin: 2 hours.
5. Embed.

The highest quality THF can be obtained from Eastman Kodak Co. and Fisher Scientific Co. *The Laboratory*, clinical edition (Fisher Scientific Co., vol. 28, no. C4, 1960) suggests that when it is necessary to repurify THF, it is simpler and safer to pass it through a column of activated alumina than to use metallic sodium.

INFILTRATING WITH PARAFFIN

Paraffin, the original substance used in this method has been replaced by refined mixes of paraffin and resin, but the title *paraffin* is still with us.

Paraffin is considered to be either soft or hard. The melting point of soft paraffin lies in the 50–52°C or 53–55°C ranges; that of the hard, in the 56–58°C or 60–68°C ranges. The choice of melting point depends upon the thickness at which the tissue is to be sectioned or upon the type of tissue—hard paraffin for hard tissues and soft paraffin for soft ones. If relatively thick sections are to be cut, use a soft paraffin; otherwise the sections will not adhere to each other in a ribbon. If thinner sections are desired ($5–7\mu$), use a paraffin in the 56–58° grade. For extremely thin sections of less than 5μ, the best results can sometimes be obtained with a hard paraffin of 60–68° melting point. The sections will retain their shape and size without excessive compression and will ribbon better than if the paraffin is softer. Temperature can also influence the choice of paraffin; hot weather will force the use of a harder paraffin than can be used in cool weather. If it is impractical to stock more than one kind of paraffin, that with a 56–58°C melting point is usually the best choice.

Paraplast and Paraplast Plus[1] are excellent embedding media, rigidly controlled mixtures of paraffin and several plastic polymers of regulated molecular weights. Any reference to Paraplast is applicable to either mixture. They are smooth, require no ice for embedding, and cut well with less compression than paraffin. I have discovered that the Paraplasts do not absorb enough water to soften hard tissues before sectioning (p. 61). This obstacle can be overcome by infiltrating with Tissue Prep[2] (56–58°) and embedding in a Paraplast.

Normally, tissues are transferred directly from clearer to paraffin. Keep the oven temperature just high enough to maintain the paraffin in a melted state, no higher. This lessens the danger of overheating tissue, which can harden and shrink. The paraffin must be fully molten to infiltrate the tissue effectively; partly melted, "mushy" paraffin penetrates poorly, if at all. Melted paraffin that has been kept in a warm oven for several days or weeks is better for infiltrating and embedding than freshly melted paraffin. After 30 minutes to 1 hour in the first bath, the tissue is removed to a fresh dish of paraffin for a similar length of time. Two changes of paraffin are sufficient for most requirements, but for some tissues that are difficult to infiltrate—such as horny skin, bone, or brain—a third change may be necessary, and the total time of infiltration may need to be extended to as much as 6 hours, or even overnight. Fortunately, such cases are rare.

However, when infiltrating friable tissues, the following preliminary step must be added. After the tissue is well cleared, begin to saturate the solution with fine shavings of paraffin until some of the paraffin remains undissolved. Leave the tissue in the saturated clearer for 4–6 hours, or overnight for large pieces. Then with a warm spatula transfer the tissue to melted paraffin.

The use of a vacuum oven for infiltrating will remove air from some tissues (lung) and eliminate holes in the final paraffin block (Luna and Ballou 1959; Weiner 1957).

All suggestions concerning the use of paraffin also apply to Tissue Prep and Paraplast.

---

[1] Curtin Matheson Scientific Co.; Fisher Scientific Co.; Scientific Products; VWR Scientific.

[2] Fisher Scientific Co.

EMBEDDING (BLOCKING) WITH PARAFFIN

As soon as the tissue is thoroughly infiltrated with paraffin, it is ready to be embedded; the paraffin is allowed to solidify around and within the tissue. The tissue is placed in a small container or paper box already filled with melted paraffin, and the whole is cooled rapidly in water. Before transferring the tissue, warm the instruments that manipulate it. This will prevent congealing of paraffin on metal surfaces. Handle the tissue and paraffin as rapidly as possible to prevent the paraffin from solidifying before the tissue is oriented in it. Orientation is important. If the tissue is placed in a known position and carefully marked with a slip of paper in the hardening paraffin (Fig. 5–1, a) it is easy to determine the proper surface for sectioning. If a paper box is used, the orienting mark may be made directly on one of its flaps.

Each technician eventually adopts his or her own pet embedding mold or container. A few suggestions: petri dishes, syracuse watch glasses, shallow stender dishes, test tubes to concentrate solid contents of tissue or body fluids, and a neat little dish for tiny pieces—a miniature syracuse watch glass (watch glasses, U.S. Bureau of Plant Industry Model, 20 mm inside diameter, 8 mm deep).[3] Lightly coat the insides of glass dishes with glycerol; then the solidified paraffin block loosens readily. Cast lead L's[4] when placed on a small flat metal rectangle can be adjusted to almost any size for embedding, and, being metal, they cool the paraffin more quickly than glass. Also they break loose immediately from the paraffin. The Pop-out Embedding Mold[5] opens to release the paraffin block.

Paper or foil boxes may be fashioned as shown in Figure 3–1. Perhaps the one advantage of these is that they and the data recorded on them can be left on blocks during storage. The Peel-A-Way Disposable Embedding Molds[6] are made of lightweight plastic, are available in five sizes, and are easily broken at the corners so the sides can be peeled down and the mold pulled away. Perforated tabs to fit can be purchased. Various devices, such as refrigerator trays with dividers, can be used for embedding a number of large pieces of tissue. Sections of aluminum (see footnote 4 above) are designed to interlock and form 20 or 30 compartments. Embedding rings and molds are parts of the Tissue-Tek[7] system. Tims[8] of Lab-Line is still another embedding container. The literature and catalogs are full of ideas; the above are just a few of them.

---

[3] Arthur H. Thomas Co. no. 9850.

[4] Lipshaw Manufacturing Co. nos. 334, 358, 359; Scientific Products.

[5] Lipshaw Manufacturing Co. no. 335 A–D; Scientific Products.

[6] Ames Co.; Scientific Products.

[7] Ames Co.; Fisher Scientific Co.; Scientific Products.

[8] Curtin Matheson Scientific Co.

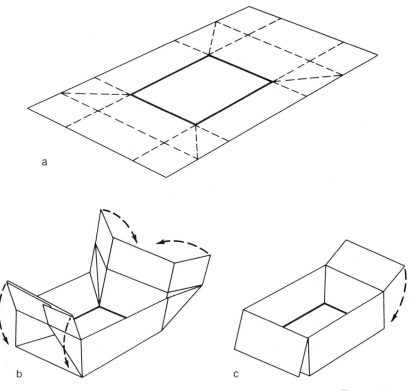

Figure 3-1
Folding bond paper or aluminum foil to make a box for paraffin embedding.

When small amounts of paraffin are ready to be solidified, they can be cooled immediately in water, preferably at a temperature of 10–15°C, or even at cool room temperature. (Paraplast does not require this treatment.) Make certain that a solidified scum has formed over each potential block before dipping it below the surface of the water. Water colder than 10°C causes the block to contract too strongly, and it may crack. Furthermore, normal crystals may form in the center, but abnormal crystals may form in the periphery of the block. The perfect block is one in which the paraffin crystals are contiguous, and the paraffin appears clear and homogeneous. Paraffin may contain 7–15% air dissolved in it and will appear clear when that air is distributed evenly through its mass, but pockets of air produce milky spots, a condition called *crystallization*. Slow hardening of the block in the air or too rapid cooling may cause this effect, particularly in large blocks. Quick hardening of outer surfaces will trap the air. When these problems arise, cool the blocks from the bottom, and with a hot instrument keep the upper surface melted for a short time, thus releasing some of the air. Then blow across the top until a scum forms, and ease the block into water. This treatment also prevents excessive shrinkage in the center of the block and the enclosed tissue (Dempster 1944c).

If the paraffin does crystallize, difficulty may be encountered during sectioning; the only remedy is to return the block to melted paraffin, allow it to remelt, and repeat the embedding process. Experienced technicians soon learn how fast to handle the paraffin to reduce crystallization to a minimum.

The so-called Histo Centers (Curtin Matheson Scientific Co.) or Tim-stations (Lab-Line) are available and perhaps useful for professional technicians, but they are not practical for a class of students.

Thoroughly hardened paraffin blocks can be stored indefinitely without injury to the tissue, but they must be kept in a cool place where they cannot soften.

Embedding Cellular Contents of Body Fluid "Cell Blocks"

"Cell blocks" are clusters of individual cells which have been concentrated and embedded for sectioning. The process for embedding them is as follows:

1. Collect the material in centrifuge tubes and add fixative: 1 hour, or overnight. Agitate occasionally.
2. Concentrate by centrifuging (preferably in small tubes); decant and add water or alcohol depending on requirement of fixative. Loosen material at bottom of tube and stir with small glass rod: 5–10 minutes, or longer.
3. Centrifuge and decant. Add next solution and stir thoroughly. Dehydrate in this manner, and clear: approximately 10–15 minutes for each step, depending on size of particles.
4. Add melted paraffin and place tube upright in a glass or beaker in oven. Stir slightly with warm instrument to work paraffin to bottom of tube: 30 minutes. Stir a bit once during this period.
5. Cool tube.
6. Break tube. Place paraffin block in tiny paper box in a small dish of paraffin in the oven. Leave only long enough for block to begin to soften. Cool quickly.

Clarke (1947) suggests alternate steps 5 and 6:

5. Warm tube carefully in water, slightly warmer than the paraffin. As soon as paraffin against the glass melts, tip tube and allow paraffin block to slide out.
6. Mount block in a small hole dug in a square of paraffin and blend two together with a warm needle or spatula.

Farnsworth (1956) transfers minute pieces with a pipette to a piece of lens paper lying on a blotter. The latter absorbs the fluid, then the lens paper bearing the tissue is laid flat on the surface of melted paraffin. She embeds in depression slides wiped with glycerine.

Arnold (1952) concentrates small organisms in agar and carries through with the usual processes (see also p. 515).

The Filtrator Coffee Paper method for bone marrow (p. 240) can be adapted to use with many body fluid cells.

For other methods of handling clumps of cells, see Del Vecchio et al. (1959b); DeWitt et al. (1957); McCormick (1959b); Seal (1956); Taft; and Arizaga-Cruz (1960).

TIMING SCHEDULE FOR PARAFFIN METHOD

The timing that follows is for tissue blocks ±10 mm in size—smaller pieces will require less time; larger pieces, more time.

Schedule Using Ethyl Alcohol (or Methanol)

1. Fix overnight or 24 hours. (Bouin, Gomori, Susa, Stieve, formalin, Zenker. If tissue fixed in Zenker tends to harden, do not leave it overnight in the fixative; transfer to 3–5% aqueous potassium dichromate overnight, then proceed as usual on schedule.)
2. Wash in water, running if possible: 6–8 hours or overnight. (Exceptions: Bouin, Gomori, Susa, and Stieve-fixed tissue can be transferred directly to 50% or 70% alcohol without washing.)
3. Transfer to 50% ethyl alcohol: 1 hour.
4. Transfer to 70% ethyl alcohol: 1 hour. (See special iodine treatment for mercuric chloride fixatives, p. 13.)
5. Transfer to 95% ethyl alcohol: 1 to $1\frac{1}{2}$ hours.
6. Transfer to absolute ethyl alcohol no. 1: $\frac{1}{2}$ to 1 hour.

7. Transfer to absolute ethyl alcohol no. 2: $\frac{1}{2}$ to 1 hour.
8. Transfer to toluene no. 1: $\frac{1}{2}$ to 1 hour.
9. Transfer to toluene no. 2: $\frac{1}{2}$ to 1 hour.
10. Transfer to melted paraffin no. 1: $\frac{1}{2}$ to 1 hour.
11. Transfer to melted paraffin no. 2: $\frac{1}{2}$ to 1 hour.
12. Embed.

Alternate Schedule Using Isopropyl Alcohol

1–4. Same as for ethyl alcohol.
5. Transfer to isopropyl alcohol no. 1: $\frac{1}{2}$ to 1 hour.
6. Transfer to isopropyl alcohol no. 2: $\frac{1}{2}$ to 1 hour.
7. Transfer to toluene no. 1: $\frac{1}{2}$ to 1 hour.
8. Transfer to toluene no. 2: $\frac{1}{2}$ to 1 hour.
9. Transfer to melted paraffin no. 1: $\frac{1}{2}$ to 1 hour.
10. Transfer to melted paraffin no. 2: $\frac{1}{2}$ to 1 hour.
11. Embed.

As noted in Chapter 2 (p. 32), if the tissue is well hardened by the fixative, it is not necessary to dehydrate the tissue through a number of graduated steps, such as 50%, 60%, 70%, 80%, 95%, and absolute alcohol. Even the elimination of the 50% and 60% steps is possible, when time presents a problem; also the use of isopropyl alcohol considerably shortens the schedule. (See Lampros 1962, for a rapid method (3–6 hours) that uses dioxane.)

In passing tissues from one fluid to another, use the decantation method to avoid excessive manipulation with forceps and reduce injury to the tissue. After pouring off a solution, drain the tissue bottle briefly against a paper towel or cleansing tissue to pull off as much as possible of the discarded solution. This reduces contamination of the new solution. Since 95% alcohol, absolute alcohol, clearers, and melted paraffin all contribute to hardening tissue, avoid leaving it in any of these fluids for longer than the maximum time recommended above (preferably only for the minimum period), and never overnight.

AUTOMATIC TISSUE PROCESSORS

Most large laboratories now accomplish the foregoing processes with a machine, such as the Autotechnicon.[9] The changes are controlled with a

---

[9] Technicon Co.

timing device, and the tissues are shifted automatically through a series of beakers or other types of container. The timing device can be present to handle changes during the night so that the tissues will be ready for embedding in the morning. Small metal or plastic receptacles with snap-on lids hold the tissues and labels, and are deposited in a basket that clips into the bottom of the lid of the instrument. When the time arrives for removal of the tissues to a new solution, the lid rises and rotates to lower the basket into the next container. The two final beakers are thermostatically controlled paraffin baths. A technician can set the timing device for any interval desired—15 minutes, 30 minutes, 1 or more hours, etc.—over a period of 24 hours. The newest models have a clock that can control the instrument over a weekend or for several days. If you have trouble with the temperature control on the Technicon paraffin baths, see Eisler and Polk, Stain Technology *51* (1976), p. 301.

There are several tissue processors on the market in addition to the Autotechnicon. There is a Lipshaw model, and the Tissuematon of the Fisher Scientific Company. Shandon Southern Instruments has a single-station model as well as a multiple-station type. Others are American Optical (Scientific Products) and Histomatik (Lab-Line).

Chapter 4
Microtomes and Microtome Knives

MICROTOMES

The first instrument for cutting sections was made by Cummings in 1770. It was a hand model that held the specimen in a cylinder and raised it for sectioning with a screw. In 1835, Pritchard adapted the instrument to a table model by fastening it to a table with a clamp and cutting across the section with a two-handled knife. These instruments were called cutting machines until Chevalier introduced the name *microtome* in 1839. Sliding cutting machines were developed in 1798, rotary microtomes in 1883 and 1886. The Spencer Lens Company manufactured the first clinical microtome in 1901. The large Spencer rotary microtome with increased precision became available in 1910 (Richards 1949).

The following types of microtomes are used in American laboratories:

1. The rotary microtome for paraffin sections and cryostat sectioning—the most widely used. Special rustproof rotary microtomes are manufactured for use with cryostats.
2. The sliding microtome for nitrocellulose (celloidin) and plastic sections: Not always the most practical method; slow and expensive, but often unexcelled for hard and large objects such as eyes, bone, and cartilage; also for cases in which shrinkage must be kept to a minimum.
3. The clinical (freezing) microtome for unembedded tissues: Quick, cheap; required for some processes in which certain cell components must be retained, such as fat and enzymes; used for immediate diagnosis.
4. The ultrathin-sectioning microtome for sections thinner than 1μ for

electron microscopy: Only for special techniques and not commonly found in student laboratories.

5. The base sledge microtome for exceedingly large tissues, e.g., brains and hard blocks of tissues: Only for special techniques and not commonly found in student laboratories.

6. The Faust Scientific Microtome is a small, inexpensive instrument that clamps to the table top and cuts sections of about 25μ with a razor blade.

7. The Smith and Farquhar Tissue Sectioner[1] is one of two instruments for cutting unfrozen fresh or fixed tissue with minimum disruption of fine structure and minimum loss of enzymatic activity. It has a cutting range of $5-230\mu$, it cuts 50–200 strokes per minute, and injector razor blades are used in it. The Mark II Vibratome[2] uses a vibrating blade sectioning in a liquid bath to lubricate the blade and prevent heat buildup. Smith (1970) compares the two microtomes, and it is advisable to consult his article before purchasing one of these models.

Microtomes should always be kept well oiled to prevent parts from wearing unnecessarily or sticking. Either of these defects can cause imperfect sectioning (sections of varying thicknesses). Obtain advice from an expert about the parts to be kept oiled, and consult the booklet accompanying the instrument. The best oil for this purpose is Bear Brand Household Oil, formerly Pike Oil.[3]

MICROTOME KNIVES

There are three familiar types of microtome knives:

1. The plane-edge for frozen sections and paraffin ribbons.
2. The biconcave used sometimes for paraffin work.
3. The plane-concave for celloidin and plastic, sometimes for paraffin.

Because knives seem to demand hours of attention, they often become the technician's nightmare, and keeping them in optimum condition is difficult.

Theoretically, a perfect cutting edge is the juncture of two plane smooth surfaces meeting at as small an angle as is feasible—ideally 14°, as sug-

---

[1] Du Pont Instruments/Sorvall.
[2] Oxford Labs.
[3] Behr-Manning, Troy, N.Y.

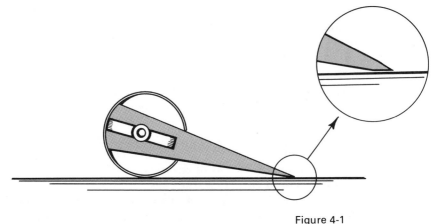

Figure 4-1
Microtome knife bevel as controlled by knife back.

gested by Dr. Lorimer Rutty (Krajian and Gradwohl 1952). These cutting surfaces are called the cutting facets. The cutting edge of a very sharp knife, when examined by reflected light under 100X magnification, appears as a fine discontinuous line. It may vary slightly in width, but it should show only a slight reflection, a narrow, straight, bright line. At a magnification of about 500 times, the edge will have a finely serrated appearance. The fineness or coarseness of these serrations depends on the degree of success in sharpening the knife. The facets are determined by the "back," which is slipped on the knife during sharpening to raise it just enough to form these facets (Fig. 4–1). Every knife must have its own back. Never interchange backs. The back must fit tightly enough so that it does not move about on the knife. When it shows considerable wear, buy a new one, because as the back wears the cutting facets are widened. Do not ram the knife into the back; hold the knife in one hand and the back in the other and work it on gradually with a rocking motion.

The importance of taking good care of a knife cannot be overemphasized. Clean it after use; materials such as blood and water, if left on the knife edge, corrode it. Clean the knife with xylene on soft cleansing tissue and wipe it dry. Do not strike the edge with hard objects, such as a section lifter or dissecting needle; the edge can be damaged or dented.

Sharpening Knives

The glass plate has had the longest and most successful use as a sharpening instrument. In 1857, von Mohl used a thick glass plate; a rotating glass plate was developed by von Apathy in 1913, and a vertical one by Long in 1927. Leather strops came into use in 1922, followed a couple of years later by Carborundum hones, and then by finer grained hones and strops of canvas or leather. Many argue against a strop, saying that a knife cannot be honed sharp enough for good sectioning. If, however, there are fine serrations, gentle stropping frequently can remove them and improve the cutting ability of the knife.

If a glass plate is used, it should be at least $\frac{3}{16}$ of an inch thick, approximately 14 inches long, and an inch or two wider than a microtome knife. Levigated alumina (approximately 1%) added to a neutral soap solution is excellent for sharpening; the polishing should be done with diamantine—a small amount. A final stropping will remove fine serrations (Uber 1936).

The Franz knife sharpener[4] uses a revolving glass disc over which the knife is swept back and forth.

Some pointers on the use and care of hones and strops may be of help to beginners. Never apply oil to hones, only water and neutral soap solution or Lava soap. Never allow the hone to dry while it is in use; always keep it wet with water or water plus soap. Two types of hones can be used, a coarse one for fast grinding of the knife edge when the nicks are deep, and a polishing hone to remove serrations and small amounts of metal left by the preliminary honing. A yellow Belgian hone is one of the best.

There are two patterns of microtome knife honing. Each has the same number of strokes. As in Figure 4–2, follow course 1, moving from bottom left to top right, roll knife over on its back, and return over same path in the opposite direction along course 2, moving from top right to bottom left; repeat both strokes three or four times. Move knife toward right into the position shown in Figure 4–3 and follow course 3, moving from bottom right to top left, and return from top left to bottom right along course 4—same number of times as in previous step. Continue this pattern of strokes on hone until deep nicks and scrates are eliminated.

When all nicks and deep serrations have been removed (check under dissecting scope), the knife is ready to be stropped, a final polishing action. Do not use a sagging hammock-like strop, unless it can be held tightly flat so that it does not round the knife edge. The best type is mounted on a felt pad on a hardwood block. This allows a bit of cushion, but on a solid surface. With a soft cloth keep the strop surface clean and free of dust. Rubbing the leather with the hand improves the texture. Do not use mineral oil on it. If the leather becomes dry, work in neat's foot oil, working over small areas at a time, not the entire strop. Buff with soft towelling at once; do not allow the oil to sit on the leather, lest it soak in or leave a gummy substance.

---

[4] Arthur H. Thomas Co.

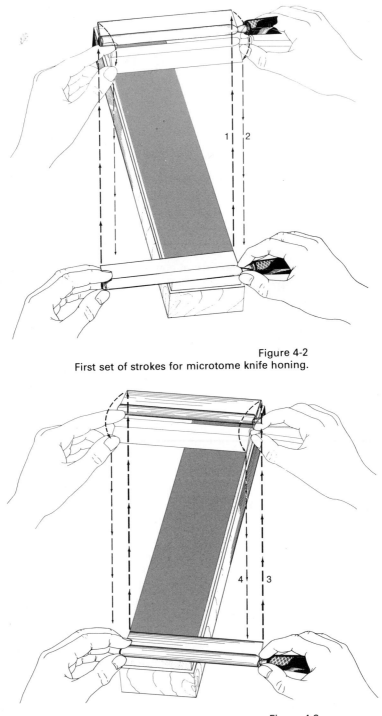

Figure 4-2
First set of strokes for microtome knife honing.

Figure 4-3
Second set of strokes for microtome knife honing.

The Lipshaw Co. offers several block-strop combinations; the revolving one is especially handy.

First use a honing strop, usually made of pigskin with a fine abrasive embedded in it. Then follow with the finishing strop, a fine-grain horsehide. The two patterns for stropping are similar to those for honing, but the cutting edge moves away from the strop (Figs. 4–4 and 4–5). Repeat these strokes on the honing strop until fine abrasions caused by the hone have been smoothed away, and then polish with a few strokes on the finishing strop. Only a dozen strokes on the latter should be necessary if the preliminary honing and stropping have been properly done. While honing and stropping, use both hands, left hand on the back and adjacent knife surface, and the right hand on the handle. Press *lightly*, guiding with both hands and with uniform pressure from both. Always use the same pressure on the forward strokes as on the return ones. Never hurry honing or stropping; use at least one full second for each stroke.

Most laboratories have replaced the "old fashioned" hone and strop for sharpening knives with mechanical knife sharpeners. All scientific supply

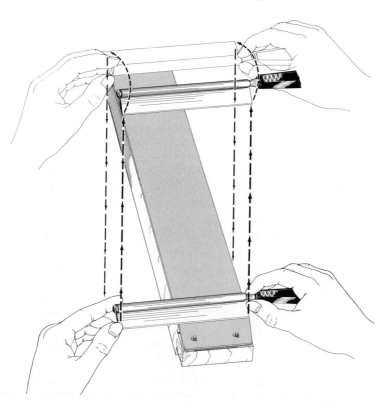

Figure 4-4
First set of strokes for microtome knife stropping.

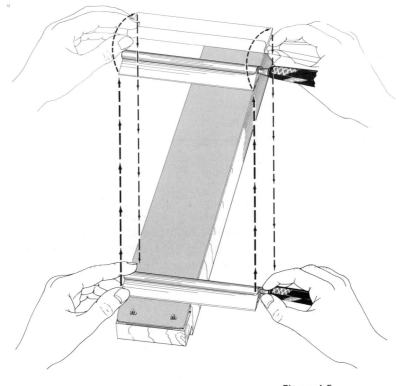

Figure 4-5
Second set of strokes for microtome knife stropping.

houses list one or more varieties of sharpeners, and some excellent sharp-eners are now available.[5] Because most technicians develop a preference for one type, I offer no recommendation, but find the American Optical type the easiest one to operate with satisfactory results.

If the cost of microtome knives and their maintenance is prohibitive for a class of beginning students, the razor blade holders[6] are useful. Students can use a razor blade once and throw it away if the edge has become dull, or sharpen it on an old knife strop.

The life of the sharp edge of a good knife can be prolonged if the technician has another knife for the preliminary trimming of paraffin

[5] American Optical Corp.; Curtin Matheson Scientific; Hacker Instruments; Lipshaw Manufacturing Co.; Scientific Products; and Shandon Southern Instruments.

[6] Spencer razor blade holder, American Optical Co.; stainless steel razor blade holder, International Equipment Co.; diSPo blades and holder, Scientific Products; disposable histo-knife, Lab-Line Biomedical Products Inc.; knife with disposable blades, Biochemical and Nuclear corporation.

blocks. After undesirable parts of the tissue block have been trimmed away, the good knife can be substituted for the old one, and the required sections collected.

Marengo (1967) describes an adjustment of the AO sharpener to produce edges of greater keenness, with a facet bevel between 25° and 30°, instead of the usual 30° to 31°.

Collins (1969) describes improved sectioning with knives etched with nitric acid.

Chapter 5
Paraffin Sectioning and Mounting

SECTIONING

The embedded blocks (Fig. 5–1a) are trimmed into squares or rectangles, depending on the shape of the tissue, with two edges parallel (Fig. 5–1b). The two short or side edges need not be parallel (sometimes with advantage, as will be indicated later). Wooden blocks or metal object discs, such as those sold by supply houses,[1] are covered with a layer of paraffin. (*Suggestion:* Use a one-inch brush which can be kept in a beaker of melted paraffin in the paraffin oven.) Then a heated instrument (spatula or slide) is held between the paraffin on the wooden block or metal disc and the undersurface of the tissue block (Fig. 5–2a). When both surfaces are melted, the instrument is withdrawn and the tissue block pressed firmly into the paraffin. Support object discs in holes in a length of wood or plastic, 1 inch thick. Make holes slightly larger than the diameter of the object disc's stem, $1\frac{1}{4}$ inches apart and approximately 1 inch deep. If wooden blocks are used, they will support themselves. After the paraffin has cooled, the tissue block is ready to be sectioned (Fig. 5–2b).

Some molds are designed to form paraffin blocks that can be clamped unmounted in the microtome.[2] Although these unmounted blocks can be used for large pieces of tissue, they will not do for small trimmed blocks of small pieces of tissue, which are more easily handled if they are mounted on an object disc or a wooden block. Trimmed blocks also create a closer alignment of the sections in a ribbon; this is particularly desirable in serial

---

[1] Lipshaw Manufacturing Co. no. 800.

[2] The Johns Hopkins Microtome Object Clamp, American Optical Corp., is used for large unmounted blocks. A Tims (Lab-Line) clamp is designed to hold Tims embedded blocks.

Figure 5-1
Untrimmed (a) and trimmed (b) paraffin tissue blocks.

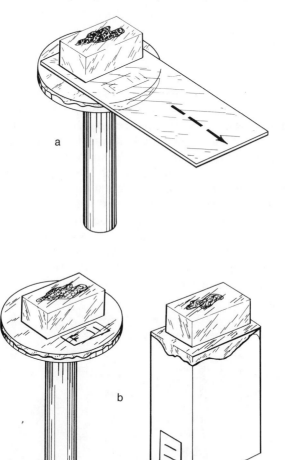

Figure 5-2
Paraffin tissue blocks: (a) mounting on an object disc;
(b) mounted on an object disc or wooden block and
ready for sectioning.

sectioning. Paraffin or Paraplast blocks may be kept in ice water until the technician is ready to begin sectioning. At least 10 to 15 minutes in ice water is necessary for thin sections. Palmer and McDonald (1963) describe a cooling device for the microtome which keeps the knife and block cold throughout the sectioning.

Set a rotary microtome for section thickness (5 or 6μ). Raise the tissue carrier and place in it the object disc with its mounted tissue block. Tighten the clamp of the tissue carrier onto the stem of the object disc. If using a wooden block, allow it to extend about $\frac{1}{4}$ inch beyond the metal clamps; this will prevent the paraffin block from breaking loose from the wooden block when the clamps are tightened.

Insert the microtome knife and tighten its clamps. The knife must be held in the clamps at a proper angle for optimal sectioning, producing a minimal amount of compression, and allowing the sections to adhere to each other. Many suggestions can be made concerning angle determination, but in most instances the technician finds the answer after a process of trial and error. The cutting facets are small, as determined by the back. When placed in the microtome, the knife must be tilted just enough so the cutting facet next to the block clears the surface of the block. If the tilt is not sufficient (Fig. 5–3a), the surface of the block is pressed down with the wedging effect of the facet, and no section results. At the next stroke of the knife, this compression increases, and finally a thick section is cut, a composite of the compressed sections. Too great a tilt of the knife makes it scrape through like a chisel, rather than cut through the tissue (Fig. 5–3b).

Turn the feed screw handle (left side of instrument), moving the feed mechanism backward or forward until the face of the tissue block barely

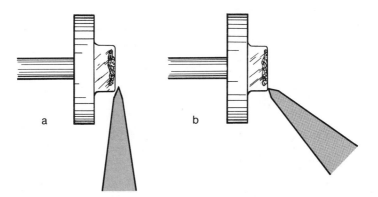

Figure 5-3
Paraffin knife tilt: (a) insufficient; (b) excessive.

touches the cutting edge of the knife. With an old knife start trimming into the block until the desired area is reached. Change to a good knife and readjust tissue carrier if necessary. Never touch the edge of the knife with anything hard. Fragments of paraffin can be flicked off with a camel's hair brush or removed with a fingertip. Scratches appearing in sections often can be remedied by rubbing the knife edge with a fingertip—*upward,* unless you like cut fingers. Also, this motion on the back of the knife will remove bits of paraffin, which can cause scratches. Cleansing tissue dipped in xylene is also helpful. (*Warning:* Discard the first section after cleaning the knife; it probably is a thick one.) When all sectioning is completed, use xylene to clean the knife; leave no corrosive material on its edge.

Section with any easy rhythm; never rush, or the result is likely to be sections of varying thicknesses. As the sections move down on the knife they form a "ribbon," each section adhering to the one that precedes it as well as the one that follows it. During the cutting of the sections, a bit of heat is generated, enough to soften the paraffin and cause the individual sections to stick to each other. As the ribbon forms, hold it away from the knife with a camel's hair brush and ease it forward so that the sections do not remain on the knife. This is advisable to prevent the sections from bunching and piling on each other. They can stack high enough to topple over the edge and get caught between the tissue carrier and the next stroke of the knife. The parallel edges of the paraffin block must be cut clean and parallel. If the edges have not been trimmed but remain as the original sides formed by the mold, a ribbon will not form.

In dry weather, static is frequently a problem during sectioning. Woolen and nylon clothing tends to create more static than cotton clothing, particularly when in contact with plastic-covered chairs. The sections stick to the knife or to parts of the microtome. They break apart and stay bunched up instead of lying flat. The friction of the knife as it crosses the paraffin block forms static electricity on its surface. An inexpensive little instrument, the Reco Neutra-Stat,[3] may be used to relieve this situation. An alpha-radiating static eliminator strip in a slot on the head of this instrument irradiates the air with harmless alpha particles which ionize the air and discharge static from the block surface. An antistatic spray can be tried. See Bryan and Hughes (1976) for easily arranged equipment to ease static formation.

The ribbons formed during sectioning can be mounted directly on slides, floated on a water bath, or laid in order in a box. It is particularly convenient to lay the ribbons in order in the box when the slides cannot be made immediately or when numerous sections—serial sections, for

---

[3] No. 61–579, Model M, E. Machlett & Son; Lipshaw Manufacturing Co.

instance—have to be cut. Shallow hosiery boxes make handy containers and, if painted black inside with India ink, provide an excellent dark background for the sections. If the shorter edges of the block have not been trimmed parallel, a serrated edge along the ribbon indicates the exact position of each individual section, minimizing the danger of cutting through one. The sections can be stored in boxes until all required slides have been mounted. Valuable sectioning time is conserved by this means.

For thin sections of 2–5μ, the Jung plane-concave knife,[4] well sharpened, is an excellent knife to use. Compression is reduced to a minimum. It is also superior for nitrocellulose sectioning. See Collins (1969) concerning an etched microtome knife for improved sectioning.

Suggestion: Use a one-inch paint brush to free the microtome of paraffin litter after sectioning.

Difficulties Encountered in Sectioning: Suggested Remedies[5]

1. Ribbons are crooked.
    a. Wedge-shaped sections caused by poor trimming; sides of paraffin block are not parallel or not parallel to edge of knife.
    b. Part of knife edge may be dull; try another part of it.
    c. Uneven hardness of the paraffin; one side may be softer than the other, or contain areas of crystallization; re-embed.
2. Sections fail to form ribbons (usually due to hardness of paraffin).
    a. Use softer paraffin (lower melting point).
    b. Blow on knife to warm it or dip it in warm water.
    c. Cut thinner sections.
    d. Place table lamp near knife and block to warm them both.
    e. Resharpen knife.
    f. Lessen tilt of knife and clean edge.
    g. Dip block in softer paraffin and retrim so a layer of this paraffin surrounds original block.
3. Sections are wrinkled or compressed.
    a. Resharpen knife; a dull knife compresses badly.
    b. Paraffin too soft; re-embed in harder paraffin.
    c. Cool block and knife.
    d. Increase tilt of knife.
    e. Clean edge of knife with finger or xylene; remove any paraffin collected there.

[4] American Optical Corp.
[5] Modified from Richards (1949).

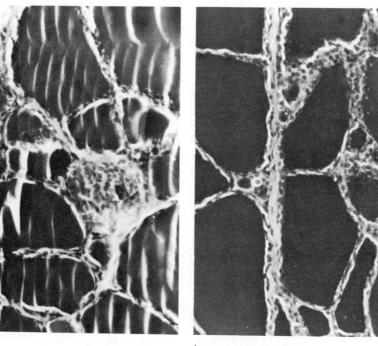

a   b

Figure 5-4
Thyroid sectioning: (a) crumbling colloid in paraffin section not soaked before sectioning; (b) same thyroid sectioned after soaking in water.

     f. Tissue is not completely infiltrated,* or too much crystallization is present.
     g. Soak tissue block in water,† $\frac{1}{2}$–1 hour or overnight if necessary (Fig. 5–4).
   4. Ribbons are split or scratched longitudinally (see Fig. 8–4).
     a. Nick in knife; move to another part of edge or resharpen knife.

---

  * Poor infiltration is usually caused by traces of water or alcohol. Correct by first removing paraffin that is present. Soak in toluene for 2 or 3 hours (or more); change twice, then place in absolute alcohol for 1 or more hours. This should remove all traces of water. Clear again in toluene (check against milky appearance); reinfiltrate and embed.
  † When soaking in water is recommended, the cut face of the tissue is exposed to tap water for $\frac{1}{2}$–1 hour, or overnight for stubborn cases and when time is of no concern. I have found this treatment completely satisfactory, but some technicians advocate the addition of glycerin (1 part to 9 parts water) or 60% ethyl alcohol instead of water (Lendrum 1944). The fluid works in through the cut tissue surface and softens tough parts. (*Exception:* Do not soak nervous system tissue at any time and lymph nodes and fatty tissue only briefly. Paraplast will not absorb water.) Diegenbach (1970) adds a fabric softener (1 : 9) to the water.

    b. Knife dirty or gritty along edge.

    c. Dirt or hard particles in tissue or in paraffin; crystals from fixing solution not adequately removed; filter stock paraffin, or decalcify or desilicify tissue.

    d. Decrease tilt of knife.

    e. Tissue too hard; soak in water.†

5. Tissue crumbles or falls out of paraffin.

    a. Poor infiltration;* reinfiltrate and re-embed.

    b. Not completely dehydrated.

    c. Not completely dealcoholized.

    d. Too long in paraffin bath or too hot while there; soak in water.†

    e. Clearing fluid made tissue too brittle; soak in water.†

6. Sections cling to block instead of knife.

    a. Knife dull or dirty.

    b. Increase tilt of knife.

    c. Paraffin too soft or room too warm; try harder paraffin or cool block.

    d. Infiltrating paraffin too hot, or too long exposure to solutions that harden; soak in water.†

7. Tissue makes scratching noise while sectioning.

    a. Tissue too hard; paraffin too hot or too long exposure to solutions that harden; soak in water.†

    b. Crystals in tissue; fixing reagents not adequately removed by washing; calcium or silicon deposits present. (See pp. 29, 495.)

8. Knife rings as it passes over tissue.

    a. Knife tilted too much or too little.

    b. Tissue too hard; soak in water.†

    c. Knife blade too thin; try a heavier one.

9. Sections curl, fly about, or stick to things, owing to static electricity from friction during cutting, especially in weather of low humidity.

    a. Increase humidity in room by boiling water in open pan.

    b. Ground microtome to water pipe.

    c. Postpone sectioning until weather is more humid; early morning sectioning often is best.

    d. See p. 59 for suggestions concerning clothing, furniture, and static eliminator.

10. Sections are skipped or vary in thickness.

    a. Microtome in need of adjustment or new parts.

    b. Tighten all parts, including knife holder and object holder clamp.

---

* See note on p. 61.

† See note on p. 61.

c. Large or tough blocks of tissue may spring the knife; soak them in water.*

d. Knife tilt too great or too little.

MOUNTING

Since the embedding medium, that is, paraffin or Paraplast, usually has to be removed from the sections to allow uniform staining throughout the cells, the tissue sections must be affixed to some solid surface or they will disintegrate during removal of the embedding medium. The sections are, therefore, mounted on glass slides which support the tissue and also permit the mounting of many sections in sequence.

Slides and Cover Glasses

Most microscopic materials (sections, whole mounts, smears, touch preparations) are mounted on glass slides and are covered with microscopic cover glasses. A noncorrosive quality of each should be purchased to insure a nonreacting surface. The slides may be obtained in the regular 3×1 inch size or in 3×2 inch for larger tissue sections. The brands labeled "Cleaned, ready for use," save many technician hours that otherwise would be devoted to cleaning slides. In addition, they permit a uniform surface for blood smears, loose cells, and touch preparations.

Cover glasses are manufactured in circles, squares, and rectangles, ranging in thickness from no. 0 to no. 3. Thickness no. 0 is used rarely, usually only when an oil immersion objective rides unusually close to the slide. This size is so thin that it fractures easily. Thickness no. 1 (about 0.15 mm) usually is adequate for oil immersion work. Thickness no. 2 (about 0.20 mm) can be used for whole mounts or when oil immersion is unnecessary. Thickness no. 3 (about 0.30–0.35 mm) is occasionally used when a tougher glass is required and for large whole mounts. Circles and rectangles may be purchased in sizes up to $\frac{7}{8}$ inch in diameter. Circles are designed for mounts for which ringing the cover is necessary in order to create a seal against evaporation, as for whole mounts in glycerol jelly or any other volatile mountant. (A ringing table can be used to make the supporting ring for the cover glass and the final seal around the edge of the cover glass.) The rectangles come in 22 and 24 mm widths, and in 30, 40, 50, and 60 mm lengths. Cover glasses will mount more efficiently if they are cleaned in absolute alcohol and wiped dry, and fewer bubbles will cling to the glass when it is pressed into place over the mounting medium.

---

* See note on p. 61.

Mounting Techniques

Before sections are mounted, a glass-marking pencil should be used to make an identifying mark on one end of the slide, such as a tissue number, a student locker number, or a research project number, to prevent confusion with other slides. In addition, this mark enables the worker to distinguish on which side of the slides the sections are mounted and thus to avoid failure during the staining process.

Types of permanent glass-marking pencils include diamond points (a real commercial diamond mounted in a handle) and steel pencils with tungsten-carbide tips.[6] A vibro-engraver or vibro-tool can be used for engraved marking on the glass. Wax china-marking pencils make only temporary inscriptions, dissolving in staining solutions. Some inks are reasonably satisfactory; one of the best is Gold Sea Laboratory Ink.[7] Sanfords *Sharpie*[8] is water resistant and can be used on slides with frosted ends, but this may be removed by xylene or alchol.

The customary means of affixing sections to slides is with egg albumen and water. (Gelatin and blood serum can also be used, p. 550.) Tissue-tac,[9] a special solution for adhering sections and cellular substances to slides, is clear and never picks up stain. (See also subbed slides, p. 550.) It is essential that slides be absolutely clean to insure adherence of the sections throughout the staining procedure. With one finger, smear a thin film of albumen fixative (adhesive, p. 548) on the slide, and with a second finger, wipe off excess albumen. This should keep the film thin enough. Thick albumen picks up stain, makes a messy looking slide, and can prevent a clear image of sharp, uncluttered tissue elements. If there is a continued tendency to get too much albumen on the slide, try a "floating solution": Dilute a couple of drops of albumen fixative with about 10 ml of distilled water and float the sections on the solution on the slide (see also Schleicher 1951) or try Haupts' adhesive (p. 549). Lillie (1965) suggests that, since water is used to float the sections on a thin coat of fixative and since the fixative constituents are water-soluble, it is doubtful that enough albumen remains on the slide to be an actual adhesive agent. He considers it probable that the albumen acts as a surface tension depressant and aids in closer attraction of sections to slide. Arn and Landolt (1975) mount Paraplast sections without an adhesive. This can be tried by experienced technicians.

If a water bath[10] is used for spreading the sections, have the temperature 5–10°C below the melting point of the paraffin being used. Excessive

---

[6] Fisher Scientific Co., no. 13–378.
[7] No. A–1408 Clay-Adams, Inc.
[8] Curtin Matheson Scientific Co.
[9] Dade Reagents; Scientific Products.
[10] Fisher Scientific Co.; Lab-Line; Lipshaw Manufacturing Co.; Scientific Products.

heat for flattening may cause shrinkage of cell structure and tearing and displacement of tissue parts. When removing the sections from the microtome knife or from the box, stretch them as flat as possible as they are placed on the water surface. After they have warmed a bit, they can be pulled to more nearly their original size with a couple of dissecting needles. With a hot needle or spatula separate the required sections from the rest of the ribbon. Dip an albumenized slide under them and with a needle hold them against the slide while removing it from the bath. When the sections are spread smoothly, the excess water can be drained off; this is facilitated by touching a piece of cleansing tissue or filter paper to the edge of the slide and around the sections. Permit slides to dry overnight on the warm plate or in an oven of comparable temperature. If they must be stained on the same day that they are mounted, dry them in one of the mechanical hot-air dryers.[11] If agar (Cordova and Polanco 1962) or gelatin is added to the water bath, unalbumenized slides can be used.

Frandsen (1964) describes mounting serial sections from a water bath, but I have always found direct mounting on slides quicker and simpler.

Sections collected (glossy side down) in a box can be separated with a spatula or razor and removed to an albumenized slide. Add enough distilled water to spread well beyond the edge of the sections. Place the slide on a slide warmer[12] to correct any compression acquired during sectioning and keep it on the warmer until the sections are stretched flat. Dense tissues will compress less than the paraffin surrounding them and often will develop folds because the paraffin does not expand enough on the warm plate to permit the tissues to flatten. Pieces with a lumen will invariably demonstrate this fault. One of the simplest ways to correct it is to break away the paraffin from the outside edge of the tissue. Do this when the ribbon is floating on the water on the slide, but not while it is warm, or the paraffin will stick to the dissecting needles. Allow the slide to cool and then split away the paraffin with care so that the tissue is not injured (Fig. 5–5). If the paraffin tends to stick to the needles during this process, wipe off the needle tips with the fingers. Clean needles will minimize this problem. A rather violent method for straightening stubborn ribbons is the addition of some acetone or alcohol to the water. Lin and Corlett (1969) use a xylene spray to relieve compression; this must be applied with great care to avoid overcompensation.

Certain tissues frequently develop cracks through the sections while drying—a tendency common to spleen, liver, lymph nodes, and nervous tissue. To prevent this, as soon as the sections are spread, drain and blot them free of as much water as possible and dry them at a higher temper-

---

[11] Technicon or Lipshaw Manufacturing Co.

[12] Fisher Scientific Co.'s Slide Warmer, no. 12–594–5; or Lipshaw Manufacturing Co.'s Slide Warmer, no. 371 or 374.

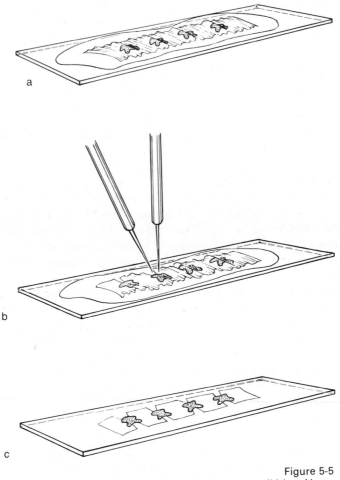

Figure 5-5
Paraffin sections: (a) compressed or folded sections; (b) breaking paraffin with dissecting needles to permit ribbon expansion without folds; (c) expanded and flattened sections.

ature (60–65°C), in the mechanical hot-air dryer, or run them rapidly through a Bunsen burner flame until the paraffin melts, but no longer (Fig. 5–6). This must be handled with care; too hot or too long in the flame impairs future staining.

Properly mounted sections will have a smooth, almost clear appearance. If they have a creamy, opaque texture, and if they reflect light when examined from the undersurface, air is caught between the glass and the sections. Tissues so mounted will float off in aqueous solutions. The cause usually lies in poorly cleaned slides or in drying the slides at too low a

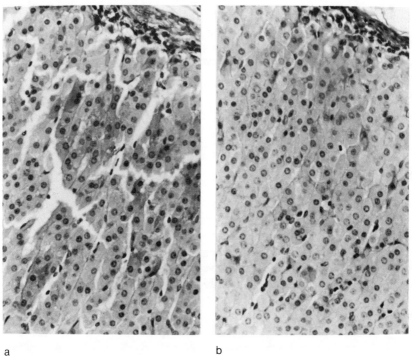

a   b

Figure 5-6
Adrenal sectioning: (a) cracks in an adrenal section caused by poor paraffin infiltration; (b) cracking of same tissue has been prevented as suggested on p. 65.

temperature or not drying long enough. If, when mounting the sections, bubbles appear under them, use distilled water that has been boiled for some time in order to release trapped air.

Several other conditions may cause tissues to loosen from slides. These include brittle sections, a large yolk or osmic acid content, and the use of strong acids or alkalis during staining. In these situations the sections will require special treatment as described below. In addition, poorly infiltrated sections and nervous tissue that have been soaked in water will usually fail to adhere, but can sometimes be saved by the following method.

After drying, and when ready to stain, dissolve out the paraffin with toluene or xylene. Place the preparations for 1–2 minutes in absolute ethyl alcohol, followed by 1–2 minutes in a dilute solution (about 0.5%) of nitrocellulose in ether–absolute alcohol (50:50). Drain off excess nitrocellulose, but do not allow it to dry completely; wave the slide in air for a second and place it in 70% or 80% alcohol. Continue with the planned staining method. Also try mounting this type of tissue on subbed slides (p. 550) instead of on albumenized slides. This is much easier than coating the mounted sections, especially for nervous tissue. I always use subbed slides in preference to nitrocellulose-coated ones. When water used in mounting sections does not puddle uniformly on subbed slides, but forms tight-up pools, sections are difficult to spread; to correct this, apply a thin coat of egg albumen over the subbed surface before adding the sections and water.

Serial Sections

With a few exceptions, the usual paraffin method is used for serial sections, and in most cases every section is mounted and in correct order. While sectioning, arrange the ribbons in a cardboard box from left to right and top to bottom (like a printed page). When the sectioning is finished, remove the sections to slides in the above order. Serially number each slide with a glass-marking pencil (do not use a grease pencil, which may wash off in staining solutions). This system affords simple sorting of slides when all processing is complete.

Chapter 6
The Microscope

THE COMPOUND MICROSCOPE

The fundamental parts of a compound microscope are the objectives and the eyepieces or oculars. Both are systems of lenses; the oculars at the top and the objectives at the bottom of a tube. The other basic parts are the stage on which the object to be viewed is placed and a mirror or system of mirrors for directing light. The ordinary laboratory microscope uses bright-field illumination with direct light amplified by a substage condenser and mirror. The condenser is also a system of lenses, the purpose of which is to concentrate light on the object to be viewed so it is seen in a bright, well-lit field.

The objectives magnify the specimen a definite amount, forming an image, which is magnified again by the ocular eyepiece. The lenses of a good microscope should not only magnify but also improve the visible detail. This is the resolving power, and it is a function of (1) the wave length of the light used; (2) the lowest refractive index (refractive index is a measurement of the refraction or bending of light rays as they pass through one medium to another at an oblique angle) between the objective and the substage condenser (p. 70); and (3) the greatest angle between two rays of light entering the front lens of the objective. Blue light increases resolving power over white light; ultraviolet increases it even more, but it cannot be seen, and the results must be photographed.

Oculars are manufactured in a variety of powers of magnification (engraved on them); 5X and 10X are the customary lenses for student and laboratory microscopes—the 5X for low power over large areas, the 10X for general work. Most of these microscopes are equipped with three objectives: 50 to 32 mm (low power); 16 mm (middle or intermediate power); and 8 to 4 mm (high power). All are engraved with their magnifying power. An additional objective will be included on the four-objective nosepiece and can be used interchangeably with the low power (or any other power) on the three-objective nosepiece. As one shifts from low power to higher powers, the magnification increases, but the size and depth of the examining field decreases. Also, the clearance of the objective above the specimen (working distance) decreases, and more light is required. The working distance is the distance between the front of the objective and the top of the cover glass when the specimen is in focus.

Achromatic objectives (producing an image essentially free of color fringes) are found on most laboratory microscopes. These objectives are corrected chromatically for the light of two wavelengths, the C (red) and F (blue) lines of the spectrum, and spherically for the light of one color, usually in the yellow greens. Achromats give a good image in the central portion of the field but not as sharp an image at the edges as can be obtained with apochromatic objectives. These are essential for critical microscopy and color photography because they are corrected for three colors, including the G (violet) lines of the spectrum, and thereby reduce color fringes even further. Lenses, if not corrected in this manner, exhibit chromatic aberration, and color fringes will confuse the border of the specimen image. For best results, apochromatic objectives should be accompanied by a corrected condenser of a numerical aperture at least equal to that of the objective aperture and compensating ocular eyepieces. A correction collar on the apochromatic objectives can be adjusted to compensate for the thickness of the cover glass on the slide. The collar can be adjusted with one hand while the fine adjustment is manipulated with the other hand.

The body tube supporting the lenses may have a fixed length, or it may be equipped with an adjustable drawtube. A scale on the drawtube is used to determine the length of the body tube for the various lenses used in it—usually 160 mm for American-made objectives. With high-power objectives this scale becomes increasingly useful in adjusting the length of the drawtube to achieve sharp detail through a thick cover glass.

At the bottom of the microscope, under the stage, there is a mirror that directs light through the diaphragm and into a condenser, which increases the illumination of the specimen. Usually there are two mirrored surfaces: one plane (flat) to direct the light, reflecting a moderate amount of light with parallel rays but no change in form; the other, concave to converge the light rays to form a cone, concentrating a large amount of light. The concave surface replaces a condenser and can be used for low magnifications, but it should not be used when a condenser is present, nor should it be used for high magnifications in place of the plane mirror.

Immediately below the microscope stage is the substage condenser, whose lenses serve to converge the parallel beam of light from the mirror so that a cone of light passes through the aperture of the stage. This cone of light is focused on the specimen under examination and then is extended to fill the back lens of the objective. Without a condenser the back lens of high-power objectives is not filled with the proper amount of light. By opening or closing the iris diaphragm mounted below the condenser, the diameter of the light entering the condenser can be controlled. A two-lens Abbe condenser is found on most laboratory and student microscopes. For critical research microscopy, a corrected condenser with a centering mount should be used. It should be carefully centered to the objective, and immersion oil should be used between it and the slide for finest detail.

The image seen in the compound microscope is inverted: it is upside down and turned right side to left. The movement of the slide will be reversed. For clarity, most images depend on color in the specimen or color added to it and/or differences in refractive indices in different parts of the specimen and the medium in which it is mounted.

Innumerable types and sources of light are used in laboratories. If a table lamp, such as that found in many student laboratories, is used, use a daylight bulb in it. Good, simple types of microscope lamps are equipped with blue ground-glass filters. Place the lamp 10 to 15 inches in front of the mirror and with all of the light directed below the stage of the microscope. The surface of the bulb or of the filter is focused on the specimen by way of the substage condenser. If the lamp is the small substage type of lamp, it may be used in front of the mirror, or, if the mirror is removed, it is placed under the condenser. A more critical illumination (Köhler method) can be obtained with better kinds of lamps provided with condensing lens and a coil or ribbon filament.

Controlled illumination is lighting by a cone of rays whose proportions are regulated by a stop at the illuminator and by the iris diaphragm of the condenser. The whole area of the aperture of the objective is used, and none of the refracted light should fall on the lens mounting or on the drawtube.

THE OPERATION OF A MICROSCOPE

Student Microscopes

Without a substage condenser

1. Place slide on stage and center the field to be examined over the opening in stage.

2. Turn low-power objective over specimen a short distance above it—one-half inch will do.
3. Adjust concave mirror, moving it back and forth, until light is as uniform as possible through the hole and onto the specimen.
4. Raise the body tube by coarse adjustment until the specimen is in focus. Check mirror, adjusting for better light.
5. Higher powers may be swung into position and light adjusted if necessary for greater magnification.

With substage condenser and lamp without lenses

1. Place slide on stage as above. Adjust objective, plane mirror, and body tube for focus, as above.
2. Remove ocular eyepiece and, while looking down the tube, open iris diaphragm of the substage condenser until it coincides with the margin of the back lens of the objective, just enough to fill it with light, no more. Focus the condenser up and down until the light is uniform through the objective.
3. Replace ocular eyepiece. Make similar adjustments for all objectives.

With substage condenser and lamp with condenser lenses

1. Place lamp 8–10 inches from microscope and direct light onto center of plane mirror.
2. Adjust lamp bulb or lamp condenser until the filament image focuses on the microscope condenser. The visibility of the filament image can be increased by closing the lamp iris.
3. Place slide on stage and focus on it. With the lamp iris closed, raise or lower the microscope condenser until the lamp iris appears in sharp focus with the slide specimen. Open the lamp iris until it disappears at the edge of the field.
4. Remove the ocular eyepiece and adjust the microscope iris by opening it until it coincides with the margin of the back lens of the objective.
5. Do not change diaphragm positions when proper adjustments have been made. Intense light can be reduced by the use of filters.

To adjust both parts of a binocular microscope, close the left eye and focus with the fine adjustment until the image is sharp for the right eye. Leave the fine adjustment alone and focus for the left eye with the focusing adjustment on the tube outside the ocular eyepiece. Between the two eyepieces is an adjustment for changing the interpupillary distance between the eyes. Turn the adjustment slowly until the right and left images blend and a single field is seen with both eyes.

Research Microscopes

Critical illumination (Köhler method) using a lamp with condensing lenses and an iris diaphragm

1. Place a slide with fine-detailed specimen in position on microscope stage. Center the condenser in relation to the stage aperture.
2. Place the lamp so the diaphragm is about 10 inches from the microscope mirror and line it up with the microscope so the light filament centers on the plane surface of the mirror. Insert a neutral filter.
3. Using a 10X ocular and 16 mm objective, adjust the mirror until the light passes through the substage condenser to the slide. Focus microscope on the specimen and center the light.
4. Rack up the substage condenser until it almost touches the slide and close the substage diaphragm. Close the lamp diaphragm about halfway and adjust the lamp condenser until an image of the filament is focused on the closed substage condenser diaphragm. This can be observed in the microscope mirror or with a small hand mirror. Partly open the substage iris diaphragm.
5. Focus with the substage condenser until a sharp image of the lamp diaphragm coincides with the specimen on the slide. With the mirror, center the diaphragm image.
6. Take out the ocular eyepiece, and, looking down the tube, observe the back lens of the objective. Open the substage diaphragm until its image edge almost disappears in the back lens, but remains just visible. The back lens should be filled with light—full cone illumination. Because the light source is imaged on the diaphragm of the substage condenser, the light source is in focus at the back focal plane of the objective, and the filament of the lamp can be seen.
7. Replace the ocular eyepiece. If the tube length has to be adjusted, do so (usually 160 mm for American objectives).

Certain precautions are imperative for good illumination and bear repetition. The back lens should always be filled with even light, and the aperture of the substage diaphragm should be closed to a minimum to furnish as wide a cone of light as the specimen will take. Closing down the diaphragm to reduce light intensity, and thereby increasing contrast in transparent objects, is poor technique. Light intensity should be controlled by the kind of bulb, the use of filters, or a rheostat or variable transformer inserted in the lamp system.

Unless you have had previous experience with a certain microscope, when changing from one objective to another of higher power (high dry or oil-immersion) do not take it for granted that the microscope is parfocal (focal planes of all objectives lie in same position on the tube). The higher power may crush a valuable specimen. Proceed as follows: Examine slide with the naked eye for approximate part to be examined. Center the area over the stage aperture. Check with low power, then middle power for the particular area of interest, then change to high power, but raise the tube $\frac{1}{4}$ to $\frac{1}{2}$ inch at least before revolving the nosepiece. Then, while watching the bottom of the objective from the side, lower the tube until the objective almost touches the cover glass. Look into the microscope and rack up the tube until the specimen is in focus. Remember, when an objective of a different numerical aperture is used, the condenser diaphragm must be adjusted (step 6). The numerical aperture (NA) number is used to compare the resolving power of the lenses: The higher the numerical aperture, the greater the resolving power and the wider the condenser diaphragm must be opened.

Use of Oil-Immersion Objective

The working distance with this objective is short, and great care must be taken to prevent crushing a cover glass. Thin cover glasses must be used, or the specimen mounted on the cover glass itself. Greater illumination is required than with the dry objectives, and a wider cone of light is needed. The space between the objective and the cover glass, also frequently between the slide and the condenser, must be filled with a suitable medium of a refractive index and dispersion like that of the glass of the lenses. This medium is provided by immersion oil (cedarwood oil, Cargille,[1] crown oil).

1. Using a dry objective, locate and center the area to be examined.
2. Raise microscope tube, swing oil-immersion objective into position.
3. Place a drop of immersion oil on the cover glass over the area to be examined. Watching (from the side) the lower edge of the objective, lower it slowly until it makes contact with the oil. (Oil may also be placed on the substage condensor lens.)
4. Observing through the microscope, very slowly lower the objective

---

[1] Cargille Laboratories.

with the coarse adjustment until the specimen is in focus. Make more critical focus with the fine adjustment.

5. Remove ocular eyepiece and adjust the light on the back lens of the objective; open the condenser diaphragm.

After use, clean oil-immersion objective; never leave oil on it. Wipe it with clean lens paper, then with a little xylene or chloroform on lens paper, finally drying it with clean dry lens paper. Do not leave xylene or chloroform on the lens.

Other Hints for Efficient Microscopy

Always keep all surfaces clean and free from dust and grease. This includes the lamp condenser, filters, mirror, lenses of the substage condenser, lenses of objectives, slides, cover glasses, binocular prisms, and lens of ocular eyepieces. Use a good grade of lens paper and a fine camel's hair brush. Omit,[2] a lint and dust remover, assists in cleaning lenses and other parts of a microscope.

If glasses are worn, protect them from scratching by mounting a rubber band around the top of the ocular. (Fingers of old rubber gloves can be cut to fit the oculars.)

If, instead of a mechanical stage, clips are present on a fixed stage, place one clip over one end of the slide. Then, with two fingers of one hand controlling the loose end of the slide, move the slide for examining, while manipulating the fine adjustment with the other hand. Individual preferences will determine which hand to use for which motion. With a mechanical stage, the right hand has to control the knobs and the left hand, the fine adjustment.

If it is convenient to have a pointer in the eyepiece, unscrew and remove the top lens of the ocular eyepiece. Inside, about halfway down its tube, is a circular shelf. Place a small drop of Permount (or other quick-drying mountant) on this shelf and, with small forceps, place an eyelash in this drop. Arrange the eyelash to lie flat and project slightly short of center of the hole formed by the circular shelf. Screw lens back in place.

When carrying a microscope, carry it upright, preferably with both hands. Never tilt it; the ocular eyepiece can fall out. When using a microscope to check staining effects or with any wet mounts, place a thin piece of glass on the stage. A lantern-slide cover glass ($3\frac{1}{4} \times 4$ inches) is usually available and is an ideal size for most stages. Clean it after use to prevent contamination of future slides or material.

---

[2] Century Laboratories.

MEASURING DEVICES USED ON A MICROSCOPE

Reading a Vernier

Most mechanical stages are equipped with a vernier—a device for the purpose of relocating the same spot previously examined on a slide (Fig. 6–1). Two numerical scales run side by side, a long one constructed on a millimeter scale, and a short one constructed on a scale of nine millimeters divided into ten equal divisions. When the area of reference on the slide is centered in the objective, read the vernier. Check the point of coincidence of the zero of the short scale against the long scale. Read the lower whole figure on the long scale, to the left of the zero on the short scale. This determines the whole number, the number to the left of the decimal point in the final reading (21 on the long scale in Fig. 6–1). *Exception:* If the zero of the short scale coincides exactly with a whole number on the long scale, then the number is recorded as a whole number with no decimals following it. The decimal is determined by the point of coincidence of the line of the short scale that perfectly coincides with any line on the long scale (.5 of the short scale in Fig. 6–1). The reading in Figure 6–1, therefore, is 21.5.

Measuring Objects with an Ocular Micrometer

The measurement of slide specimens usually is done with a micrometer disc placed in the ocular, but first the disc's actual value with respect to the magnification at which it is being used has to be calibrated against a stage micrometer. Stage micrometers usually have a 2 mm scale divided into 0.01 mm divisions, or a 0.2 inch scale divided into 0.001 inch divisions. The ocular micrometer has a 5 mm scale divided into 50 (0.1 mm) divisions or 100 (0.05 mm) divisions.

Unscrew the top lens of the ocular, place the ocular micrometer on the circular shelf inside, and replace the lens. Focus on the stage micrometer,

Figure 6-1
A vernier scale.

moving it until the zero lines of both micrometers coincide. Then a definite distance on the stage micrometer will correspond to a certain number of divisions on the ocular micrometer.

The rest is simple; for example:

The ocular micrometer has 100 divisions.

Suppose that 30 divisions (each measuring 0.01 mm) of the stage micrometer equal the 100 ocular micrometer divisions. Then 30 × 0.01 mm = 0.30 mm, the length of the 100 divisions on the ocular micrometer.

Next, 0.30 ÷ 100 = 0.003 mm (3.0μ), the value of each single division.

Therefore, when the ocular micrometer (each division measuring 3.0μ) is focused on a specimen, and the specimen requires 9 divisions to meet its length, multiply 9 by 3.0μ to equal 27.0μ (.027 mm), the total length of the specimen.

Obviously, each combination of oculars and objectives and each different tube length must be calibrated in this fashion.

SPECIALIZED MICROSCOPY

The image of a biological specimen can be formed and used in several ways. The most common form, known in all laboratories, is the use of color images, either natural color in the specimen or differential color applied to it. But unless one of a few vital stains are employed, the color image method cannot be used on live material. Most living material is transparent to light; that is, the light wave passes through it with very little loss of intensity, so other means for examining it must be employed.

Dark-Field Microscopy

In dark-field microscopy, the objects themselves turn the light into the microscope by reflecting it or scattering it, and the object appears luminous on a dark background. No light from an outside source reaches the eye. A stop in the substage condenser blocks out the central part of the solid cone of light formed in the condenser, and only oblique rays in a hollow cone of light, striking from the sides, illuminate the object.

The simplest form of dark field can be developed with a black central patch stop inserted in the carrier under the condenser. Dark-field elements in a threaded mount can be used in place of the upper lens of an

Abbe condenser. This is practical with all objectives if the numerical aperture does not exceed 0.85, and immersion oil is used with the slide.

Dark-field illumination can be modified with colored stops and outer rings below the substage condenser, instead of black stops, to produce optical staining. The central stop determines the background color and the outer ring, the color of the object, giving optical coloring to unstained objects and helping to reveal detail and structure.

Dark-field and optical staining can be useful for examining the following: (1) bacteria, yeasts, and molds; (2) body fluids, plant or animal; (3) colloids; (4) living organisms in water; (5) foods, fibers, and pigments; (6) insects and scales; (7) crystals; (8) bone, plant, and rock sections.

It is a common means of studying the results of microincineration, the investigation of minerals present in different parts of tissues. Paraffin sections of tissue fixed in a solution of formalin and alcohol are mounted on slides and placed in an electric furnace. The temperature is then slowly raised until all organic matter is burned away and only the mineral skeleton of the tissue is left. This appears white under dark-field observation.

Phase and Interference Microscopy

Although most living material is transparent to visible light, different components of tissue do alter to a different extent the phase of light waves passing through them. That is, the light velocity is altered, advanced, or retarded, and its vibration is said to be changed in phase. When two waves come together and are in phase, brightness increases, but if they are out of phase and are of equal amplitude, interference occurs and the eye sees black not light. Waves of intermediate difference in phase produce a series of grays. It is the aim of the phase microscope to change slight phase differences into amplitude differences and to produce a variation in intensity from light to dark contrast observable in the specimen under the microscope.

Both phase and interference microscopes follow this principle of interference phenomenon—combining light waves that are out of phase with each other to produce combined waves of greater and lesser amplitude. The means of applying the principle is different.

Phase microscopy

The cost of phase microscopy is not excessive, and the parts can be adapted to any bright-field microscope. The system requires phase contrast objectives (achromatic objectives with fixed-phase plates and a two-lens Abbe condenser with an iris diaphragm and a revolving ring below the condenser to carry four annular ring diaphragms, which are stops that produce different-sized cones of light). The correct annular ring diaphragm must be centered to the phase ring of the correct objective and must match its numerical aperture. The annular diaphragm causes the light to strike the object in the shape of a hollow cone and gives rise to two types of waves: one type passes straight through the object (undiffracted), and the other type is diffracted into a different course. Different components in the tissue diffract differently, producing the differences in intensity.

In phase microscopy, images exhibit a "halo" as a result of the diffraction of light at the phase-changing annulus (annular ring). The light that has passed straight through the specimen is made to interfere with the light diffracted sideways by it. Only the refracting structures are observed, and these edges and abrupt changes of refractive index produce the halo around the images.

Interference microscopy

A more sensitive and accurate instrument is the interference microscope, which is better adapted to measuring the refractive indices of a specimen. It produces an image of extremely high contrast with clearly defined boundaries as pseudo-3-dimensional images. It reveals colorless or homogeneously colored structures which would be invisible or not pronounced enough to be viewed in the conventional light microscope.

In this microscope the light splitting and recombining is carried out externally to the specimen. Birefringent plates (doubly refracting) are cemented to the top lens of the condenser and to the front lens of the objective. One set of rays passes through the object, the other set passes through a clear region at one side, and they are then recombined. Any phase differences between them remain constant and can interfere to give light or dark. A refractile object in one beam causes a change in light intensity. The two rays can be arranged to bring out a black background and a bright object or vice versa, and also to give high contrast between the object and inclusions in it. The lateral separation of the two rays is only a few microns. The Nomarski prism avoids the need for two prisms (one in the objective and one in the substage condensor) and can be used with all objectives. The prism is made of two cemented components of a uniaxial birefringent crystal mounted in an interference contrast slide in the substage condensor. It can be used with either reflected or transmitted light.

Phase versus interference microscopy

Both operate on the same principle—the interference phenomenon, which changes phase difference into amplitude difference. However, the special features of each should be considered in choosing between the two.

Phase: (1) Light passing through the specimen is made to interfere with light diffracted sideways by it. (2) It only shows up diffracting structures in the specimen; produces a halo. (3) Apparatus is simple, reasonable in cost, easy to operate, and can be added to any conventional microscope. (4) It can be used to study living material, cytoplasm, cell inclusions, nucleus, action of physical and chemical agents on living cells. (5) It is adequate for routine examination.

Interference: (1) Light splitting and recombining is carried out on outside of specimen and under control of experimenter. (2) No halo; variations in the optical path through the object are easily interpreted. Phase change can be measured. (3) Apparatus is expensive and complicated; requires constant checking and adjusting. The Nomarski prism simplifies this. (4) It can be used on living material to determine dry mass, for example, changes in mass during cell activity; protein distribution, both in cytoplasm and nucleus.

Polarizing Microscopy

Closely related to the above types of observation is the use of polarizing attachments. These may be used on most types of microscopes, but for continued use a polarizing microscope is preferable. When a ray of plane-polarized light (vibrating in one plane) falls on the object, it is split into two rays: one obeys laws of refraction, and the other passes through the object with a different velocity. After emerging from the object, the two rays are recombined, but because their velocity is different, they will be out of phase. This phase difference is the quantity measured in a polarizing microscope.

A polarizing prism in the fork substage (instead of a condenser) or a polaroid disc in the condenser slot polarizes the light so that it vibrates in one plane only. Part of the light (ordinary ray) is reflected to the side of the prism or disc and does not illuminate the object; the other part (extraordinary ray) continues straight through the prism to emerge as polarized light. (There is obviously a loss of light, so plenty of it must be used.) If the polarizer is a prism, it usually can be rotated 360° and the amount of rotation in the field checked.

In or above the ocular is fitted another polarizing prism, the analyzer, whose vibration direction is set at 90° to that of the polarizer. The extraordinary ray from the polarizer becomes the ordinary ray in the analyzer and is reflected out of the field, unless an anisotropic substance (doubly refracting when placed between the analyzer and polarizer) rotates the plane of polarization and interferes with the path of light. Such a substance divides the light from the polarizer into two beams, one of which passes through the analyzer, making the object appear to glow against a dark background. Isotropic substances (singly refracting) do not polarize and therefore do not divide the beam of light and do not glow.

Some uses of polarizing microscopy

1. To determine whether an object is isotropic or anisotropic (if it rotates the plane of polarization).
2. To determine differences in physical properties in different directions, such as those occurring in the study of mitotic spindles.
3. To reveal molecular orientation of structures, chemical constitution, chemical and physical intervention in the cell, pressures, or tensions; all can produce anisotropic effects.
4. To determine refractive indices.
5. Can be used on natural and artificial fibers; cellulose fibrils; lamellar plasma differentiation; pseudopodia; spindles and asters; nerve fibers, muscle fibers, chromosomes; chemical and mineral crystals; crystallized hormones and vitamins; dust counts; starch grains; and horn, claw, and bone sections.
6. Can be used on fresh unfixed material.

X-Ray Microscopes

Matter reflects X-rays too feebly to make it possible to construct refracting objectives for forming enlarged X-ray images of objects. Reflecting objectives have been made, but the resolution achieved by them is less than that obtained with a light microscope. Two alternative methods for obtaining an X-ray image are X-ray diffraction and X-ray shadow microscopy. X-ray diffraction is a complex technique; the position and amplitudes of the spectra diffracted by the specimen are determined experimentally, and the object which produces the spectra is reconstructed mathematically. Individual atoms in molecules can be resolved. In X-ray shadow microscopy magnified images may be produced by using X-rays, and good resolution can be obtained. The short wave length of X-rays makes them easily absorbed by biological material, and the laws of X-ray absorption can then be used to obtain quantitative information.

Ultraviolet Microscopy

Short waves beyond the visible spectrum have a profound effect upon living matter and are useful for the physical analysis of such matter. Some kinds of living material, under the influence of ultraviolet radiation, visibly radiate, glow, and fluoresce; some kinds remain dark. A microscope designed for this type of observation has optical glass in the objectives but is equipped with a quartz substage condenser. A quartz lamp condenser is part of the system to help concentrate the maximum amount of near-violet on the specimen. The object itself should not fluoresce but will absorb, partially absorb, or transmit the ultraviolet and thereby reveal structural differences in a photographic image without the use of stains. The specimen must be mounted in a nonfluorescent medium: water, glycerol, or mineral oil. Because glass fluoresces and will reduce contrast and possibly obscure some detail in the specimen, quartz cover glasses and slides should be used for best results, as should fluorescent-free immersion oil.

Resolution is increased over that of the conventional microscope, and differences in structure in the specimen are enhanced by the ultraviolet absorption. The technique is neither complicated nor expensive.

For greater resolution and higher selective absorption, a quartz ultraviolet microscope is better than the conventional ultraviolet microscope. It is, however, more difficult to handle and to focus, and it is more costly. The quartz microscope is equipped with fused quartz objectives, crystalline quartz eyepieces, a quartz substage condenser, and a quartz right angle prism. Quartz slides and cover glasses should be used. The light source must be ultraviolet.

Some cells are immune to ultraviolet, and others are only mildly affected; these can be examined with excellent results under this microscope. But in those that are killed by ultraviolet, formation of artifacts should be expected. Because of the possibility of damage to these cells, focusing can be done with visible radiation and then ultraviolet used only during photography or photoelectric measuring. Tissue cannot be fixed for ultraviolet (this decreases absorption), but ice-solvent or freeze-dry techniques can be used. Ultraviolet photomicrographs are difficult to interpret at times and their interpretation requires considerable experience.

The ultraviolet microscope has become a useful instrument for measuring the selective absorption of cellular components (measured at specific wavelengths). Its chief uses have been for the measurement of cellular concentration and localization of nucleic acids, RNA (ribonucleic acid), and DNA (deoxyribonucleic acid) content, and for observations on normal and neoplastic (tumoring) tissue (in which the nuclear RNA content is high). Some substances can interfere with such observations. Perhaps among the more interesting of these substances are the barbiturates, which absorb heavily and should not be used for sedation or anesthesia in animals to be prepared for this technique.

Fluorescent Microscopy

An object fluoresces when it absorbs ultraviolet light reflected on it or transmitted through it and then emits the energy as visible light of a specific violet, blue, green, yellow, orange, or red color. Secondary fluorescence can be induced by the use of fluorochromes (strongly fluorescent dyes or chemicals) applied to the specimen. Fluorochromes can induce fluorescence in objects that do not fluoresce naturally, and will intensify fluorescence in these that do. As with all ultraviolet preparations, nonfluorescent solutions, mounting media, and oil immersion must be used. Quartz slides and cover glasses may be used but are not essential for the techniques included in this book.

Apparatus required for fluorescent microscopy

Fluorescent microscopes can be purchased from several of the optical companies (Reichert, Zeiss-Winkel, American Optical). In addition to a compound microscope with a focusable condenser, minimum equipment includes a mercury lamp and filter and yellow check filters for each ocular eyepiece. Although some technicians feel that a front-surfaced aluminized mirror should be used, it is not essential. A monocular microscope gives a brighter image than a binocular microscope.

Keep all equipment free from grease, which is opaque to ultraviolet light.

Fluorescent microscopy differs from ultraviolet microscopy in that the fluorescent microscopy observes the specimen directly by innate autofluorescence or by secondary fluorescence induced by fluorochromes; ultraviolet microscopy records specific absorption of the ultraviolet light by certain structures in the specimen.

Uses for fluorochromes

Alkaline phosphatase (Burstone 1960*b*)
Amyloid and connective tissue (Vasquez and Dixon 1956; Vassar and Culling 1959, 1962)
Bacteria (Braunstein and Adriano 1961; Carter and Leise 1958; Moody et al. 1958; Yamaguchi and Braunstein 1965)
Blood vessels and lymphatics (Schlegel 1949)

Bone marrow (Werth 1953)

Cancer (Umiker and Pickle 1960; Vinegar 1961; von Bertalanffy and von Bertalanffy 1960)

Chromosomes (Schiffer and Tiju 1962)

Counterstaining (Smith et al. 1959)

Elastin (Shelley 1969)

Fat (Metcalf and Paton 1944)

Fungi (Clark and Hench 1962)

Glycogen (Burns and Neame 1966)

Granules of islets of Langerhans (Hartroft 1951)

Malarial parasites and other blood parasites (Ingram et al. 1961; Rothstein 1958; Rothstein and Brown 1960)

Motor nerve endings (Cole and Mielcarek 1962)

Mucin (Lev and Stoddard 1969; Sidney 1961)

Mucopolysaccharides (Kuyper 1957; Saunders 1962, 1964; Takeuchi 1962)

Nerve cells (Zeiger et al. 1951)

Nucleic acids and nucleoproteins (Armstrong 1956; Armstrong et al. 1957; de Bruyn et al. 1953)

Rickettsiae (Anderson and Greiff 1964)

Zinc (Sternberg et al. 1964)

See text for other uses.

Electron Microscopy

Among the methods used by investigators in biological and medical fields, one of the newest and fastest growing is electron microscopy. Work on electron microscopes began in the 1930s, and by the 1940s electron microscopes had been developed in different parts of the world. Knoel and Ruska in Germany had a practical, marketable research scope by 1939. England began work in 1936, and RCA in the United States began the development of one in 1938 and had one on the market in 1941. The electron miscrosope has a higher resolving power than the light microscope and reveals submicroscopic structures that were previously invisible. It uses electrons as the illuminating beam, and focuses the beam on the object by the use of magnets. As the beam passes through the object under observation, some of the electrons are scattered by the object, causing shadows of the scattered electrons on photographic film.

Electron beams must be handled in vacuum, therefore no wet or living tissue ordinarily can be observed. Sections must be thin $(0.01–0.05\mu)$ to prevent appreciable electron absorption, because the image depends on the differential scattering rather than absorption. The specimen is supported on a metal grid and is placed in the column at the electron-gun side of the object–lens aperture.

Special fixation, embedding, and sectioning are required to prepare ultrathin sections for use in the electron microscope. To facilitate the scattering of electrons, the density of the specimen must be increased. Because the usual fixing solutions do not meet this requirement, the so-called electron stains must be used in their place. These are compounds of heavy metals (tungsten and osmium), such as phosphotungstic acid and osmic acid. Potassium permanganate (Luft 1956), acrolein (Luft 1959), and, more recently, aldehydes (Sabatini et al. 1963, 1964) have been advocated. Only plastic embedding permits ultrathin sectioning. Conventional microtomes have been modified for thin sections, but specially designed microtomes are available. Glass or diamond knives are essential for ultrathin sectioning; steel knives are rapidly dulled by cutting plastic.

Three-dimensional images may be obtained by metal shadowing, and this is finding favor with many researchers. Metals, such as silver, chromium, palladium, platinum, and others, are evaporated on the specimen at a small angle. A thin film of the metal covers the specimen except for a "shadow" caused by the specimen being in the way of passage of the metal. From the shape and dimension of the shadow, the height and contour of the specimen can be observed. Of the many references, important ones are Glauert (1972, 1974); Hayat (1970, 1972, 1975); Kay (1965); Koehler (1973); Mercer and Birbeck (1961); Parsons (1970); Pease (1964); and Schultze (1969).

Further references for microscopy are as follows: Barer (1956, 1959); Beck (1938); Belling (1930); Bennett (1950); Cosslett and Nixon (1960); Davies (1958); Dempster (1944a, 1944b); Engström (1956, 1959); Gage (1943); Ham (1957); Munz and Charipper (1943); Needham (1958); Nurnberger (1955); Oster (1955); Popper and Szanto (1950); Richards (1954); Ruch (1955); Scott (1955); Shillaber (1944); Vickers (1956); and Walker (1958).

I recommend the excellent booklets about special types of microscopes that are easily obtained from microscope suppliers and are descriptive of parts and their use and care. These are on file next to my microscope at all times.

Chapter 7
Stains and Staining Action

Because most tissues do not retain enough color after processing to make their components visible under a bright-field microscope, it is expedient to add colors to tissues by staining them. Correctly chosen stains aid in identifying tissues and their elements, and in diagnosing pathological conditions. Knowledge about the structure and action of the chemicals and stains used for tissue identification must be reviewed and understood. The subject is extensive, and the student is advised to consult specialized references for detail that cannot be included here. One of the best is H. J. Conn's *Biological Stains* (1969).

NATURAL DYES

Cochineal and Carmine

These are members of a group of dyes called natural dyes. Unlike other natural dyes, cochineal and carmine are derived from an animal source—a minute insect, *Coccus cacti,* which lives on spineless cacti. The dye is present as a purple sap in the females, which are harvested, dried, and pulverized to produce cochineal. This dye by itself has little affinity for tissue unless iron, aluminum, or some other metal is present. With the salt of one of these metals as a mordant (see p. 88), staining will result. Alum cochineal, a commonly used form of this dye with mordant, is an efficient nuclear stain.

The dye carmine is derived from cochineal by boiling the cochineal with a salt, usually alum, to produce a precipitate. This precipitate is insoluble in water, and before it can be used as a stain it must be converted into a soluble compound such as ammoniacal carmine or acetocarmine, in a process that will be described in the section on mordants (p. 88).

Hematoxylin

In many respects, hematoxylin can be regarded as the most important of the natural dyes. It was one of the first histological dyes and remains one of the most widely known and used dyes. Hematoxylin is extracted from the heartwood of longwood trees from South and Central America and the West Indies. The tree is *Caesalpina campechianum,* one of the legumes similar to acacia or cassia trees. The crude material is exported as logs, as chips, or as dried aqueous extract of the heartwood. It is then extracted with ether in a continuous extraction apparatus evaported to dryness, dissolved in water, filtered, and crystallized out of solution. All of these steps are slow and difficult to handle and require costly apparatus, thus making hematoxylin one of our most expensive dyes.

In the crystallized condition it is not yet a dye, and its color must be allowed to develop by oxidation into hematein (color acid—no relation to hematin, the colored constituent of red blood cells). Oxidation may be accomplished in either of two ways: "naturally" (a slow process of exposure to air for 3 to 6 weeks, as in Heidenhain hematoxylin) or artificially by the use of mercuric oxide, hydrogen peroxide, or another oxidizing agent (a more rapid process, as in Harris hematoxylin). Used alone, hematein is only a weak and diffuse dye with little affinity for tissues. A weak acid will not combine with nuclear elements in sufficient quantity to produce efficient staining. Some form of mordanting is required to form a base from this dye, which will then stain the acidic nuclear elements. The most commonly used mordants are alum salts of aluminum, potassium, or iron.

When properly oxidized, hematoxylin is an exceedingly powerful dye with various shades of staining from purples, through blues, and into blue blacks. The iron-mordanted form is one of the most valuable dyes for mitotic study, and it gives to the chromatin a precise black or blue-black color. This black color is the result of the presence in hematoxylin of some tannin, and the latter in combination with iron salts produces a lasting black color.

Hematoxylin staining is discussed further in Chapter 9.

Other Natural Dyes

Other natural dyes are saffron from stigmas of *Crocus;* indigo from plants of the genus *Indigofera;* berberine from barberry; orcein and litmus from the lichens *Lecanora* and *Rocella;* and brazilin, chiefly from a few species of *Caesalpina,* a tropicopolitan genus of leguminous trees (Mohr 1950). Orcein, a specific dye for elastin (present in elastic fibers), is prepared by boiling the plants in water. The lecanoric acid in them splits to produce orcinol—a resorcinol with a methyl group attached to it. Orcinol with NH_3 (ammonia) and atmospheric oxygen forms orcein.

MORDANTS

Metallic mordants have been used for over one hundred years, but the exact nature of their action still is somewhat mysterious.

A mordant is a metallic salt or hydroxide that combines with a dye radicle to form an insoluble compound. Such a compound is called a lake. For example, carmine dissolved in a solution of aluminum sulfate, becomes positively charged and acts as a highly basic dye. The term *mordant* should not be used for all substances that increase staining action, but only for salts and hydroxides of di- and trivalent metals.

The use of mordants in staining tissues has many advantages. Once the mordant–dye compound has formed in the tissue, the dye is relatively permanent, is insoluble in neutral solutions, and can be followed by many other forms of staining. Dehydration will not decolorize the tissue. Blocking and extraction methods have very little effect on mordant dyes. There is considerable flexibility in mordant–dye procedures, as the mordant can be used before the dye, together with the dye, or following the dye (rare).

Carmine and hematoxylin are two dyes often used with mordants for staining tissue. For these dyes, the mordants commonly used are aluminum, ferric and chromium salts, and alums (potassium alum, ammonium alum, iron alum, and chrome alum). A solution of ferric chloride can be used in place of iron alum, and Cole (1933) recommends the use of a phosphate-ripened hematoxylin with it. The ferric chloride causes the tissues to stain more rapidly and more intensely than does the iron alum. Iron alum solutions are quite acid; the greater the amount if iron alum present, the more acid the solution. Adding even a weak acid, such as acetic acid, to an alum hematoxylin makes the solution more selective for nuclei.

For long-lasting solutions of combined mordant and dye, mordants with little or no oxidizing action must be used: ammonium alum, potassium alum, phosphotungstic acid, phosphomolybdic acid, and iron alum–ferrous sulfate. If a long life is not required, it is possible to use mordants with a vigorous oxidizing action: ferric chloride, ferric acetate, and ferric alum (Cole 1943). Since the usefulness of these solutions lasts only a matter of hours, they must be prepared immediately before use. In iron-hematoxylin solutions that are premixed, a high mordant-to-dye ratio tends to produce nuclear staining, while a high dye-to-mordant ratio favors myelin staining. Though the term *iron hematoxylin* is widely used, it is actually chemically incorrect; hematoxylin cannot form a lake, but first must be oxidized to hematein (Puchtler and Sweat 1964).

Hematein forms blue-green lakes with copper, lilac with nickel, red with tin, dark brown with lead, and purple with bismuth.

When using two separate solutions, a mordant of any kind can be used if followed by a well-ripened hematoxylin solution. If the solution is unripened, the mordant should include a substance of considerable oxidizing power, a ferric or chromium salt. The value of two solutions lies in the fact that the dye can be preceded by a salt, which cannot be used in combination with the dye in a single solution. For example, ferric chloride, when added to an ammonia-ripened hematoxylin, will throw down a precipitate of ferric hydroxide. Double mordanting can be profitable; a mordant followed by a solution of hematoxylin containing a mordant gives excellent results. The mordants for separate use are ammonium or potassium alum, ferric ammonium alum (2–3 drops of HCl increases contrasts), and ferric chloride (HCl increases contrast). They yield the following colors with hematoxylin: ammonium alum, bright blue nuclei; potassium alum, lilac or violet; chrome alum, cold gray blue; iron alum, blue to black (Cole 1943).

When used with hematoxylin, some mordants are more effective than others for staining particular types of material. The following is a list of tissue elements and the metals effectively used on them (Mallory 1938).

Nuclei: aluminum, iron, tungsten
Myelin sheaths: chromium, iron, copper
Elastic fibers: iron
Collagen: molybdenum
Fibroglia, myoglia, neuroglia, epithelial fibers: tungsten
Axis cylinders: lead
Mucin: iron
Fibrin: tungsten

Progressive and Regressive Staining

Lakes, formed by mordant and dye, can be used progressively, but when mordant and dye are used separately, regressive staining usually is more effective. In progressive staining the stain is added to the tissue until the correct depth of color is reached. In regressive staining the sections are overstained, and excessive amounts are removed by one of the following methods.

Method I: Excess Mordant

With an excess of free mordant present outside the tissue, the mordant–dye complex in the tissue is broken up, and, since the amount of mordant in the tissue is smaller than that in the differentiating fluid, the dye moves out of the tissue into the fluid. (If the tissue is left long enough in the fluid, it is conceivable that most of the dye could move out of it into the excess mordant, causing the sections to become colorless.) Because the nuclei hold considerably more dye than the cytoplasm, most of the dye is lost from the cytoplasm, but some still remains in the nuclei. When enough dye has been extracted from the cytoplasm, and when the correct intensity is left in the nuclei, remove the slides from the mordant and wash them thoroughly, usually in running water, to remove excess mordant. Remaining traces can cause the stain to fade in time.

Method II: Acids

Acids are effective differentiators for some dyes, but there is no completely adequate explanation for their action.

Method III: Oxidizers

Oxidizers furnish a third method of regressive staining, and by this method the dye can be oxidized to a colorless condition. Oxidizers are slow in action; the parts of the cell that hold only a small amount of dye will be bleached before those containing greater quantities of it. A complete explanation for this is not available. Picric acid, a commonly used oxidizer, has both a moderate oxidizing and a weak acidic action.

Accentuators and Accelerators

Not to be confused with mordants are accentuators and accelerators. Accentuators are substances that, contrary to the action of mordants, do not become a part of the dye complex or lake. Instead they increase the selectivity or stainability of the dye, e.g., phenol in carbolfuchsin. Accelerators, as their name implies, accelerate the action of the dye—usually of importance in silver impregnation—e.g., chloral hydrate.

SYNTHETIC DYES

Natural dyes had no competition until the middle of the nineteenth century when William Perkin worked out the processes for making aniline or coal-tar dyes. These were the first synthetic dyes.

Synthetic dyes, like natural dyes, can be used either progressively or regressively. An acid solution is often used to remove excess basic dye; an alkaline solution is used to remove excess acidic dye. In some cases alcohol can be used as a differentiator, particularly for basic dyes; but in general a sharper differentiation is achieved by using an acid.

The real importance of synthetic dyes lies in their use for double and triple staining—the use of two or more stains on the same slide. The dyes, if properly chosen, will stain histologically; that is, each dye, because of a known specificity, will stain only specific parts of the cells. This type of staining has obvious advantages. Owing to their chemical nature, synthetic dyes make this kind of staining possible. If the dyes are synthesized, their formula can be controlled, and the significant part of the dye is either anionic (acid) or cationic (basic) in action. Actually, dye powder as purchased is a salt, but the salts of the so-called basic dyes give up OH^- ions and act as cations, and the acidic dyes give up H^+ ions and act as anions. Therefore, an acid, or anionic, dye is the salt of a color acid, usually a sodium salt; a basic, or cationic, dye is the salt of a color base, usually a chloride. (See structure of synthetic dyes, p. 93.) The terms anionic and cationic are more appropriate than acidic and basic and are becoming more widely used. Basic dyes have an affinity for nuclei, which are basophilic (readily stained by basic dyes), and acidic dyes have an affinity for the cytoplasm, which is acidophilic (readily stained by acidic dyes).

When the sodium salt of an acid dye and the chloride of a basic dye are mixed, there is a tendency for ions to interchange. The so-called neutral dyes are formed and give results differing from those obtained with ordinary double staining using separate acid and base dyes. Polychroming is a process in which a dye forms other dyes spontaneously—the basis of the development of modern blood staining (Romanovsky stains). Methylene blue in solution is oxidized into one or more compounds of lower methylation. Therefore, all methylene blue solutions, on standing, contain lower homologs, primarily azure A, B, and C, and the methylene blue is now called polychrome methylene blue. The azures are more violet in color and more selective in their action than unpolychromed methylene blue and account for the differential action of blood stains in the differentiation of white blood cells. When eosin is added, it enters into chemical combination with the above basic dyes to form an insoluble precipitate, which, when redissolved, is the basis of Wright, Leishman, and May-Grunwald stains. The pure azures may now be purchased. Pure azure A-eosinate is equal to a traditional Giemsa stain. Azure C is not as good. Azure B is best and, as an eosinate, can be substituted for May-Grunwald staining.

The dyes are not stable in aqueous solutions and precipitate out; the dye powder must therefore be dissolved in alcohol. Since alcoholic solutions do not stain well, different methods of using them have been employed in the Romanovsky stains. Sometimes they are used immediately after mixing (Giemsa), or, after applying the alcoholic solution, the stain is diluted with water (Wright). How the stains act is uncertain. Probably certain parts of the cells have an affinity for the neutral stain, others have an affinity for the basic dyes and break up the stain to acquire the basic part; other cell parts do the same with the acidic portion. These actions by the cell parts are termed, respectively, neutrophilic, basophilic, and oxyphilic. The cells are stained in a solution of the proper pH, or they are overstained in a neutral solution and differentiated to the proper coloration of their parts. (The Maximow method uses alcohol; the Giemsa method, dilute acetic acid; and the Wolbach method, rosin.)

Polychroming must not be confused with another type of staining, metachroming, in which certain substances are stained in one color and others in another color by the same dye. The definition of metachromasia given by Paddy (1970) is "a characteristic reversible color change that any dye may undergo by virtue of a change in its environment not involving chemical reaction of the dye."

The explanation for the reaction to the dye Thionin may be as follows: Thionin stains chromatin blue; it stains mucus, ground substance of cartilage, and granules of mast cells red. The dye seems to exist in aqueous solution in two forms: (1) the normal color, blue; and (2) the metachromatic color, red. Both forms are always present, but the red is in a polymerized form of the blue, and color change may be due to some interaction between the chromophores present. The blue form is favored by an increase of temperature, a lowering of pH, a decrease of dye concentration, or the addition of salts, alcohol, or acetone. The red form is favored by a decrease of temperature, a raising of pH, or an increase in concentration of dye (Bergeron and Singer 1958).

Tissue components showing metachromasia are made up of large anionic molecules containing sulfate, phosphate, or carboxylic acid radicals in abundance, and the groups are close enough to permit secondary bonding between the dye molecules, thus permitting polymerization of the bound dye. Mucin, the ground substance of cartilage, and the granules of mast cells contain substances of this nature and therefore take up the metachromatic form.

Some other metachromatic stains are methyl violet, brilliant cresyl blue, azure A, B, and C, toluidine blue, new methylene blue, methylene blue, crystal violet, safranine, bismarck brown, and basic fuchsin.

The majority of dyes do not stain metachromatically, but are orthochromatic in action. This means that their action is direct and predictable under normal conditions; if it is a blue dye, it stains blue; if it is a green dye, it stains green; and so forth.

Some dyes are not stable in solution and will gradually give rise to other colored agents, which should be considered impurities. The presence of such agents is called allochromasia. Orthochromatic dyes may give rise to metachromatic impurities by allochromasy. Nile blue solutions, an example of allochromasia, contain cation of the dye (blue), anion of the dye (sulfate), Nile red (red or rose), and imino base of Nile blue (orange yellow).

A chromotrope is a substance that can alter the color of a metachromatic dye. Examples of chromotropic tissue substances are the matrix of cartilage, granules of mast cells, and secretions of some mucus glands. Bathochromic (Baker 1958) means a shift in the peak of the absorption curve of a dye in solution toward longer wavelengths. Hypsochromic means a shift toward the shorter wavelengths.

Structure of Synthetic Dyes

The synthetic (coal tar or aniline) dyes are derivatives of benzene, all built on the benzene ring:

There are an infinite number of ways in which this ring can combine with other radicals to produce compounds of various degrees of complexity. Certain chemical groups called chromophores ($C=C$, $C=O$, $C=S$, $C=N$, $N=N$, $N=O$, and NO_2) are associated with color. When these groups are attached to a benzene derivative, the compound acquires the property of color and is known as a chromogen. A chromogen, however, is not yet a dye. It has no affinity for tissues, will coat them only mechanically, and can be easily removed by mechanical means. The compound must also contain a group that gives it the ability to form a union with some tissue end group, either directly or through chelate action of a mordant metal. The commonest type group is one which permits electrolytic dissociation (the formation of anions and cations). This auxiliary group, known as an auxochrome, furnishes combining properties to the compound.

A simple demonstration of the principle of synthetic dye formation is picric acid. The nitro group ($-NO_2$) is a chromophore, and three of these can displace three hydrogen atoms on the benzene ring to produce trinitrobenzene:

This is yellow, a chromogen, but not yet a dye. Replace one or more H^+ with an OH^-, the auxochrome, and produce dissociable picric acid:

Since the auxochromes are the salt-forming groups, they usually determine the action of the dye; the more basic or acidic groups present, the more basic or acidic is the action of the dye. If both a basic and an acidic group are present, the basic group predominates but is weakened in action by the acidic group.

Most commercial dyes are sold as salt. An acid dye is the salt of a color acid, usually sodium salt; the basic dye is the salt of a color base, usually a chloride. The basic chromophores are: azo group ($-N=N-$), azin

group $\left(\begin{array}{c} N \\ | \\ N \end{array} \right)$ and indamine group ($-N=$). The acid chromophores

are: nitro group ($-NO_2$) and quinoid benzene ring:

Simpler compounds can be converted into more complex ones by substituting radicals for the H^+ atoms present. As more H^+ atoms are replaced by radicals, the deeper the color becomes. The simplest dyes tend to be yellow. As methyls or other groups are added, the colors pass through reds, violets, blues, and greens. Using ethyl groups instead of methyl groups deepens the color.

Nomenclature of stains has no absolute conformity. For the names of some stains, the color is used (orange G, Martius yellow); for others, a chemical term (methyl green). If the term is followed by a letter or numeral (Sudan III, IV; ponceau 2R, 4R), one dye is being distinguished from a related one. B indicates a more bluish color, Y or G, a more yellowish color; WS means water-soluble; A, B, and C distinguish among azures.

NATURE OF STAINING ACTION

Biologists and biochemists disagree about the nature of staining action—is it chemical, physical, or a combination of both? If chemical, this can mean that some parts of the cells are acid and others alkaline, the acid parts tending to combine with cations and the alkaline with anions. Absorption and diffusion of the dye occur, penetrating the cellular elements and combining with them. This action can be combined with physical action, where there is an adsorption of the dye, an attraction of plus and minus charges for each other, and a condensation of the dye on the surface of the cell parts. Minute particles of the dye are deposited on the surface of the tissue by selective adsorption and then enter into combination with the tissue. The proteins, nucleic acids, and other components of the protoplasm proceed along lines of chemical laws by exchanging ions. Stearn and Stearn (1929, 1930) maintain that the confusion lies in the term adsorption, that it may be either a chemical or a physical force, and that the adsorbent can form ions and then proceed along chemical lines. The development of physical chemistry has helped to resolve the question. There appears to be no doubt that both physical and chemical factors take part in the mechanism of salt formation—the basic action of stains.

In any case, the staining properties depend on three factors:

1. Strength of dye.
2. Rate of ionization of tissue proteins and dyes.
3. pH value of dye solution and tissue proteins.

In addition, staining can be affected by other conditions:

1. Alcoholic or aqueous solution of dye.
2. Low or high temperature during reaction.
3. Simple or multiple combinations of dyes.
4. Strong or weak concentration of dye in solution.
5. Permeability of tissues and dyes.

STANDARDIZATION OF STAINS

In the early days of tissue staining, it was difficult to secure reliable dyes. The textile dye industry was the sole source of dyes, and products received were often unsatisfactory. There was no adequate standardization of color, and no standardization of chemical content was undertaken. There were variations in color and quality and many impurities in the dyes. Grübler in Germany, the first to try to standardize dyes, built a highly specialized business in this field. He did not actually manufacture dyes, but bought up batches from other firms and tested them for technical use. After the beginning of the twentieth century, certain events changed Grübler's hold on the business of standardization. Perhaps of greatest influence were the two world wars, causing Germany to lose its monopoly in the dye industry. No country could afford to remain dependent on another country for its source of dye.

The lack of German dyes led to a new form of standardization in the United States. A body called the Commission on Standardization of Biological Stains was organized, and it later became the Biological Stain Commission. The object of this commission is to work with the manufacturers, showing them what the biologists require, testing their products, and permitting approved batches to be put on the market with the stamp of the commission on them. Specifications of the most important dyes have been drawn up by the commission. The specifications are partly chemical and partly spectrophotometric; they contain detailed instructions for testing the dyes and the results to be expected from the tests.

Batches of dyes approved by the commission bear a special label furnished by the commission and known as a certification label. On it is a C.I. (Color Index) number, indicating the certification number of that particular batch. This number means that (1) a sample of that batch was submitted to the commission for testing, and a portion of it is on file; (2) the sample proved true to type of spectrophotometric tests; (3) its dye content met specifications and is correctly indicated on the label; (4) it was tested by experts in the procedure named on the label and found satisfactory; and (5) no other batch can be sold under the same certification number. Any description of the use of the stain should be followed by its C.I. number or the letters C.C., indicating that it is Commission Certified. C.I. numbers used in this text are from the new revised list.

The actual dye content of commercial dyes may vary from 32% to 99%, and some batches may contain colored compounds with staining properties different from those of the actual dyes. Even different batches of a dye can vary. Rosenthal et al. (1965) developed a simple paper chromatography method for testing dyes that requires only simple laboratory equipment.

Unless otherwise specified, dyes can be purchased from most of the scientific supply houses with chemical outlets: Curtin Matheson Scientific Co., Hartman-Leddon Co., Fisher Scientific Co., Allied Chemical Corporation (National Aniline Division), and others.

Foreign stains can be obtained from the following sources:

Esbe Laboratory Supplies (outlet for Gurr Stains).
Edward Gurr.
G. T. Gurr.
Searle Diagnostic, G. D. Searle.
British Drug Houses.
Pfatz and Bauer, Inc.
Roboz Surgical Instrument Company (American agent for Chroma-Gesellschaft [Dr. K. Holborn and Söhne, G. Grübler]).
Biomedical Specialties (Gurr stains).

For greater detail concerning stains and their actions, read: Baker (1945, 1958); Cole (1933, 1943), Conn (1946, 1948, 1961, 1969); Gurr (1960); Holmes (1929); Singer (1952); and Stearn and Stearn (1929, 1930). Concerning the present status of dyes read Stain Technology *50* (1975), pp. 65–81.

Chapter 8
Mounting and Staining Procedures

Certain standard principles usually apply to the staining of tissue on slides, but there are many exceptions and variations.

Paraffin sections first must be deparaffinized, because most stains are applied in either aqueous or alcoholic solutions and will not penetrate efficiently through paraffin-infiltrated tissues. The customary solvent for paraffin is xylene. This is followed by the removal of the xylene with absolute alcohol because stains rarely can be applied successfully in a xylene medium. After the removal of the xylene, the general rule is to transfer the slides to a medium comparable to the solvent of the dye being used. That is, if the dye is a water solution, the slides are hydrated through a series of decreasing alcoholic and increasing aqueous dilutions, such as 95%, 80%, 70%, and 50% alcohol, and finally they go into water. If the dye is dissolved in a 50% alcoholic solution, then the slides are carried only to 50% alcohol before going into the staining solution.

During the hydration process, undesirable pigments or other materials (mercuric chloride crystals, formalin pigment, etc.) are removed and the slides are washed well to remove the reagent. Counterstains (background color) or other special treatments are applied in their proper sequence, to allow each dye or chemical to maintain its specific effect on the tissue elements. Improper sequences of staining, decolorizing, or other solutions can result in a poorly stained slide. Hematoxylin-eosin staining (p. 119) is a simple example: if the eosin (counterstain) is applied before the hematoxylin (nuclear) stain, the eosin stain will be completely removed by the action of the hematoxylin stain.

MECHANICAL AIDS

If slides are to be transported in quantity, rather than individually, several types of holders on the market are useful. One type holds 50 slides;[1] one holds six slides.[2] Another model holds five slides and fits as a cap on coplin jars.[3] Some are baskets; others are clips that fit into special staining dishes. The tissue-processing machines are equipped with slide carriers to fit the instrument. Phosphor-bronze spring wire, 0.05 inch diameter, can be fashioned into coils of $\frac{3}{4}$ inch in diameter and cut into any length to hold from 3 to 20 slides (Fig. 8-1). This combination of slides and coil can be used in rectangular staining dishes with the slides resting on their long edges, or standing upright in tall stender dishes.[4]

Cover glass mounts can be stained by using Chen's (1942) staining rack[5] (Fig. 8-2), which holds 12 22 mm cover glasses, or racks[5] that hold either 5 or 30 cover glasses. Cover-glass staining dishes are far less convenient to use than racks and lead to breakage and wasted time in handling; each cover glass has to be processed individually with forceps.

PROCESSING SLIDES FOR MOUNTING

A final processing of slides is necessary to make permanent preparations for examination and storage without deterioration. All alcohol and water must be extracted (with certain exceptions), and a medium applied that maintains the tissues in a clear and transparent condition, does not alter the color or intensity of the stains, and holds a cover glass in place. The water is removed through increasing concentration of alcohol until absolute alcohol is reached. The final reagent is xylene (or a similar solvent), which removes the alcohol and makes the sections lose their opacity and become clear. Finally, a mounting medium (mountant) is applied and the cover glass lowered into place, completely covering the sections. (The solvent of the mounting medium is usually either toluene or xylene.)

COVER GLASS MOUNTING

The tidiest method for mounting a cover glass is shown in Figure 8-3: Apply a thin streak of medium on the cover glass; turn cover glass over

---

[1] Arthur H. Thomas Co.
[2] Ward's Natural Science Establishment.
[3] Peel-A-Way Scientific Co.; Fisher Scientific Co.; Scientific Products.
[4] VWR Scientific.
[5] Lipshaw Manufacturing Co.

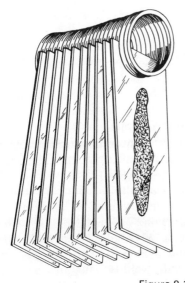

Figure 8-1
A staining coil carrying multiple slides.

Figure 8-2
T. T. Chen's cover glass staining rack in a tall slender dish. Handle for rack lies beside dish.

and rest it on edge on the slide, beside the sections; ease cover glass into place slowly to allow air to escape from under cover glass; press firmly from center outward to distribute the medium evenly. An alternate method is to apply mounting medium along one edge of sections on slide; rest one edge of cover glass adjacent to mounting medium; lower gradually to ease out air without bubble formation; press gently in place.

If, during microscopic examination of stained and mounted slides, dull black spots replace nuclear detail (Fig. 8-4b), the clearing solution was partly evaporated out of the sections before the cover glass was in place. Return such slides to xylene, dissolve all mounting medium to allow the air to leave the sections, and remount. If the slides appear dull, almost milky, instead of crystal clear, water is present. Remove all mounting medium in xylene and return slides to absolute alcohol—preferably a fresh solution—clear, and remount.

MOUNTING MEDIA (MOUNTANTS)

Natural Resins

Formerly, natural resins were used as mounting media: Canada balsam, composed of terpenes, carboxylic acid, and esters; gum damar, composed

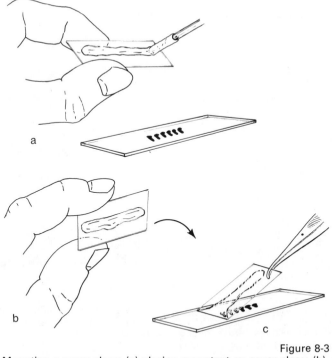

Figure 8-3
Mounting a cover glass: (a) placing mountant on cover glass; (b) turning cover over; and (c) lowering it onto slide.

of unsaturated resin acids and a little ester; and gum sandarac, an unsaturated acid resin. These dried slowly, were variable in composition, and unpredictable in behavior. Some developed acidity, faded stains, and would turn yellow and crack after a few years. One still used, particularly for blood smears, is Euparal (refractive index 1.483), a eutectic (melts at low temperature) mixture of oil of eucalyptus, gum sandarac, salol, paraldehyde, menthol, and camphor. Euparal Vert (green because of copper salt content) is claimed to intensify hematoxylin stains. Dried smears may be mounted directly in Euparal, but sections must be carried into 95% alcohol and then into Euparal essence before mounting. (If alcohol is carried into the Euparal, sections will fade.) Drain off excess essence and apply Euparal.

Synthetic Resins

The synthetic resins now available are superior to natural resins in most respects; their composition can be controlled; they are stable and inert;

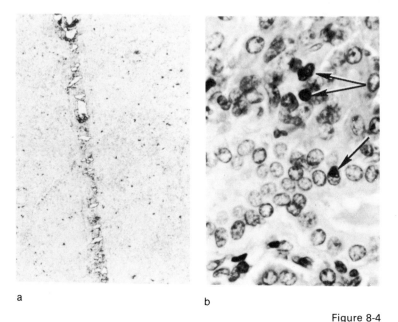

a   b

Figure 8-4
Common imperfections in sections: (a) knife scratch in paraffin section as seen under microscope before deparaffinization; (b) black areas caused by drying of sections before cover glass was applied.

they dissolve readily in xylene or toluene, do not require long drying, and adhere tightly to glass; and they have the correct refractive indices, are pale in color, and do not yellow with age. The resins should have refractive indices of 1.53 to 1.54 or better (Lillie et al. 1953).

The most widely used synthetics are the β-pinene polymers (terpene resins), such as Permount; Piccolyte, refractive index 1.51 to 1.52; HRS (Harleco Synthetic Resin), refractive index 1.5202; Biloid Synthetic Resin, refractive index 1.5396; Kleermount; Histoclad, refractive index 1.54; Pro-Texx; Technicon Mounting Medium, refractive index 1.5649; Namount; and Preservaslide.[6] There is no reason to recommend one of the above products over the others; all are equally efficient. They are soluble in xylene, toluene, aromatic hydrocarbon solvents, and in chlorinated hydrocarbons such as chloroform, but not in dioxane.

Lillie et al. (1950) recommend Lustron 2020[7] for good preservation of

---

[6] These products are manufactured by the following companies, respectively: Fisher Scientific Co.; General Biological Supply Co.; Hartman-Leddon Co.; Will Corp.; Carolina Biological Supply Co.; Clay Adams Co.; Scientific Products; Technicon Co., Allied Chemical Corp.; and Curtin Matheson Scientific Co.

[7] Mosanto Chemical Corp.

Prussian blue mounts, which tend to fade in some resins. Lustron is a polystyrene, refractive index 1.59, and is water resistant and soluble in varying degrees in aromatic and chlorinated hydrocarbons. Lillie et al. use it in diethylbenzene as solvent. I have had excellent results using toluene or xylene without the formation of air bays as the xylene evaporated. If air bays do form, diethylbenzene is recommended; Lillie et al. suggest adding 5 ml dibutylphthalate to 70 ml xylene and 25 g polystyrene. This does not delay hardening.

There are also the German synthetics, Caedax, Eukitt,[8] and Rhenohistol; and the British synthetics, Xam, Cristalite, and Clearmount. These are similar in most respects to American products.

Most synthetic mountants are allowed to air dry, but leaving synthetic resin mounts on a slide warmer overnight will make them relatively firm. Slides can be cleaned, marked with ink, and stored without dislodging cover glasses.

Epoxy Resin Mounting Medium (Ors 1971)

Solution A

Epon 812[9]	62.0 ml
DDSA[9]	100.0 ml

Solution B

Epon 812[9]	100.0 ml
NMA[9]	89.0 ml

Store both solutions at 5°C and bring to room temperature before mixing. Mix A and B in equal parts.

To each 10 ml add:

DMP 30[9]	0.15 ml

Stir thoroughly and let stand until bubbles disappear. Ready for use and can be stored for several months at room temperature. Harden mounted slide overnight at 60°C. Refractive index 1.488.

---

[8] Calibrated Instruments.
[9] Polysciences Inc.

Aqueous Mounting Media

Aqueous mounting media are indispensable for the preservation of tissue elements that are soluble in alcohol or hydrocarbons or are demonstrated by the use of dyes soluble in these fluids. Nearly all media that have been proposed use (1) gelatin and gum arabic as solidifying agents with water; (2) sugars and salts for increasing the refractive index; and (3) glycerols and glycine as plasticizing agents. Gum arabic slowly hardens by drying, gelatin sets by cooling, and the glycerin keeps the tissues from cracking or overdrying. The addition of phenol, thymol, merthiolate, or Zephiran prevents mold growth.

Kaiser Glycerin Jelly

water	52.0 ml
gelatin	8.0 g
glycerin	50.0 ml

Add as a preservative either (1) carbolic acid (phenol), 0.1 g; (2) merthiolate, 0.01 g; or (3) Zephiran, 0.1 ml.

Allow gelatin to soak for 1–2 hours in water, and add glycerin and preservative. Warm for 10–15 minutes (not above 75°C) and stir until mixture is homogeneous. This keeps well in covered jar in refrigerator. If heated above 75°C, the gelatin may be transformed into metagelatin, which will not harden at room temperature.

Von Apathy Gum Syrup

gum arabic (acacia)	50.0 g
sucrose (cane sugar)	50.0 g
distilled water.............................	50.0 ml
formalin	1.0 ml

Dissolve lumps of gum arabic in warm water; add sucrose. When dissolved, filter and allow to cool. Add formalin. Do not use powdered form of gum; it makes a milky solution, whereas the lump form produces a clear medium. The cover glass does not have to be sealed as it does with Kaiser glycerin jelly above.

Farrants Medium

distilled water..............................	40.0 ml
gum arabic (acacia)	40.0 g
glycerin	20.0 ml
carbolic acid (phenol)	0.1 g
(or merthiolate or Zephiran)	

This solution can be purchased from Amend Drug and Chemical Company.

Gray and Wess Medium (PVA) (1952)

PVA (polyvinyl alcohol) 71–24 powder[10]	2.0 g
70% acetone	7.0 ml
glycerin	5.0 ml
lactic acid	5.0 ml
distilled water............................	10.0 ml

Make a paste of the dry alcohol with acetone. Mix half of water with glycerin and lactic acid, stir into paste. Add rest of water drop by drop while stirring. Solution will be cloudy but become transparent if warmed in water bath about 10 minutes.

Hoyer Medium (Beek 1955)

gum arabic	30.0 g
glycerin	16.0 ml
chloral hydrate (a drug)	200.0 g
distilled water............................	50.0 ml

Dissolve gum in water; a little heat helps. Add chloral hydrate, then glycerin.

PVA Mounting Medium (polyvinyl alcohol)

PVA 71–24 powder	15.0 g
water	100.0 ml

---

[10] E. I. du Pont de Nemours and Co.

Add PVA powder slowly to cold water. Heat in water bath (80°C) and stir until the solution becomes as viscous as thick molasses. Filter out undissolved lumps through two layers of cheesecloth. Solution will appear milky, but after standing for several hours it will clear. Spread a thin film over stained blood smears when cover glasses are undesirable.

The following solutions are recommended for small whole mounts, insects, worms, tiny invertebrates, etc.

Berlese Mounting Medium

Gray formula (1952)

water	10.0 ml
glacial acetic acid	3.0 ml
dextrose syrup	5.0 ml
gum arabic (acacia)	8.0 g
chloral hydrate (a drug)	75.0 g

Mix water with acid and syrup; dissolve gum in this. This requires a week or more. Stir at intervals. When solution is complete, add chloral hydrate. This is one of the best media for insects.

Thick formula

distilled water............................	40.0 ml
saturated aqueous solution of chloral	
hydrate (a drug)	30.0 ml
gum arabic (acacia)	25.0 g
glycerin	5.0 ml

Dissolve gum arabic in water, add chloral hydrate and then glycerin. Filter.

Thin formula

distilled water............................	100.0 ml
gum arabic (acacia)	60.0 g
glycerin	40.0 ml
saturated aqueous solution of chloral	
hydrate	100.0 ml

Prepare like thick formula.

Lactophenol Mounting Medium

melted phenol (carbolic acid)	3 parts
lactic acid	1 part
glycerin	2 parts
distilled water...........................	1 part

Turtox Mounting Media

General Biological Supply House offers nonresinous mounting media that permit the mounting of either living or preserved organisms from water or alcohol. Small specimens containing little air can be mounted directly, but larger ones must be mounted from water or alcohol. The CMC series (9, 9AB, 9AF, and 10), refractive index 1.38 to 1.39, are fast-drying media used primarily for mounting small arthropods and diptera, such as fleas, ticks, mites, lice, and mosquito adults, larvae, and pupae. The CMCP series (9, 9AB, 9AF, and 10) has a higher refractive index, 1.40 to 1.41, a more rapid killing and relaxing action that the CMC series, and can be used for mounting parasites and free-living worms.

Yetwin Mounting Medium for Nematodes and Ova (1944)

10% bacto-gelatin, granular, Difco, aqueous ...	150.0 ml
glycerin	50.0 ml
1% chromium potassium sulfate aqueous (chrome alum)	100.0 ml
phenol (carbolic acid), melted...............	1.0 ml

Dissolve gelatin in boiling water, add glycerin. After mixing, add chrome alum solution and phenol. The medium liquefies in 15 minutes at 65°C. Organisms may be transferred from glycerin or formalin directly into medium. The gelatin hardens to form a permanent mount.

Abopon

Abopon[11] is water-miscible and can replace glycerin jelly in several special staining techniques. It can be purchased in the form of lumps or solution. Dissolve the lumps, approximately 50 g, in 25 ml distilled water. The fluid form is usually too thick; dilute with distilled water.

[11] Glyco Chemicals.

Aquamount

Aquamount[12] is an excellent, water-miscible, quick-drying, neutral mountant. It does not require ringing. I use it in preference to glycerin jelly for mounting acid phosphatase preparations, and for fat stains on frozen sections.

Diaphane

Diaphane[13] is available in two forms, colorless (refractive index, 1.4777) and green (refractive index, 1.4792). Slides can be mounted directly from absolute, 95%, or 70% alcohol.

Fluorescent Mounting Medium (Rodriquez and Deinhardt 1960)

Elvanol 51–05 (Du Pont)	20.0 g
0.14 sodium chloride buffered with 0.01 M	
$KH_2PO_4 - Na_2HPO_4 \cdot 12H_2O$ (pH 7.2)	80.0 ml
glycerin	40.0 ml

Agitate frequently during a period of 16 hours. Remove undissolved particles of Elvanol by centrifuging at 12,000 rpm for 15 minutes. The pH should be between 6 and 7.

Herr's $4\frac{1}{2}$ Fluid (Perry et al. 1975; Blackburn and Christophel 1976)

85% lactic acid	2 parts
chloral hydrate (a drug)	2 parts
phenol	2 parts
clove oil..................................	2 parts
xylene	1 part

The above are determined by weight. Sections 30–50 μ can be cut, transferred to slides, stained, washed, blotted partially dry, and mounted in Herr's with a cover slip. The fixed tissue can even be surrounded with hot paraffin to support it while sectioning. The sections will fall clear of the paraffin for mounting on the slides. This produces excellent clearing for phase and interference microscopy.

---

[12] Edward Gurr.
[13] Will Corp.

AQUEOUS MOUNTING TECHNIQUES

Aqueous mounts—sections or whole mounts—are removed from the water, placed on a slide, and covered with a drop of mounting medium. The cover glass, held in a horizontal position, is placed directly on the medium. Do not drop it from a slanted position as this may push the object to one side. In many cases it is not necessary to press the cover glass into place; its own weight is sufficient. Since the sections or objects are not attached to the slides, too much pressure may result in disarranged and broken material.

Using Two Cover Glasses

The material is mounted out of glycerin into glycerin jelly between two cover glasses, one of which is smaller than the other. Clean away excess jelly and air dry overnight. Invert the pair of cover glasses on a drop of resinous medium, such as Permount or Piccolyte, on a slide. This will permanently seal the mount, and it can be treated like resin mount.

Ringing Cover Glasses

If the mountant contains a volatile substance like water and the slides are to be relatively permanent, the cover glass must be sealed with a ringing material. Ringing cements, such as Turtox Slide Ringing Cement[14] are sold by supply houses. Others easily obtainable are Duco Cement, colorless nail polish, gold size, and asphaltum. Lustron 2020[15] dissolved in xylene can be used.

Conger (1960) recommends Dentists' Sticky Wax.[16] It is solid and slightly tacky at room temperature, but flows easily when melted. It adheres well and does not leak, but will crack off cleanly if frozen with dry ice or liquid air. Good for acetocarmine and aceto-orcein preparations.

A firm and reasonably permanent ringing cement is cover glass cement Kroenig from Riedel-de-Häen AG, Seelze-Hannover, Germany.[17]

For temporary mounts, melted paraffin can be used to ring a cover glass. Because it is susceptible to temperature damage and will crack away from the cover, it is not recommended for slides subject to hard usage or for a long time.

[14] General Biological Supply House.
[15] Monsanto Chemical Co.
[16] Kerr Co.
[17] Roboz Surgical Instrument Co.

If the cover glasses are round, a turntable rotating on a steel pin or ball bearing facilitates the ringing operation.[18] The turntable spins rapidly, and, with the ringing cement on a brush, a neat seal can be made by following concentric guidelines.

A liquid cover glass can be used for blood smears and the like without a glass cover. Diatex, Flo-texx, or Coverbond are available from Scientific Products, Trycolac, from Lab-Line.

---

[18] Watson & Sons.

Chapter 9
Hematoxylin Staining

The strength of hematoxylin as a staining agent depends on its proper oxidation and the use of a mordant, as discussed in Chapter 7. A number of methods have been developed for oxidizing hematoxylin.

Baker (1958) recommends oxidation with sodium iodate for preparing hematoxylin solutions. Start with a wholly unoxidized hematoxylin dye powder, not hematein powder. Solutions started with the hematein powder tend to lose their strength by flocculation (sedimentation) of the products of oxidation. When in solution, the hematoxylin dye should be only partly oxidized by the sodium iodate. Use less chemical than would be required for complete oxidation; about one-fourth to one-half of the full amount of oxidizer is adequate. The solution will then continue to combine with atmospheric oxygen and thereby maintain its strength. Such solutions are allowed to ripen slowly for six weeks or more. (Mayer hematoxylin is an example of this type.) They will produce brilliant staining for many months.

The choice of oxidants can be one of convenience; all give equivalent results. Iodine is used as a 1% alcoholic solution, the following as 0.1% aqueous: sodium iodate ($NaIO_3$), sodium metaperiodate ($NaIO_4$), potassium permanganate ($KMnO_4$).

The rate of oxidation is also affected by the type of solvent used. A neutral aqueous solution forms hematein in a few hours; an acid solution does this more slowly, and an alkaline solution, more rapidly. Alcoholic solutions are slow, and the addition of glycerin retards them even more. The function of the additives is not always clear. Some, such as chloral hydrate, thymol, and salicylic acid, may be preservatives; alcohol can prevent molds; and glycerin can stabilize against overoxidation. The latter action probably forms the precipitate coating on the inside walls of the storage bottle. Overoxidation also can be prevented by a minimum of air space in the bottle. Color changes that take place in a stock solution indicate its efficiency. The changes, with no mordant present, are from water white through lilac, bright purple, deep purple, red, orange red, orange brown, to brown. At the purple stage the solution is most vigorous; at the red stages, less so; at the brown stage it is no longer useful. The lifetime of alcoholic solutions is five times greater than that of aqueous ones (Cole 1943; Clark 1974, 1975).

SINGLE SOLUTIONS

Delafield Hematoxylin (Carlton 1947)

Dissolve 4 g hematoxylin in 25 ml absolute ethyl alcohol. Mix gradually into 400 ml ammonia alum ($NH_4Al(SO_4)_2 \cdot 12H_2O$), saturated aqueous (approximately 1 part alum to 11 parts distilled water). Leave exposed to light in a flask with a cotton plug for 3–5 days. Filter. To the filtrate add 100 ml glycerin and 100 ml methyl alcohol. Allow to ripen for at least 6 weeks. The ripened solution will keep for years in a stoppered bottle.

Hance and Green (1961) bubbled pure oxygen for 4 hours into 3 liters of freshly prepared Delafield solution. This reduced the oxidation time from weeks to hours. The solution thus oxidized stained more rapidly than solutions oxidized by atmospheric oxygen and could be diluted to half strength for use.

Ehrlich Hematoxylin (Gurr 1956)

hematoxylin	2.0 g
ammonia alum ($NH_4Al (SO_4)_2 \cdot 12H_2O$)	3.0 g
alcohol, methyl or ethyl	100.0 ml
glycerin	100.0 ml
distilled water	100.0 ml

Ripens in 6–8 weeks or may be ripened for immediate use with 0.24 gm sodium iodate.
Add 100 ml glacial acetic acid. Keeps for years.

Harris Hematoxylin (Mallory 1944)

Dissolve 1.0 g hematoxylin in 10 ml ethyl alcohol. Dissolve 20 g potassium alum $(KAl(SO_4)_2 \cdot 12H_2O)$ or ammonia alum $(NH_4Al(SO_4)_2 \cdot 12H_2O)$ in 200 ml water and boil. Add hematoxylin and boil $\frac{1}{2}$ minute. Add 0.5 g mercuric oxide. Cool rapidly. Add a few drops of acetic acid to keep away metallic luster and brighten nuclear structure. Does not keep longer than a month or two. (*Warning:* Mix with care; the reaction can be explosive.)

Mayer Hematoxylin (Mallory 1944)

Add 1 g hematoxylin to 1 liter distilled water. Heat gently and add 0.2 g sodium iodate and 50 g potassium alum $(KAl(SO_4)_2 \cdot 12H_2O)$. Heat until dissolved; add 1 g citric acid and 50 g chloral hydrate. Allow to ripen, preferably for 6–8 weeks, although it can be used within 1–2 weeks. Chloral hydrate is a drug and may be difficult to obtain. In that case use Lillie's modification: hematoxylin, 5 g in 700 ml distilled water. Add 300 ml glycerin, 50 g ammonia alum, 0.2–0.4 g sodium iodate, and 20 ml glacial acetic acid.

Gill's Hematoxylin (1974)

hematoxylin	4.0 g
sodium iodate	0.4 g
aluminum sulfate	35.2 g
distilled water	710.0 ml
ethylene glycol	250.0 ml
glacial acetic acid	40.0 ml

The ready-for-use solution can be purchased from Polysciences and Lerner Laboratories. Staining time: 3–4 minutes.

DOUBLE SOLUTIONS

Weigert Iron Hematoxylin (Lillie Modification 1968)

Solution A

ferric chloride $(FeCl_3 \cdot 6H_2O)$	2.5 g
ferrous sulfate $(FeSO_4 \cdot 7H_2O)$	4.5 g
hydrochloric acid	2.0 ml
distilled water	298.0 ml

Solution B

hematoxylin	1.0 g
95% ethyl alcohol	100.0 ml

Mix A and B. The solution turns black at once and is usable for 2–3 weeks or until it turns brown. Stain progressively 3–5 minutes.

Groat Variation of Weigert Hematoxylin (1949)

In this variation a single solution is used.

distilled water.............................	50.0 ml
sulfuric acid (spgr 1.84, 94% H_2SO_4)	0.8 ml
ferric alum ($FeNH_4(SO_4)_2 \cdot 12H_2O$)	1.0 g
95% ethyl alcohol	50.0 ml
hematoxylin	0.5 g

Mix in order given, at room temperature. Filter. Stain progressively 9–10 minutes.

Krutsay Iron Hematoxylin (1962*b*)

hematoxylin	1.0 g
potassium alum ($KAl(SO_4)_2 \cdot 12H_2O$)	50.0 g
potassium iodate (KIO_3)	0.2 g
hydrochloric acid..........................	5.0 ml
distilled water.............................	1000.0 ml

This can be used like Mayer hematoxylin or converted to an iron hematoxylin like Weigert's by adding 8 volumes of the solution below to 100 of above hematoxylin just prior to use.

ferric alum ($FeNH_4(SO_4)_2 \cdot 12H_2O$)	2.0 g
hydrochloric acid..........................	0.5 ml
distilled water.............................	100.0 ml

A fresh solution is brownish violet; if it turns orange, it is no longer usable. Staining time: 5 minutes.

Slidder Iron Hematoxylin (1968)

Solution A

hematoxylin	1.0 g
95% ethyl alcohol	100.0 ml
aluminum chloride hydrate ($AlCl_3 \cdot 6H_2O$)	10.0 g

Solution B

ferrous sulfate hydrate ($FeSO_4 \cdot H_2O$)	10.0 g
distilled water	100.0 ml

Combine A and B and add:

hydrochloric acid, concentrated	2.0 ml
sodium iodate, 9% aqueous	2.0 ml

Allow to stand 48 hours, ready for use. Stain 5–10 minutes, wash and differentiate with 0.5% HCl in 70% alcohol, rinse and neutralize with 2% aqueous potassium acetate or other mildly alkaline solution.

Cole Hematoxylin (Clark et al. 1973 modification)

Stock Solutions

> iodine, 1% in 95% ethyl alcohol
> potassium aluminum sulfate $KAl(SO_4)_2 \cdot 12H_2O$,
>   1.2% aqueous
> hematoxylin, 10% in absolute alcohol

Working Solution

1% iodine	2.0 ml
10% hematoxylin	1.0 ml
1.2% alum	100.0 ml

The stock solutions are stable for years. Place the freshly mixed working solution in a paraffin oven (55–60°C) overnight. Cool, ready for use. Staining time: 5 minutes.

Double Solutions That Are Not Mixed Before Use

Heidenhain Iron Hematoxylin

Solution A

ferric alum ($FeNH_4(SO_4)_2 \cdot 12H_2O$)	4.0 g
distilled water.............................	100.0 ml

Keep in refrigerator to prevent precipitation on sides of bottle.

Solution B

hematoxylin	10.0 g
95% ethyl alcohol	100.0 ml

Let stand until a deep wine-red color; 4–5 months is not too long. Add 4–5 ml of this stock solution to 100 ml distilled water; this gives a practically aqueous solution and is already ripe. Saturated aqueous lithium carbonate—3 drops—added to the working solution improves color.

A and B are not usually mixed. A is a mordant solution, and precedes B. If they are mixed for use, the solution deteriorates rapidly.

Testing Hematoxylin Solutions

Add several drops of the solution to tap (not distilled) water. If it turns bluish purple immediately, it is still satisfactory; but if it changes slowly, stays reddish or brownish, it has weakened or broken down and should be discarded.

SUBSTITUTES FOR HEMATOXYLIN SOLUTIONS

Gallocyanin (Berube et al. 1966)

gallocyanin ..:............................	0.15 g
chrome alum ($CrK(SO_4)_2 \cdot 12H_2O$), 15%	
aqueous	100.0 ml

Boil 10–20 minutes. Cool. Filter. Restore volume to 100 ml by washing the precipitate. Keeps about a week, then deteriorates slowly. An iron lake may be prepared by substituting 5% aqueous iron alum for the

chrome alum (Proescher and Arkush 1928; also see Terner and Clark 1960*a*,*b*).

Hematein (Hemalum) (Kornhauser 1930)

hematein	0.5 g
95% ethyl alcohol	10.0 ml
potassium alum ($KAl(SO_4)_2 \cdot 12H_2O$),	
5% aqueous	500.0 ml

Grind hematin with alcohol in glass mortar, and add to the aqueous potassium alum. Immediately ready for use.

Iron Alizarine Blue S (Meloan and Puchtler 1974)

alizarine blue S (C.I. 67415), 2% aqueous	100.0 ml
ferric chloride, 40% aqueous	5.0 ml

Mix well and let stand 15 minutes. Filter. Stain for 15 minutes. Wash 10–15 minutes and proceed to counterstain. Fisher Scientific commercial ferric chloride, 10–11% Fe^{+++}, can be substituted for the above. Do not boil the stain; it is good for 2–3 weeks.

Celestin Blue B

celestin blue B (C.I. 51050)	2.5 g
ferric ammonium sulfate, 5% aqueous	500.0 ml

Bring to a boil for 3 minutes. Cool and filter. Add 70 ml glycerin. Good for 6–12 months.

This is an excellent stain and can be followed by acidic stains of the Mallory or Masson type. It can precede hematoxylin; the iron alum in the solution acts as a mordant for the hematoxylin. Use a Mayer-type hematoxylin for 3–6 minutes. Many prefer this combination over a Weigert iron hematoxylin. Lillie et al. (1973) use iron gallein for hematoxylin.

COUNTERSTAINS (PLASMA STAINS) FOR HEMATOXYLIN, GALLOCYANIN, AND HEMATEIN

Eosin

eosin Y (C.I. 45380)	1.0 g
70% ethyl alcohol	1000.0 ml
glacial acetic acid	5.0 ml

Dilute with equal volume of 70% alcohol for use and add 2–3 drops of acetic acid.

Eosin (Putt 1948)

eosin Y (C.I. 45380)	1.0 g
potassium dichromate	0.5 g
saturated aqueous picric acid	10.0 ml
absolute ethyl alcohol	10.0 ml
distilled water	80.0 ml
acetic acid (optional)	1 drop

Eosin–Orange G

1% eosin Y (C.I. 45380) in 95% ethyl alcohol ..	10.0 ml
orange G (C.I. 16230), saturated solution in 95% ethyl alcohol (approximately 0.5 g per 100 ml)	5.0 ml
95% ethyl alcohol	45.0 ml

Orange G

orange G	1.0 g
phosphotungstic acid	5.0 g
90% ethyl or methyl alcohol	100.0 ml

Other Acceptable Counterstains

1. Acid fuchsin (C.I. 42685), 5% aqueous (slight acidity improves stain). If overstained, rinse with tap water.
2. Orange G (C.I. 16230), saturated in 95% ethyl alcohol.
3. Van Gieson or substitute (p. 147).

4. Bordeaux red (C.I. 16180), 1% aqueous.
5. Biebrich scarlet (C.I. 26905), 1% aqueous; a good counterstain.
6. Additional "eosins"
   a. Eosin Y (C.I. 45380) or eosin B (C.I. 45400), 0.1–0.5% in 95% ethyl alcohol.
   b. Erythrosin B (C.I. 45430), 0.1–0.5% in 95% ethyl alcohol.
   c. Phloxine B (C.I. 45410), 0.5% aqueous, plus a few drops of acetic acid (p. 241).
7. Congo red (C.I. 22120), 0.5% aqueous.
8. Light green SF, yellowish (C.I. 42095), 0.2–0.3% in 95% ethyl alcohol
9. Aniline blue WS (C.I. 42780) (pp. 137, 243).
10. Fast green FCF (C.I. 42053), similar to light green or aniline blue.
11. Wool green (C.I. 44090), 0.5% aqueous.

HEMATOXYLIN STAINING PROCEDURES

Delafield (or Harris) Hematoxylin, I (Progressive Method)

FIXATION

Any general fixative or one specific for nuclear detail.

SOLUTIONS

Hematoxylin, see p. 112.
Counterstain, see p. 118.

PROCEDURE

The slides are passed through a "down" series of jars, a process often termed running down slides to water (or hydration), because a series of alcohols of decreasing strength is used (from left to right, top row of Fig. 9–1). Never at any time during this procedure allow the slides to dry.

1. Xylene (or toluene): 2–3 minutes or longer.

   Two changes may be necessary to insure removal of paraffin.

2. Absolute alcohol: 2–3 minutes or longer.
3. 95% alcohol: 2–3 minutes or longer.
4. 70% alcohol: 2–3 minutes or longer.

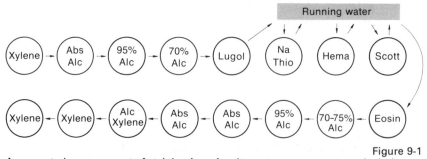

Figure 9-1

A suggested arrangement of staining jars. An alternate arrangement can include two jars of xylene in the "down series" (left to right) when many slides are being stained. An absolute alcohol–xylene in "up series" (right to left) is optional.

If mercuric chloride was absent from fixative, skip steps 5 through 7 and proceed to step 8.

5. Lugol solution (p. 544): 3 minutes.
6. Running water: 3 minutes.
7. 5% sodium thiosulfate, $Na_2S_2O_3$: 2–3 minutes.
8. Running water: 3–5 minutes.
9. Delafield hematoxylin: 2–5 minutes; check after 1 minute for stain intensity. Fresh solutions stain faster than old ones. If not dark enough, return slides to stain. Rinse off stain in tap water before checking under microscope. If slide becomes too dark, convert to regressive type of stain.
10. Running water: 3–5 minutes.
11. Scott solution (p. 545): 3 minutes.
12. Running water: 3–5 minutes.
13. Counterstain: 1 or more minutes, depending on stain used.

Transfer slides through "up" series, termed running up slides (or dehydration), using a series of alcohols of increasing strength. Steps 14 and 15 can control intensity of the counterstain. Watch timing carefully in these solutions (from right to left, bottom row, Fig. 9–1).

14. 70% alcohol: 1 or more dips.
15. 95% alcohol: few dips.
16. Absolute alcohol: 3 minutes or longer.
17. Absolute alcohol: 2–3 minutes or longer.
18. Absolute alcohol–xylene (1 : 1), optional: 2–3 minutes or longer.
19. Xylene: 2–3 minutes or longer.
20. Xylene: 2–3 minutes or longer.

21. Mounting medium: keep sections moist with xylene during this process; they must not dry. Add cover glass.

RESULTS

nuclei—deep blue
cytoplasmic structure—pink, rose, etc., depending on counterstain

COMMENTS

1. Slides may be left in stronger alcohols and xylene longer than scheduled (indicated by "or longer"), but do not leave them much longer in any solutions weaker than 80% or 95% alcohol. Weaker alcohols and water can loosen the sections from the albumen.
2. Isopropyl alcohol can be substituted for ethyl alcohol in this schedule, but cannot be used as a stain solvent.
3. Two to five minutes for Delafield's (step 9) is an approximate time only. This may have to be varied according to the tissue used. Tissue sections are individualistic, often due to the type of fixative employed on them, and a certain amount of trial and error may be required to develop the exact timing schedule.
4. Poor staining can result from improper fixation, leaving gross tissues too long in alcohol or iodine, faulty processing during preparation for embedding, or careless handling of the slides in preparation for staining. In many instances no amount of trial and error will produce a perfect stain. The only correctable faults are those made during deparaffinization and hydration of the slides. Then the slides should be transferred in a reversed process through the solutions, back to xylene. Change to fresh solutions and try again.
5. Sometimes too many slides have been taken through the fluids and the fluids become contaminated. If more than 20 slides are being stained, use two changes of xylene in step 1 and follow step 20 with a third change. Prevent further contamination by draining slides properly. On removing them from a solution, touch the edges against the inner surface of the staining jar, and then against paper towels. But do not be too thorough and let the slides dry; merely drain off excess liquid.
6. Hematoxylin staining must be left in a "blued" condition. The original pinkish color of the hematoxylin, like Delafield's, must be converted to a blue color by alkalinity. In the progressive method this is accomplished in running water to remove excess dye and start the bluing action. Then use Scott solution (step 11), a couple of drops of ammonia in a final wash, or a weak solution of sodium bicarbonate to

insure the alkaline condition. In the regressive method (below) the same applies, and, as a precaution against neutralization caused by carrying one solution over into the other (step 5), after several exchanges of slides add a drop or two of acid and base to them. If there is any pinkness left in the nuclei, they are not adequately blued.

Delafield (or Harris) Hematoxylin, II (Regressive Method)

PROCEDURE

1. Deparaffinize and run slides down (hydrate) to water, removing $HgCl_2$ if present.
2. Stain in Delafield hematoxylin, until slides are well overstained: 15–20 minutes.
3. Wash in running water: 3–5 minutes.
4. Transfer to 70% alcohol.
5. Destain in ±0.5% HCl (2–3 drops in 60.0 ml of 70% alcohol) for a few seconds. Remove to 70% alcohol, with drop of ammonia added, until sections turn blue. Examine under microscope. If nuclei are too dark, repeat above procedure. When nuclei stand out as a sharp blue against colorless background, blue slides thoroughly in alkaline solution: 3–5 minutes. 70% alcohol, rather than water, is recommended for these procedures because the water tends to loosen the sections from the slides.
6. Transfer to 70% alcohol.
7. Counterstain.
8. Dehydrate as for progressive staining.
9. Clear and mount.

RESULTS

nuclei—deep blue, sharper than in the progressive method
other elements—colors of counterstain

Mayer Hematoxylin

FIXATION

Any general fixative or one specific for nuclear detail.

SOLUTIONS

Hematoxylin, see p. 113.
Counterstains, see p. 118.

PROCEDURE

1. Deparaffinize and run slides down (hydrate) to water, removing $HgCl_2$ if present.
2. Stain in Mayer hematoxylin: 11 minutes.
3. Wash in running water: 3 minutes.
4. Blue in Scott solution: 3 minutes.
5. Wash in running water: 3–5 minutes.
6. Counterstain.
7. Dehydrate quickly through 70% and 95% alcohols.
8. Finish dehydration, absolute alcohol.
9. Clear and mount.

RESULTS

nuclei—deep blue
other elements—colors of counterstain

COMMENTS

Mayer hematoxylin is my favorite alum hematoxylin. It keeps well, and a single solution will stain several hundred slides over a period of two weeks without losing its intensity. To mordant with celestin blue B see p. 117.

Heidenhain Iron Hematoxylin

FIXATION

Any general fixative, preferably one containing $HgCl_2$.

SOLUTIONS

Hematoxylin and iron alum, see p. 116.
Counterstains, see p. 118.

PROCEDURE

1. Deparaffinize and run slides down (hydrate) to water, removing $HgCl_2$.
2. Mordant in 4% iron alum: 15–30 minutes (or overnight, see comments below).
3. Wash in running water: 5 minutes.
4. Stain in hematoxylin: 15–30 minutes (or overnight).
5. Wash in running water: 5 minutes.
6. Destain in 2% aqueous iron alum or saturated aqueous price acid (see comments below) until nuclei stand out sharp and clear against colorless background. Before examining the slides, dip them in water containing a couple of drops of ammonia. *Warning:* too much ammonia will loosen sections.
7. Wash in running water: 15–30 minutes or longer.
8. Counterstain, if desired.
9. Run slides up through alcohols (dehydrate), clear, and mount.

RESULTS

nuclei—blue black to black
muscle striations—blue gray
other elements—colors of counterstain, or very light gray to colorless if no counterstain used

COMMENTS

The shorter staining schedule produces blue-black nuclei and the overnight staining a truer black. The overnight staining is recommended for mitochondria. Picric acid usually perfects a more sharply differentiated nucleus than destaining with iron alum. Prolonged destaining with iron alum can leave a tan or yellow tinge in the cytoplasm. Yellow color left in the tissue after picric acid is due to insufficient washing following the destaining. Striations of muscle and some protozoan structures are better differentiated by iron alum. Thorough washing is essential after either method lest the sections continue to destain slowly and fade (picric acid destaining, Tuan 1930).

Destaining agents act as both an acid and an oxidizer, reacting in two ways: (1) As an acid they extract stain faster from the cytoplasm than from the nuclei; (2) as an oxidizer they bleach the stain uniformly. If an acid alone is used, the staining of the nuclei is favored, and if an oxidizer alone is used, the staining of the cytoplasm is enhanced.

Iron alum mordanting can precede a hematoxylin such as Mayer's and produce a blacker (iron-type) nuclear staining. Destain, if necessary, in the same manner as for Heidenhain's.

Hutner (1934) recommendation for nuclei

1. Mordant in 4% iron alum: 1 hour.
2. Stain hematoxylin: 1 hour.
3. Destain in saturated aqueous picric acid.
4. Wash and blue by adding 1–2 drops ammonia to 70% alcohol.

Hutner (1934) recommendation for cytoplasm

1. Mordant in iron alum: 30 minutes.
2. Stain in hematoxylin: 30 minutes.
3. Destain in freshly made 95% alcohol, 2 parts to Merk's superoxol (30% H_2O_2), 1 part.
4. Rinse in 1 change 70% alcohol: 10 minutes.

For brevity, intermediate steps have been omitted from the above procedures. Oxidizer and acid can be combined to approximate iron alum destaining effects:

0.25% HCl in 95% alcohol, 2 parts; superoxol, 1 part.
Wash in 70% alcohol and blue.

Cell membranes and various cytoplasmic structures stain well with Heidenhain hematoxylin, probably because of the use of iron alum. The metal is first bound to the tissues and then combines with the hematoxylin. It is likely that proteins are most frequently the bearers of iron-binding residues and are responsible for the precise staining action of Heidenhain hematoxylin (Puchtler 1958).

Acidified Hematein (Puchtler and Sweat 1964)

Dissolve 0.1 g hematoxylin in 2 ml 95% ethanol. Add 18 ml distilled water and 2 ml of a 1% aqueous solution of sodium iodate. Ripen for 15 minutes or longer. Add 60 ml of 10% acetic acid. This solution is not stable and should be prepared shortly before use.

Acidified hematein may be used in step 4 of Heidenhain method. Mordant (step 2) for 1 hour, wash thoroughly, and stain with acidified hematein for 1 hour. Wash thoroughly. Counterstain if desired. Dehydrate, clear, and mount. No differentiation is required.

For double iron hematoxylin solutions and their use, see p. 113.

Phosphotungstic Acid–Hematoxylin (Puchtler et al. 1963)

FIXATION

Any general fixative, preferably one containing $HgCl_2$; if tissue is formalin-fixed, mordant sections before staining in saturated aqueous mercuric chloride, rinse, and treat with Lugol's.

SOLUTIONS

Hematoxylin solution

hematoxylin	1.0 g
phosphotungstic acid	20.0 g
distilled water.............................	1000.0 ml

Dissolve hematoxylin and phosphotungstic acid in separate portions of water. Heat hematoxylin solution until no more color change is noted. Cool to room temperature and combine with phosphotungstic acid solution. Add 0.177 g potassium permanganate. Let stand one month.

Langeron iodine solution

iodine	1.0 g
potassium iodide	2.0 g
distilled water.............................	200.0 ml

PROCEDURE

1. Deparaffinize and run slides down (hydrate) to water.
2. Treat with Langeron iodine solution: 10 minutes.
3. Rinse in distilled water and transfer to 5% sodium thiosulfate: 2 minutes.
4. Wash in running water: 5 minutes.
5. Transfer to hematoxylin solution (room temperature) and place in paraffin oven (approximately 60°C): 1 hour.
6. Rinse briefly in two changes 95% alcohol.

7. Dehydrate in 3 changes absolute alcohol and clear in 2 changes xylene. Mount.

RESULTS

fibrin, nuclei, muscle fibers, keratin, terminal bars—dark blue
collagen, reticulin, basement membranes, amyloid, cartliage, thrombocytes—shades of red

COMMENTS

A freshly prepared solution must stand one month to be sufficiently ripened, otherwise the staining will be of low contrast. The staining properties improve with age and the solution keeps for many months, even up to a year. Do not preheat the dye solution, and store it at room temperature. If kept in the oven, it will deteriorate within 1–2 days. This is an excellent stain for muscle and can be used with PAS and Luxol fast blue.

HEMATOXYLIN SUBSTITUTE PROCEDURES

Gallocyanin (Berube et al. 1966; Einarson 1951)

Gallocyanin is an excellent substitute for hematoxylin and can be used in most procedures designed for hematoxylin. The solution is made in a few minutes and ready for immediate use; also an iron lake can be prepared. No differentiation is required, and it is better than hematoxylin for tissues of the central nervous system. It can replace thionin for ganglia and glia cells, but not for myelin sheaths. It is good for Negri bodies. The preferred fixatives are acetoformol, Zenker (acetic or formol), or formalin—fixatives that preserve chromophilic substance.

SOLUTION

Gallocyanin, see p. 116.

PROCEDURE

1. Deparaffinize and hydrate to water.
2. Leave in stain overnight, or 3 hours at 56°C.
3. Wash thoroughly and proceed to counterstain.
4. Dehydrate, clear, and mount.

RESULT

nuclei—blue

Hematein (Hemalum) (Kornhauser 1930)

FIXATION

Any general fixative, preferably one containing $HgCl_2$.

SOLUTIONS

Hemalum, see p. 117.
Counterstains, see p. 118.

PROCEDURE

1. Deparaffinize and hydrate slides down to water, remove $HgCl_2$.
2. Stain hemalum, progressively: about 5 minutes.
3. Wash in running water: 5–10 minutes.
4. Counterstain.
5. Dehydrate, clear, and mount.

RESULTS

nuclei—blue
other elements—color of counterstain

Eriochrome Cyanin (Chapman 1977).

FIXATION

A good general fixative.

SOLUTION

eriochrome cyanin R (C.I. 43820)[1]	0.2 g
ferric chloride ($FeCl_3$), anhydrous	0.9 g
HCl......................................	5.0 ml
distilled water............................	100.0 ml

---

[1] Curtin Matheson Scientific Co.

PROCEDURE

1. Deparaffinize and hydrate sections to water; remove $HcCl_2$ if present.
2. Rinse in 0.5 N HCl: few seconds.
3. Stain: 1–2 minutes.
4. Rinse in 0.5 N HCl: 2 seconds.
5. Wash in running water: 30 seconds.
6. Blue in 0.1% sodium bicarbonate: 1 minute.
7. Wash in running water: 5 minutes.
8. Counterstain, eosin or the like, if desired.
9. Dehydrate, clear, and mount.

RESULTS

nuclei—blue
matrix of hyalin cartilage, some mucus, and granules—pale blue
other structures—shades of pink and red

RED NUCLEAR STAINING

Darrow Red (Powers and Clark 1963; Powers et al. 1960)

FIXATION

Any general fixative.

SOLUTION

Darrow red[2]	50.0 mg
0.2 M glacial acetic acid (pH 2.7)	200.0 ml

Boil briefly. Cool and filter. Stable for 1 month.

PROCEDURE

1. Use frozen sections in water, or deparaffinize paraffin sections and hydrate to water.
2. Stain in Darrow red: 20–30 minutes.
3. Rinse in distilled water.

---

[2] Curtin Matheson Scientific Co.

4. Differentiate and dehydrate through 50%, 70%, and 95% alcohol, not so slowly as to remove too much dye from nuclei, but slowly enough to decolorize background.
5. Completely dehydrate in *n-butyl* alcohol, clear, and mount.

RESULTS

nuclei—red
other elements—depending on other stains applied

COMMENTS

This is a good nuclear stain for contrast with blue cytoplasmic stains—following Luxol fast blue for example. It is a good Nissl stain at pH 2.7. For glial nuclei, use acetic acid–acetate buffer at pH 3.5 to prepare the staining solution. Stain 10–15 min. For celloidin sections, use 25 mg of dye in 200 ml of acetic acid–acetate buffer, 20 minutes. Dehydrate overnight in terpineol-xylene, 1 : 3.

Scarba Red (Slidders et al. 1958)

FIXATION

Any general fixative.

SOLUTIONS

Scarba red

melted phenol (carbolic acid)	2.0 g
neutral red	1.0 g

Mix thoroughly, allow to cool, dissolve in:

95% ethyl alcohol	15.0 ml

Add:

2% aniline in water	85.0 ml
glacial acetic acid	1.0–3.0 ml

Mix well and filter. Keeps for at least 6 months.

Differentiator

70% ethyl alcohol	85.0 ml
formalin	15.0 ml
glacial acetic acid	15.0 drops

PROCEDURE

1. Deparaffinize and hydrate slides to water (remove $HgCl_2$ if present).
2. Stain in Scarba red: 5 minutes.
3. Rinse in distilled water.
4. Transfer to 70% alcohol: 2–3 minutes.
5. Differentiate until nuclei are clear and sharp against colorless background.
6. Dehydrate in 95% and absolute alcohol.
7. Clear in xylene and mount.

RESULT

chromatin and calcium—red

For Restoration of Basophilic Properties see p. 565.

Part II
SPECIFIC STAINING METHODS

Chapter 10
Staining Connective Tissue and Muscle

Stains for connective tissue and muscle are used to demonstrate various components and features, such as collagen, reticulum, elastin, hyalin, fibrin, and striations in muscle cells. (Hemopoietic tissues are covered in Chapter 16, p. 219.)

MALLORY STAINING

Mallory Triple Connective Tissue Stain

Innumerable modifications of this method have appeared in the literature, particularly those using phosphomolybdic and/or phosphotungstic acid. Holde and Isler (1958) support the use of phosphomolybdic acid if selective staining of connective tissue is desired; the acid diminishes background and nuclear staining and stains connective tissue blue. They feel that the staining of connective tissue is due to the action of the acid on the fibers. Baker (1958) suggests that phosphomolybdic acid acts as a "colorless acid dye" in the tissue, chiefly on collagen. It acts as a dye excluder with acid dyes such as acid fuchsin, excluding them from collagen. Then the aniline blue stains the collagen selectively, but is excluded from other tissues.

It is true that various dyes (acid and basic triphenylmethane) form lakes with phosphomolybdic acid; such compounds are used in textile dying (Puchtler and Sweat 1963b). Basic dyes and some amphoteric dyes are bound by interaction between the acid groups of phosphomolybdic acid and accessible basic groups of the dye (Smith et al. 1966). If no phosphomolybdic acid is present, aniline blue stains more strongly and with very little selectivity. The action of phosphomolybdic and phosphotungstic acid has been called a mordant action, but this is disputed. Everett and Miller (1974), in a series of experiments, showed that most histological structures stain deeply by aniline blue; after phosphotungstic acid treatment only the connective tissue stains. Their preliminary conclusions are that the poly acids act not as mordants but as selective blocking agents against all tissues except connective tissue. Quintarelli et al. (1971) and Scott (1976) believe phosphotungstic acid combines by electrostatic bonds with cationic groups. This is directly related to the pH, and a salt-type linkage is involved between the acid and organic cations.

In regard to the use of phosphotungstic acid, it may be of interest to note that certain tissue elements react more specifically to certain metals than to others: collagen to molybdenum; and fibroglia, myoglia, neuroglia, and epithelial fibers to tungsten. Lillie (1952) writes that phosphotungstic acid intensifies plasma staining, and phosphomolybdic acid, fiber staining. The acids may be used separately or together, depending on the desired final effect.

Oxalic acid is often used. Mallory (1944) claims that it makes the aniline blue stain more rapidly and intensely. Baker (1958) seems to agree when he suggests that oxalic acid lowers the pH, seeming thereby to aid staining with aniline blue and also with orange G.

No polychroming or metachroming takes place in the combination of dyes; they react orthochromatically in varying intensities of their own colors. In Mallory's combination of stains, aniline blue (acidic) stains connective tissue and cartilage; orange G (acidic) stains blood cells, myelin, and muscle; acid fuchsin (acidic) stains the rest of the tissue (including the nuclei) in shades of pink and red. This is not, however, an efficient method for staining nuclei, and they will fade in a few years. Modifications of Mallory's using basic fuchsin, carmine, or azocarmine result in more permanent nuclear staining, since the basic stain reacts more reliably with the nuclei than the acidic one. This method may also be preceded by hematoxylin staining for a brilliant permanent nuclear stain.

The final staining may be followed by an acetic acid wash (0.5–1%): 3–5 minutes or longer, which produces more transparent sections without altering the color.

Suggestion: Try Sirius supra blue in place of aniline blue (Sweat et al. 1968). Aniline blue WS was formerly a mixture of water blue 1 and methyl blue; that mixture now appears to be obsolete. Aniline blue ordered from some manufacturers is water blue; from others, methyl blue. The dyes are apparently considered interchangeable with each other as well as with the old mixture. However, I find the new dyes inferior in color to the old mixture and in most cases substitute Sirius supra blue. Treat sections for 5 minutes with 1% phosphomolybdic acid; wash with 3 changes of distilled water: 10 seconds each. Stain for 5 minutes in the following solution:

Sirius supra blue FGL–CF[1]	2.0 g
distilled water	100.0 ml
glacial acetic acid	2.0 ml

Wash in 2–3 changes of distilled water: 10 seconds each. Dehydrate, clear, and mount.

Note: Do not combine Sirius supra blue and phosphomolybdic acid as can be done with aniline blue; the combination is not successful with Sirius supra blue.

Pantin Method (1946)

FIXATION

Any general fixative, preferably one containing $HgCl_2$.

SOLUTIONS

Mallory I

acid fuchsin (C.I. 42685)	1.0 g
distilled water	100.0 ml

Phosphomolybdic acid

phosphomolybdic acid	1.0 g
distilled water	100.0 ml

Mallory II

aniline blue WS (C.I. 42780)	0.5 g
orange G (C.I. 16230)	2.0 g
distilled water	100.0 ml

---

[1] Roboz Surgical Instrument Co.

PROCEDURE

1. Deparaffinize and hydrate slides to water; remove $HgCl_2$. If $HgCl_2$ is absent from fixative, mordant in saturated aqueous $HgCl_2$, plus 5% acetic acid: 10 minutes. Wash, treat with Lugol's and sodium thiosulfate; wash and rinse in distilled water.
2. Stain in Mallory I: 15 seconds.
3. Rinse in distilled water to differentiate reds: 10 or more seconds.
4. Treat with phosphomolybdic acid: 1–5 minutes.
5. Rinse briefly in distilled water.
6. Stain in Mallory II: 2 minutes.
7. Rinse in distilled water.
8. Differentiate aniline blue in 90% ethyl alcohol.
9. Dehydrate in absolute alcohol, clear, and mount.

RESULTS

nuclei—red
muscle and some cytoplasmic elements—red to orange
nervous system—lilac
collagen—dark blue
mucus, connective tissue, and hyaline substance—blue
chitin—red
yolk—yellow to orange
myelin and red blood cells—yellows and orange
dense cellular tissue (liver)—pink with red nuclei
bone matrix—red

COMMENTS

From distilled water (step 7) the slides can be run through the "up" series of alcohols to control the blue color. Aqueous-alcoholic solutions differentiate the amount of Mallory II left in the various parts of the tissue. After the slides have remained in the absolute alcohol for a couple of minutes, they should have changed from a muddly purple to a clear blue and red. An acetic acid rinse following step 7 contributes to the transparency of the sections.

For better nuclear detail, Mallory I can be preceded by alum hematoxylin staining (*results:* nuclei—blue). Lendrum and McFarlane (1940) recommend the addition of celestin blue. Krichesky (1931) proposes the following for Mallory II.

Solution A

aniline blue WS (C.I. 42780)	2.0 g
distilled water.............................	100.0 ml

Solution B

orange G (C.I. 16230)	1.0 g
distilled water.............................	100.0 ml

Solution C

phosphomolybdic acid	1.0 g
distilled water.............................	100.0 ml

Keep solutions A, B, and C in separate bottles because the mixture deteriorates on standing. When ready to use, mix equal parts of each.

This modification is useful for instructors of microtechnique classes because if large stocks of Mallory II components are kept as individual solutions and freshly mixed before use, a more brilliant stain is achieved than from a single solution that has been stocked with all the components already combined.

Azan* Stain, Mallory-Heidenhain (Koneff Modification 1938)

FIXATION

Zenker formol; other general fixatives fair, but fixation is improved by mordanting slides overnight in 3% potassium dichromate.

SOLUTIONS

Azocarmine

azocarmine G (C.I. 50085)...................	0.2–1.0 g
distilled water.............................	100.0 ml
glacial acetic acid	1.0 ml

Boil azocarmine in water 5 minutes, cool, filter, and add acetic acid.

---

* Azan from first syllables of azokarmin B and anilinblau W.

Aniline alcohol

aniline	1.0 ml
85–90% ethyl alcohol	1000.0 ml

Acid alcohol

glacial acetic acid	1.0 ml
90–95% ethyl alcohol	100.0 ml

Phosphotungstic acid

phosphotungstic acid	5.0 g
distilled water.............................	100.0 ml

Aniline blue stain

aniline blue WS (C.I. 42780)	0.5 g
orange G (C.I. 16230)	2.0 g
oxalic acid................................	2.0 g
distilled water.............................	100.0 ml
5% phosphotungstic acid (above)............	1.0 ml

Acidified water

glacial acetic acid	1.0 ml
distilled water.............................	100.0 ml

PROCEDURE

For strikingly beautiful results, particularly for pituitary, Koneff's longer method is recommended. Extra steps and changes in timing are indicated in parentheses.

1. Deparaffinize and hydrate slides down to water; remove $HgCl_2$.
2. (Treat with aniline alcohol: 45 minutes.)
3. (Treat with acid alcohol: 1–2 minutes.)
4. Stain in azocarmine, 56°C: 1 hour (or 2 hours). Check temperature carefully; if too high, azocarmine differentiates poorly.
5. Rinse in distilled water.
6. Differentiate in aniline alcohol. Check under microscope for brilliant red nuclei and very light red cystoplasm.
7. Treat with acid alcohol: 1–2 minutes.
8. Transfer to phosphotungstic acid: 2–3 hours (4 hours).

9. Rinse in distilled water.
10. Stain in aniline blue solution: 1–2 hours (4 hours).
11. Rinse in distilled water.
12. Treat with phosphotungstic acid: 3–5 minutes.
13. Rinse in distilled water.
14. Rinse in acidulated water: 1–2 minutes or longer.
15. Rinse in 70% alcohol: briefly dip.
16. Dehydrate in 95% alcohol, several dips in each of 2 changes.
17. Dehydrate in absolute alcohol, clear, and mount.

RESULTS

nuclei—brilliant red
collagen and reticulum—blue
muscle—red and yellow
basophilic cytoplasm—light blue
acidophilic cytoplasm—orange red
chromophobes—colorless or light gray

COMMENTS

The azan method is particularly recommended for pituitary and pancreas but is also outstanding for connective tissue. If a tissue does not "take" the azocarmine, it may be because of formalin in the fixative. To correct this condition, mordant the sections overnight in 3% aqueous potassium dichromate.

Poblete et al. (1976) have developed a fast form (less than 2 hours) of this method, but the quality is inferior to the one above.

Romeis (1948) suggests the substitution of acid alizarine blue for azocarmine; this shortens the time and does not require heat.

PROCEDURE

1. Stain in acid alizarine blue: 5 minutes.

aluminum sulfate ($Al_2(SO_4)_3$)	10.0 g
acid alizarine blue (C.I. 58610)	0.5 g
distilled water	100.0 ml

Bring to boil: 5–10 minutes. Let cool. Fill to 100 ml with distilled water. Filter. The solution will be a red-violet color.
2. Wash in running water.

3. Treat with phosphotungstic or phosphomolybdic acid, 5% aqueous: 30 minutes.
4. Wash in distilled water.
5. Proceed to aniline blue stain.

RESULTS

After phosphotungstic acid, the alizarine is red; it stays in nuclei and muscles but is removed from collagen. After phosphomolybdic acid, the nuclei and muscle are blue.

TRICHROME STAINING

Masson Trichrome Stain (Gurr modification 1956)

When Masson (1928) developed this method, he called it a trichrome stain, although it includes four dyes. Since then the trichrome name has been applied to many modifications of Masson's method that use many combinations of different dyes.

FIXATION

Any general fixative.

SOLUTIONS

Iron alum

ferric ammonium sulfate	4.0 g
distilled water.............................	100.0 ml

Hematoxylins, see p. 112.

Acid fuchsin

acid fuchsin (C.I. 42685)	1.0 g
distilled water.............................	100.0 ml
glacial acetic acid	1.0 ml

Ponceau de xylidine

ponceau de xylidine (C.I. 16150)	1.0 g
distilled water.............................	100.0 ml
glacial acetic acid	1.0 ml

Lillie (1940) suggests alternate stains for ponceau de xylidine: ponceau 2R; nitazine yellow; Biebrich scarlet; azofuchsin 3B, G, and 4G; Bordeaux red; chromotrope 2R; chrysoidin; eosin Y; orange G; and crocein.

Fast green

fast green FCF (C.I. 42053)	2.0 g
distilled water.............................	100.0 ml
glacial acetic acid	2.0 ml

Phosphomolybdic acid

phosphomolybdic acid	1.0 g
distilled water.............................	100.0 ml

Acidified water

glacial acetic acid	1.0 ml
distilled water.............................	100.0 ml

PROCEDURE

1. Deparaffinize and hydrate slides to water, remove $HgCl_2$. Mordant formalin-fixed tissue in saturated aqueous $HgCl_2$: 5 minutes. Wash in running water: 5 minutes. Treat with Lugol's and sodium thiosulfate.
2. Mordant in iron alum: $\frac{1}{2}$ hour.
3. Wash in running water: 5 minutes.
4. Stain in hematoxylin (Delafield or like): $\frac{1}{2}$ hour.
5. Wash in running water: 5 minutes.
6. Differentiate in saturated aqueous picric acid.
7. Wash thoroughly in running water: 10 minutes or longer.
8. Stain in acid fuchsin: 5 minutes.
9. Rinse in distilled water until excess stain is removed. Check under microscope.
10. Stain in ponceau de xylidine: 1–5 minutes.
11. Rinse in tap water, control under microscope for proper intensity of acid fuchsin and ponceau de xylidine.
12. Differentiate in phosphomolybdic acid: 5 minutes.
13. No rinse; transfer directly into fast green: 2 minutes.
14. Differentiate fast green in acidified water and dehydrating alcohols.
15. Dehydrate in absolute alcohol, 2 changes: 3 minutes each.
16. Clear and mount.

RESULTS

 nuclei—deep mauve blue to black
 cytoplasmic elements—varying shades of red and mauve
 muscle—red
 collagen, mucus—green

COMMENTS

Among the many modifications of Masson stain, this is one of the best because it offers good control of both red dyes by the use of two separate solutions. In most cases the full 5 minutes in the acid fuchsin is advisable; time in the ponceau de xylidine is not quite so critical.

Pollak Rapid Method (1944)

FIXATION

Any general fixative. Formalin-fixed tissues are improved by treating slides overnight with saturated aqueous $HgCl_2$.

SOLUTIONS

 Hematoxylins, see p. 112.

 Trichrome stain

acid fuchsin (C.I. 42685)	0.5	g
ponceau 2R (C.I. 16150)	1.0	g
light green SF, yellowish (C.I. 42095)	0.45	g
orange G (C.I. 16230)	0.75	g
phosphotungstic acid	1.5	g
phosphomolybdic acid	1.5	g
glacial acetic acid	3.0	ml
50% ethyl alcohol, up to	300.0	ml

Add acetic acid to alcohol; split solution in 4 parts. Dissolve acid fuchsin and ponceau in one part, light green in the second part, orange G and phosphotungstic acid in the third part, and phosphomolybdic acid in the fourth part. Mix four solutions and filter. Keeps well.

Acidified water

glacial acetic acid	0.2 ml
distilled water...........................	100.0 ml

PROCEDURE

1. Deparaffinize and hydrate slides to water; remove $HgCl_2$.
2. Stain in Mayer hematoxylin: 10 minutes. (Pollak uses Weigert: 10–20 minutes.)
3. Wash in running water: 10 minutes.
4. Stain in trichrome stain: 7 minutes.
5. Rinse briefly in distilled water.
6. Differentiate in acidified water: only a few seconds. Check under microscope if necessary.
7. Dip a few times in 70% alcohol.
8. Dehydrate in 95% alcohol, 2 changes: several seconds each.
9. Dehydrate in absolute alcohol, clear, and mount.

RESULTS

nuclei—dark blue
muscle, elastin—red
fibrin, calcium—purple
hyalin—pale blue
collagen, mucus—blue green

Toren (1963) combines this stain with Giemsa (p. 231) for mast cell staining.

Gomori Method (1950*b*) (Sweat et al. Modification 1968)

FIXATION

Any general fixative. Formalin-fixed tissues are improved by treating slides overnight with saturated aqueous $HgCl_2$.

SOLUTIONS

Bouin fixative

picric acid, saturated aqueous	1 part
formalin, concentrated	5 parts
glacial acetic acid	1 part

Mix shortly before use. Picric acid in solution is not dangerous, but it is explosive when dry. Remove used Bouin from the oven and discard.

Trichrome solution

chromotrope 2R (C.I. 16570)	0.6 g
aniline blue WS (C.I. 42780)	0.6 g
phosphomolybdic acid	1.0 g
dissolve in distilled water	100.0 ml
add hydrochloric acid	1.0 ml

Allow to stand 24 hours in refrigerator before use. Store in refrigerator and use cold. Do not filter. The solution is stable until the reds begin to fade.

Acidified water

glacial acetic acid	0.2 ml
distilled water	100.0 ml

PROCEDURE

1. Deparaffinize and hydrate slides to water; remove $HgCl_2$.
2. Treat with Bouin solution, 56°C: 1 hour.
3. Wash in running water: 5 minutes, or until decolorized.
4. Stain in trichrome mixture: 1 minute.
5. Rinse briefly in distilled water to remove some of excess stain.
6. Rinse in acidified water: 30 seconds.
7. Dehydrate in 95% alcohol, 2 changes. Be careful not to remove too much green.
8. Dehydrate in absolute alcohol, clear, and mount.

RESULTS AND COMMENT

There have been a number of attempts to combine the Azan and Masson methods into a quicker method. Gomori considered Pollak's to be a disappointingly dull color scheme, lacking real red shades. Almost any combination of an acid triphenylmethane dye with a sulfonated azo dye in the presence of phosphomolybdic or phosphotungstic acid will give good results. The phosphomolybdic acid favors the green and blue shades; phosphotungstic acid, the reds. A short staining time produces more red; prolonged staining, more green and blue. Rinsing in tap water weakens the reds.

The Sweat et al. modification improves the colors of the fine connective tissue fibers, so that they closely resemble those obtained with the Mallory and Mallory–Heidenhain procedures. Pretreatment with Bouin improves the affinity for stains, offers more uniform results, and reduces variations due to different fixatives. Staining with hematoxylin has been eliminated; very little of its color remains after staining in the acid trichrome mixture. If an iron hematoxylin is used, it turns the clear red of the chromotrope to a murky color.

The staining of connective tissue fibers is affected by the pH of the solution, best around pH 1.3. The Gomori solution has a pH of 2.5–2.7; by replacing the acetic acid with hydrochloric acid, the pH is lowered to approximately 1.3.

The rinse in acetic acid does not change the colors, but makes them more delicate and transparent.

I have had good results with most of the trichrome stains and can only recommend that the selection of a routine stain be made by each individual. All of the short trichrome methods offer clearer and sharper returns on thin (5–7 μ) sections than on thicker ones. For the thick sections, the Mallory, Masson, or Azan stain should be used.

All of the many trichrome staining methods cannot be included, but following are references for some excellent ones.

1. Buffalo back staining, Conn et al. (1960); Lillie (1945).
2. Churg and Prado trichrome stain (1956).
3. Greenstein simplified five-dye stain (1961).
4. Koneff aniline blue stain (1936).
5. Kornhauser quad stain (1943, 1945).
6. Movat pentachrome stain (1955).
7. Menzies modification of picro-Gomori (1959).
8. Chlorazol black E, Levine and Morrill (1941).
9. PAS-phosphomolybdic acid–Sirius supra blue FGL–CF, Sweat, Puchtler, and Sesta (1968).
10. Allochrome procedure, Lillie (1951c).
11. Sweat, Puchtler, and Woo, modification of no. 10 (1964).
12. Mallory–Heidenhain, Cason (1950).
13. Whipf's Polychrome, Wilson and Fairchild (1977).

COLLAGEN AND ELASTIN STAINING

Picro-Ponceau with Hematoxylin (Gurr 1956)

(Van Gieson Substitute, Nonfading)

FIXATION

Any general fixative.

SOLUTIONS

Hematoxylins, see p. 112.

Picro-ponceau

ponceau S (C.I. 27195), 1% aqueous	10.0 ml
picric acid, saturated aqueous	86.0 ml
acetic acid, 1% aqueous	4.0 ml

PROCEDURE

1. Deparaffinize and hydrate slides down to water, remove $HgCl_2$.
2. Overstain in hematoxylin: 5–15 minutes (Delafield or Harris type).
3. Wash thoroughly in running water until slides are deep blue: 10 minutes or longer.
4. Stain in picro-ponceau: 3–5 minutes. (This may be too strong for some tissues; in addition to staining, the stain acts as a destaining agent on the hematoxylin.) Rinse for a few seconds in distilled water and check under microscope. Continue to stain and destain or differentiate in water until nuclei are sharp.
5. Dip several times in 70% alcohol.
6. Dehydrate in 95% alcohol, 2 changes, to insure complete removal of excess picric acid. Only that which has acted as a dye must be left in the tissue.
7. Dehydrate in absolute alcohol, clear, and mount.

RESULTS

nuclei—brown to brownish or bluish black
collagenous and reticular fibers—red
elastic fibers, muscle fibers, erythrocytes, epithelia—yellow

COMMENTS

A Weigert type of hematoxylin is excellent for this method. Delafield (or similar) hematoxylin may be preceded by mordanting. After step 1: mordant with iron alum: 10 minutes; follow by washing in running water. Then proceed to step 2.

This method is so superior to so-called Van Gieson stain, which uses acid fuchsin instead of ponceau S, that the Van Gieson method has been omitted. The colors are identical but Van Gieson is unsatisfactory because it fades rapidly.

Katline (1962) recommends the following procedure if the picric solution tends to fade the hematoxylin: treat the sections prior to the picric staining with a solution of equal parts of 2.5% aqueous phosphotungstic and phosphomolybdic acid, 0.5–1 minutes; wash briefly; and proceed to the picro-ponceau solution.

Sweat, Puchtler, and Rosenthal (1964) substitute 0.1% Sirius red F3BA[2] in aqueous picric acid for the acid fuchsin in Van Gieson method.

Puchtler et al. (1973) distinguish between elastin and collagen with a fluorescent method.

Verhoeff Elastin Stain (Mallory 1944)

FIXATION

Any general fixative.

SOLUTIONS

Verhoeff stain

On electric hot plate, dissolve 3 g hematoxylin in 66 ml absolute ethyl alcohol. Cool, filter, and add 24 ml of 10% aqueous ferric chloride and 24 ml Verhoeff iodine solution. Usefulness is limited to 1–2 weeks.

Verhoeff iodine solution

potassium iodide (KI)	4.0 g
distilled water	100.0 ml

Dissolve and add:

iodine	2.0 g

Ferric chloride solution, 10%

ferric chloride ($FeCl_2$)	10.0 g
distilled water	100.0 ml

---

[2] Verona Dyestuffs.

Ferric chloride solution, 2%

10% ferric chloride	20.0 ml
distilled water.............................	100.0 ml

Picro-ponceau solution

ponceau S (C.I. 27195), 1% aqueous	10.0 ml
picric acid, saturated aqueous	86.0 ml
acetic acid, 1% aqueous	4.0 ml

PROCEDURE

1. Deparaffinize and run slides down to 70% alcohol. Removal of $HgCl_2$ not necessary.
2. Stain in Verhoeff stain: 15 minutes.
3. Rinse in distilled water.
4. Differentiate in 2% ferric chloride: a few minutes. Elastic fibers should be sharp black; nuclei brown. If destained too far, return slides to Verhoeff for another 5–10 minutes.
5. Transfer to sodium thiosulfate, 5% aqueous: 1 minute.
6. Wash in running water: 5–10 minutes.
7. Counterstain in picro-ponceau: 1 minute.
8. Differentiate in 95% ethyl alcohol, 2 changes: a few seconds each.
9. Dehydrate in absolute alcohol, clear, and mount.

RESULTS

elastic fibers—brilliant blue black
nuclei—blue to brownish black
collagen—red
other tissue elements—yellow

COMMENTS

If elastic fibers stain unevenly or not at all, Verhoeff solution is too old.

Verhoeff can be combined with Perl staining (p. 246) to show both iron and elastin (Pickett and Klavins 1961).

Miller's (1971) method using Victoria blue 4R, new fuchsin, and crystal violet also stains elastin black. The solution is stable.

Iron Gallein Elastin Stain (Churukian and Schenk 1976)

Solution A

gallein (C.I. 45445)	1.0 g
ethylene glycol	20.0 ml

Add 80 ml absolute alcohol and mix.

Solution B

ferric chloride (FeCl$_3$ · 6H$_2$O), 29% aqueous ...	4.0 ml
distilled water	95.0 ml
HCl, concentrated	1.0 ml

Mix equal parts of A and B just before use.

PROCEDURE

1. Deparaffinize and hydrate to water removing HgCl$_2$.
2. Stain: 30 minutes.
3. Wash in running water: 5 minutes.
4. Differentiate in 2% FeCl$_3$: 2 minutes.
5. Wash in running water: 15 minutes.
6. Procede with counterstain of choice.

RESULTS

elastin—black

Basic Fuchsin Stain (Horobin et al., 1974)

Solution

2 g basic fuchsin (C.I. 42500), bring to boil in 200 ml distilled water. Add 25 ml 30% aqueous ferric chloride and boil for 3 minutes. Filter and dry precipitate in incubator. Dissolve precipitate in 200 ml 95% ethyl alcohol using low heat until dissolved. Add 4 ml concentrated HCl and make up volume to 200 ml with 95% alcohol. Stored at 4°C this solution is good for at least 18 months. The dry precipitate also is stable.

PROCEDURE

1. Hydrate to water as usual.
2. Stain: 30 minutes.
3. Rinse off excess stain in 70% alcohol.
4. Differentiate in 1% HCl in 70% alcohol up to 5 minutes if required.
5. Dehydrate, clear, and mount.

RESULTS

elastin—deep purple
strongly basophilic substances (mast cells, cartilage)—purple

Orcein (Romeis 1948)

FIXATION

Any general fixative.

SOLUTION

orcein	1.0 g
70% ethyl alcohol	100.0 ml
hydrochloric acid (assay 37–38%)	1.0 ml

PROCEDURE

1. Deparaffinize and hydrate slides to water; remove $HgCl_2$.
2. Stain in orcein: 30–60 minutes.
3. Wash briefly in distilled water.
4. Dehydrate in 95% alcohol: 2 minutes.
5. Differentiate in absolute alcohol until background is almost colorless and elastin fibers are isolated.
6. Rinse in fresh absolute alcohol, clear, and mount.

RESULT

elastin—red

COMMENTS

Lillie et al. (1968) stain in orcein, 4 hours, and differentiate in ferric chloride hexahydrate (0.1 ml of 10% $FeCl_3 \cdot 6H_2O$ in 50 ml of 70%

ethanol). Dehydrate and clear. This changes the elastin color to black or reddish black.

Roman et al. (1967) combine orcein, hematoxylin, ferric chloride, and iodine in a single solution and stain for two hours. Differentiation is controlled in a ferric chloride solution, and the elastic fibers are stained a deep purple.

Humason and Lushbaugh (1969) combine silver nitrate, orcein, and Sirius supra blue stains to show reticulum, elastin, and collagen in the same section.

Some disagreement appears in the literature concerning the pH at which orcein is most effective. Weiss (1954) writes that orcein stains only from an acid alcoholic solution between pH 3 and 8 and that this puts orcein in a category between basic and acidic dyes, which generally operate at the extremes of alkaline and acid pH. He suggests that there is a formation of hydrogen bond between orcein and elastin. Since this reaction takes place in acid alcohol, it is probably due to a uniquely low positive charge of elastin in such solutions. The use of alcoholic solutions is necessary to stabilize the positively charged orcein fractions. Dempsey and Lansing (1954) agree with Weiss.

Darrow (1952), experimenting with orcein, found that pH 1–2.4 is best for elastin staining, and that above pH 2.6 collagen stains as well. A dye content of 0.4% is adequate for elastin staining; a higher concentration adds to collagen staining. For a specific elastin reaction when collagen is present, therefore, it is advisable to check the pH of the acid alcohol used for the solution and to reduce the dye content to 0.4 g per 100 ml.

SUBCUTANEOUS TISSUE STAINING

Subcutaneous tissue, areolar tissue, and omentum (membranes) can be easily fixed for staining. A simple way is to spread a piece on a slide with dissecting needles, allow to dry slightly, and immerse in fixative. The tissue, in most cases, will adhere to the slide throughout staining procedures.

If histologic rings are not available for spreading tissue, cut thick filter paper into circles of a desired diameter. Cut a hole in the center of the circle, leaving an outer edge of 6–7 mm in width. Place the circle of filter paper under the omentum or other tissue to be spread and press against the tissue. Cut out a ring of the tissue, just beyond the outer edge of the paper circle, and place in fixative. Carry filter paper and tissue through staining procedures; connective tissue stains can be applied as usual. Finally, remove the paper by peeling it off when the tissue is being placed on a slide preparatory to mounting with cover glass.

Bits of loose connective tissue that are difficult to handle may be dehydrated and infiltrated with nitrocellulose. Carefully spread the tissue in a film of nitrocellulose on a slide, allow to dry slightly, and harden in 70% alcohol. For more uniform staining, stain the tissue in place on the slide and carry it through solutions like a nitrocellulose section. This method gives more uniform staining than the dry method above.

BONE STAINING

Decalcified Sections (Romeis 1948)

FIXATION

Formalin, followed by decalcification (pp. 29–32).

SOLUTIONS

Thionin

thionin (C.I. 52000) saturated in 50% ethyl alcohol	10.0 ml
distilled water	100.0 ml

If the color does not set correctly, add 1–2 drops of ammonia to staining solution.

Carbol-xylol, see p. 543.

PROCEDURE

1. Deparaffinize and hyrate slides to water. If sections tend to loosen, treat with 0.5–1% nitrocellulose (p. 67) or try subbed slides (p. 550).
2. Stain in thionin solution: 10 minutes.
3. Wash, distilled water: 20 minutes, change several times.
4. Treat with picric acid, saturated aqueous: 1 minute.
5. Rinse in distilled water.
6. Differentiate in 70% alcohol: 5–10 minutes or more, until no more color comes off.
7. Dehydrate in 95% alcohol, 2 changes: 3 minutes each.
8. Dehydrate and clear in carbol-xylol: 5 minutes.
9. Clear in xylene and mount.

RESULTS

   lacunae and canaliculi—bordered with bluish black
   background—yellow

COMMENTS

Since paraffin sections occasionally loosen from the slides, some technicians prefer nitrocellulose embedding, but this takes more time. Subbed
slides or Haupt's adhesive are reliable enough for my use.

Hand-Ground, Undecalcified Sections (Enlow 1954)

PROCEDURE

1. If it is necessary to remove organic materials and fat, treat in following manner:
   a. Boil in soap solution: 3–4 hours; and wash in running water: 3–4
      hours.
   b. Suspend over ether or chloroform: 36–48 hours.
   c. Allow to dry thoroughly.
2. Cut slices as thin as possible without cracking them. A jeweler's
   saw is recommended.
3. Grind on sharpening hones with finely powdered Carborundum or
   household cleansing powder. Keep surface wet with water. Optical
   or metallurgical grinding and polishing equipment, if available, can
   be used. The grinding can be done between two stones, or the
   section can be held on a wet rubber or cork stopper and ground
   against the hone. If this becomes difficult to manage, glue the section on a slide with Duco Cement, and continue to grind. When the
   section is almost thin enough, loosen it from the slide with acetone,
   turn it over, glue it down, and grind the opposite side.
4. Polish off both sides with a fine leather strop.
5. Place in 95% alcohol: 5 minutes.
6. Air dry.
7. Place in plastic solution; agitate to liberate air bubbles.

   parloidin .... 28.0 g
   butyl (or amyl) acetate .... 250.0 ml

   Let stand until parloidin is completely dissolved. Stir thoroughly.

8. Transfer to slide with a drop of solution.
9. Dry thoroughly; do not add more solution.
10. When completely dry, add mounting medium and cover glass.

COMMENTS

To bring out the density and distribution of the mineral in undecalcified bone, Frost (1959) recommends methods with basic fuchsin, silver nitrate, alizarine red S, and others.

Dowding (1959) uses methyl methacrylate as a plastic embedding medium and grinds the bone down as thin as 30μ.

Hause (1959) describes a block for holding the bone while sawing it and recommends a razor saw blade no. 35–ST[3] as a good cutting instrument.

Kropp (1954) describes a plastic-embedding method using heat and pressure.

Yaeger (1958) uses freeze drying and vacuum infiltration with butyl methacrylate–ethyl methacrylate.

Norris and Jenkins (1960) describe a method using epoxy resin for preparation of bone for radioautography. The resolution is good; there are no chemicals in resin to produce artifacts in nuclear emulsions; the medium does not warp or chip when machined or abraded. (Polyesters and methacrylate do distort when machined.) They include a design for a metal and lucite mold for embedding. Sections can be made with a microtome ($6-10\mu$), with a circular saw ($50-100\mu$), or ground to give sections as thin as $10-50\mu$. Kwan (1970) uses sticky wax and a circular diamond blade. Enlow (1961) describes staining of ground and decalcified sections. Villanueva et al. (1964) stain ground sections with fast green FCF, orange G, celestin blue, and basic fuchsin.

Alizarine Red S Method for Embryos (Bone Formation)

FIXATION

A hardening action. Hollister (1934) suggests 70% alcohol for several days for fish; removal of scales desirable. Hood and Neill (1948) use 95% alcohol: 3 days. Organism preferably should be free of hair or feathers. Richmond and Bennett (1938) use 95% alcohol: 2 weeks. Some specimens require decolorization. The best method is to lay specimen in 95% alcohol in white tray. Place in direct sunlight for 24 hours each side. A sunlamp

---

[3] Lipshaw Manufacturing Co.

was used by Hollister on sunless days. Youngpeter (1964) bleaches specimens in 30% hydrogen peroxide: 1–2 hours, after step 2 in procedure below. Then he immerses in KOH until specimen is transparent.

SOLUTIONS

Potassium hydroxide, 2%

potassium hydroxide, white sticks............	20.0 g
distilled water.............................	1000.0 ml

Alizarine stock solution (Hollister 1934)

alizarine red S (C.I. 58005), saturated in 50% acetic acid...............................	5.0 ml
glycerin	10.0 ml
chloral hydrate, 1% aqueous (a drug)	60.0 ml

Alizarine working solution

alizarine stock solution.....................	1.0 ml
1–2% potassium hydroxide in distilled water ..	1000.0 ml

Make up at least 500 ml; use at room temperature.

Clearing solutions (Youngpeter 1967)

Solution A

glycerin	20 parts
4% potassium hydroxide	3 parts
distilled water.............................	77 parts

Solution B

glycerin	50 parts
4% potassium hydroxide	3 parts
distilled water.............................	47 parts

Solution C

glycerin	75 parts
distilled water.............................	25 parts

PROCEDURE

1. From hardening solution, rinse few minutes in distilled water.
2. Leave in 2% potassium hydroxide until skeleton shows through musculature: 2–4 hours for small embryos, 48 hours or longer for larger forms. St. Amand and St. Amand (1951) warmed solution to 38°C for quicker action.
3. When clear, transfer to alizarine working solution: 6–12 hours or longer depending on specimen. Skeleton should be deep red. Large specimens may require fresh changes of dyé solution.
4. Transfer directly to 2% potassium hydroxide: 1 day or longer until soft tissues are destained. The KOH may be 1.0–0.5% for small specimens. Sunlight or lamp speeds up process.
5. Transfer through the three clearing solutions: 24 hours each.
6. Transfer to pure glycerin with thymol added as preservative.
7. Store in sealed tubes or bottles. Mount on glass rods and seal in museum jars, or embed in plastic.

COMMENTS

Cumley et al. (1939) gradually replace the glycerin with 95% alcohol, absolute alcohol, and finally toluene. Then the specimens are transferred to toluene saturated with naphthalene and stored in anise oil saturated with naphthalene. This method is supposed to produce greater clarity than the glycerin storage.

Crary (1962) adds alizarine red S to the KOH solution and clears in a mixture of glycerin, 70% ethanol, and benzyl alcohol.

Alizarine is the most specific stain for calcium. The red coloring depends on the presence of a Ca base; the only other substance staining red is strontium.

Taylor (1967) uses pancreatic enzymes to clear small vertebrates.

Ojeda et al. (1970) stain the cartilage of whole chick embryos with Alcian blue.

Plastic mounts can be made of these preparations. Excellent directions are supplied in: *Embedding Specimens in Transparent Plastic*, Turtox Service Leaflet #33 (General Biological Supply House) and *How to Embed in Bio-Plastic* (Ward's Natural Science Establishment). For other methods, also for cartilage staining, see Wassersug 1976, cartilage and bone; Ojeda et al. 1970, bone; McCann 1971, and Love and Vickers 1972, cartilage.

MUSCLE STAINING

Milligan Trichrome Stain (1946)

FIXATION

10% formalin in normal saline (I have used Gomori 1–2–3 fixative with excellent staining results).

SOLUTIONS

Mordant

Solution A

potassium dichromate........................	3.0 g
distilled water..............................	100.0 ml

Solution B

hydrochloric acid, concentrated..............	10.0 ml
95% ethyl alcohol	100.0 ml

Mix 3 parts A with 1 part B; use within 4 hours.

Acid fuchsin

acid fuchsin (C.I. 42685)	0.1 g
distilled water..............................	100.0 ml

Phosphomolybdic acid

phosphomolybdic acid	2.0 g
distilled water..............................	200.0 ml

Use half of solution for orange G solution below.

Orange G

orange G (C.I. 16230)	2.0 g
1% phosphomolybdic acid	100.0 ml

Fast green

Stock solution

fast green FCF (C.I. 42053)	10.0 g
2% acetic acid (2 ml/98 ml distilled water)	100.0 ml

Working solution

fast green stock solution	10.0 ml
distilled water............................	90.0 ml

Aniline blue may be substituted for fast green.

PROCEDURE

1. Deparaffinize and transfer slides through absolute alcohol into 95% alcohol (remove HgCl$_2$ if present).
2. Mordant in potassium dichromate–hydrochloric acid solution: 5–7 minutes.
3. Rinse in distilled water.
4. Stain in acid fuchsin: 5–8 minutes.
5. Rinse in distilled water.
6. Fix stain in phosphomolybdic acid solution: 1–5 minutes.
7. Stain in orange G: 5–10 minutes.
8. Rinse in distilled water.
9. Treat with 1% aqueous acetic acid (1 ml/99 ml distilled water): 2 minutes.
10. Stain in fast green: 5–10 minutes.
11. Treat with 1% acetic acid: 3 minutes.
12. Rinse in 95% alcohol, transfer to second 95% alcohol: 5 minutes.
13. Finish dehydration in absolute alcohol, 2 changes: 3 minutes each.
14. Clear and mount.

RESULTS

nuclei, muscle—magenta
collagen—green (blue with aniline blue)
red blood cells—orange to orange red

COMMENTS

This is a beautiful, precise stain showing strong contrast between muscle and connective tissue. Smooth muscle cells stand out sharply and clearly.

Milligan reports that this stain's weak point is the nuclear stain, which is improved by using Gomori 1–2–3 fixative.

Galigher[4] recommended pinning out strips of muscle on a smooth piece of wood. Allow just enough tension to hold the muscle straight without stretching it. Fix in Susa fixative (p. 17): 3–6 hours. Stain the sections by the Azan method (p. 139). The bands of the striations appear in red, blue, or yellow. Galigher also found phosphotungstic acid–hematoxylin to be a useful stain for muscle.

Iron hematoxylin staining (p. 123) can produce beautiful striation staining if it is followed by careful destaining with iron alum instead of picric acid. Wash thoroughly (at least 30 minutes) in running water after destaining; do not counterstain.

Most of the general fixatives are satisfactory for muscle fixation; however, Zenker is best only for the preservation of intercalated discs.

See Poley and Forbes (1964) for staining to show muscle infarction.

See Puchtler et al. (1969) for a PAS–myofibril stain that is excellent for muscle fiber bands.

---

[4] Personal communication.

Chapter 11
Silver Impregnating Reticulum

SILVER IMPREGNATION

According to Baker (1958) methods under this heading depend upon the local formation within tissues of a colored substance that is not a dye. Impregnation applies to the condition which develops when an unreduced metal (silver, etc.) is taken up from a solution of salt or other compound and deposited in a colloidal state on a tissue element. Following impregnation, the tissue is removed to a reducing solution of a photographic type and the metal is reduced to the elementary state, probably in the form of a black deposit. Thus the tissue itself does not reduce the metal, but some extraneous reducer is required to perform the reaction.

Silver staining started in 1843 when Krause tried small pieces of fresh tissue in silver nitrate. Golgi, an Italian, in 1872 fixed nerve tissue for a long period of time in a dichromate solution. Then he tried soaking the same tissue in silver nitrate solution and found that silver dichromate was deposited selectively, leaving the impregnated components in sharp relief against an almost colorless background. Ramón y Cajal (1903, 1910) saw the possibilities of the silver method and experimented with it further, including ways of reducing it. In 1906, he and Golgi were jointly awarded the Nobel Prize in physiology and medicine. By continuing to improve the use of silver and gold impregnations, Ramón y Cajal and del Río-Hortega, one of his pupils, were able to systematically investigate the histology of the nervous system. In the early 1900s Protargol (Bodian method) and other organic silver compounds were introduced as substitutes for silver nitrate. The albumen fraction in the organic compounds is considered to act as a protective colloid that prevents too rapid reduction by formaldehyde, thereby inducing the formation of finer-grained deposits of silver. Subsequent experimentation with silver on tissue sections led to many modifications of the methods of Ramón y Cajal, Bielschowsky, and others. Silver impregnation can be very complicated because of the many factors involved to make it specific for various tissue elements.

Ammoniacal silver is the familiar complex and, when reduced by formaldehyde to metallic silver, forms a colloidal solution containing negatively charged particles. These may be precipitated out by oppositely charged surfaces, which can be changed to repel or to absorb the silver. Thus the negatively charged silver (formed by reduction) is deposited on positively charged surfaces and allows selective impregnation of neurofibrils, reticulum, Golgi, and so on. It is probable that fixation and dehydration coagulate the proteins and leave them positively or negatively charged; this would account for the difference in charges of different tissue elements. The pH of the silver solution is a strong factor in determining charges, and it depends on whether an ammonium or sodium hydroxide or sodium or lithium carbonate is used (Ramón y Cajal and del Río-Hortega methods).

As already mentioned, protective colloids, such as gum arabic and mastic, and the use of Protargol slow down the reduction of the silver to produce a finer grain (Liesegang method). Dilute reducing reagents, combined with the above, can have the same effect (Ramón y Cajal and Bielschowsky methods). Temperature is a factor, because it increases the kinetic energy of the particles and permits a greater number of collisions of the particles against the tissue surfaces. In some methods, copper is added to the silver solution, supposedly to speed up impregnation by initiating the reduction to metallic silver. Too heavy a deposit is prevented by removal of some of the silver from the solution. Thus various applications of the principles can be used to control the impregnation of different kinds of tissue elements (Silver 1942). Chemical properties of tissues and their responses to these conditions all help to determine the place and amount of deposition.

The types of silver impregnation (all ammoniacal) can be classified in the following manner: (1) ammoniacal silver nitrate, (2) ammoniacal silver hydroxide, and (3) ammoniacal silver carbonate. In all of these solutions the silver is present largely in the form of a complex silver ammonia cation $[Ag(NH_3)_2]^+$. In (1), ammonia alone is used to form the precipitate, the chief product in solution being silver diammino nitrate. In (2), the Bielschowsky method (also modified by Ramón y Cajal), sodium hydroxide is used to form the precipitate and ammonia to redissolve it, the chief product being silver diammino hydroxide. The difference between the Bielschowsky and the Ramón y Cajal methods lies in the way the silver is applied and the reduction performed. The Ramón y Cajal method uses a single ammoniated silver impregnation followed by reduction in Pyrogallol, hydroquinone, or one of the aminophenols. The Bielschowsky method uses double impregnation (silver nitrate followed by ammoniated silver) and reduction in formalin. In (3), the del Río-Hortega method, sodium carbonate (sometimes lithium carbonate) is used to form the precipitate and is followed by ammonia, the chief product being silver diammino carbonate. (Kubie and Davidson 1928.)

The reactions of these solutions to various conditions should be understood:

1. The ammoniacal silver nitrate is most stable, least sensitive to light, and least readily reduced, and it combines least easily with tissues. During formalin reduction, a cloud of finely divided gray dust slowly develops, and the staining is slow. This method is rarely used in preference to others.

2. Ammoniacal silver hydroxide is the least stable, the most sensitive to light, and the most readily reduced, and it combines most easily with tissues. Almost instantly a heavy black cloud appears. It combines almost at once with the tissue, the solution darkens quickly, and a precipitate begins to form. Since silver nitrate is not reduced in an acid solution, but reduces readily in an alkaline solution, it stands to reason that the ammoniacal silver hydroxide solution is the most sensitive of the three.

3. The properties of the ammoniacal silver carbonate solution are between those of the above two. Its precipitate forms more promptly than that of ammoniacal silver nitrate (1), but not as fast as that of ammoniacal silver hydroxide (2). Its precipitate is darker than that of (1) but not as dark as that of (2). The solution's reaction begins within 5 to 10 minutes and reaches optimum color before the solution begins to darken. Although solution (2) and this one are used interchangeably, this solution has the advantage in that its hydroxide $(OH)^-$ ion concentration is not high enough to render it as unstable and oversensitive as solution (2), and the presence of buffer salts makes the reduction proceed steadily and evenly. As the acid (HNO_3) is formed during the reduction, the buffer absorbs it and blocks its effect; thus the reduction is not lessened. In addition, the presence of CO_3 ion buffers the formalin and prevents formation of formic acid, also an effective stop to further reduction. Foot (1929) buffered his formalin and prolonged the reducing action, making his results darker and more intense. He warned, however, that the buffer must be kept to a minimum so the reaction would not become too intense. Equi-molar silver solutions produce the most uniform results. In most cases, it is wise not to dissolve completely the precipitate that is first formed since this can cause inferior results.

The use of pyridine, a fat solvent, precedes some methods; removes cephalin, lecithin, myelin, mitochondria, galactolipids, etc.; and makes subsequent penetration of the silver easier. (This is particularly true for connective tissue.) Toning with gold chloride is optional in many methods, but it may yield a more desirable color and improve contrasts. The timing in gold chloride apparently is variable. It need be only long enough to make the desired change in color, usually a few seconds. If the reaction is slow, the solution has weakened. The final fixing in sodium thiosulfate (hypo) is necessary to remove all unreduced silver.

Corked bottles should be avoided, since cork extractives may disrupt selective impregnation of tissue elements (Deck and DeSouza 1959). In all silver methods, take care that metal instruments do not come in contact with the silver solution; a black precipitate may dribble down the surface of slides handled in such a manner. Coat forceps with paraffin, or use horn or wooden instruments. All glassware for metallic stains (silver, gold, etc.) must be acid-clean. Soak in cleaning solution (p. 570), wash thoroughly in running water to remove cleaning solution, and rinse 4 or 5 times with distilled water. Keep one set of glassware for silver stains only. If aqueous solutions of silver nitrate only (no other chemical present) are milky, the distilled water is at fault; glass distilled is recommended. Use silver solutions in the dark and protected from dust.

The loosening of sections from slides is a common problem during silver impregnation. Davis and Harmon (1949) use a rinse (0.5 ml acetic acid, 2% aqueous, in 50.0 ml of water) before reduction in sulfite and hydroquinone in the Bodian method (p. 188). Many technicians find that Masson gelatin fixative (p. 549) is superior to Mayer albumen (p. 548) for affixing sections for silver processes. Transferring unmounted paraffin sections through all solutions and mounting and deparaffinizing at the conclusion of the method is feasible. I use subbed slides (p. 550).

Prepare ammoniacal silver solutions just before use. Smith (1943) warns that ammoniacal silver hydroxide solutions will become explosive if they stand for too long a time. Explosions have occurred in several laboratories as the result of the formation of explosive silver amide in ammoniacal silver hydroxide solutions that have been stored for some time. (See Laboratory Safety, p. 572.)

For complete discussions concerning silver impregnation, see Baker (1958); Beech and Davenport (1933); Bensley (1959); Foot (1929); Kubie and Davidson (1928); Long (1948); and Silver (1942).

SILVER IMPREGNATION FOR RETICULUM

For reticulum the silver method probably depends on the local reduction and selective precipitation of the silver by the aldehydic groups of the carbohydrates in the reticulum. The silver is reduced to a dark-brown lower oxide and precipitated on the fibers. The formalin (sodium sulfite, hydroquinone) then reduces the precipitate to black metallic silver.

Among the numerous impregnation methods for reticulum, the following, when properly handled, have never failed to produce precise results for me, and usually with a minimal loss of sections. Subbed slides (p. 550) can be used to reduce loss of sections.

Naoumenko and Feigin Method (1974)

FIXATION

Any good general fixative

SOLUTIONS

Silver solution

To 35 ml distilled water add, in order, 7 ml 8% aqueous ammonium nitrate, 8 ml 4% aqueous sodium hydroxide, and 3.8 ml 10% aqueous silver nitrate.

Potassium permanganate

0.25% potassium permanganate, aqueous	45.0 ml
0.67% acetic acid, aqueous	0.5 ml

Mix fresh; not stable.

Oxalic acid

oxalic acid................................	1.0 g
distilled water............................	100.0 ml

Formalin solution

formalin	0.2 ml
distilled water............................	100.0 ml

Gold chloride

gold chloride.............................	1.0 g
distilled water............................	100.0 ml

PROCEDURE

1. Deparaffinize and hydrate slides to water, remove $HgCl_2$.
2. Treat with potassium permanganate: 2 minutes.
3. Wash in distilled water: 15 seconds.
4. Bleach in oxalic acid: 2 minutes.
5. Wash in 2 changes distilled water: 1 minute each.
6. Impregnate with silver solution: 6 minutes or longer, depending on following steps.

   From step 6 process slides individually.

7. Dip, one slow dip, in 70% alcohol.
8. Reduce in formalin solution, 2 changes, over a period of 2 minutes. Briefly with agitation in first, and remaining time in second.
9. Wash in distilled water: 1 minute.
10. Gold tone in gold chloride: 1 minute.
11. Rinse in distilled water: 1 minute.
12. Fix in 5% sodium thiosulfate: 1 minute.
13. Rinse in distilled water: 1 minute.
14. Counterstain if desired.
15. Dehydrate, clear, and mount.

RESULTS

Reticulum—black

Gridley Method (1951)

FIXATION

Any good general fixative.

SOLUTION

Ammoniacal silver hydroxide

To 20.0 ml of 5% silver nitrate (1 g/20 ml distilled water) add 20 drops of 10% sodium hydroxide. Add fresh 28% (reagent) ammonia drop by drop until precipitate which forms is almost redissolved. Add distilled water up to 60.0 ml.

PROCEDURE

1. Deparaffinize and hydrate slides to water. Remove $HgCl_2$ if present.
2. Treat with 0.5% periodic acid (0.5 g/100 ml water): 15 minutes.
3. Rinse in distilled water.
4. Treat with 2% silver nitrate (2 g/100 ml water): 30 minutes, room temperature.
5. Rinse in 2 changes of distilled water.
6. Impregnate in ammoniacal silver solution: 15 minutes, room temperature.
7. Rinse rapidly in distilled water.
8. Reduce in 30% formalin (30 ml/70 ml water): 3 minutes. Agitate gently.
9. Rinse in 3 or 4 changes of distilled water.
10. Tone in gold chloride (10 ml 1% stock solution/40 ml water) until yellow-brown color has changed to lavender gray.
11. Rinse in distilled water.
12. Fix in 5% sodium thiosulfate (5 g/100 ml water): 3 minutes.
13. Wash in running water: 5 minutes.
14. Counterstain if desired.
15. Dehydrate, clear, and mount.

RESULTS

reticulum fibers—black
other tissue elements—depends on counterstain

COMMENTS

Gridley uses periodic acid for oxidation in prefence to potassium permanganate and oxalic acid because the latter combination frequently causes
the sections to detach from the slides.

Gomori Method (Mallory 1944)

FIXATION

10% neutral formalin.

SOLUTIONS

Ammoniacal silver solution

To 20 ml of 10% silver nitrate solution (3 g/30 ml distilled water) and 4–5 ml of a 10% potassium hydroxide solution (0.5 g/5 ml distilled water), add 28% ammonia water, drop by drop, shaking the flask continuously, until the precipitate that forms is completely dissolved. Carefully add silver nitrate solution, drop by drop, until the precipitate that forms disappears when the solution is shaken. Make up the solution with distilled water to twice its volume. Always use acid-clean glassware.

Potassium permanganate

potassium permanaganate	0.5 g
distilled water.............................	100.0 ml

Potassium metabisulfite

potassium metabisulfite	2.0 g
distilled water.............................	100.0 ml

Ferric ammonium sulfate

ferric ammonium sulfate	2.0 g
distilled water.............................	100.0 ml

Formalin

formaldehyde, concentrated	20.0 ml
distilled water.............................	80.0 ml

Gold chloride solution

gold chloride stock solution (1 g/100 ml water) .	20.0 ml
distilled water.............................	80.0 ml

Sodium thiosulfate

sodium thiosulfate	2.0 g
distilled water.............................	100.0 ml

PROCEDURE

1. Deparaffinize and hydrate slides to water.
2. Oxidize in potassium permanganate solution: 1 minute.
3. Wash in tap water: 2 minutes.
4. Decolorize in potassium metabisulfite: 1 minute.
5. Wash in tap water: 2 minutes.
6. Sensitize in ferric ammonium sulfate: 1 minute.
7. Wash in tap water: 2 minutes. Rinse in 2 changes of distilled water: 30 seconds each.
8. Impregnate in silver solution: 1 minute.
9. Rinse in distilled water: 20 seconds.
10. Reduce in formalin solution: 3 minutes.
11. Wash in tap water: 3 minutes.
12. Tone in gold chloride: 10 minutes. Sections turn purplish gray.
13. Rinse in distilled water.
14. Reduce in potassium metabisulfite: 1 minute.
15. Fix in sodium thiosulfate: 1 minute.
16. Wash in tap water: 2 minutes.
17. Counterstain, if desired; dehydrate, clear, and mount.

RESULT

reticulum fibers—black

COMMENTS

This is a good reliable method; it is quick, and I have had no loosening of sections with this procedure.
Also see the Nassar and Shanklin method, 1961.

Wilder Method (1935)

FIXATION

Any good general fixative.

SOLUTIONS

Phosphomolybdic acid, 10%

phosphomolybdic acid	10.0 g
distilled water	100.0 ml

Uranium nitrate

uranium nitrate	1.0 g
distilled water...........................	100.0 ml

Ammoniacal silver nitrate

Add ammonia (28% reagent), drop by drop, to 5 ml of 10% silver nitrate (0.5 g/5 ml water) until precipitate that forms is almost dissolved. Add 5 ml of 3.1% sodium hydroxide (3.1 g/100 ml water). Barely dissolve the resulting precipitate with a few drops of ammonia. Make the solution up to 500 ml with distilled water. Use immediately. Glassware must be acid-clean.

Reducing solution

distilled water...........................	50.0 ml
formalin	0.5 ml
uranium nitrate, 1% (above)	1.5 ml

Make up fresh each time.

Gold chloride

gold chloride stock solution (1 g/100 ml water)	10.0 ml
distilled water...........................	40.0–80.0 ml

PROCEDURE

1. Deparaffinize and hydrate slides to water; remove $HgCl_2$.
2. Wash thoroughly in distilled water.
3. Treat with phosphomolybdic acid: 1 minute.
4. Wash in running water: 5 minutes.
5. Treat with uranium nitrate: 5 seconds or less.
6. Rinse in distilled water.
7. Impregnate with ammoniacal silver nitrate solution: 1 minute.
8. Dip quickly in 95% alcohol and immediately into reducing solution: 1 minute.
9. Wash in distilled water: 2–3 minutes.
10. Tone in gold chloride until yellow colors turn purplish gray.
11. Brief rinse in distilled water.
12. Fix in sodium thiosulfate, 5% (5 g/100 ml water): 3–5 minutes.
13. Wash in running water: 5 minutes.
14. Counterstain if desired.
15. Dehydrate, clear, and mount.

RESULTS

   reticulum fibers—black
   other tissue elements—depends on counterstain

COMMENTS

Phosphomolybdic acid replaces potassium permanganate as an oxidizer; the phosphomolybdic acid shows less tendency to loosen sections. Sensitization with uranium nitrate reduces the time and eliminates the heat required by some reticulum methods. Lillie (1946) disagrees that uranium nitrate is a sensitizer and claims that it is an oxidizer. (For a combination reticulum, collagen, and elastin stain, see Humason and Lushbaugh 1967.) Fitzgerald and Pohlman (1969) adapted a method for reticulum using silver proteinate reduced by hydroquinone.

Chapter 12
Silver Impregnating and Staining Neurological Elements

For nervous tissue, frozen sections are best; alcohol and xylene embedding may remove lipids, which can be an essential part of the tissue. The loss of lipids may result in no impregnation of oligodendroglia and weakened microglia, since their impregnation depends on lipid complexes. Periodic and chromic acid oxidation also weaken the reaction of oligodendroglia and microglia. (The reverse is true for connective tissue. Then pyridine, periodic, or chromic acid treatment is desirable to suppress the impregnation of nervous tissue elements.)

Myelin, because of its high lipid content (cholesterol, cerebroxide, and phospholipids) is soluble in fat solvents, and most of it dissolves away in dehydrating and clearing solutions. Empty spaces remain at the former sites of myelin. However, special fixatives can be used to preserve myelin through dehydration, clearing, and paraffin embedding; osmic acid fixes it efficiently and at the same time colors it black. Overnight chromation of tissue blocks in 3% aqueous potassium dichromate preserves formalin-fixed myelin for paraffin processing.

If tissues from the central nervous system are to be embedded in paraffin, they require long periods of fixation to harden them; they also require extended treatment with alcohol, clearing solutions, and paraffin. If the tissue is not sufficiently fixed and hardened, it will compress badly during sectioning, and the sections may not adhere to the slides during staining. The softness and almost jelly-like consistency of nervous tissue is due to the absence of supporting tissues like those of ordinary connective tissue which has tough intercellular substances, such as elastin and collagen. Only a delicate and ectodermally derived cellular substance—neuroglial cells and fibers—supports the nervous tissue elements by lying between and binding together the nerve fibers and blood vessels. (Connective tissue wrappings—meninges—do cover the brain and spinal column.)

Ordinary stains, such as hematoxylin and eosin, do not demonstrate the innumerable processes of neuroglial cells, because only the nuclei stain. Silver nitrate methods impregnate the fibers and aid in classifying them. When impregnating tissues from the central nervous system, bear in mind these facts: If the hydroxide method is used, astrocytes and microglia are not specially impregnated; if the carbonate method is used, the opposite effect takes place. For impregnation, do not heat the solutions above normal body temperature.

Neurological techniques, as has perhaps become evident, so often necessitate highly specialized methods that many technicians prefer to avoid them. Precise attention to all details, however, can produce beautiful and exciting slides. Carefully follow directions for fixation, such as the solution composition, the duration of fixation, and whether fixation is or is not followed by washing. Always wash in distilled water unless tap water is specified. For making silver and other special solutions, use double (glass) distilled water if possible and glassware cleaned in cleaning solution (p. 570), washed well in running water, and rinsed 4 or 5 times in distilled water. All chemicals should be at least reagent grade. Use no corks in containers and no metal instruments in silver solutions. If in doubt about the age of solutions, make fresh ones. If, while being toned with gold chloride, the tissue retains a yellow or brownish hue, the gold chloride is weakened. Prepare a new solution.

Artifact precipitates are difficult to avoid, but strict adherence to procedure details will reduce them to a minimum.

The method of embedding and sectioning will depend on the impregnating or staining technique to follow. Paraffin and nitrocellulose methods are used at times, but, as mentioned above, frozen sections are usually more satisfactory. Sections of neurological tissue are frequently thicker than for other tissues—7 to 20μ, or even more. Mount sections on subbed slides (p. 550).

When fixing an entire brain, do not allow it to rest on the bottom of the container. Carefully insert a cord under the circle of Willis on the underside of the brain and support the two ends of the cord on the sides of the container. The brain, hanging upside down, should be free of the bottom of the vessel but should remain completely submerged in fixative. The spinal cord can be supported in a graduated cylinder filled with fixative. Run a thread through one end of the spinal cord and tie the thread around an applicator stick supported on the edges of the cylinder. For a perfusion method for the brain, see p. 27.

Note: In neurological techniques, some methods employ section staining, and some, block staining. The letters S and B will be used in this chapter to indicate which type of staining is being described.

GLIA

Holzer Method for Glia Fibers—S

FIXATION

10% formalin, preferably buffered.

SOLUTIONS

Phosphomolybdic acid

phosphomolybdic acid, 0.5% aqueous, freshly mixed	10.0 ml
95% ethyl alcohol	20.0 ml

Alcohol-chloroform

absolute ethyl alcohol	20.0 ml
chloroform	80.0 ml

Crystal violet

crystal violet (C.I. 42555)	5.0 g
absolute ethyl alcohol	20.0 ml
chloroform	80.0 ml

Potassium bromide

potassium bromide	10.0 g
distilled water	100.0 ml

Differentiating Solution

aniline oil	30.0 ml
chloroform	45.0 ml
ammonia	5 drops

PROCEDURE

1. Hydrate sections to water.
2. Treat with phosphomolybdic acid: 3 minutes.
3. Drain and flush sections with alcohol-chloroform.

4. Drain off alcohol-chloroform, but keep sections moist.
5. Cover with crystal violet: 30 seconds.
6. Drain and add potassium bromide: 1 minute.
7. Drain and blot dry.
8. Add differentiating solution: 30 seconds.
9. Drain, clear with several changes of xylene, and mount.

RESULTS

glia fibers—blue

COMMENTS

This is a rapid and beautiful stain. If it is preceded with a PAS stain, the attachment of astrocytes to capillary walls can be shown.

See Proescher's modification of Heidelberger Victoria blue (1934) for glia cells and fibrils, and Waldrop and Puchtler (1975) using Levafix red violet.

Penfield Modification of del Río-Hortega Silver Carbonate Method (McClung 1950)—S

FIXATION

10% formalin at least one week (longer fixation also will give excellent results); or fix in:

formalin	14.0 ml
ammonium bromide	2.0 g
distilled water	86.0 ml

SECTIONING

Frozen sections, 20μ.

SOLUTIONS

Globus hydrobromic acid

40% hydrobromic acid	5.0 ml
distilled water	95.0 ml

Silver carbonate solution

Combine 5.0 ml of 10% aqueous silver nitrate and 20.0 ml of 5% aqueous sodium carbonate. Add ammonium hydroxide, drop by drop, until precipitate is just dissolved. Add distilled water up to 75 ml. Filter. The solution keeps for long periods if stored in a dark bottle.

PROCEDURE

1. Place sections in 1% formalin or distilled water.
2. Transfer to distilled water plus 1% of ammonia. Cover to prevent escape of ammonia and leave overnight.
3. Transfer to hydrobromic acid, 38°C: 1 hour.
4. Wash in 3 changes of distilled water, 2 minutes in each.
5. Mordant in 5% aqueous sodium carbonate: 1 hour. Tissue may remain in this solution 5–6 hours with no ill effect.
6. Impregnate in silver solution: 3–5 minutes or until sections turn a smooth gray when transferred to reducer. Try single sections at 3 minutes, 5 minutes, or longer.
7. Reduce in 1% formalin. Agitate during reduction: 2 minutes.
8. Wash in distilled water: 1 minute.
9. Tone in gold chloride solution (1 g/500 ml water) until bluish gray.
10. Fix in 5% sodium thiosulfate: 3 minutes.
11. Wash in running water: 5 minutes. Dehydrate, clear, and mount.

RESULT

oligodendroglia and microglia—dark gray to black

COMMENTS

If only oligodendroglia are to be shown, shorten fixation time to 2 days. Long fixation tends to increase the staining of the microglia and astrocytes.

Procedure for oligodendroglia

1. Follow fixation by treatment with 95% alcohol: 36–48 hours.
2. Wash, freeze, and cut sections.
3. Stain in a stronger silver solution by diluting the silver carbonate solution to only 45 ml: 15 minutes or more, until the sections begin to turn brown.

4. Wash for a few seconds in distilled water.
5. Reduce with agitation, wash, and tone as above.
6. Dehydrate, clear, and mount.

The oligodendroglia will stain black.

ASTROCYTES

Ramón y Cajal Gold Chloride Sublimate Method (Mallory 1944)—S

FIXATION

Fix thin slices of tissue in:

formalin	15.0 ml
ammonium bromide	2.0 g
distilled water	85.0 ml

Maintain at 37°C: 1 day.

SECTIONING

Frozen sections, 15μ thick.

SOLUTIONS

Gold chloride sublimate solution

mercuric chloride crystals (*not* powder)	0.5 g
gold chloride, 1% aqueous	35.0 ml
distilled water	6.0 ml

Prepare fresh, using acid-clean glassware. Pulverize mercuric chloride; add to distilled water. Heat gently. When dissolved, add gold chloride, and filter.

Del Río-Hortega carbol-xylol-creosote mixture

creosote	10.0 ml
phenol (carbolic acid), melted	10.0 ml
xylene	80.0 ml

PROCEDURE

1. Place sections in 1% formalin.
2. Wash quickly in distilled water, 2 changes.
3. Place in gold chloride. Flatten sections, keep in dark at room temperature: 4–6 hours. When sections appear purplish, examine them for astrocytes.
4. Wash in distilled water: 5–10 minutes.
5. Fix in 5% aqueous sodium thiosulfate: 5–10 minutes.
6. Wash in distilled water, several changes.
7. Float on subbed slide, blot gently, dehydrate with 95% alcohol: 3 minutes.
8. Clear with del Río-Hortega solution.
9. Blot and mount.

RESULTS

astrocytes and processes—black
background—unstained or light brownish purple
nerve cells—pale red
nerve fibers—unstained

COMMENTS

After being fixed in hypo, extra sections can be preserved in 1% formalin for long periods.

For reliable results staining should be done within three weeks of initial fixation. If tissues have been in formalin for a long period, treat the frozen sections as follows:

1. Place in ammonium hydroxide, 1 part to 9 parts water: 24 hours.
2. Rinse in 2 changes of distilled water.
3. Treat in 40% hydrobromic acid (HBr), 1 part to 9 parts water: 2–4 hours.
4. Rinse in 2 changes of distilled water with 0.5% of dilute ammonia (step 1) added.

Phosphotungstic Acid Hematoxylin (Mitchell 1975)—S

FIXATION

10% formalin, preferably buffered.

SOLUTIONS

Susa fixative, see p. 17.

Potassium permanganate

       potassium permanganate .... 0.25 g
       distilled water .... 100.0 ml

Oxalic acid

       oxalic acid .... 1.0 g
       distilled water .... 100.0 ml

Phosphotungstic acid–hematoxylin, see p. 126.

PROCEDURE

1. Hydrate sections to water.
2. Mordant in Susa fixative (p. 17): 60 minutes.
3. Wash in distilled water: 5 minutes.
4. Oxidize in potassium permanganate: 5 minutes.
5. Rinse off excess permanganate in distilled water.
6. Bleach in oxalic acid till colorless.
7. Wash in several changes of distilled water: 5 minutes.
8. Stain in phosphotungstic acid hematoxylin: 2–4 hours.
9. Dehyrate briefly through 2 changes 95% alcohol.
10. Complete dehydration in 2 changes absolute alcohol, clear, and mount.

RESULTS

astrocyte fibers—blue
nuclei—blue
collagen—red
myelin—blue

NISSL SUBSTANCE

Cresyl Violet (Powers and Clark 1955)—S

Nissl substance is characteristic of fixed nerve cells. It is found in a granular form distributed throughout the cytoplasm and stains brilliantly with basic aniline dyes.

FIXATION

Bouin recommended; others satisfactory: Zenker not recommended.

SECTIONING

Paraffin method, 10μ; thinner sections are of no advantage.

SOLUTIONS

Cresyl violet

cresyl violet–acetate	0.2 g
distilled water.............................	150.0 ml

Buffer solution, pH 3.5

0.1 M (approx.) acetic acid (6 ml/1000 ml water)...................................	94.0 ml
0.1 M (approx.) sodium acetate (13.6 g/1000 water)...................................	6.0 ml

Working solution

buffer solution	100.0 ml
cresyl violet solution......................	6.0–12.0 ml

PROCEDURE

1. Deparaffinize and hydrate sections to distilled water.
2. Stain 20 minutes in working solution. Use solution only once.
3. Rinse quickly in 70% ethyl alcohol and 95% ethyl alcohol.
4. Dehydrate in isopropyl alcohol, 2 changes: 3–4 minutes each.
5. Clear and mount.

RESULT

Nissl substance—purple

COMMENTS

Banny and Clark (1950) recommend Matheson Coleman and Bell cresyl violet for Nissl staining.

Manns (1960) stains with lithium hematoxylin for myelin, and with cresyl fast violet for Nissl substance.

Thionin (Clark and Sperry 1945)—S

FIXATION

Bouin recommended; others satisfactory; Zenker not recommended.

SECTIONING

10μ; thinner sections of no advantage.

SOLUTIONS

Lithium carbonate, 0.55%

lithium carbonate	5.5 g
distilled water	1000.0 ml

Thionin

thionin (C.I. 52000)	0.25 g
0.55% lithium carbonate	100.0 ml

PROCEDURE

1. Deparaffinize and hydrate slides to water; remove $HgCl_2$.
2. Treat with lithium carbonate solution: 5 minutes.
3. Overstain in thionin: 5–10 minutes.
4. Rinse in distilled water.
5. Dip in 70% alcohol: few seconds.
6. Dehydrate in butyl alcohol, 2 changes: 2–3 minutes in each.
7. Clear and mount.

RESULT

Nissl substance—bright blue

COMMENTS

If differentiation is necessary, briefly rinse slides in 95% ethyl alcohol (following step 5) and place in aniline, then in lithium carbonate saturated in 95% alcohol. Proceed to step 6.

Gallocyanin—S

FIXATION

Any general fixative.

SECTIONING

10μ.

SOLUTION AND PROCEDURE

See pp. 116, 127.

RESULT

Nissl substance—blue

NERVE CELLS, PROCESSES, AND FIBRILS

Ramón y Cajal Method (Favorsky Modification 1930)—**B**

FIXATION

Cut slices perpendicular to organ surface, about 5 cm thick. Place in 70% alcohol, plus 0.5% glacial acetic acid: 6 hours.

PROCEDURE

1. Transfer to 80% alcohol: 6 hours.
2. Treat with ammoniacal alcohol: 24–36 hours.

   a. For cerebrum, cerebellum, spinal cord, or ganglia, add 4 drops
      of ammonia to 50 ml of 95% alcohol.
   b. For medulla, add 9 drops of ammonia to 50 ml of 95% alcohol.
 3. Wash in distilled water, several changes, until pieces sink.
 4. Treat with pyridine: 1–2 days.
 5. Wash in running water: overnight; then wash in distilled water,
    several changes.
 6. Blot on filter paper and place in relatively large volume of 1.5%
    aqueous silver nitrate: 5 days in dark, 38°C.
 7. Rinse in distilled water and place in following fluid: 24 hours.
    Ramón y Cajal's reducing fluid:

pyrogallic acid or hydroquinone	1.0 g
distilled water	100.0 ml
neutral formalin	15.0 ml

 8. Rinse in distilled water, several changes over a period of at least 1
    hour.
 9. Dehydrate and embed in paraffin, celloidin, or double embed.
10. Make sections perpendicular to surface of organ and about 15 or
    more μ thick.
11. Affix to slides, dry, remove paraffin with xylene, and mount. Lay
    celloidin sections on slides and mount in resin.

RESULTS

neurofibrils—black
background—brownish yellow

Rapid Golgi Method (Kemali 1976)—B

FIXATION

Fix small brains or pieces of large brains in a dark container in the following solution: 6–12 hours at room temperature.

SOLUTIONS

Solution A

potassium dichromate	3.0 g
distilled water	100.0 ml

Solution B

sodium barbitol (a drug, may be unobtainable)	0.61 g
sodium chloride	0.29 g
calcium chloride	0.22 g
distilled water...........................	100.0 ml

Add 2 g osmic acid and adjust the pH to 7.2 if necessary with HCl. Stock solutions keep well for several months at 4°C.

Working solution

Solution A................................	85.0 ml
Solution B	15.0 ml

PROCESSING

Remove from fixing solution and wash several times in an aqueous solution of 0.75% silver nitrate. Impregnate in a fresh silver solution (0.75%) in darkness, room temperature: 6–12 hours or more. A thin slice taken from the block of tissue will help determine the depth of impregnation. Dehydrate, embed, and section. Deparaffinize sections, clear, and mount.

RESULTS

nerve cells and processes—deep black

COMMENTS

Impregnation will be best at cut surfaces and outer structures of whole brains. Also see Tunturi 1973.

Ramón y Cajal Pyridine-Silver Method (Davenport et al. Modification 1934)—**B**

FIXATION

As soon as possible in:

absolute ethyl alcohol......................	98.0 ml
ammonia, concentrated	2.0 ml

Fix for 1–6 days, preferably no longer.

PROCEDURE

1. Treat tissue blocks in 5% aqueous pyridine: 24 hours. This is recommended, but Davenport et al. say it can be optional.
2. Wash in distilled water: 2–6 hours; change every half hour.
3. Impregnate in 1.5–2.0% aqueous silver nitrate, 37°C: 2 to 3 days or longer, depending on size of tissue blocks. A minimum time yields the best differentiated tissue. The longer the tissue is left in silver nitrate, the greater the tendency for everything to stain. The time will have to be determined by trial and error.
4. Wash in distilled water: 20 minutes to 1 hour. This is a critical step. The amount of silver washed from the tissue depends on whether the water is changed often or whether the tissue is shaken in the water; change every 10 minutes or use a large volume of water and shake every 10 minutes. Equal staining of the periphery as well as central parts of the tissue determines correct washing. A light periphery indicates too much washing.
5. Reduce in 4% aqueous Pyrogallol: 4 hours.
6. Dehydrate, clear, embed, and section.
7. Mount on slides, deparaffinize, and clear. Add mountant and cover glass.

RESULTS

nerve—yellow to brown
neurofibrils—brown to black
axis cylinders of myelinated fibers—yellow to brown
axis cylinders of nonmyelinated fibers—black

COMMENTS

If the preparation is too light, reduce the time in pyridine and washing. If the preparation is too dark, omit the pyridine and wash 48 hours after fixation, or reduce the concentration of the silver nitrate solution and wash longer between impregnation and reduction.

Bielschowsky Method (Davenport et al. Modification 1934)—**B**

FIXATION

10% formalin: 2 days. For embryos add 0.5% of trichloracetic acid to 10% formalin.

Add 5.0 ml of concentrated ammonia to 40.0 ml of 2% aqueous sodium hydroxide. Mix well. Add slowly from a burette or pipette 8.5% aqueous silver nitrate until opalescence remains in the solution (about 40.0 ml). Shake the hydroxide solution while adding the silver nitrate. Add 0.5–1.0 ml of ammonia. Dilute with about 5 parts of distilled water: dilution is not critical.

PROCEDURE

1. After fixation, wash: 1 hour.
2. Transfer to 50% aqueous pyridine: 1–2 days.
3. Wash in distilled water: 2–6 hours, depending on size of block. Change every half hour.
4. Impregnate with 1–1.5% aqueous silver nitrate: 3 days, 37°C.
5. Wash in distilled water: 20 minutes to 1 hour, depending on size. Change every 10 minutes or use a large volume of water and shake every 10 minutes. Periphery and central portions must be equally stained.
6. Impregnate in ammoniated silver solution: 6–24 hours.
7. Wash in distilled water: 15 minutes to 1 or 2 hours, depending on size.
8. Reduce in 1% formalin: 6–12 hours.
9. Wash in running water: 10–15 minutes.
10. Dehydrate, clear, and embed. Section and mount.
11. Deparaffinize, clear, and cover.

Alternate method (if gold toning is desired):
  a. Following step 9, rinse in distilled water.
  b. Gold tone and fix.
  c. Return to step 10.

RESULTS

nerve fibers, neurofibrils—brown to black (no gold toning)
                    —gray to black (with gold toning)

Fluorescent Method (Zeiger et al. 1951)—S

FIXATION

95% ethyl alcohol.

Paraffin method.

PROCEDURE

1. Deparaffinize and hydrate slides to water.
2. Stain in 0.1% aqueous acridine orange (C.I. 46005): 6 minutes.
3. Differentiate in 95% ethyl alcohol: 2 seconds.
4. Blot with filter paper and mount in fluorescent mountant.

RESULTS

nonmyelinated fibers—bluish gray
myelinated fibers—brownish orange

COMMENTS

Fresh tissue can be frozen, cut, and stained in acridine orange made up in physiological saline or in Ringer solution.

Bodian Method (Russell Modification 1973)—S

FIXATION

Williams (1962) recommends:

formalin	40.0 ml
glacial acetic acid	10.0 ml
80% ethyl alcohol	100.0 ml
picric acid	2.0 g

Not suitable are chromates, chromic and osmic acid, or mercuric chloride. 10% formalin causes excessive staining of connective tissue.

SECTIONING

Paraffin method.

SOLUTIONS

Protargol solution

Protargol[1]	1.0 g
distilled water............................	100.0 ml

Sprinkle Protargol on surface of water in a wide dish or beaker. Do not stir—this is critical. When the granules are dissolved, pour the solution into a coplin jar containing 6 g of copper shot. This prevents the surrounding tissue from becoming impregnated with silver and obliterating some cellular detail.

Reducing solution

hydroquinone	1.0 g
formalin	5.0 ml
distilled water............................	100.0 ml

Make up fresh.

Gold chloride

gold chloride.............................	1.0 g
distilled water............................	100.0 ml

Oxalic acid

oxalic acid...............................	2.0 g
distilled water............................	100.0 ml

PROCEDURE

1. Deparaffinize and hydrate slides to water.
2. Impregnate in Protargol solution, 37°C: 12–24 hours.
3. Wash in distilled water, several changes.
4. Reduce: 15 minutes.
5. Rinse in distilled water, 6 changes: 1 minute total.
6. Tone in gold chloride: 4 minutes.
7. Rinse in distilled water, 6 changes: 1 minute total.

---

[1] Roboz Surgical Instrument Co., imported "Silver Protein."

8. Develop in 2% aqueous oxalic acid. Check under microscope until background is gray and fibers are sharply defined: approximately 3 minutes.
9. Wash in distilled water, 6 changes: 1 minute each.
10. Fix in 5% aqueous sodium thiosulfate: 5 minutes.
11. Wash in running water: 5–10 minutes.
12. Rinse in distilled water.
13. Counterstain if desired.
14. Treat with acidified water: 5 minutes.
15. Dehydrate, clear, and mount.

RESULTS

nerve fibers—black
background colors—depending on counterstain

COMMENTS

Foley (1943) considers counterstaining essential, claiming that it serves as contrast between nervous and nonnervous tissue and adds to transparency of the sections. To counterstain by his method follow step 11 above with the following:

12. Stain in gallocyanin (p. 116): overnight.
13. Wash thoroughly in running water: 5–10 minutes.
14. Mordant in 5% aqueous phosphotungstic acid: 30 minutes.
15. Transfer directly to dilution (20 ml/30 ml water) of following stock solution: 1 hour.

aniline blue WS (C.I. 42780)	0.1 g
fast green FCF (C.I. 42053)	0.5 g
orange G (C.I. 16230)	2.0 g
distilled water	92.0 ml
glacial acetic acid	8.0 ml

16. Differentiate through 70% and 95% alcohols.
17. Dehydrate, clear, and mount.

Consistent selective results with silver impregnation have plagued technicians. Loots et al. (1977) attacked this problem for nervous tissue and concluded Protargol yields the best results. To prevent darkening of the solution during incubation (probably due to reduction of the silver), a bit of oxidizing agent has been added. Such a solution can be used more than once and is not excessively light sensitive. Instead of metallic copper, add copper nitrate.

Dissolve 1 g Protargol (or Merck's albumin-silver) in 100 ml distilled water. When dissolved add in order with agitation:

1% copper nitrate aqueous	2.0 ml
1% silver nitrate aqueous	2.0 ml
30% hydrogen peroxide (undiluted)	2–4 drops

Substitute for step 2 above, 3–5 days in dark, 37°C, and proceed to step 3 as usual.

See Herr et al. (1976) for a Bodian method for frozen sections.

Nauta and Gygax Method (1951)—S

FIXATION

10% formalin: 2 weeks to 6 months.

SECTIONING

Frozen sections, 15–20μ.

SOLUTIONS

Silver solution A

silver nitrate	1.5 g
distilled water..............................	100.0 ml
pyridine	5.0 ml

Silver solution B

Dissolve 0.45 g silver nitrate in 20.0 ml distilled water. Add 10.0 ml of 95% ethyl alcohol. From calibrated pipettes add 2.0 ml ammonia (concentrated) and 2.2 ml of 2.5% aqueous sodium hydroxide. Mix thoroughly and keep the container covered to prevent escape of ammonia.

Reducing solution

10% ethyl alcohol	45.0 ml
10% formalin	2.0 ml
1% citric acid	1.5 ml

PROCEDURE

1. Demyelinate sections in 50% ethyl alcohol plus 1.0 ml ammonia per 100 ml: 6–12 hours. A longer time has no ill effect.
2. Wash in distilled water, 3 changes: few seconds each.
3. Impregnate in silver solution A: 12–24 hours.
4. With no washing, transfer into silver solution B: 2–5 minutes.
5. Transfer directly into reducing solution until the sections turn gold in color.
6. Transfer to 2.5% aqueous sodium thiosulfate: 1–2 minutes.
7. Wash in distilled water, at least 3 changes.
8. Dehydrate rapidly, clear, and mount.

RESULTS

nerve fibers and endings—black
cells—pale yellowish brown

COMMENTS

This method is nonselective and stains normal as well as degenerating axons. For degenerating axons see p. 203.

Nauta and Gygax sections may be counterstained with cresyl echt violet. Follow step 7 with the cresyl echt violet stain: 6 minutes in preheated solution, 57°C, just before use.

cresyl echt violet (cresyl violet acetate)	1.0 g
distilled water	100.0 ml

Just before using add 15 drops of 10% glacial acetic acid.

Differentiate in 90% alcohol, dehydrate in absolute alcohol, clear, and mount.

Glees Method (Novotny Modification 1974, 1977)

FIXATION

Perfuse with buffered formalin and store at least 1 week. Embed and section. Use subbed slides.

SOLUTIONS

Silver nitrate

silver nitrate	20.0 g
distilled water	100.0 ml

Reducing solution 1

distilled water	400.0 ml
95% ethyl alcohol	45.0 ml
10% formalin	13.5 ml
1% acetic acid	13.5 ml

Ammoniacal silver solution

silver nitrate	5.0 g
80% ethyl alcohol	100.0 ml

Add drop by drop 25% NH_4OH until precipitate formed just redissolves. A few grains can remain. Add 3 more drops. This must be freshly prepared. Keep tightly covered during use and use clean glassware. Avoid use of metal instruments.

Reducing solution 2

10% formalin	400.0 ml
95% ethyl alcohol	50.0 ml
1% acetic acid	20.0 ml

The reducing solutions are stable and store well.

Luxol fast blue, see p. 197.
Cresyl echt violet, see p. 192.

PROCEDURE

1. Deparaffinize and hydrate sections to distilled water.
2. Place in silver solution: 2 hours room temperature. Sections should appear light brown.
3. Treat with reducing solution 1: 10 minutes.
4. Place in ammoniacal silver solution: 15 minutes. Sections should appear orange brown.

5. Rinse thoroughly in absolute alcohol.
6. Rinse in reducing solution 2.
7. Treat with fresh reducing solution 2: 10 minutes.
8. Wash in running water: 10 minutes.
9. Fix in 5% sodium thiosulfate: 3 minutes.
10. Wash in running water: 10 minutes.
11. Dehydrate to 95% alcohol.
12. Stain in Luxol fast blue: overnight, 56°C.
13. Allow to cool and rinse slides in absolute alcohol.
14. Rinse in distilled water.
15. Treat with 0.05% lithium carbonate: 3–7 minutes.
16. Differentiate in 70% alcohol until neurophil is yellow; nerve cells, orange brown; and myelin, still blue.
17. Rinse in 1% acetic acid to stop differentiation.
18. Stain with cresyl echt violet, see above, p. 192.
19. Differentiate in 95% alcohol, dehydrate in absolute alcohol, clear, and mount.

RESULTS

axons—black or dark brown
myelin—blue
Nissl bodies and glial nuclei—violet
neurophil—yellow

COMMENTS

The method may be halted at end of step 11 and finished by dehydrating, clearing, and mounting. Step 2 is not critical; the duration may be overnight. The reducing solutions are not critical, but step 5 (absolute alcohol) is—it helps prevent formation of precipitate on slides—but do not leave longer than the recommended time. Impregnation can be lost.

Cole Method (1946) (Modified: Whole Mounts)

FIXATION

None; carry fresh tissue directly into step 1.

PROCEDURE

1. Tease striated costal muscle into strips 1 mm in diameter and a few millimeters in length, and place in either of following solutions (Zinn and Morin 1962): (1) 1 part commercial lemon juice and 1 part distilled water; or (2) 0.01 M citric acid. Minimum time is 10 minutes; maximum time, 30 minutes.

Use a separate clean glass container for each step that follows.

2. Wash in several changes of distilled water: 5 minutes.
3. Transfer to 1% aqueous gold chloride. Make up this solution the day before it is to be used. Keep in dark: 60 minutes, or until tissue turns dark yellow. (If pieces are wider than 4 mm, a longer time is required.)
4. Wash in several changes of distilled water: 5 minutes.
5. Transfer to 20% formic acid (20 ml/80 ml water): 10–20 hours in dark. Do not use metal forceps; coat forceps with paraffin or use glass rods.
6. Rinse in tap water.
7. Transfer to 95% methyl alcohol–glycerin (1 : 1): several hours. Then remove top of container and allow alcohol to evaporate.
8. Transfer to pure glycerin.
9. To mount: Place a piece of muscle in a very small drop of glycerin (usually amount carried over by the piece is sufficient) on a round cover glass. Lay a smaller size cover glass over it. Spread the muscle fiber to single fiber thickness by using gentle pressure and strokes at right angles to the fibers. Turn cover glasses over and mount in Permount (or the like) on a glass slide. (Double cover glass mounting, p. 109.)

RESULTS

muscle fibers—red blue
motor end plates and medullated axons—black
muscle fiber nuclei—unstained

COMMENTS

Carey (1941) uses undiluted fresh filtered lemon juice in place of the citric acid. He claims that if the tissue requires more than 12 hours in the formic acid (step 5), the gold chloride technique is faulty. The color of the tissue should be gold, not brown.

Boyd (1962) fixes in fresh filtered lemon juice and formic acid, 3 : 1 : 2–10 minutes.

Cole and Mielcarek (1962) outline a fluorescent method.

Cavanagh et al. (1964) stain with Sudan black B.

Pyridine-Silver Method (Gladden 1970)

PROCEDURE

1. Fix muscle in absolute ethanol, 4.5 ml; distilled water, 5 ml; and concentrated nitric acid, 0.1 ml: 24 hours.
2. Transfer into absolute ethanol, 10 ml, and ammonia, 0.1 ml: 24 hours.
3. Wash in distilled water: 30 minutes.
4. Treat in pyridine: 2 days.
5. Wash in distilled water, 5–8 changes: 24 hours.
6. Place in 2% aqueous silver nitrate in dark, 25°C: 3 days.
7. Reduce in 5% aqueous formic acid, 10.0 ml, and Pyrogallol, 0.4 g: 6–24 hours.
8. Wash in distilled water and store in glycerin.
9. Tease muscle to show nerve endings and mount in glycerin.

MYELIN

Luxol fast blue B was first used by Klüver and Barrera (1953) for staining myelin sheaths. Margolis and Pickett (1956) combined Luxol fast blue MBSN with other methods to differentiate various neurological elements. Salthouse (1962) stained myelin sheaths a deep blue with Luxol fast blue ARN, which he said had a greater affinity for phospholipids than Luxol fast blue MBSN. In 1964 he reported Luxol fast blue G, with which he had stained myelin a blue-black color, to be superior to ARN. If dissolved in isopropyl alcohol, the Luxol dyes bind to more phospholipids than if dissolved in ethyl alcohol. The Margolis and Pickett and the Salthouse solutions are listed below: they can be used interchangeably. All Luxol fast blue solutions keep indefinitely.

Luxol Fast Blue—S

FIXATION

Salthouse (1962) recommends 10% formalin or calcium-formalin, but the following solution is best:

calcium chloride	10.0 g
distilled water.............................	900.0 ml

Dissolve and add:

cetyltrimethylammonium bromide	5.0 g

Dissolve and add:

formalin, concentrated	100.0 ml

EMBEDDING

Paraffin method.

SOLUTIONS

Luxol fast blue MBSN (Margolis and Pickett 1956)

Luxol fast blue MBSN[2]	0.1 g
95% ethyl alcohol	100.0 ml
acetic acid, 10% aqueous	0.5 ml

Luxol fast blue ARN (Salthouse 1962)

Luxol fast blue G[3]	1.0 g
95% ethyl alcohol	1000.0 ml
glacial acetic acid	0.2 ml

Luxol fast blue G (Salthouse 1964)

Luxol fast blue G[3]	1.0 g
95% isopropyl alcohol	1000.0 ml
glacial acetic acid	0.2 ml

---

[2] E. I. du Pont de Nemours and Company.
[3] Curtin Matheson Scientific Co.

Periodic acid, 0.5%

periodic acid	0.5 g
distilled water	100.0 ml

Lithium carbonate, 0.05%

lithium carbonate	0.5 g
distilled water	1000.0 ml

Schiff reagent, see p. 210.

Sulfurous acid

sodium metabisulfite, 10% aqueous	6.0 ml
distilled water	100.0 ml
1 N HCl (see p. 542)	5.0 ml

Cresyl echt violet (cresyl violet acetate)

cresyl echt violet	0.1 g
distilled water	100.0 ml

Just before use add 15 drops 10% acetic acid. Filter and preheat solution to 57°C.

PROCEDURE

1. Deparaffinize and hydrate slides to water. (Remove $HgCl_2$ if present.) Transfer to 95% alcohol: 2 minutes.
2. Stain in Luxol fast blue MBSN, 60°C: overnight; or Luxol fast blue ARN or G, 35–40°C: 2–3 hours.
3. Rinse off excess stain in 95% alcohol.
4. Rinse in distilled water.
5. Dip in lithium carbonate: 15 seconds.
6. Differentiate in 70% alcohol: 20–30 seconds.
7. Rinse in distilled water.
8. Dip in lithium carbonate, second solution: 20–30 seconds.
9. Differentiate in 70% alcohol: 20–30 seconds.
10. Rinse in distilled water. If differentiation is not complete, repeat steps 8 and 9.
11. Oxidize with periodic acid: 5 minutes.
12. Wash in distilled water, 2 changes: 5 minutes total.

13. Treat with Schiff reagent: 15–30 minutes.
14. Treat with sulfurous acid, 3 changes: 2 minutes each.
15. Running water: 5 minutes.
16. Stain in cresyl echt violet: 6 minutes, preheated.
17. Differentiate in 95% alcohol.
18. Dehydrate in absolute alcohol, clear, and mount.

RESULTS

myelin—blue green
PAS positive elements—rose to red
nuclei—blue
Nissl granules— deep blue purple

COMMENTS

Luxol is a trade name of the DuPont Company. The Luxol fast blue used by Klüver and Barrera was the amine salt of a sulfonated copper phthalocyanin. Phthalocyanin-amine dyes may have one to four sulfonyl groups on each molecule and one of several kinds of bases. Because of the copper phthalocyanin, Luxol fast blue methods are sometimes called copper phthalocyanin methods. Luxol fast blue ARN, however, is a diaryl quanidine salt of a sulfonated azo dye with no copper present, and it has a greater affinity for phospholipids than the MBS salt. Klüver and Barrera thought the reaction was due to the affinity of the copper phthalocyanin for the porphyrin present in myelin, but the superior reaction of the ARN salt proves that it does not depend on the copper. Pearse (1960) suggests that Luxol fast blue staining of fixed tissues is due to the presence of lipoproteins rather than lipids, that it is an acid–base reaction to form a salt, with the lipoprotein base replacing the phthalocyanin. Lycette et al. (1970) consider the phosphate group of phospholipids to be essential for the reaction. Whatever the explanation, the sensitivity of the dye for myelin is excellent. Also, stores of phospholipids can be clearly shown by this method. Salthouse (1962) demonstrates that the ARN salt forms complexes with many phospholipids.

Hale et al. (1960), using Luxol fast blue MBSN, differentiate abnormal from normal collagen; abnormal collagen did not stain, but normal collagen stained an intense blue.

Snodgrass and Lacey (1961) modify the Luxol fast blue method to differentiate degenerating myelinated fibers.

Lockard and Reers (1962) use it with neutral red to include Nissl staining.

Margolis and Pickett (1956) also describe methods for following Luxol fast blue with phosphotungstic acid–hematoxylin to differentiate neuroglia from myelin, and with oil red O to distinguish degenerating myelin from normal myelin.

Dziabis (1958) describes a method for staining gross brain sections.

Luxol Fast Blue–Holmes Silver Nitrate (Margolis and Pickett 1956)—S

FIXATION

10% formalin preferred.

EMBEDDING

Paraffin method.

SOLUTIONS

Silver nitrate, 20%:

silver nitrate	20.0 g
distilled water	100.0 ml

Boric acid

boric acid	12.4 g
distilled water	1000.0 ml

Borax

borax	19.0 g
distilled water	1000.0 ml

Impregnating fluid

In a 500 ml cylinder, mix 55 ml boric acid solution and 45 ml borax solution. Dilute to 494 ml with distilled water. With pipette add 1 ml of 1% silver nitrate (1 g/100 ml water). With another pipette add 5 ml of 10% pyridine (10 ml/100 ml water). Mix thoroughly.

Reducer

hydroquinone	1.0 g
sodium sulfite (crystals)	10.0 g
distilled water.............................	100.0 ml

Can be used repeatedly, but only for a few days.

Luxol fast blue, see p. 197.

PROCEDURE

1. Deparaffinize and hydrate sections to water.
2. Treat with silver nitrate in dark, room temperature: 1 hour. Prepare impregnating fluid.
3. Wash in distilled water, 3 changes: 10 minutes.
4. Impregnate, 37°C: overnight.
5. Shake off superfluous fluid and place in reducer: 2 minutes.
6. Wash in running water: 3 minutes, rinse in distilled water.
7. Tone in 0.2% aqueous gold chloride: 3 minutes.
8. Rinse in distilled water.
9. Treat with 2% aqueous oxalic acid: 3–10 minutes; when axons are thoroughly black, remove.
10. Rinse in distilled water.
11. Fix in 5% aqueous sodium thiosulfate: 3 minutes.
12. Wash in running water: 10 minutes.
13. Rinse briefly in 95% alcohol.
14. Stain in Luxol fast blue: overnight.
15. Rinse in 95% alcohol; rinse in distilled water.
16. Treat with 0.05% aqueous lithium carbonate: 15 seconds.
17. Differentiate in 70% alcohol: 20–30 seconds.
18. Rinse in distilled water.
19. Repeat steps 16 and 17 if necessary. (Three repeats are usually necessary for sharp differentiation.)
20. Dehydrate, clear, and mount.

RESULTS

axis cylinders—black
myelin sheaths—green blue

COMMENTS

Margolis and Pickett write that slides may be left in the distilled water (step 6) until it is time to prepare them for Luxol fast blue staining. However, leaving the sections in water tends to detach them from the slides. I have lost no sections by carrying the slides through to step 13. Transfer them into absolute alcohol and coat them with nitrocellulose. Harden the nitrocellulose and store the slides in 70–80% alcohol until you are ready to stain them. Slides can be left in this condition for a weekend with completely satisfactory results. The nitrocellulose protection also permits freer agitation of slides in the differentiating fluids. Subbed slides can be used.

Chromic Acid–Hematoxylin (Lillie and Henderson, 1968)

FIXATION

10% formalin.

SOLUTIONS

Chromic acid

chromic acid	0.2 g
distilled water	100.0 ml

Acetic hematoxylin

1% stock alcoholic hematoxylin	10.0 ml
distilled water	89.0 ml
glacial acetic acid	1.0 ml

or

1% aqueous acetic acid	100.0 ml
hematoxylin	100.0 mg

Borax-ferricyanide

borax	1.0 g
potassium ferricyanide	1.0 g
distilled water	100.0 ml

PROCEDURE

1. Cut frozen sections, 10–15μ, transfer to 1% aqueous gelatin, and pick up on slides.
2. Drain and blot firmly with filter paper. Immerse in 1% aqueous gelatin: 1–2 minutes. Drain.
3. Fix gelatin in formalin vapor: 30 minutes. Transfer to 5% formalin until ready to use.
4. Wash in running water: 10 minutes.
5. Mordant in chromic acid: 4 hours.
6. Wash in running water: 10 minutes.
7. Stain in acetic hematoxylin: 2 hours.
8. Rinse in distilled water.
9. Differentiate in borax-ferricyanide: approximately 1 hour. Check under microscope.
10. Dehydrate, clear, and mount.

RESULTS

myelin—blue black
cells and gray substance—yellow to colorless

COMMENTS

This method is recommended over the Ora or Mahon methods because the myelin sheaths remain a more normal size in relation to other structures in this method, whereas paraffin embedding may cause shrinkage in thin sheaths.

For a Mahon method for celloidin embedded tissue, see Metz (1976). Angulis and Sepinwall (1971) use gallocyanin as a myelin stain.

DEGENERATING AXONS

Nauta and Gygax Method (modified by Powell and Brown 1975)—S

FIXATION

Perfusion method seems to be preferred.

1. Perfuse with 0.9% sodium chloride (intracardiac and intra-aortic cannulation) until escaping fluid is clear.

2. Perfuse with 500 ml of 5% aqueous potassium dichromate and 2.5% aqueous potassium chlorate. (This step is peculiar to Anderson (1959); others perfuse with a formalin solution.)
3. Remove tissue and fix in 10% formalin neutralized with carbonate: 1 week or longer. Nauta and Gygax recommend 1–3 months as best.

EMBEDDING AND SECTIONING

1. If embedding is necessary, use gelatin. Wash slices (5–10 mm thick) in running water: 24 hours. Incubate in 25% gelatin, 37°C (cover tightly): 12–18 hours. Wipe off excess gelatin and immerse in cool formalin: 6 hours or longer.
2. Frozen sections: Cut sections 15–25μ and place in 10% formalin, room temperature or cooler. Can be stored in formalin, preferably in refrigerator. Process only a few sections (6–10) at one time.

Adey et al. (1958) prefer a dual-freezing block method, using dry ice evaporated in 70% alcohol, over CO_2 freezing. This produces more even freezing. Rapid freezing and thawing with CO_2 may rupture some of the fibers; the use of Freon equipment will sometimes prevent this.

SOLUTIONS

Bleach

1% hydroquinone, aqueous; mix fresh	1 part
1% oxalic acid, aqueous	1 part

Silver solution A

0.5% uranyl nitrate, aqueous	30.0 ml
2.5% silver nitrate, aqueous	36.0 ml
distilled water	56.0 ml

Silver solution B

0.5% uranyl nitrate	40.0 ml
2.5% silver nitrate	60.0 ml

Silver solution C

2.5% silver nitrate	60.0 ml
ammonium hydroxide, fresh, concentrated	4.0–5.0 ml
2.5% sodium hydroxide aqueous	2.8–3.0 ml

Mix in above order and use immediately.

Reducing solution

distilled water.............................	405.0 ml
95% alcohol	45.0 ml
1% citric acid, aqueous	13.5 ml
10% formalin, aqueous......................	13.5 ml

Gelatin-alcohol

Dissolve 6.0 g gelatin in 400.0 ml hot distilled water. Cool and add 400.0 ml 80% alcohol.

PROCEDURE

1. Rinse sections thoroughly in 4 changes of distilled water: 2–5 minutes each. Agitate.
2. Soak in 0.05% potassium permanganate aqueous: 2–6 minutes.
3. Rinse briefly in distilled water.
4. Bleach with agitation in freshly mixed solution: 30 seconds to 2 minutes.
5. Rinse thoroughly in distilled water. Sections can remain here overnight.
6. Place in silver solution A, freshly mixed: 30 minutes.
7. Transfer directly to silver solution B, also freshly mixed: 30 minutes.
8. Rinse thoroughly in distilled water.
9. Transfer into silver solution C, freshly mixed: 2 minutes. Do not crowd the sections or overuse the solution. Change every 8–12 sections. 2 minutes usually is adequate, but check the intensity under the microscope after the first 30 seconds.
10. Transfer directly into reducing solution: 2 minutes in each of 2 changes.
11. Fix in 0.5% sodium sulfate: 1 minute or more.
12. Rinse thoroughly in distilled water.

13. Place in gelatin-alcohol: 5 minutes or overnight.
14. Mount on slide; blot gently but do not dry.
15. Dehydrate in 95% and absolute alcohol (2 changes): 3 minutes each.
16. Clear in 2 changes xylene: 5 minutes each, and cover.

RESULTS

degenerating axons—black
normal axons—various shades of brown

COMMENTS

See also Nauta and Gygax (1951); Nauta and Ryan (1952); Wall (1950); White (1960); and additional references included therein.

The Nauta-Gygax impregnation of degenerating fibers is related to unsaturated lipids. Since the Nauta reaction is abolished by prior bromination, this suggests that the basic mechanism of the impregnation may involve unsaturated bonds, the ethylene bonds of cholesterol esters. The presence of these esters is attributed principally to the breakdown of the myelin. It is possible that degeneration of the axis cylinders is also part of the reaction (Giolli 1965).

Hamlyn (1957) and Guillery et al. (1961) describe methods for paraffin-embedded sections that are stained after mounting.

Marchi Method—B

FIXATION

10% formalin plus 1% of potassium chlorate: 24–48 hours, no longer (Swank and Davenport 1934a).

SOLUTION
Marchi fluid (Poirier et al. 1954)

osmic acid, 0.5% aqueous	11.0 ml
potassium chlorate, 1% aqueous	16.0 ml
formalin	3.0 ml
acetic acid, 10% aqueous	3.0 ml
distilled water	67.0 ml

PROCEDURE

1. Cut tissue into thin slices, about 3 mm for easier impregnation.
2. Chromate in 2.5% aqueous potassium dichromate, in a dark enclosure: 7–14 days, change twice.
3. Transfer directly to Marchi fluid, a volume 15–20 times that of tissue: 1–2 weeks depending on size of tissue. Turn tissue over every day to improve penetration.
4. Wash in running water: 24 hours.
5. Dehydrate and embed. If using celloidin, keep embedding steps to a minimum. If using paraffin, avoid xylene and its solvent action on osmic acid; chloroform is safer and keeps steps to a minimum.
6. Deparaffinize slides with chloroform and mount in chloroform-resin.

RESULTS

degenerating myelin—black
background—brownish yellow
neutral fats—black

COMMENTS

The principle behind Marchi method is that the myelin of medullated nerves oxidizes more easily than degenerating myelin. Normal myelin is oxidized by chromating and will not react with osmic acid. Degenerating myelin contains oleic acid, which does not oxidize during chromating, therefore reduces the osmic acid, and stains black.

For additional reading about neurological staining, see Culling (1957); Lillie (1954b); Mettler (1932); Mettler and Hanada (1942); Poirier et al. (1954); and Swank and Davenport (1934a,b; 1935a,b).

For phospholipids see p. 289.

Chapter 13
PAS and Feulgen Techniques, and Related Reactions

Feulgen and PAS techniques involve two chemical reactions: (1) the oxidation of a-amino alcohol and/or 1,2 glycol groups to aldehydes; and (2) the reaction of resulting aldehydes with Schiff reagent (p. 209) to form a purple-red color. The Schiff reaction sequence for detecting the aldehydes is complicated, and two interpretations have been suggested: (1) the formation of an amino sulfinic acid followed by an additional reaction of the aldehyde (Wieland and Scheuing 1921); (2) the formation of an amino alkylsulfonic acid (Hörmann et al. 1958; Rumpf 1935). Hardonk and van Duijan (1964a,b,c) and Nauman et al. (1960) confirm the second interpretation. The two oxidizers most commonly used are chromic and periodic acids. The periodic acid breaks the carbon chains of the polysaccharides containing the 1,2 glycol groupings and oxidizes the broken ends into aldehyde groups. Chromic acid is a weaker oxidizer whose action is limited almost exclusively to glycogen and mucin (the principle of the Bauer method). If necessary, glycogen and starch can be demonstrated to the exclusion of other reactants by iodine or Best carmine (p. 294), and mucin can be demonstrated by Alcian blue (p. 297) or metachromatic methods (p. 316). If it is desirable to prevent the reaction of glycogen or starch, the saliva or diastase treatment is simple and effective (p. 211).

Among the nucleic acids, oxidation will not form aldehydes, and acid hydrolysis is required (see pp. 213, 272).

SCHIFF REAGENT

A century ago Hugo Schiff carried out extensive research on the reactions of amines with aldehydes and reported that the addition of a few drops of aldehyde would restore a red-violet color to a rosanilin (fuchsin, magenta) dye solution that had been decolorized by SO_2. His name has been applied to the colorless derivative formed by the action of SO_2 on basic fuchsin, or any of its component dye moieties.

Feulgen (1914) discovered that hydrolysis of fixed tissues exposed the deoxypentose of the nucleus in an aldehyde form. Then he and Rossenbeck (1924) described the Feulgen reaction for DNA, that mild acid hydrolysis followed by Schiff reagent gave a reddish-purple color to DNA-containing structures.

In an acid solution and with an excess of SO_2 present, basic fuchsin (a mixture of several related phenyl methane dyes, rosanilins, and pararosanilins built on the quinoid ring) is reduced to form a colorless N-sulfinic acid (fuchsin sulfurous acid). This reagent, with the addition of aldehydes, forms a new phenyl methane dye, slightly different from basic fuchsin since the color is more purple red than pure red. The chemical reaction is not wholly understood. Thus areas rich in DNA show deep coloration after hydrolysis. Schiff is stable as long as an excess of SO_2 with high acidity is present. Anything removing these conditions restores the original dye and produces a pseudoreaction. But when regenerated by an aldehyde, the dye becomes extremely resistant to such agents. Schiff reagent is not a dye; it lacks a chromophore and is, therefore, colorless. Baker (1958) considers this an example of localized synthesis of a dye; when Schiff reagent comes in contact with an aldehyde, the chromophore of the triarylmethane (quinoid ring structure) dye is reconstituted. The additive compound of the aldehyde with the Schiff could be called a dye.

Schiff reagent deteriorates rapidly at temperatures above 40°C; but at 0–5°C, if kept tightly stoppered, the deterioration is slow. Always keep it in the refrigerator when not in use. Under these conditions it will keep as long as six months.

Chung and Chen (1970) restore exhausted Schiff reagent by adding 0.5 g $NaHSO_3$ to 100 ml of the reagent. This can be repeated 2 or 3 times as the solution becomes exhausted, and is advisable even when the solution is kept in the refrigerator. If the solution shows a red tinge, it is partially exhausted; if it shows a purple tinge it is completely exhausted. In either case it can be recovered by this method.

For references concerning the Schiff reaction, see Atkinson (1952); Baker (1958); Bensley (1959); Glegg et al. (1952); Gomori (1952); Kasten (1960); Lhotka and Davenport (1949); Lillie (1951a,b; 1953a,b; 1954a,b); Lodin et al. (1967); McManus (1961); and Stowell (1945, 1946).

SCHIFF REACTIONS

PAS Technique, Aqueous

FIXATION

Any general fixative, but if glycogen or other soluble polysaccharides are to be demonstrated, fixation and washing should be done in alcoholic fluids of no less that 70% alcoholic content.

SOLUTIONS

Periodic acid

periodic acid (HIO_4)	0.6 g
distilled water.............................	100.0 ml
nitric acid, concentrated	0.3 ml

Schiff reagent (Lillie 1951*b*)

basic fuchsin (C.I. 42500)	0.5–1.0 g
distilled water.............................	85.0 ml
sodium metabisulfite ($Na_2S_2O_5$)	1.9 g
N HCL	15.0 ml

Place in bottle with approximately 50–60 ml of free air space. Shake at intervals for at least 2 hours or overnight. Add 200 mg activated charcoal: 1 minute; shake occasionally. Filter. If solution is not water-white, the charcoal is old. Try a fresh batch and refilter. Store Schiff reagent in a bottle with a minimum of air space above the solution and keep in refrigerator. This will decrease loss of SO_2 (Elftman 1959*c*). It can be stored frozen, but never at room temperature.

Kasten (1960) recommends the use of less than 50 mg of charcoal per 100 ml of solution, claiming that too much charcoal abstracts some of the discolored dye and reduces its sensitivity. Do not leave the charcoal in the solution too long, about 45 seconds, and filter.

Sodium bisulfite

sodium metabisulfite ($Na_2S_2O_5$)	0.5 g
distilled water.............................	100.0 ml

PROCEDURE

1. Deparaffinize and hydrate slides to water; remove $HgCl_2$.
2. Treat with periodic acid, aqueous: 5 minutes.
3. Wash in running water: 5 minutes.
4. Treat with Schiff reagent: 10 minutes.
5. Transfer through sulfite solutions, 3 changes: $1\frac{1}{2}$–2 minutes each (see comment 2).
6. Wash in running water: 5 minutes.
7. Counterstain, if desired (see comment 4).
8. Dehydrate, clear, and mount.

RESULTS

fungi—red
nuclei and other tissue elements—color of counterstain
Many tissues give positive PAS reactions: glycogen, starch, cellulose, mucins, colloid of thyroid, cartilage matrix, chitins, reticula, fibrin, collagen—rose to purplish red (see comment 6).

COMMENTS

1. When preparing slides, avoid excessive use of egg albumen, which contains sufficient carbohydrate to react with the Schiff.
2. Some technicians eliminate the sulfite rinses, but high chlorination of the water makes this a questionable practice. The reagent must be removed, and the sulfite solutions insure this. Renew them often.
3. *Control slides:* To remove glycogen, run the slides down to water and subject them to the saliva test. Saliva contains a diastatic enzyme which dissolves glycogen and starch. Human saliva can be negative for diastase and should be tested on a known starch before its use on control slides. A reliable substitute is a 1% solution of diastase of malt[1] in a phosphate buffer (p. 556), pH 6.0, 37°C: 1 hour.

   To remove mucin, treat slides with lysozyme, 0.1 mg to 10.0 ml of Sorensen $M/15$ phosphate buffer (p. 556), pH 6.24, room temperature: 40–60 minutes.

   To remove RNA, treat slides with 0.01 mg RNAse in 10.0 ml $0.2\,M$ acetate buffer (p. 552), pH 5.0, at room temperature: 10–15 minutes (or tris-HCl buffer, $0.5\,M$, pH 7.5, 56°C: 2 hours).

---

[1] Fisher Scientific Co.

4. *Counterstains:* Nuclei—hematoxylin; for glycogen—fast green, FCF, as in Feulgen method; mucin or acid polysaccharides—an acid dye (fast green); other polysaccharides—a basic dye (malachite green).

5. If it is necessary to remove the PAS from the tissue, treat it with potassium permanganate until all the color is removed; then bleach the permanganate with oxalic acid.

6. A number of carbohydrate and carbohydrate–protein substances give PAS reaction: acid mucopolysaccharides, glycolipids, mucoproteins, glycoproteins, pituitary gonadotropins, neutral mucopolysaccharides, unsaturated lipids, phospholipids, and others. Among the polysaccharides that give PAS reaction are glycogen, starch, and cellulose; all of these have 1,2 glycol groups. Cartilage has a polysaccharide compound that makes this tissue react positively. Among the mucoproteins, the mucins are carbohydrates and give PAS reaction. Certain other structures that are of unknown chemical composition, but that contain polysaccharides, will show positive reactions: striated and brush borders and reticulin fibers are examples. Fats may become colored with Schiff's after application of periodic acid if they are glycolipids. Unsaturated fats color, probably because of the oxidation of C=C sites to 1,2 glycols. Simple sugars are always lost, no matter what technique is employed. To retain complex sugars, avoid use of water solutions (McManus 1961).

7. The PAS technique can be combined with many other procedures, such as silver, aldehyde-fuchsin, Luxol fast blue, Sudan black, alkaline phosphatase (Elftman 1963; Himes and Moriber 1956; Lazarus 1958; and Moffat 1958). For black periodic and black Bauer methods, see Lillie et al. (1961).

PAS Technique, Alcoholic (Bedi and Horobin 1976)

FIXATION

Same as for aqueous technique.

SOLUTIONS

Periodic acid

periodic acid (HIO$_4$)	1.0 g
90% ethyl alcohol	100.0 ml

Keep in dark; solution is unsatisfactory if it turns brown.

Schiff reagent

Prepare a fresh solution of de Tomasi's Schiff. Dissolve 1 g basic fuchsin in 200 ml boiling distilled water. Shake for 5 minutes. Cool to exactly 50°C. Filter and add to filtrate 20 ml N HCl. Cool to 25°C and add 1 g sodium (or potassium) metabisulfite. Keep in dark 14–24 hours. Add 2 g activated charcoal. Shake for 1 minute. Filter and add an equal volume of 1% phosphotungstic acid aqueous. Shake. A white precipitate of phosphotungstic acid–Schiff complex is formed. Centrifuge down and discard supernatant. Redissolve precipitate in same volume of absolute alcohol as was used to prepare the complex. The precipitate can be dried and frozen at 0°C and stored for several months.

PROCEDURE

1. Deparaffinize sections and remove $HgCl_2$ if present.
2. Hydrate sections to 70% alcohol and briefly rinse in distilled water.
3. Treat with periodic acid: 60 minutes.
4. Wash in absolute alcohol, 2–3 changes: 3 minutes each.
5. Treat with Schiff complex: 30 minutes.
6. Wash in several changes absolute alcohol: 20 minutes.
7. Clear and mount.

COMMENTS

Some polysaccharides, mucosubstances, and fatty acids are extremely water soluble and demonstrating them by aqueous solutions becomes difficult. Alcoholic solutions are therefore recommended as more reliable for such substances.

Feulgen Reaction

FIXATION

A fixative containing $HgCl_2$ is preferred.

SOLUTIONS

5 N hydrochloric acid, see p. 542.

Schiff reagent, see p. 210.

Bleaching solution (sulfurous acid)

1 N HCl (p. 542)	5.0 ml
potassium bisulfite ($K_2S_2O_5$) or sodium bisulfite	
($Na_2S_2O_5$), 10% aqueous	5.0 ml
distilled water............................	100.0 ml

or

HCl, concentrated	1.0 ml
potassium or sodium bisulfite	0.4 g
distilled water............................	100.0 ml

For best results, make up bleach fresh each time.

Fast green

fast green FCF (C.I. 42053)	00.5 g
95% ethyl alcohol	100.0 ml

PROCEDURE

1. Deparaffinize and hydrate slides to water; remove $HgCl_2$. (Leaving slides in 95% alcohol overnight will remove lipids that might cause a plasmal reaction.)
2. Rinse at room temperature in 5 N HCl: 2 minutes.
3. Hydrolyze in 5 N HCl, room temperature: 1 hour (see comment 1).
4. Rinse at room temperature in 5 N HCl; rinse in distilled water.
5. Stain in Schiff reagent: 2 hours in dark.
6. Drain and transfer quickly into bleaching solution, 3 changes: $1\frac{1}{2}$–2 minutes each.
7. Wash in running water: 10–15 minutes.
8. Rinse in distilled water.
9. Counterstain in fast green: 10 seconds.
10. Dehydrate, clear, and mount.

RESULTS

DNA-containing substance—red violet
other tissue elements—shades of green

COMMENTS

1. The specificity of the Feulgen reaction has been attacked periodically, but it seems evident that a properly applied Feulgen reaction can be reasonably specific. Because the results of the reaction can vary with conditions, optimum conditions are essential. For instance, excessive hydrolysis will allow the degraded nucleic acids to diffuse from the tissue. Depending on the fixative used, the following times of hydrolysis are recommended (Pearse 1968):

Carnoy	6 minutes	Formalin sublimate ...	8 minutes
Carnoy, formula B ...	8 minutes	Helly	8 minutes
Carnoy-Lebrun	6 minutes	Regaud	14 minutes
Champy	25 minutes	Susa	18 minutes
Chrome acetic	14 minutes	Zenker	5 minutes
Flemming	16 minutes	Zenker-formol	5 minutes
Formalin	8 minutes		

The conventional hydrolysis temperature has been 60°C, but Deitch et al. (1968) and Decosse and Aiello (1966) agree that $5 N$ HCl at room temperature is preferable. The Feulgen reaction time is less critical at room temperature, and Feulgen values may be from 5 to 30% higher than when hydrolysis is done at 60°C.

2. Kasten and Burton (1959) make the following Schiff reagent. It is quickly prepared and colorless. It does not stain the hands and does not require refrigeration. It can be made more sensitive by boiling it for 1 minute.

basic fuchsin (C.I. 42500)	0.05 g
distilled water	300.0 ml
sodium hyposulfite ($Na_2S_2O_4$)	6.0 g

Solution should decolorize immediately. Filter if necessary. Ready for immediate use.

3. Pink solutions of Schiff reagent may have lost their potency. Test by pouring a few drops into 10 ml of 40% formalin. A good solution changes rapidly to reddish purple, but, if the color changes slowly and becomes blue purple, the solution is breaking down. Older batches of reagent tend to fade more than new batches (de la Torre and Salisbury 1962).

4. Chen (1944b) uses the following weak Flemming fixative for avian parasites. It gives beautiful results on any kind of smear preparation and on very small pieces of tissue, which should be fixed for 1–4 hours.

chromic acid, 1% in normal saline	25.0 ml
acetic acid, 1% in normal saline	10.0 ml
osmic acid, 2% in normal saline	5.0 ml
normal saline	60.0 ml

Wash smears for 1 hour in running water; proceed to step 2, hydrolysis. Wash pieces of tissue: overnight before embedding for sectioning.

5. Kasten and Lala (1975) observed false positives after glutaraldehyde fixation and blocked free aldehydes by reducing them with a fresh solution of 0.5% sodium borohydride ($NaBH_4$) in 1% sodium phosphate, monobasic ($NaH_2PO_4 \cdot H_2O$) aqueous solutions for 1 hour, room temperature. Hydrolyze, wash, and treat with Schiff.

It is always advisable to wash well after an aldehyde fixative like formalin, but it is more difficult with glutaraldehyde. The latter presents more opportunities for the introduction of free aldehydes into fixed cells.

6. The first of the series of three bleaching solutions (step 6) will begin to accumulate Schiff reagent and turn pink. Then it is advisable to remove that solution, shift steps 2 and 3 to steps 1 and 2, and add a new step 3 solution.

7. Swartz and Nagy (1963) found less fading of Feulgen stain in tissues that were mounted in Eukitt[2] than in those mounted in Clearmount or Permount.

8. Block and Godman (1955) hydrolyzed in 1 N trichloracetic acid and stained with a trichloracetic acid–Feulgen. The trichloracetic acid was substituted for hydrochloric acid, mole for mole in the hydrolyzing solution and the Feulgen reagent. In this way the histone proteins were left intact.

Deoxyribonucleic Acid (DNA) Fluorescent Technique (Culling and Vasser 1961)

FIXATION

10% formalin for sections; methyl alcohol for smears. (Other fixatives may require a different time for hydrolysis.)

---

[2] O. Kindler; Calibrated Instruments.

SOLUTIONS

Fluorescent Schiff reagent

acriflavine dihydrochloride	1.0 g
potassium metabisulfite	2.0 g
distilled water............................	200.0 ml
N hydrochloric acid	20.0 ml

Dissolve acriflavine and potassium metabisulfite in distilled water; add hydrochloric acid. Keep overnight before using.

Periodic acid

periodic acid.............................	1.0 g
M/15 sodium acetate	10.0 ml
absolute ethyl alcohol	90.0 ml

Acid alcohol

70% ethyl alcohol	99.0 ml
hydrochloric acid, concentrated	1.0 ml

PROCEDURE

1. Deparaffinize and hydrate sections to water. Smears can be carried directly into next step.
2. Transfer to preheated N hydrochloric acid, 60°C: sections, 10 minutes; smears, 3–4 minutes (or 1% periodic acid: 10 minutes for PAS).
3. Wash briefly in distilled water.
4. Transfer to fluorescent Schiff reagent: 20 minutes.
5. Wash in acid alcohol: 5 minutes.
6. Wash in fresh acid alcohol: 10 minutes.
7. Dehydrate in absolute alcohol: 1 minute.
8. Clear in xylene: 1 minute, and mount.

RESULTS

DNA—bright golden fluorescence
other elements—dark green to black

COMMENTS

The advantage of this method over the conventional Feulgen reaction is that smaller amounts of dye molecules are more easily observed.

Culling and Vasser warn that previously heated hydrochloric acid is important; cold hydrochloric acid can produce negative results. Also, timing is important; the reaction may weaken if slides are left in hydrolysis too long.

See Betts (1961) and Kasten and Glover (1959) for other fluorescent Schiff-type reagents. Also see Armstrong et al. (1957); Keeble and Jay (1962); Metcalf and Paton (1944); and Nash and Plaut (1964) for fluorescent method for DNA and RNA.

Chapter 14
Staining Hematologic Elements and Related Tissues

Hemopoietic (blood-forming) tissues is connective tissue specialized to produce blood cells and to remove worn-out blood cells from the bloodstream. There are two varieties of hemopoietic tissue: (1) myeloid, which produces erythrocytes, granular leukocytes, and platelets (thrombocytes); and (2) lymphatic, which produces most of the nongranular leukocytes.

Blood cells and fluid may become parasitized in various ways: for example, with malaria, trypanosomes, inclusion bodies of various diseases, such as rickettsia and psittacosis. Most of these foreign elements are best demonstrated with a Romanovsky type of stain.

See Chapter 21, p. 401, for techniques for demonstrating enzymes found in hemopoietic tissue.

BLOOD SMEARS

Preparation for Thin Smears

Slides must be clean for a uniform smear. Handle slides at the edges, keeping fingers off the clean surface. Prick the finger and, when a small drop of blood appears, wipe it away. Touch the next drop of blood to the clean surface of the right end of the slide. Place the narrow edge of another slide at a 20–30° angle on the first slide and to the left of the drop of blood. Pull to the right until the slide touches the blood. As soon as the blood has spread along the line of contact, push the right hand toward the left. Push steadily until all the blood disappears or the other end of the slide is reached. Move the hand rapidly; if the smear is spread too slowly, the leukocytes concentrate along the edges and in the tail of the smear (Christophers 1956). This method drags the blood but does not run over it and crush some of the cells. The hand can be kept from shaking by resting it on the table. Also, do not use a slide with a rough edge, which will leave streaks in the smear. If the blood seems thick, reduce the 20° angle to feed it out at a slower rate. For thin blood increase the angle (Fig. 14–1). See Pequeno (1960) for a clever method of using a pen to make thin smears.

Dry the slides rapidly in the air; waving them facilitates drying and prevents crenation (notching or scalloping of edges) of the red cells.

For cover-glass preparations (Fig. 14–2), place a small drop of blood in center of cover glass; place a second cover glass directly over the top of the drop at a slight angle to the bottom glass. Allow the blood to spread to edges of cover glasses. Quickly slide cover glasses apart. Air dry. See p. 402 for a similar method of spreading blood on slides.

Preferably, blood smears should be stained immediately or within 24 hours. If they must be stored, place them in a tight box away from dust and flies.

Blood smears are commonly stained with a Romanovsky type of stain, or neutral stain (p. 91). Neutrality is essential, and therefore dilution is

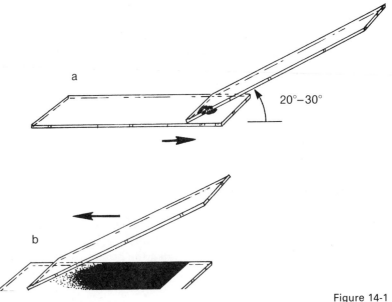

Figure 14-1
Preparation of thin smears: (a) Drop of blood is placed near one end of slide and second slide pulled against it; (b) second slide pulls blood across first slide.

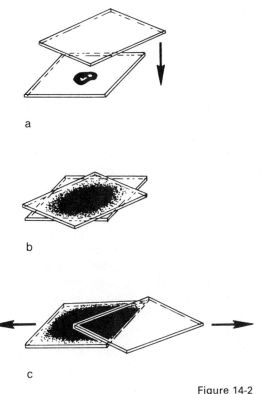

Figure 14-2
Preparation of cover-glass films: (a) A square cover
glass is placed on top of blood on another square
cover glass; (b) blood is allowed to spread; (c) top
cover glass is pulled across and off bottom cover
glass.

usually made with a buffer solution of a known pH. Distilled water is often
too acid, and tap water, too alkaline. The smears require no fixation in the
usual sense, since Wright stain (1902) includes both fixing agent (methyl
alcohol) and stain.

A common practice is to leave stained blood smears uncovered, because some mounting media tend to fade Romanovsky stains after a
period of time. The fading takes place slowly and is an inconvenience only
if slides are to be stored for years. Plastic coatings have been recommended, but they tend to be uneven in thickness and are easily scratched.
See Steinman (1955) for preparation of a plastic coating. Hollander (1963)
adds 1% of 2,6-di-*tert*-butyl-*p*-cresol to the synthetic mountant Permount
to prevent the fading of blood smears. Try plastic mountants, see p. 103.

Thin Smear Method

SOLUTIONS

Wright stain

The stain may be purchased in three forms: (1) dry powder, by the bottle, 1, 10, 25, or 100 g; (2) dry powder in capsule form, 0.1 g per capsule; or (3) in solution, 1, 4, 8, or 16 ounces.[1]

The powder (bulk or capsule) is ground up with methyl alcohol (0.1 g to 60.0 ml). The alcohol must be labeled *neutral* and *acetone free*. Grind thoroughly in a glass mortar, and pour off supernatant (surface-floating) liquid. If undissolved powder remains, pour back the liquid and grind again.

A slower but perhaps easier method is satisfactory. Add 0.3 g stain to 100.0 ml methyl alcohol and 3 ml glycerin is a stoppered bottle. Shake occasionally. After 24 hours, the solution is usually ready for use. To check whether a staining solution is satisfactory, with a pipette release a small drop of the solution onto filter paper. A good stain forms a deep blue spot, but a poor stain spreads into a broad pink halo around the central blue spot.[2]

For longer life store blood stains made up in methanol in the refrigerator.

Buffer solution, pH 6.4

monobasic potassium phosphate (KH_2PO_4)	6.63 g
anhydrous dibasic sodium phosphate	
(Na_2HPO_4)	2.56 g
distilled water	1000.0 ml

Bohorfoush diluent (1963)

sodium thiosulfate ($Na_2S_2O_3$)	0.1–0.2 g
distilled water	1000.0 ml

Allow to stand 1 hour. Bohorfoush claims good uniform staining with this solution.

---

[1] Scientific Products sells Camco Quik Stain (no. B4130), a single solution of Wright stain with built-in buffering agent. A dip of only 30 seconds is required. Other quick stain kits can be obtained from Harleco and Hemal Stain Co.

[2] From "Lab Hint: A Quick Check for Wrong Wright's," *Clinical Reporter,* Fisher Scientific Co.

PROCEDURE

1. With wax pencil draw 2 marks across slide delineating region to be stained (length of 40 mm cover glass).
2. Cover with 10–12 drops of Wright stain: 1–2 minutes.
3. Add an equal amount of buffer or diluent: 2–4 minutes.
4. Rinse in distilled water or buffer solution, 1 or 2 dips. Precipitate deposit on the slide can be avoided by flushing the slide with a pipette, or carry the stain with the slide into the wash water; do not drain it off first.
5. Blot with 2 sheets of filter paper; press but do not rub.
6. Allow slide to thoroughly dry before applying cover glass.

ALTERNATE PROCEDURE

1. Fix dried film in methyl alcohol: 1–2 minutes. This reduces red cell distortion and artifacts.
2. Dilute Wright stain with an equal amount of buffer. About 10 drops of each is sufficient for one slide. Use immediately.
3. Cover slide with diluted stain: 4 minutes.
4. Rinse and dry as in above procedure.

RESULTS

erythrocytes—pink
nuclei—deep blue or purple
basophilic granules—deep purple
eosinophilic granules—red to red orange, bluish cytoplasm
neutrophilic granules—reddish brown to lilac, pale-pink cytoplasm
granules of monocytes—azure
lymphocyte granules—larger and more reddish than monocyte granules, sky-blue cytoplasm
platelets—violet to purple

COMMENTS

Precipitate formation can be troublesome; the dark granules obscure the blood cells and are confusing in malarial smears. In addition to poor washing in step 4, another cause of precipitate may be evaporation during exposure to undiluted stain. Methyl alcohol is highly volatile and readily lost by evaporation. In dry warm weather use more Wright stain or shorten the time a bit, or cover the slides with a petri dish. Rapid evaporation is easily detected, and more stain can always be added. I always cover the slides in any type of weather.

Proper rinsing in step 4 is one of the most important steps in this procedure. Never use tap water; it is too alkaline. Distilled water is usually satisfactory, but for best results rinse in the buffer, pH 6.4.

The longer the washing with water or buffer, the more stain is removed from the white cells. Only a dip or two is usually sufficient, but if the white cells are overstained, differentiate them by longer washing.

If the slides are overstained, the erythrocytes are too red, the white cells too pale, or the stain has precipitated, the slides can be recovered: (1) Cover the entire slide almost to excess with additional Wright stain: 15–30 minutes; (2) rinse with distilled water or buffer; (3) dry (Morrison and Samwick 1940).

Wright Stain for Cold-Blooded Vertebrates (Heady and Rogers 1962)

PROCEDURE

1. Cover with Wright stain: 2 minutes.
2. Add equal portion of phosphate buffer, pH 6.7: 3 minutes (p. 556).
3. Wash in distilled water or buffer solution and blot as above.

Wright Stain for Birds (Santamarina 1964)

SOLUTIONS

Wright stain

Wright stain powder	3.3 g
absolute methyl alcohol	500.0 ml

This is stronger than the solution for human blood. Do not use one that is too dilute.

Formalin solution

formalin, concentrated	0.25 ml
distilled water	500.0 ml

Adjust to pH 6.8 by adding either 0.25% sodium carbonate or 0.25% hydrochloric acid as needed.

PROCEDURE

1. Cover dried smears with Wright solution: 8 minutes.
2. Slowly add formalin solution with a medicine dropper. Do not allow solution to overflow.
3. When a metallic sheen covers entire surface of fluid, pour off and flush immediately with distilled water adjusted to pH 6.8 as above.
4. Blot
5. Differentiate in ether–absolute methyl alcohol (1 : 1). Dip slides up and down 6 to 10 times. Examine them under microscope; if they are too bluish, differentiate further.
6. Clear and mount.

RESULTS

erythrocytes—yellow to pinkish cytoplasm, purple chromatin
thrombocytes—gray-blue cytoplasm, purple chromatin
lymphocytes—blue granules, purplish-red chromatin
monocytes—light-blue granular cytoplasm, purple chromatin
heterophils—yellowish to brownish-red rods, light-purple chromatin
eosinophils—similar to heterophils
basophils—dark-purple granules

Preparation of Thick Blood Films

With this type of film, a relatively large quantity of blood is concentrated in a small area, thereby increasing the possibilities of finding parasites. The concentration and timing of staining are adjusted so the action is stopped at the point when the leukocytes have stained, some hemoglobin has been dissolved, and the red cell membranes have not yet begun to stain. At this point the leukocytes, platelets, and protozoa are stained and lie on an unstained or very lightly stained background, which is yellowish from the remaining hemoglobin. Freshly prepared films stain better than films one or more days old.

Puncture the skin deeply enough to form a large drop of blood. On a slide, cover a space the size of a dime with enough blood (about 3–4 average drops) to spread easily. Too much blood will crack and peel off when dry. Smear it by circling the slide under the finger without making actual contact. Some find it easier to swirl the blood with a dissecting needle or the corner of a slide. Burton (1958) taps the slide against the table top, so the drop of blood flows transversely across the slide for $\frac{3}{8}$ to $\frac{3}{4}$ inch. Practice will determine the best method and how much blood to use.

Too thin a film has no advantage over a thin smear. An ideal film is several cell layers thick in the center, tapering off to one cell thickness at the periphery.

Allow the slide to dry in a horizontal position; if tilted, the blood will ooze to one edge of the film. Protect from dust and flies, and do *not* fix by flaming or alcohol.

Field-Wartman Thick Film Method (Field 1941; Wartman 1943)

SOLUTIONS

Solution A

methylene blue (C.I. 52015)	0.8 g
azure A (C.I. 52005)	0.5 g
dibasic sodium phosphate, anhydrous	5.0 g
monobasic potassium phosphate	6.25 g
distilled water.............................	500.0 ml

Solution B

eosin Y (C.I. 45380)	1.0 g
dibasic sodium phosphate, anhydrous	5.0 g
monobasic potassium phosphate	6.25 g
distilled water.............................	500.0 ml

Dissolve phosphate salts first, then add stains. (The azure will go into solution by grinding in glass mortar with a small amount of phosphate solvent.) Set aside for 24 hours. Filter. Store in refrigerator.

PROCEDURE

1. Dip slides in solution A: 1 second.
2. Rinse gently in clean distilled water: few seconds, or until stain ceases to flow from film.
3. Dip in solution B: 1 second. Use this solution with care; it tends to decolorize the leukocytes stained with methylene blue–azure, and accelerates the dissolving of hemoglobin.
4. Rinse gently in distilled water: 2–3 dips.
5. Dry, do not blot. Stand slides on end and allow to air dry.
6. When completely dry, add mounting medium and cover glass.

RESULTS

There will be a thicker, central area of partially laked blood. This may not be well suited for examination. But the surrounding area, and especially at the edge toward which the hemoglobin has drained, will be creamy, sometimes mottled with pale blue. This is the best area for study.

Leukocytes
cytoplasm—pale blue, poorly defined
nuclei—dark blue, well defined
eosinophilic granules—bright red, large, well defined
neutrophilic granules—pale purple pink, small, indistinct
basophilic granules—deep blue with reddish cast.

Platelets—pale purple or lavender

Parasites
cytoplasm—blue
chromatin—purplish red or deep ruby red
pigment—unstained yellow granules of varying intensity

Wilcox Thick Film Method (1943)

SOLUTIONS

Wright-Giemsa stock solution

Giemsa powder	2.0 g
glycerin	100.0 ml

Heat in water bath, 55–60°C: 2 hours, stirring at intervals. Avoid absorption of moisture by covering mouth of flask with double thickness of paper secured with rubber band.

Add:

aged Wright staining solution (2.0 g/1000.0 ml methyl alcohol)	100.0 ml

Let stand overnight; add in addition:

        Wright staining solution ....   800.0 ml

Filter; ready for use.

Working solution

        Wright-Giemsa stock solution ....   1 part
        neutral distilled water (buffer) ....   9 parts

Buffer, pH 7.0

Solution A

        dibasic sodium phosphate, anhydrous ....   9.5 g
        distilled water ....   1000.0 ml

Solution B

        monobasic potassium phosphate ....   9.7 g
        distilled water ....   1000.0 ml

Working solution

        Solution A ....   61.1 ml
        Solution B ....   38.9 ml
        distilled water ....   900.0 ml

PROCEDURE

Short method
1. Stain film: 10 minutes.
2. Flush scum from top of solution with buffer to avoid picking up precipitate on slides.
3. Remove slides to neutral distilled water: 1 minute.
4. Dry slides standing on end. Do not blot. Mount.

Long method (This method produces a more brilliant stain.)

1. Stain: 45 minutes in dilution of 1 to 50 of buffer.
2. Flush off scum and transfer slides to buffer: 3–5 minutes.
3. Dry as above.

Thin smear and thick film can be stained on same slide. Fix thin portion in methyl alcohol for 1–2 minutes, taking care that thick film does not come in contact with alcohol. Stain as above (short method), but shorten washing time to 2 to 3 dips so thin smear is not too light; this shorter washing time will leave a deeper background in the thick film.

COMMENTS

Modifications of this method have been made by Fenton and Innes (1945); Manwell (1945); and Steil (1936).

Old smears will stain more brilliantly if treated 5–10 minutes in alcohol-acetic solution—10 drops glacial acetic acid to 60 ml absolute alcohol.

BLOOD TISSUE ELEMENTS AND INCLUSION BODIES[3]

Giemsa Stain (Cramer et al. 1973)

FIXATION

Any good general fixative, but Zenker preferred.

SOLUTIONS

Azure-eosinate stock solution

azure II-eosin[3]	2.0 g
azure II[3]	1.0 g
azure B–eosin[3]	1.0 g
azure A–eosin[3]	0.5 g

Dissolve in equal parts of glycerin and methanol, 250 ml each, and allow to stand at room temperature overnight. Shake well 5–10 minutes and store unfiltered in a brown, well-capped bottle at room temperature. Stable for months.

---

[3] Curtin Matheson Scientific Co.

Working solution A (2 hour stain)

      stock solution ....   5.0 ml
      distilled water.............................   65.0 ml

pH should be 4.8–5.2. If greater than 5.2, add 1–2 drops of 1% acetic acid.

Working solution B (overnight stain)

      stock solution ....   3–5 drops
      distilled water.............................   65.0 ml

pH should be 6.5–6.8; adjust with 1% acetic acid if needed.

PROCEDURE

1. Deparaffinize and hydrate sections to distilled water.
2. Stain in working solution A: 2 hours; in B: overnight.
3. Dip quickly in 1% acetic acid.
4. Blot immediately and rinse several times in absolute alcohol until only slightly bluish tint appears in alcohol running off slides.
5. Clear and mount.
6. If too blue, return through xylene and absolute alcohol into several very quick dips in 0.5–1% acetic acid. Blot, and return through absolute to xylene.

RESULTS

    nuclear chromatin—dark blue
    eosinophil granules—red orange
    mast cell granules—dark purple
    erythrocytes—pink to red
    connective tissue—pink to light purple
    lymphocytes—blue cytoplasm

COMMENTS

Acetic acid removes the blue colors rapidly; this has to be carefully controlled. The red tones are removed rapidly by alcohol. Too red a color, therefore, means overdifferentiation. Try restaining. *Warning:* Such slides may have to be discarded.

Jenner-Giemsa Stain for Malarial Parasites (McClung 1939)

FIXATION

Smears may be fixed during staining process. Fix tissue sections in any good general fixative, preferably one containing mercuric chloride and alcohol.

SOLUTIONS

Jenner solution

Jenner stain	0.2 g
methyl alcohol, neutral, acetone-free	100.0 ml

Giemsa stock solution

Giemsa powder	3.8 g
methyl alcohol, neutral, acetone-free	75.0 ml
glycerin	25.0 ml

Work stain into glycerin, warm for 2 hours, 60°C oven. Add methyl alcohol.

Working solution

Giemsa stock solution	10.0 ml
distilled water...........................	100.0 ml

PROCEDURE

1. Deparaffinize sections and run down to 50% alcohol. Remove $HgCl_2$, if present.
2. *a*. Sections—flood with ample amount of Jenner solution and add an equal amount of distilled water, or mix equal amounts of Jenner solution and water and flood slides or stain slide in coplin jar: 4 minutes.
   b. Smears—flood with Jenner solution: 3 minutes. Add equal amount of distilled water: 1 minute.
3. Pour off Jenner; no rinse.
4. Flood with diluted Giemsa solution or place in coplin jar of solution: 15–20 minutes.

5. Rinse off stain with distilled water and continue to differentiate with distilled water. If too blue, eosin color can be brought out by rinsing in 0.5–1% acetic acid in water.
6. Dehydrate sections in following sequence of acetone and xylene.

> acetone 95 parts: xylene 5 parts
> acetone 70 parts: xylene 30 parts
> acetone 50 parts: xylene 50 parts
> acetone 30 parts: xylene 70 parts
> acetone 5 parts: xylene 95 parts

Smears may be blotted and air-dried before mounting.

Suggestion: In place of the above solutions, use 3 changes isopropyl alcohol, xylene–isopropyl alcohol (1 : 1). Do not use ethyl or methyl alcohol, which extract the stain.

7. Clear in xylene and mount.

RESULTS

chromatin of parasite—red or purplish red
cytoplasm of parasite—blue (other elements—color similar to Wright stain)
pigment—yellow brown to black
Schüffner's stippling—red

COMMENTS

A beautiful clear stain for any blood picture, particularly blood parasites.

Brooke and Donaldson (1950) recommend the use of Triton X-30, a nonionic liquid detergent to prevent the transfer of malarial parasites between films when mass staining.

For a fluorescent antibody technique, see Ingram et al. (1961) and Sodeman and Jeffery (1966). For a fluorescent stain see Hansen et al. (1970). An excellent lab manual on malaria techniques is Shute and Maryon (1966).

Giemsa Staining for Cold-Blooded Vertebrates (Pienaar 1962)

1. Fix dried smears in undiluted Wright (p. 222) or Jenner (p. 231) stock stains: 3–5 minutes. (Smears may be previously fixed in absolute methyl alcohol.)

2. Add equal volume of distilled water buffered to pH 6.2–6.8: 5–6 minutes (pH is important).
3. Pour off solution, but do not rinse. Stain in diluted Giemsa stock (p. 231) (1 ml/10 ml distilled water): 20–30 minutes.
4. Flush with distilled water to prevent stain precipitate on slide.
5. Differentiate, if necessary, in distilled water, pH 6.2.
6. Rinse well in distilled water. Blot, dry, and mount.

Pienaar recommends the following modification for snake eosinophils, which are distorted in the above method.

1. Jenner solution: 1–3 minutes.
2. Buffered water dilution: 10 minutes.
3. Dilute Giemsa: 10–15 minutes.
4. Rinse briefly in distilled water. Dehydrate rapidly through acetone, xylene–absolute alcohol (1 : 1), xylene. Mount.

Giemsa Staining for Birds (Lucas and Jamroz 1961)

1. Fix dried smears in Wright stain: 4 minutes.
2. Add distilled water: 4–5 minutes.
3. Wash in running water: 2 minutes.
4. Stain in dilute Giemsa (15 drops/10 ml water): 15–20 minutes.
5. Wash in distilled water. If precipitate is present, run Wright stain over the slide. Wash off with running water.
6. Stand slides on edge to dry. Do not blot.

Reticulocyte Staining (Brecher 1949)

SOLUTION

new methylene blue N (C.I. 52030)	0.5 g
potassium oxalate	1.6 g
distilled water............................	100.0 ml

PROCEDURE

1. Mix equal amounts of stain and fresh or heparinized (or oxalated) blood on a slide. Draw mixture into a capillary pipette and allow to stand: 10 minutes.

2. Expel mixture on a slide and mix again.

3. Make thin smears. Air dry.

RESULTS

reticulum—deep blue
red cells—light greenish blue

COMMENTS

If slides fade, fix in methanol and restain with Wright stain. The slides can be stained with Wright immediately following the air drying in step 3 if desired.

Heinz bodies are demonstrated by this stain, pale to deep blue against the pale green background of the red cells (Thompson 1961). Simpson et al. (1970) recommend Rhodanile blue as a selective stain for Heinz bodies.

Lucas and Jamroz (1961) use the following for birds: Mix a drop of blood with a drop of 1% brilliant cresyl blue in 0.85% NaCl or Ringer avian solution (p. 546). Let stand 2 minutes and smear. Dry, fix in methyl alcohol, and stain with Wright.

For cold bloods (Pienaar 1962), mix a few drops of blood and stain together in equal quantity in a small covered dish, and leave 5 minutes. Make up the stain, 1% in Ringer or Locke solution for cold bloods (p. 546). Make smears on slightly warm slides. Dry, fix in methyl alcohol, and stain with Wright.

HEMOGLOBIN STAINING

Buffalo Black for Hemoglobin (Puchtler et al. 1964)

FIXATION

Any general fixative.

SOLUTIONS

Tannic acid

tannic acid .... 5.0 g
distilled water........................... 100.0 ml

Phosphomolybdic acid

phosphomolybdic acid	1.0 g
distilled water.............................	100.0 ml

Buffalo black

Saturate a solution of buffalo black NBR (C.I. 20470) in methanol–glacial acetic acid (9 : 1). Let stand 48 hours. Do not filter. Keeps several months. Filter the used portions back into the storage bottle.

PROCEDURE

1. Deparaffinize and hydrate to water.
2. Treat with tannic acid: 10 minutes.
3. Transfer through 3 changes of distilled water.
4. Treat with phosphomolybdic acid: 10 minutes.
5. Transfer through 3 changes of distilled water.
6. Stain in buffalo black: 5 minutes.
7. Differentiate in methanol–acetic acid (9 : 1), 2 changes: 5 minutes total.
8. Place in absolute ethanol or propanol, clear, and mount.

RESULTS

hemoglobin—dark blue to purplish black
other tissues—yellow

COMMENTS

Do not use freshly prepared tannic acid. The tannic acid should be more than 48 hours old; it keeps indefinitely.

Puchtler and Sweat (1963*a*) combine hemoglobin staining with a ferrocyanide staining for hemosiderin.

Cyanol Reaction (Dunn, 1946)

FIXATION

Formalin, preferably buffered to pH 7; fix no longer than 48 hours. Dried smears may be methyl-alcohol fixed and then taken to step 2 below.

SOLUTIONS

Cyanol reagent stock solution

cyanol	1.0 g
distilled water............................	100.0 ml
zinc powder, CP	10.0 g
glacial acetic acid	2.0 ml

Bring to boiling point. In a short time the solution is decolorized. Stable for several weeks.

Working reagent

Filter 10 ml of stock reagent, and add:

glacial acetic acid	2.0 ml
hydrogen peroxide, commercial, 3%	1.0 ml

PROCEDURE

1. Deparaffinize and hydrate slides down to water.
2. Treat with working reagent: 3–5 minutes.
3. Rinse in distilled water.
4. Counterstain in red nuclear stain.
5. Dehydrate, clear, and mount.

RESULT

hemoglobin—dark blue to bluish gray

COMMENTS·

Gomori (1952) uses 1.0 g acid fuchsin instead of cyanol, and produces a red hemoglobin. Counterstain with hematoxylin or other blue nuclear stain.

The benzidine reaction for hemoglobin has been removed since benzidine is considered to be a carcinogen.

Fetal Hemoglobin (Modified from Kiossoglou et al. 1963)

FIXATION

Air dry blood smears for 1 hour. Fix in absolute methanol: 5 minutes; follow by 80% ethanol: 5 minutes.

SOLUTIONS

Buffer

McIlvaine citric acid–phosphate, pH 3.2–3.4 p. 557.

Stain

eosin Y (C.I. 45380)	0.5 g
distilled water.............................	100.0 ml

or

ponceau S (C.I. 27195).....................	0.5 g
1% aqueous acetic acid	100.0 ml

PROCEDURE

1. Rinse fixed slides in tap water. Follow by a rinse in distilled water.
2. Treat in buffer, 37°C: 5 minutes. Slides must be in vertical position.
3. Rinse in tap water.
4. Stain: 3 minutes.
5. Rinse in tap water: few seconds. Dry standing in vertical position.

RESULTS

cells containing fetal hemoglobin—deep pink to red
adult cells—colorless or pale-staining ghosts

COMMENTS

The smears can be stained with Mayer hematoxylin for 5 minutes before staining with eosin or ponceau S. Also they may be cleared and mounted. Count 500 cells and express the number of F-containing cells as a percentage of the total. Smears must be fresh; if left unstained overnight they can give false determinations.

BONE MARROW STAINING

Maximow Eosin-Azure Stain (Block et al. 1953)

FIXATION

Zenker–neutral formalin: 30 minutes. Neutralize formalin by adding 2 g lithium or magnesium carbonate to 500 ml of formalin. Excess of carbonate should be present. Wash in running water 1–24 hours before preparation for embedding (see comment 1).

SOLUTIONS

Solution A

eosin Y (C.I. 45380) or eosin B (C.I. 45400) . . .	0.1 g
Wright buffer	100.0 ml

Solution B

azure II	0.1 g
Wright buffer	100.0 ml

Working solution

Wright buffer	2.0 ml
distilled water	40.0 ml
Solution A	8.0 ml

Stirring vigorously, add gradually

Solution B	4.0 ml

Fresh solutions are best; their action deteriorates after 3 or 4 weeks. Working solution should appear deep violet in color, and a precipitate should not form in it for an hour or more. If a precipitate forms on the slides, the stain mixture was improperly made; solution B was added to solution A too rapidly or without stirring. If the eosin loses its brilliance, solutions are old. Store solutions A and B in the refrigerator.

PROCEDURE

1. Deparaffinize and hydrate slides to water; remove $HgCl_2$. If tissue was fixed in a fixative without potassium dichromate, chromate slides overnight in 2.5%–3% aqueous potassium dichromate. Wash thoroughly in running water: 15 minutes. Proceed to step 2.
2. Stain in Mayer hematoxylin: 30–45 seconds, no more.
3. Wash in running water: 5–10 minutes.
4. Wash in distilled water: 5 or more minutes.
5. Stain in eosin-azure overnight.
6. Differentiate in 95% alcohol. In stubborn cases of differentiation (old solutions) a brief immersion in colophonium alcohol (p. 563) may help to sharpen the colors. Differentiation must be done under the microscope.
7. Dehydrate in absolute alcohol, clear, and mount. If, after mounting, the stain still appears undifferentiated, remove the cover glass and mountant with xylene. Transfer to absolute alcohol: 1–2 minutes. Transfer for a brief period to 95% alcohol. Examine the slide under a microscope immediately, since the tissue destains rapidly after this treatment. Dehydrate, clear, and mount.

RESULTS

nuclei—dark purple blue
erythrocytes—light pink
eosinophilic granules—brilliant red
cytoplasm—pale blue

COMMENTS

1. After 30 minute fixation, cells in center are not as well-fixed as those at periphery, but longer fixation produces granular cytoplasm in eosinophilic erythrocytes and erythroblasts. If 30 minute fixation is not adequate for fixing bone marrow contained in bone, follow Zenker-formol fixation by overnight treatment with 3% aqueous potassium dichromate. Also crack the bone and break away a section of the bone to permit penetration of the fixative.
2. Block[4] embeds in methacrylate. This is recommended for thin sections of $1–3\mu$ with a minimum of shrinkage in the tissue and with precise cellular detail (see p. 453).

---

[4] Personal communication.

3. Lillie (1965) stains 1 hour in the following eosin–azure solution:

azure A, 0.1%	4.0 ml
eosin B, 0.1%...........................	4.0 ml
0.2 M acetic acid	1.25 ml
0.2 M sodium acetate	0.75 ml
acetone	5.0 ml
distilled water...........................	25.0 ml

The solution should be made up immediately before use. Transfer to acetone, 2 changes; acetone-xylene (1 : 1) and xylene. Lillie alters the amount of acetic acid and sodium acetate in the solution. The above mixture is for material fixed in Zenker. The proportions of acetic acid and sodium acetate should be adjusted following other fixatives (Lillie).

4. Bone marrow when collected should have an anticoagulant added to it. Gardner (1958) and many other workers mix the bone marrow in a tube containing 0.5 mg heparin powder. Kniseley[5] wets his syringe with d-potassium EDTA as an anticoagulant. Smears can be made with some of the aspirated (drawn out by suction) marrow. The remaining material is poured into a small funnel lined with Filtrator Coffee Paper (Zbar and Winter 1959). Rinse marrow, while in funnel, several times with saline, or until the marrow has lost most of its color. This also helps to wash the material into one mass at the apex of the funnel. Partially clotted marrow will be broken up and washed free of blood, leaving excellent clear particles of marrow. Fold in the filter paper, place in tissue receptacle and carry through fixation, washing, dehydration, and infiltration without removing from filter paper. No marrow will be lost. Minute amounts of marrow may prove difficult to recover from the filter paper without loss of material. I allow the paraffin around the marrow (on the filter paper) to congeal slightly, then carefully scrape paraffin and marrow together, and remove to melted paraffin in mold. The congealed paraffin containing the marrow will sink to the bottom of the mold, or it can be pressed down into desired position. This method keeps the marrow concentrated reasonably well. For other methods for handling bone marrow, see Berthrong and Barhite (1964); Conrad and Crosby (1961); Leach (1960); and Raman (1955).

5. Gude and Odell (1955) recommended for dilution Vinisil (Abbott Laboratories), a 3.5% solution of polyvinylpyrrolidone (PVP) in

[5] Ralph M. Kniseley, Oak Ridge Associated Universities. Personal communication.

normal saline. Endicott (1945) used plasma serum as a diluting fluid to thin smears. The serum can be kept on hand for several months if stored in refrigerator.

6. For a fluorescent method, see Werth (1953).

Phloxine–Methylene Blue (Thomas 1953)

FIXATION

Any general fixative, Zenker preferred; or mordant slides in Zenker after hydration.

SOLUTIONS

Phloxine solution

phloxine B (C.I. 45410)	0.5 g
distilled water.............................	100.0 ml
glacial acetic acid	0.2 ml

A slight precipitate will form. Filter before use. (I have found this unimportant; the precipitate settles at the bottom of the bottle and does not collect on the tissue.)

Methylene blue–azure solution

methylene blue (C.I. 52015)	0.25 g
azure B (C.I. 52010)	0.25 g
borax	0.25 g
distilled water.............................	100.0 ml

PROCEDURE

1. Deparaffinize and hydrate slides to water; remove $HgCl_2$.
2. Stain in phloxine: 1–2 minutes.
3. Rinse well in distilled water.
4. Stain methylene blue–azure: $\frac{1}{2}$–1 minute.
5. Partially destain in 0.2% aqueous acetic acid.
6. Complete differentiation in 95% alcohol, 3 changes.
7. Dehydrate in absolute alcohol; clear and mount.

RESULTS

nuclei—blue
plasma cell cytoplasm—blue
other tissue elements—shades of rose and red

COMMENTS

Thomas substitutes azure B for azure II of other methods because of the uncertain composition of the latter. It is unnecessary to use colophonium alcohol as in other methods. I recommend Thomas' method over all others I have tried. It is practically foolproof. See Greenstein's (1957) modification. He uses an alcoholic phloxine solution and mixes equal parts of separate solutions of methylene blue and azure B.

Delez and Davis (1950) make up their phloxine with oxalic acid; 1% phloxine in 0.05% aqueous oxalic acid.

This staining method is frequently used for general staining in place of a hematoxylin-eosin method.

STAINING FOR FIBRIN

Ledrum Acid Picro-Mallory Method (Culling 1957)

FIXATION

Any good general fixative.

SOLUTIONS

Celestin blue

celestin blue B (C.I. 51050)	0.25 g
ferric alum, 5% (5 g/100 ml water)	50.0 ml

Boil 3 minutes. Cool and filter. Add:

glycerin	7.0 ml

Keeps for several months.

Mayer hematoxylin, see p. 113.

Picro-orange

picric acid saturated in 80% alcohol	100.0 ml
orange G (C.I. 16230)	0.2 g

Acid fuchsin

acid fuchsin (C.I. 42685)	1.0 g
distilled water.............................	100.0 ml
trichloracetic acid	3.0 g

Phosphotungstic acid

phosphotungstic acid	1.0 g
distilled water.............................	100.0 ml

Aniline blue

aniline blue WS (C.I. 42780)	2.0 g
distilled water.............................	100.0 ml
glacial acetic acid	2.0 ml

PROCEDURE

1. Deparaffinize and hydrate slides to water; remove $HgCl_2$.
2. Stain in celestin blue: 3–5 minutes.
3. Rinse in tap water.
4. Stain in Mayer hematoxylin: 5 minutes.
5. Wash in running water: 5 minutes.
6. Rinse in 95% alcohol.
7. Stain in picro-orange: 2 minutes.
8. Stain in acid fuchsin: 5 minutes.
9. Rinse in distilled water.
10. Dip in equal parts of picro-orange and 80% alcohol: few seconds.
11. Differentiate in phosphotungstic acid: 5–10 minutes, until colors are clear.
12. Rinse in distilled water.
13. Stain in aniline blue: 2–10 minutes.
14. Rinse in distilled water: 2–3 minutes.
15. Dehydrate, clear, and mount.

RESULTS

fibrin—clear red
erythrocytes—orange
collagen—blue
nuclei—blue black

COMMENTS

This method is specific for fibrin, setting it off sharply from other tissue elements. See also phosphotungstic acid–hematoxylin (p. 126), the Gram-Weigert method (p. 352), and the Rosindole method (Puchtler et al. 1961). The latter is also considered specific for fibrin.

Chapter 15
Staining Pigments and Minerals

Pigments are a heterogeneous group of substances containing enough natural color to be visible without staining. Sometimes, however, color is added to give them more intense differential staining. Some pigments are artifacts, such as formalin pigment (p. 7). Others, exogenous pigments, are foreign pigments that have been taken into the tissue in some manner. Carbon is a common pigment found in the lungs of city dwellers, particularly of people from coal-burning cities. Endogenous pigments are found within the organism and arise from nonpigmented materials. Iron-containing hemoglobin can be broken down into iron-containing pigment, hemosiderin, and a brown pigment containing no iron, called hematoidin, or bilirubin, which can be oxidized to biliverdin (green). Normal hemoglobin, when not broken down into hemosiderin, will not show a positive Prussian blue reaction (p. 24). The iron is masked or occult, and the organic part of the hemoglobin molecule must be destroyed in order for the iron to be unmasked.

Melanin (brown or black pigment) is found normally in the skin, hair, and eye, but may occur pathologically anywhere in the body. The pigment is formed from tyrosine by a special enzyme, tyrosinase. Lipofuscin, sometimes known as the "wear and tear" pigment, can be found in the heart muscle, adrenals, ganglion cells, and liver. The lipofuscins are derived from lipid or lipo-protein sources, at least partly by oxidation. Several enzymes are found in them, nonspecific esterases and acid phosphatase in particular. Melanin and lipofuscin are brownish pigments stainable by fat dyes and some basic aniline dyes such as fuchsin. They are metachromatic (p. 314) with methyl green and give a positive Schiff reaction after periodic acid treatment (p. 210).

Calcium may be present in tissues in a number of different forms, but the methods here are applicable to the insoluble forms—deposits of calcium phosphate and calcium carbonate (Ham 1957).

STAINING FOR IRON

According to Baker (1958), the iron reaction is an example of local formation of a colored substance which is not a dye. It is a type of histochemical test wherein a tissue is soaked in a colorless substance. Certain tissue elements react with the substance and become colored. This well-known test is called the Berlin blue, Prussian blue, or Perl reaction: The iron is dissolved from hemosiderin by hydrochloric acid, and then reacts with potassium ferrocyanide to form the Berlin blue precipitate, ferric ferrocyanide.

Sometimes fading occurs in slides due to the reduction of the Berlin blue to colorless ferro ferrocyanide. The mounting resin probably takes up oxygen while drying and deprives the sections of oxygen, thereby reducing them to the colorless condition. If this takes place and it is essential to recover the slides, treat them with hydrogen peroxide. The newer synthetics are not as prone to reduce Berlin blue as former resins; also Lustron 2020 is recommended (p. 102).

Iron Reaction (Hutchison 1953)

FIXATION

Hutchison recommends:

sodium sulfate	12.0 g
glacial acetic acid	33.0 ml
formalin	40.0 ml
distilled water to make	200.0 ml

Go directly into 70% alcohol from fixative.

SOLUTIONS

Ferrocyanide–hydrochloric acid solution

Solution A

potassium ferrocyanide	2.0 g
distilled water.............................	50.0 ml

Solution B

hydrochloric acid..........................	2.0 ml
distilled water.............................	48.0 ml

Prepare the two parts of the solution separately, mix them together, warm slightly, and filter. Place in paraffin oven, 56°C, a short time before using.

Safranin

safranin O (C.I. 50240).....................	0.2 g
distilled water.............................	100.0 ml
glacial acetic acid	1.0 ml

Kernechtrot

Kernechtrot (nuclear fast red)[1]	0.1 g
5% aluminum sulfate (5 g/100 ml water).......	100.0 ml

Dissolve with heat. Cool and filter.
Add a crystal of thymol as preservative.

PROCEDURE

1. Deparaffinize and hydrate slides to water.
2. Wash in distilled water: 3 minutes.
3. Treat with ferrocyanide–hydrochloric acid, 56°C: 10 minutes.
4. Rinse, several changes distilled water: 5 minutes.
5. Counterstain with safranin: 2–5 minutes, or Kernechtrot: 5–8 minutes.
6. Rinse in 70% alcohol.
7. Dehydrate, clear, and mount.

---

[1] Roboz Surgical Instrument Co.

RESULTS

iron pigment—brilliant greenish blue
nuclei—red
other tissue elements—shades of red and rose

COMMENTS

Hutchison claims that warming the solution is the most important step in this method. Do not leave the slides in the solution longer than 10 minutes. If they have been well washed in distilled water, precipitate seldom forms on the sections. This method produces deep, brilliant colors.

For masked or occult iron (Glick 1949), pretreat slides with 30% hydrogen peroxide alkalized with dilute ammonia (1 drop/100 ml peroxide): 10–15 minutes. Wash well and proceed to step 3. If the hydrogen peroxide tends to form bubbles under the sections and loosen them, keep the solution and slides cool in the refrigerator during the treatment.

I prefer 3% nitric acid in 95% alcohol overnight. Wash in 90% alcohol and proceed as usual. This method does not unmask iron in hemoglobin.

Past (1961) decalcifies osseous tissue in 5% EDTA (ethylenediaminetetraacetic acid) adjusted to pH 12 with N NaOH before staining for iron.

Tartakowsky Method[2]

FIXATION

Place small pieces (3 × 4 mm) 12–24 hours in:

70% ethyl alcohol	95.0 ml
ammonium sulfide $(NH_4)_2S$	5.0 ml

PROCEDURE

1. Transfer pieces to 20 ml absolute ethyl alcohol plus 2 drops ammonium sulfide: 6–12 hours.
2. Rinse very briefly in distilled water.
3. Place in 1.5% aqueous potassium ferrocyanide: 30 minutes, room temperature.
4. Transfer to 0.5% aqueous HCl: 10 minutes.
5. Wash thoroughly with distilled water, 4 changes: 10 minutes each.

---

[2] Personal communication, Dr. P. B. van Weel, University of Hawaii.

6. Dehydrate, embed in Paraplast, and section.
7. Deparaffinize and hydrate to water.
8. Stain in Kernechtrot (p. 247): 10 minutes.
9. Rinse briefly in 10% alcohol. Blot lightly and transfer to absolute alcohol.
10. Complete dehydration, clear, and mount.

RESULTS

iron—blue
nuclei—dark red
cytoplasm—pale red

COMMENTS

Although cytoplasm fixation is not ideal, the treatment is extremely sensitive. Even diffuse iron is stained, but its blue color will fade in time.

Turnbull Blue Method for Ferrous Iron (Pearse 1968)

FIXATION

10% buffered formalin; other general fixatives satisfactory if acid is absent.

SOLUTIONS

Ammonium sulfide

A saturated solution of $(NH_4)_2S$, 20–30% content, analytical reagent.

Potassium ferricyanide

potassium ferricyanide $(K_3Fe(CN)_6)$	20.0 g
distilled water............................	100.0 ml

Hydrochloric acid

hydrochloric acid (sp gr. 1.188–1.192, 37–38% HCl)	1.0 ml
distilled water............................	99.0 ml

Safranin or Kernechtrot (see p. 247).

PROCEDURE

1. Deparaffinize and hydrate slides to water; remove $HgCl_2$.
2. Wash in distilled water.
3. Treat with yellow ammonium sulfide: 1–3 hours.
4. Rinse in distilled water.
5. Treat with equal parts of potassium ferricyanide and hydrochloric acid, freshly mixed: 10–20 minutes.
6. Rinse in distilled water.
7. Counterstain with safranin: 2–5 minutes, or Kernechtrot: 5–8 minutes.
8. Rinse in 70% alcohol.
9. Dehydrate, clear, and mount.

RESULTS

ferrous iron and ferric iron coverted to ferrous iron—deep blue
other tissue elements—shades of rose and red

COMMENTS

The dinitroresorcinol used by Humphrey (1935) gives a disappointingly dull slide. Hukill and Putt's (1962) technique with 3,7-diphenyl-1, 10-phenanthroline (Bathophenanthroline) also gives disappointing results.

Kutlík (1970) stains for 2 hours at room temperature in chlorate hematoxylin (1% hematoxylin in 7.3% aqueous potassium chlorate, filtered). Wash well and counterstain. This forms a black iron hematein lake where iron compounds are present. I have not tried this method, but feel it has promise, particularly when photographs are desired. The solution will keep for 6 months.

BILE PIGMENT (BILIRUBIN) STAINING

Glenner Method (1957)

FIXATION

Use fresh tissue quickly frozen by the Adamstone-Taylor cold knife method (p. 395), or fix tissue in formalin for 6 hours and prepare Carbowax sections.

SOLUTION

> potassium dichromate, 3% aqueous .... 25.0 ml
> buffer, pH 2.2 (0.1 N HCl, 8.0 ml; 0.1 N potas-
>   sium dihydrogen phosphate, 17.0 ml) .... 25.0 ml

PROCEDURE

1. Treat sections with dichromate solution, room temperature: 15 minutes.
2. Wash in running water: 5 minutes.
3. Counterstain with hematoxylin, safranin, or Kernechtrot.
4. Dehydrate rapidly through 95% alcohol: 5–10 seconds.
5. Finish dehydration in absolute alcohol: 1 minute.
6. Clear through absolute alcohol–xylene and xylene.
7. Mount in Permount or the like.

RESULT

bilirubin—emerald green

COMMENTS

The pH of 2.2 is crucial; it must be low enough to result in complete oxidation of all bilirubin, but not so acid that it causes loss of the pigment from the sections. Glenner also includes a method for the demonstration of bilirubin, hemosiderin, and lipofuscin in the same section.

This reaction is recommended in preference to Stein's iodine method (Glick 1949), which cannot be considered reliable at all times because the reactants are diffusible and the final color can spread from the original site. The test is based on oxidation of the bile pigment (bilirubin) to green biliverdin by the iodine solution. The Gmelin method using nitric acid is even less permanent.

Kutlík (1958) and Lillie and Pizzolato (1968) describe argentaffin methods (p. 305). For definite blackening of the bile pigment, Lillie and Pizzolato recommend 5–16 hours exposure to diammine silver at 25°C. The solution is made as follows: to 2.0 ml of 28% NH_4 OH add 35–40 ml of 5% $AgNO_3$, rapidly at first, with swirling to dissolve the precipitate that forms. Then add smaller amounts of $AgNO_3$ until only a faint turbidity persists.

See Lillie and Pizzolato (1968) for a more complete discussion of bile pigments.

Azo-Coupling Method

FIXATION

Formalin.

SOLUTIONS

Buffer solution

Solution A

sodium acetate	1.17 g
sodium barbital (a drug)	2.94 g
distilled water	100.0 ml

Solution B

hydrochloric acid	0.84 ml
distilled water	100.0 ml

Working solution, pH 9.2

solution A	5.0 ml
solution B	0.25 ml
distilled water	17.75 ml

Staining solution

fast red B salt[3]	50.0 mg
buffer, pH 9.2	50.0 ml

PROCEDURE

1. Deparaffinize and hydrate sections to water or cut frozen sections.
2. Stain: 30 seconds.
3. Wash off excess stain in tap water.
4. Stain in hematoxylin: 5 minutes.
5. Wash in tap water: 5 minutes.
6. Dehydrate, clear, and mount.

---

[3] Sigma Chemical Co.

RESULTS

bile pigments—orange red
nuclei—blue

MELANIN AND LIPOFUSCIN STAINING

Lillie (1956*a,b*) **Nile Blue Method**

FIXATION

Any good general fixative.

SOLUTION

Nile blue A (C.I. 51180) (Nile blue sulfate) .... 0.05 g
sulfuric acid, 1% (1 ml conc. H_2SO_4/99 ml
distilled water........................... 100.0 ml

PROCEDURE

1. Deparaffinize and hydrate slides to water: remove $HgCl_2$.
2. Stain in Nile blue A solution: 20 minutes.
3. Wash in running water: 10–20 minutes.
4. Mount in glycerin jelly.

RESULTS

lipofuscins—dark blue or blue green
melanin—pale green
erythrocytes—greenish yellow to greenish blue
myelin—green to deep blue
nuclei—poorly stained

Lillie Alternate Nile Blue Method

PROCEDURE

Steps 1 and 2 as above.
3. Do not wash in water. Rinse quickly in 1% sulfuric acid to remove
excess dye.

4. Dehydrate at once in acetone, 4 changes: 15 seconds each.
5. Clear in xylene and mount.

RESULTS

cutaneous, ocular, meningeal melanins—dark green
mast cells—purple red
lipofuscins—unstained but appear yellow to brownish
muscle, myelin, erythrocytes—unstained
nuclei—greenish to unstained

COMMENTS

Nile blue sulfate stains basophilic tissue and acidic lipids, such as free fatty acids, phospholipids, and certain lipid components of some pigment deposits of a lipogenic character.

Lipofuscins stain with Nile blue by two mechanisms: a fat solubility mechanism operating at pH below 1.0; and an acid–base mechanism operating at levels above pH 3.0. When stained by the second mechanism, they retain a green stain after acetone or brief alcoholic extraction, but when the first mechanism is used, they are promptly decolorized by acetone or alcohol (Lillie 1956b, see also Lillie 1965).

Melanins stain with basic dyes at pH levels below 1.0 and retain the stain when dehydrated and mounted (Lillie 1955).

Ferro-Ferricyanide Method (Lillie 1965)

FIXATION

Avoid chromate fixatives; others are satisfactory.

SOLUTIONS

Ferrous sulfate

ferrous sulfate ($FeSO_4 \cdot 7H_2O$)	2.5 g
distilled water	100.0 ml

Potassium ferricyanide

potassium ferricyanide ($K_3Fe(CN)_6$)	1.0 g
distilled water	99.0 ml
glacial acetic acid	1.0 ml

PROCEDURE

1. Deparaffinize and hydrate slides to water.
2. Treat with ferrous sulfate: 1 hour.
3. Wash in distilled water, 4 changes: total 20 minutes.
4. Treat with potassium ferricyanide: 30 minutes.
5. Wash in 1% aqueous acetic acid: 1–2 minutes.
6. Counterstain if desired: picro-ponceau satisfactory; do not use hematoxylin.
7. Dehydrate, clear, and mount.

RESULTS

melanins—dark green
background—faint greenish or unstained; with picro-ponceau, collagen stains red, and muscle and cytoplasm, yellow and brown

COMMENTS

Lillie says this method is highly selective. No other pigments react in this procedure, except occasionally hemosiderin.

Indophenol Method (Alpert et al. 1960)

FIXATION

Any general fixative, or use fresh-frozen sections.

SOLUTION

sodium-2,6-dichlorobenzenone-indophenol	0.1 g
50% ethyl alcohol	100.0 ml

Prepare just before use. Filter and add 1% aqueous hydrochloric acid until color of solution is red (approximately 1 : 5) or until pH is 2.0.

PROCEDURE

1. Deparaffinize and hydrate slides to water.
2. Treat with HCl-indophenol: 5 minutes.
3. Dip in tap water several times until background loses color.
4. Mount in glycerin jelly or Von Apathy (p. 104).

RESULTS

lipofuscins—red
erythrocytes—blue
other tissue elements—colorless to blue or slightly pink

COMMENTS

Avoid alcohol; it removes color immediately. Stain fades in about 1 day.

STAINING FOR CALCIUM DEPOSITS

Von Kossa Method (Mallory Modification 1944)

FIXATION

10% formalin or alcohol, although other fixatives give reasonably good results.

SOLUTIONS

Silver nitrate

silver nitrate (AgNO₃)	5.0 g
distilled water.............................	100.0 ml

Fresh solution always best; never use one more than 1 week old.

Safranin or Kernechtrot, see p. 247.

PROCEDURE

1. Deparaffinize and hydrate slides to distilled water; remove $HgCl_2$.
2. Treat with silver nitrate in dark: 30 minutes.
3. Rinse thoroughly in distilled water.
4. Expose slides (in distilled water) to bright light (75–100W bulb satisfactory): 1 hour. Lay them over a white background to expedite reaction.
5. Wash thoroughly in distilled water.
6. Counterstain in safranin: 2–3 minutes, or Kernechtrot: 5–8 minutes.
7. Rinse in 70% alcohol.
8. Dehydrate, clear, and mount.

RESULTS

    calcium deposits—dark brown to black
    nuclei and other tissue elements—shades of red and rose

COMMENTS

If the exposure to light in step 4 produces too much brown or yellow background, try developing in a photographic developer: 5 minutes (step 4).

Step 5. Rinse in distilled water and fix in hypo: 5 minutes. Wash well and continue to step 5 above.

If iron blocks out the calcium, treat 10–15 minutes with .005 M sodium EDTA in normal saline. Wash in distilled water and proceed to step 2 (McGee-Russell 1958).

This reaction actually demonstrates phosphates and carbonates rather than calcium itself, but since soluble phosphates and carbonates are washed out, the calcium phosphates and calcium carbonates remain to react with the silver (Lillie 1954b, 1965), and the test can be regarded as sufficiently specific.

Renaud (1959) demonstrates that a high percentage of alcohol (at least 80% content) in the fixing fluid is necessary in order to detect calcium in some tissues (heart and coronary vessels). Water and even buffered formalin can dissolve out some small deposits of calcium salts.

Pizzolato and McCrory (1962) demonstrated that neither chemical reduction nor exposure to illumination was necessary in the Von Kossa technique.

Pickett and Klavins (1961) followed step 5 with an elastin stain (preferably aldehyde-fuchsin) to demonstrate calcium and elastin simultaneously.

Carr et al. (1961) describe a method using chloranilic acid (red brown) to differentiate calcium deposits from carbon deposits.

Calcium can be present in areas of calcification before it can be demonstrated by the Von Kossa metal substitution method. But there are reagents that are considered capable of complexing with calcium, such as antraquinone dyes. See Chaplin and Grace (1976) and the alizarine red S method below.

Alizarine Red S (Puchtler et al. 1969)

FIXATION

Alcohol or Carnoy fixative should be used for best preservation of calcium. Formalin (even buffered) or Zenker-formol remove calcium deposits.

SOLUTION

alizarine red S (C.I. 58005)	0.5 g
phosphate or barbital buffer, pH 9	100.0 ml

PROCEDURE

1. Deparaffinize and hydrate slides to water.
2. Stain in alizarine red: 30–60 minutes.
3. Rinse in buffer, pH 9: 5 seconds.
4. Dehydrate in 3 changes absolute ethanol: 2 minutes each.
5. Clear and mount.

RESULT

calcium—orange red

COMMENTS

The recommendations made by Puchtler et al. (1969) must be observed to obtain reliable results with this method. Proper fixation is important; calcium staining is most intense after alcohol and Carnoy fixation. Formalin and Zenker produce weakly or moderately colored deposits and staining decreases after storage in formalin, disappearing completely after 2–3 weeks. Alizarine red S stains red more reliably in a buffer solution of pH 9 than at other pHs. It stains red without a yellow tinge within 5 minutes, but intensity of color increases with staining times up to an hour. The alizarine dyes in aqueous solution of pH 7 or below become only slightly dissociated with little tendency to form calcium salts. If the solutions are made at pH 9, then many of the 2-OH groups of alizarine become dissociated and the formation of monoalizarate yields the red color for calcium. At a higher pH of 12, almost all of the 2-OH and about half of the 1-OH groups are dissociated and a blue alizarate forms. The sulfonic acid group of alizarine red S also forms a salt with calcium. Alizarine, however, cannot be considered specific for calcium alone; other metals also react with the stain. It is sensitive toward ions of uranium, titanium, bismuth, thallium, zirconium, hafnium, and thorium. The Puchtler article should be read for a better understanding of this stain and its reaction.

REMOVAL OF PIGMENTS

Melanin Pigment

This will appear as brown, grayish, or almost black granules.

Permanganate Method of Removal (Lillie 1965)

1. Hydrate slides to water.
2. Immerse slides in 0.5% aqueous potassium permanganate: 12–24 hours.
3. Wash in running water: 5 minutes.
4. Immerse slides in 1% aqueous oxalic acid: 1 minute.
5. Wash in running water: 10 minutes, and proceed to stain.

Performic or Peracetic Acid Method of Removal

Lillie (1954*b* and 1965) claims that the permanganate method can be unpredictable and that the best bleaching is done with performic or peracetic acid: 1–2 hours:

Performic acid

90% formic acid, aqueous	8.0 ml
30% H_2O_2 (undiluted reagent)	31.0 ml
H_2SO_4, concentrated	0.22 ml

Keep at or below 25°C. About 4.7% performic acid is formed within 2 hours, but it deteriorates after a few more hours.

Peracetic acid

glacial acetic acid	95.6 ml
30% H_2O_2 (undiluted reagent)	259.0 ml
H_2SO_4, concentrated	2.2 ml

Let stand 1–3 days. Add 40 mg disodium phosphate as stabilizer. Store at 0.5°C. Keeps for months in refrigerator.

Chlorate Method of Removal

Treat sections 24 hours in 50% alcohol plus small amount of potassium chlorate and a few drops of HCl (concentrated). Wash 10 minutes before staining.

Bromine Method of Removal

1% bromine in water: 12–24 hours. Wash well before staining.

Chromic Acid Method of Removal

Mixture of equal parts of 1% chromic acid and 5% calcium chloride, aqueous: 8–12 hours. Wash well.

Peroxide Method of Removal

10% H_2O_2: 24–48 hours. Wash well before staining.

The peroxide method is considered by some to be the best; it is specific for melanin; other pigments resist longer than 48 hours. The permanganate and chromic acid methods allow no differentiation between pigments.

Formalin Pigment

Brown and black crystalline granules and artifacts, produced by formalin, are found in and around blood, and are considered a hematein derivative. The crystals are formed when the tissue is fixed with formalin at an acid pH. They usually will not form when the pH of the formalin fixative is above 6.0.

Barrett Method of Removal (1944)

Saturated solution of picric acid in alcohol: 10 minutes to 2 hours.

Malarial Pigment

This appears as amorphous brown-black granules. Its histochemical properties apparently are the same as those of formalin pigment.

Gridley Method of Removal (1957)

Method 1

1. Hydrate slides down to water.
2. Bleach for 5 minutes in:

acetone	50.0 ml
3% H_2O_2	50.0 ml
28–29% ammonia, concentrated	1.0 ml

or overnight in 5% aqueous ammonium sulfide (diluting 20% analytical reagent 4 : 1).
3. Wash thoroughly in running water: 15 minutes or longer.

Method 2

Bleach in 3% H_2O_2: 2 hours.

Carbon

Carbon (opaque black) usually appears in the lungs and adjacent lymph nodes, sometimes in the spleen and liver. It is possible to distinguish it from malarial pigment, iron, or some other pigment or precipitate, since the carbon is black and is insoluble in concentrated sulfuric acid in which other pigments will dissolve.

Hemosiderin

This yellowish, brownish, or greenish brown pigment resists bleaching, does not dissolve in alkalis or acids, and is not argentaffin. It can be identified by Perl's test, p. 246.

Bile Pigments

These are yellowish green, resist bleaching, and are argentaffin. They can be converted by H_2O_2, Lugol solution, nitrous acid, or dichromates into emerald green biliverdin.

Chapter 16
Staining Proteins and Nucleic Acids

PROTEIN STAINING

The proteins found in cells may be classified as either (1) simple proteins, which on hydrolysis yield *a*-amino acids and their derivatives; or (2) conjugated proteins, which yield, in addition to amino acids, nonprotein materials. The simple proteins may be fibrous or globular proteins. Collagens, reticulins, keratin, and fibrin are simple fibrous proteins and insoluble in most aqueous media. The globular proteins—albumins, globulins, globins, and histones—are soluble in aqueous media. Among the conjugated proteins are nucleoproteins, mucoproteins, lipoproteins, and glycoproteins.

Because of their omnipresence in cells, proteins can be disappointing subjects for precise and brilliant localization. Sometimes, however, it does become necessary to ascertain the protein or nonprotein nature of a granulation or another kind of locale in cells. "Protein" techniques exist, but few are specific for proteins as such. Most of these techniques demonstrate a component of the protein (such as one of the amino acid groups), or they are specific for the aromatic nuclei, the phenolic functions, or the guanidine grouping (Gomori 1952; Pearse 1968).

The Millon reaction, one of the classical methods, is not specific for proteins, but for aromatic groups found in the amino acid tyrosine. The method, therefore, is specific for tyrosine, which is in most proteins. Two Millon variations follow.

Millon Reaction (Gomori 1952)

FIXATION

Any good general fixative.

SOLUTIONS

Solution A

mercuric acetate	5.0 g
distilled water............................	100.0 ml
tricloracetic acid	15.0 g

Solution B

sodium nitrite, 1% aqueous	10.0 ml

PROCEDURE

1. Run sections down to water. (Cut sections at least 10μ thick.)
2. Incubate sections in solution A, 30–37°C: 5–10 minutes.
3. Add solution B; incubate another 25 minutes.
4. Rinse in 70% alcohol.
5. Dehydrate, clear, and mount.

RESULT

tyrosine—pink to brick red

Millon Reaction (Romeis 1948)

SOLUTION

Dilute 400 ml of concentrated nitric acid with distilled water to make 1 liter. Let stand 48 hours. Dilute again, 1 part of this solution with 9 parts of distilled water. Add an excess of mercuric nitrate and allow to stand several days with frequent shaking to complete saturation. To 400 ml of filtrate, add 3 ml of the 40% original solution and 1.4 g sodium nitrite.

PROCEDURE

1. Fix, section, and hydrate to water as in the procedure for the first Millon reaction above.
2. Immerse sections in reagent: 3 hours.
3. Give the slides a short dip in 1% nitric acid.
4. Transfer quickly through 70%, 95%, and absolute alcohol, and xylene; mount.

RESULT

tyrosine—rose or red

Mercuric Bromphenol Blue Method (Chapman 1975)

FIXATION

Any general fixative.

SOLUTION

mercuric chloride	1.0 g
sodium bromphenol blue	0.05 g
2% acetic acid aqueous	100.0 ml

PROCEDURE

1. Deparaffinize sections (5μ or less) and hydrate to water.
2. Transfer to bromphenol blue solution: 15 minutes, room temperature.
3. Rinse in 2 changes 0.5% acetic acid, aqueous: 20 minutes total.
4. Blot and dehydrate rapidly in 2 changes absolute alcohol; agitate several times.
5. Transfer to xylene: 2 minutes.
6. Treat with xylene plus 0.5 ml n-butylamine per 100 ml until section turns blue.
7. Clear in 2 changes xylene and mount.

RESULT

This appears to be limited to proteins and peptides that are not removed by washing—blue.

COMMENTS

I cannot claim that this method is valid histochemically, but bromphenol blue continues to be used for proteins.

See Kaniwar (1960) about the uncertain specificity of this method.

The identification of the individual amino acids in tissue sections involves a number of procedures, some of which are controversial and unreliable as to specificity, and some of which are complex. Available techniques include methods for amino groups, carboxyl groups, tyrosine, tryptophane, histidine, the SS and SH groups, and arginine. Only a small number of these can be included here.

Ninhydrin Reaction (Serra 1946)

Ninhydrin (triketo-hydrindene-hydrate) reacts with the a-amino groups, oxidizing them to carbon dioxide, ammonia, and an aldehyde. The reaction is not stable and its products are diffuse, but the aldehydes that are formed will react with Schiff reagent to produce a red color. Glenner (1963) questions the value of this method, since it reacts with both free and bound a-amino acids, but is not limited to them. See also Kasten (1962) and Puchtler and Sweat (1962).

FIXATION

10% formalin.

SOLUTION

phosphate buffer, pH 6.98 (p. 556) .... 50.0 ml
ninhydrin, 0.4% aqueous (triketo-hydrindene-
   hydrate .... 50.0 ml

PROCEDURE

1. Deparaffinize and hydrate slides to water, or use frozen sections.
2. Place slides on a rack over boiling water, cover with stain, and steam for 1–2 minutes.
3. Drain stain off the slide and mount in glycerin jelly.

RESULT

Blue or violet color indicates the presence of amino acids, peptides, and proteins.

COMMENTS

Slides should be examined at once because the color diffuses readily and begins to fade within a day or two.

Ninhydrin-Schiff Reaction (Yasuma and Ichikawa 1953)

FIXATION

Any general fixative, but 10% formalin, Zenker, or absolute alcohol are recommended.

SOLUTIONS

Ninhydrin

ninhydrin.................................	5.9 ml
absolute ethyl alcohol......................	100.0 ml

Schiff reagent, see p. 210.

PROCEDURE

1. Deparaffinize sections. (Use 10μ sections.)
2. Transfer to absolute ethyl alcohol, 2 changes: 2 minutes each.
3. Incubate in ninhydrin solution, 37°C: 5–24 hours.
4. Wash in tap water: 3–5 minutes.
5. Proceed as in Feulgen reaction, p. 213.

RESULT

protein or peptide—red or purple

COMMENTS

See Van Duijn (1961) for acrolein-Schiff reaction, which is based on the reaction of acrolein with tissue compounds. All proteins except arginine-rich protamines are positive. Alloxan also may be used (Yasuma and Ichikawa 1953).

Staining of Histones

Histones are the basic protein of chromatin.

Alfert and Geschwind Method (1953)

FIXATION

10% formalin: 3–6 hours. Zenker, Susa, and Carnoy do not permit specific staining. Wash overnight in running water.

SOLUTIONS

Trichloracetic acid (TCA)

trichloracetic acid	5.0 ml
distilled water	100.0 ml

Fast green

fast green FCF (C.I. 42053)	0.1 g
distilled water	100.0 ml

Adjust pH to 8.0–8.1 with a minimum of NaOH. Prepare fresh.

PROCEDURE

1. Deparaffinize and hydrate to water.
2. Immerse in TCA solution in boiling water bath to remove nucleic acids: 15 minutes.
3. Wash out TCA with 2 changes of 70% alcohol: 10 minutes each. Then rinse in distilled water.
4. Stain in fast green: 30 minutes.
5. Wash in distilled water: 5 minutes.
6. Transfer directly to 95% alcohol, absolute alcohol, and xylene; mount.

RESULT

histone—green

COMMENTS

This is a specific chromosomal stain, and all experiments indicate that histones or protamines are responsible for the histological stain. The stain is not retained unless nucleic acids are removed. The histones occur in a constant quantitative ratio to the DNA in cell nuclei, and neither histones nor DNA vary in amount when cells undergo physiological changes.

Ammoniacal Silver Method (Black and Ansley 1964)

FIXATION

Buffered formalin (p. 15).

SOLUTION

Add 10% aqueous silver nitrate, drop by drop, to 3–4 ml concentrated ammonium hydroxide until the first turbidity occurs. Stir constantly. Ap proximately 38 ml $AgNO_3$ is required for 4 ml of NH_4OH. Prepare fresh.

PROCEDURE

1. Deparaffinize and hydrate to water.
2. Treat with buffered formalin: 15 minutes.
3. Wash in 5 changes of distilled water: several seconds each.
4. Stain in silver solution with gentle agitation: 5–10 seconds.
5. Wash in 5 changes of distilled water: several seconds each.
6. Treat with 3% neutral formalin with agitation: 2 minutes.
7. Wash in 3 changes of distilled water: several seconds each.
8. Dehydrate, clear, and mount.

RESULT

histone—black

COMMENTS

Black et al. (1964) have written an excellent article concerning the cell specificity of histones. Also see Vidal et al. (1971) for a silver stain.

Arginine Reaction

The arginine reaction uses the reaction of naphthol with guanidine derivatives in the presence of hypochlorite or hypobromite to produce an orange-red color. The reaction takes place in guanidine derivatives when one hydrogen atom of one of the amino groups is replaced by alkyl, fatty acid, or cyano radical. Proteins containing arginine give the color. In tissue fixed by ordinary methods the reaction is specific for protein-bound arginine and other guanidine derivatives such as glycocyamine or agmatine. Guanidine, urea, creatine, creatinine, and amino acids other than arginine do not give the color.

Sakaguchi Reaction (Modified by Deitch 1961)

FIXATION

Absolute ethyl alcohol–glacial acetic acid (3 : 1). Formalin fixation may block amino and guanidyl groups and reduce the reaction. Smears may be air-dried and fixed in methyl alcohol.

SOLUTIONS

Barium hydroxide

barium hydroxide	4.0 g
distilled water	100.0 ml

Filter just before use.

Hypochlorite solution

Clorox, freshly obtained (5% solution)	1.0 ml
distilled water	4.0 ml

Naphthol solution

2,4-dichloro-*a*-naphthol[1]	0.15 g
tertiary butyl alcohol	10.0 ml

_____

[1] Eastman Organic Chemicals.

PROCEDURE

1. Deparaffinize and hydrate sections to water. Use sections of 4μ or thinner.
2. Blot and place slides in an empty coplin jar.
3. Pour 5 parts of barium hydroxide solution in a flask. Add 1 part hypochlorite solution and then 1 part of dichloronaphthol, shaking flask during each addition. Pour immediately on slides in jar. Room temperature: 10 minutes.
4. Transfer slides rapidly through 3 changes of tertiary butyl alcohol containing 5% of tri-N-butylamine.[2] Move each slide vigorously in first change for about 5 seconds, transfer through next 2 changes: 30–60 seconds each. Change first solution frequently.
5. Transfer rapidly through 2 changes of xylene containing 5% tri-N-butylamine: 30–60 seconds each.
6. Drain rapidly and mount in Cargille's[3] (formerly Shillaber's) oil containing 10% tri-N-butylamine.

RESULT

arginine—orange to red

COMMENTS

Always keep slides moist. Drying causes a crust of barium carbonate to form. The precipitate on the underside of the slide can be wiped off with dilute acetic or hydrochloric acid.

Deitch has substituted barium for sodium or potassium hydroxide for better tissue morphology, and the sections do not tend to fall off. The addition of an organic amine to the solutions and mountants reduces the fading problem of former methods. See also Baker (1947); Carver et al. (1953); Liebman (1951); and Thomas (1950).

The following references are only a few of the many excellent papers concerning proteins and amino acids. For proteins: Deitch (1955) and Spicer and Lillie (1961). For tyrosine: Glenner and Lillie (1959) and Lillie (1957d). For histidine: Landing and Hall (1956); Pauly (1964); and Reaven and Cox (1963). For histamine: Enerbäck (1969) and Lagunoff et al. (1961). For tryptophane: Adams (1957); Bruemmer et al. (1957); and Glenner and Lillie (1957b). For SH groups: Barrnett and Seligman (1952); Bennett (1951); Bennett and Watts (1958); and Chévremont and Fréderic

---

[2] Eastman Organic Chemicals.
[3] Cargille Laboratories.

(1943). For SS groups: Adams (1956) and Adams and Sloper (1955, 1956). For arginine: Notenboom et al. (1967). For fluorescent labeling of proteins: Rinderknecht (1960).

NUCLEIC ACID AND NUCLEOPROTEIN STAINING

The two types of nucleic acids are: (1) DNA (thymonucleic acid, deoxyribonucleic acid, and deoxypentose ribonucleic acid), the principal components of nuclei; and (2) RNA (ribonucleic acid, and plasmonucleic acid), which is found in the nucleolus and ribosomes of the cytoplasm. Nucleoproteins are combinations of basic proteins and nucleic acids in various highly polymerized forms. The nucleolus is composed largely of concentrated protein, with variable amounts of DNA. Its function is part of the synthetic processes of the cell (Love and Rabotti 1963; Vincent 1955).

On hydrolysis, the nucleic acids yield: (1) purine and pyrimidine bases, (2) carbohydrates, and (3) phosphoric acid. DNA contains deoxyribose (deoxypentose) sugar, and RNA contains ribose (pentose) sugar.

Because the nucleic acids exist in many states of polymerization, fixation will cause some redistribution. Organic solvents, formalin, and most acid fixatives, however, will not cause significant losses from the cell. Pearse (1968) prefers alcohol and acetic acid to formalin for nucleic acids and nucleoproteins, but Kurnick (1955a) says neutral (pH 7.0) 10% formalin appears to be satisfactory. Pearse warns against long exposure if Carnoy fixative is used, since it causes the extraction of first RNA and then DNA. He recommends Lillie's acetic-alcohol-formalin, 4°C: 24 hours.

40% formalin, concentrated	10.0 ml
glacial acetic acid	5.0 ml
absolute ethyl alcohol	85.0 ml

Pure nucleic acids can be stained with basic dyes, but nucleic acids in living cells are bound with protein and will not react with basic dyes. Inactivation of the protein-bound amino groups by formalin or deamination (p. 280) will increase the basophilia of the nucleic acids by releasing increasing numbers of nucleic acid phosphoryl groups to bind the basic dye. By using different exposures to these reagents, Walsh and Love (1963) demonstrated several nucleoproteins. (See also Love 1957, 1962; Love and Liles 1959.)

For specific staining of nucleic acids, the Feulgen technique is considered specific for DNA (p. 213). It is thought to depend on the uncovering of aldehyde groups of deoxyribose by the hydrolysis of purine-deoxyribose bonds by hydrochloric acid. The released aldehyde gives the Schiff reaction. Proper hydrolysis is critical. Two actions occur almost simultaneously: (1) The purine bases are rapidly removed and the aldehyde groups uncovered; (2) the histones and nucleic acids are progressively removed. At the beginning, the first action predominates; as hydrolysis proceeds, the second action begins to take over and staining with Schiff will decrease. Overhydrolysis can produce a negative Feulgen reaction. The optimum time is in the area of 8 minutes; the range is 6–12 minutes.

In methyl green–pyronin staining for nucleic acids, the two dyes are used to distinguish between polymerization of the nucleic acids, not between the acids as such. Methyl green is readily bound by highly polymerized DNA; the binding is acquired at two sites, two amino groups of the dye with two phosphoric acid groups of the DNA. Depolymerizing treatment, such as heat and certain fixatives (Zenker, picric acid), must be avoided. If formalin is used, it must be accurately adjusted to pH 7.0 and fixation must be brief, so that the solution does not become acid. Pyronin, for RNA staining, stains low polymers of both nucleic acids, but it does permit methyl green to compete effectively for the highly polymerized DNA to the exclusion of the pyronin. Nuclei will stain red with pyronin alone, but they will stain green when the two dyes are used together. Control slides should be used with all methyl green–pyronin staining. If one slide is exposed to ribonuclease and one is not, and then both are stained with methyl green–pyronin, the material stained red with pyronin and removable with ribonuclease is considered RNA, and that which is not removable is not RNA. (See Kasten et al. 1962, Tepper and Gifford 1962.)

The pages of literature are full of techniques, theories, controversies, and rationalizations for the demonstration of the nucleic acids and related components of the cell. Obviously they cannot all be included here, but references to some of them are listed below.

Spicer (1961) reports that Bouin fixes RNA as well as or better than any of a number of fixatives and yields DNA without acid hydrolysis. By staining at controlled pH (RNA stains more strongly at low pH) with Schiff reagent for DNA and methylene blue for RNA, the chromatin stains red and the cytoplasmic RNA stains blue.

Kurnick (1952) uses 1% toluidine blue in 95% alcohol as a rough screen test to demonstrate areas of high concentration of nucleic acids. By comparing an unextracted slide with a preparation made after extraction of the nucleic acids, one may differentiate the basophilia caused by nucleic acids from that caused by other components of the cell.

Daoust (1964) uses 0.1% toluidine blue in veronal buffer, pH 5.0 (5 minutes); dehydrates in 3 changes of tertiary butyl alcohol (1 minute each); clears; and mounts.

Other stains used for nucleic acids and nucleoproteins are these: azure A (Anderson 1961; Flax and Pollister 1949); Biebrich scarlet (Spicer 1962); cresyl violet (Ritter et al. 1961); methenamine silver (Korson 1964); methylene blue (Deitch 1964); fluorescent methods (Armstrong 1956; Armstrong et al. 1957; deBruyn et al. 1953); Feulgen-oxidized tannin-azo technique (Malinin 1961); trimethylthionin basic fuchsin (Menzies 1963a).

Aceto-orcein and acetocarmine are popular chromosome stains (p. 474); the orcein probably depends on DNA and the carmine on protein. Hematoxylin used with an alum mordant probably stains chromatin specifically and may depend on the state of polymerization of DNA.

Tagging with ^{14}C and ^{32}P can be used for autoradiography.

Nucleoli can be "stained" with silver (Tandler 1955).

For general discussion about nucleic acids and nucleoproteins, see Gomori (1952); Kurnick (1955b); and Pearse (1968).

Pyronin–Methyl Green for Nucleic Acids (Kurnick 1952)

FIXATION

Absolute alcohol, Carnoy, or cold acetone. If formalin is used, it must be adjusted to pH 7.0 and used only briefly before the solution turns acid and produces a faint green staining of cytoplasm. Picric acid depolymerizes DNA and shows no green chromatin.

SOLUTIONS

Methyl green

methyl green (C.I. 42590)	0.2 g
0.1 M acetate buffer (pH 4.2) or distilled water	100.0 ml

Before making the solution, purify the methyl green by extraction with chloroform. Add approximately 10 g methyl green to 200 ml of chloroform in a 500 ml Erlenmeyer flask and shake. Filter with suction and repeat with smaller amounts of chloroform until the solution comes off blue green instead of lavender (usually requires at least 3 extractions). Dry and store in stoppered bottle. Purified dye is stable.

To detect crystal violet contaminant, place a drop of methyl green solution on a piece of filter paper. Hold the stained area over ammonia solution. The green color disappears, and violet color, if present, will remain (Kasten and Sandritter 1962).

Pyronin

Pyronin Y saturated in acetone, or for lighter color in 10% acetone. The British Pyronine Y for RNA (G. T. Gurr or Edward Gurr) is recommended.

PROCEDURE

1. Deparaffinize and hydrate slides to water.
2. Stain in methyl green solution: 6 minutes.
3. Blot and immerse in *n*-butyl alcohol: several minutes in each of 2 changes.
4. Stain in pyronin: 30–90 seconds. Shorten to a few dips if stain is too dark.
5. Clear in cedarwood oil, xylene, and mount.

RESULTS

RNA containing cytoplasm—red
nucleoli—red
chromatin—bright green
erythrocytes—brown
eosinophilic granules—red
cartilage matrix—green
osseous matrix—pink with trace of violet

COMMENTS

These two dyes distinguish between states of polymerization of nucleic acids, not between the acids themselves. Highly polymerized nucleic acid stains with methyl green; low polymers of both DNA and RNA stain with the pyronin. The usual techniques result in too pale a methyl green, or, if excess staining is tried, water rinses remove the methyl green and acetone removes the pyonin. Dehydrating with ethyl alcohol removes most of the color; isopropyl and tertiary butyl are no improvement. Kurnick developed the above pyronin–methyl green method when he found that *n*-butyl alcohol differentiated the methyl green but removed the pyronin.

Since the pyronin was lost anyway, he therefore decided to leave it out of the first solution and tried following the methyl green with pyronin saturated in acetone. (Because the acetone must be free of water, always use a fresh solution.)

Pyronin–Methyl Green for Nucleic Acids (Elias 1969, 1971).

FIXATION

Carnoy.

SOLUTION

methyl green GA[4]	0.5 g
Walpole acetate buffer, pH 4.1 (p. 553)	100.0 ml
pyronin GS or Y	0.2 g

PROCEDURE

1. Hydrate slides to water.
2. Stain, 37°C: 1 hour.
3. Rinse in ice-cold distilled water: 1–2 seconds.
4. Blot.
5. Rinse, agitating vigorously in tertiary butanol.
6. Dehydrate in 2 changes of tertiary butanol: 5 minutes each.
7. Clear and mount.

COMMENTS

The Elias method gives a consistent, qualitative, nucleic-acid differentiation. The ice water differentiation rinse is critical, and tertiary butanol is superior to *n*-butanol.

The critical factors in this method are the use of Carnoy fixative, a staining solution of low pH, and ice-cold distilled water rinse, followed by tertiary butanol dehydration.

Ahlqvist and Andersson (1972) recommend a low pH (3.8), formalin fixation, and purify both the methyl green and pyronin. Their stain solution is made up in a Walpole buffer at pH 3.8.

See also d'Ablaing et al. (1970)

---

[4] Chroma Gesellschaft. Roboz Surgical Instrument Co.

Thionin–Methyl Green for Nucleic Acids (Roque et al. 1965)

FIXATION

Fix thin (2 mm) pieces in 4% formaldehyde in 1% sodium acetate: 3 hours. Dry smears fixed in methyl alcohol can be used.

SECTIONING

Cut 2–4μ paraffin sections.

SOLUTION

methyl green (C.I. 42590)	0.1	g
thionin (C.I. 52000)	0.0165	g
citrate buffer, 0.01 M, pH 5.8 (0.01 M HCl,		
42 ml; 0.01 M sodium citrate, 58 ml)	100.0	ml

Dissolve the thionin in a little water and add the buffer and methyl green. Shake well and filter. This solution is best when freshly mixed. The methyl green should be chloroform-extracted (p. 274).

PROCEDURE

1. Deparaffinize and hydrate sections to distilled water. Smears can go directly into water.
2. Stain in methyl green–thionin, 40°C: 30 minutes.
3. Rinse in distilled water.
4. Dehydrate in 3 changes of alcohol mixture; 3-butyl alcohol, 80.0 ml; absolute ethyl alcohol, 20 ml: 30 seconds in first change, 3 minutes each in second and third changes.
5. Rinse in absolute ethyl alcohol, clear in xylene, and mount.

RESULTS

chromatin—green or blue green
nucleolar and cytoplasmic basophilic substance—red or reddish purple

COMMENTS

Objection to the specificity of the pyronin–methyl green method centers around the unsuitable staining properties of pyronin. Roque et al. tried various basic dyes in an effort to find a suitable substitute for pyronin.

Thionin gave the best results; it did not obscure the green coloring of chromatin and was less readily extracted by aqueous and alcoholic rinses. Also it stained more selectively than pyronin.

Use a short fixation time, no more than 3 hours. Zenker, Helly, and Carnoy can be used, but the staining is less intense, although the color contrasts are satisfactory.

A range of pH of 6–4 is satisfactory, but the color varies with the type of buffer, becoming blue when acetate or phosphate buffers are used, and green when a citrate buffer is used.

Thin sections, 2–4μ, should be cut for best color contrasts; if the RNA content is very low, sections of 4–6μ may be necessary.

Nucleic acids are stained deep blue with gallocyanin–chrome alum, p. 127. Also see Dutt (1974).

See Mundkur and Brauer (1966) for selective staining of nucleoli.

CONTROL SLIDE TECHNIQUES

In some techniques the reaction is so specific that controls are unnecessary. But in others controls are essential in order to separate a genuine reaction from one similar in appearance but not specific in nature. An essential ingredient may be omitted from a solution used in the control. This is the common procedure in histochemistry, in which the specific substrate is omitted from the incubating mixture used for the control. In other methods, inactivators, inhibitors, or extractives are used.

Extraction Techniques

In the demonstration of the specificity of a reaction of nucleic acids, the extraction of one or both of the nucleic acids is frequently essential. The original preparation can then be compared with another preparation that has been subjected to extraction techniques.

Removal of Both Nucleic Acids

Hot 5% trichloracetic acid, 90–98°C: 30 minutes.

Removal of RNA

1. 5% trichloracetic acid, 60°C: 30 minutes.
2. 10% perchloric acid, 4°C: 12 hours. The cold temperature is critical (Aldridge and Watson 1963; Kasten 1965).
3. 1 N hydrochloric acid, 60°C: 6–12 minutes (Feulgen reaction, p. 213).
4. Ribonuclease (RNAse).

RNAse (crystallized, saltfree)[5]	1.0 mg
distilled water...........................	1.0 ml

Incubate sections, 40°C: 4 hours.

RNAse usually removes nucleolar and cytoplasmic RNA, but chromosomal RNA may be resistant.

Removal of DNA (Deoxyribonuclease [DNAse] method, Daoust 1964)

DNAse (crystallized)[4]	0.05 mg
0.1 M tris maleate buffer, pH 6.5, containing	
$MgSO_4 \cdot 7H_2O$ at a concentration of 0.2 M ..	1.0 ml

Incubate, 37°C: 24 hours.

Amano (1962) claims the results are poor if the slides are incubated in a horizontal position with a few drops of the enzyme. He recommends putting the slides in a coplin jar with 40 ml of DNAse solution.

Tandler (1974) method

1. Transfer slides from water into a 10% solution of formalin in saturated picric acid (10 ml concentrated formaldehyde: 90 ml saturated aqueous picric acid): 12–24 hours. (Bouin fixative can be used.)
2. Wash in 2 changes distilled water: 2 minutes total.
3. Treat with 25% acetic acid aqueous: 2 minutes.
4. Treat with 10% aniline in 25% acetic acid: 1 hour.
5. Wash in 2 changes 25% acetic acid: 2 minutes total.
6. Wash well in distilled water and proceed to stain.

---

[5] Worthington Biochemical Corporation.

Blocking Methods

A blocking method may become an essential part of a technique, not because it stains a component, but because it prevents its staining and thereby verifies that a certain chemical group is responsible for the staining reaction.

Blocking Protein End Groups, Reactions for Amines

Deamination

In this reaction nitrous acid blocks amino groups, most rapidly the *a*-amine groups; it converts amino groups into hydroxyl groups. Van Slyke reagent (Lillie 1954*b*) is best for this.

concentrated sodium nitrite (6 gm/10 ml water)	10.0 ml
glacial acetic acid	5.0 ml
distilled water..............................	25.0 ml

Use at room temperature: 1–12 hours (10 minutes for blood films).

Acetylation (Lillie 1954*b*, 1958, 1964*a,b*)

Acetylation produces acyl derivatives of primary and secondary amines and amine compounds which prevent 1,2 glycols from reacting with PAS. Acetylation and deacetylation are most commonly used for controls with PAS for mucopolysaccharides, even if the validity of the control is not absolutely infallible.

1. Take sections to absolute alcohol.
2. Dip in pyridine.
3. Transfer to following solution, 25°C: 1–24 hours, or 58°C: 30 minutes–6 hours.

acetic anhydride	16.0 ml
anhydrous pyridine	24.0 ml

   (The acetylation of glycols is progressive but slow, and it may require 24–48 hours for a complete blockage of PAS. It can be somewhat accelerated by adding 0.4–0.5% by volume of concentrated sulfuric acid.)
4. Wash in 2 changes each of absolute, 95%, and 80% alcohols. Proceed to method for end groups under investigation.

Rapid acetylation and complete blockage of PAS can be accomplished by use of the following for 4 minutes:

acetic anhydride	25.0 ml
glacial acetic acid	75.0 ml
sulfuric acid	0.25 ml

Alcohol acetylation of phenolic groups (Lillie 1964b) can be done in a mixture of equal volumes of acetic anhydride and absolute ethyl alcohol. Use at 23–25°C: 30 minutes.

Benzoylation (Lillie 1954b; Barnard 1961)

Benzoylation is similar to acetylation, and steps 1 and 2 are the same.

3. Transfer to following, 25°C: 1–24 hours; or 58°C: 30 minutes–6 hours.

benzoyl chloride	2.0 ml
anhydrous pyridine	38.0 ml

4. Wash in alcohols as in step 4 above and proceed to method for group under investigation.

Barnard's method requires anhydrous conditions throughout benzoylation, room temperature, in a desiccator: 3 hours.

dry acetonitrile	50.0 ml
benzoyl chloride	4.2 ml
anhydrous pyridine	2.2 ml

Deacetylation (Saponification)

Acetylation blockage is almost completely reversed by saponification, which is effected by placing sections in the following for 20–30 minutes:

potassium hydroxide	1.0 g
70% ethyl alcohol	100.0 ml

and then treating for 16–24 hours in

absolute ethyl alcohol	70.0 ml
distilled water	10.0 ml
ammonia, 28% concentrated	20.0 ml

Keep solution in closed container during action. Wash in 80% alcohol and water and proceed to method under investigation.

Blocking Aldehydes

Aniline–acetic acid (Lillie and Glenner 1957), room temperature: 20–30 minutes.

glacial acetic acid	45.0 ml
aniline	5.0 ml

If this fails to block a Schiff reaction, aldehydes are excluded.

If aldehydes are made to combine with sodium sulfite or phenylhydrazine, the reaction with Schiff after hydrolysis (Feulgen) or oxidation (PAS) can be blocked (Pearse 1960).

Phenylhydrazine

phenylhydrazine	5.0 ml
glacial acetic acid	10.0 ml
distilled water to make	50.0 ml

Treat at 60°C: 2–3 hours.

Sodium bisulfite

absolute ethyl alcohol	10.0 ml
50% aqueous sodium bisulfite	40.0 ml

Treat at room temperature: 2–8 hours.

Blocking Carboxyl Groups or Other Acid Groups

This reverses metachromasia of most components and blocks basophilia of nucleic acids; it abolishes reactivity of phosphoryl, carboxyl, sulfate groups of polysaccharides, and carboxyl groups of proteins.

Methylation (Lillie 1954*b*, 1958)

0.1 N HCl (8 ml/1000 ml) made up in methyl alcohol, 25°C: 2–3 days. Use screw-cap jars for the procedure. A temperature of 58°C is required to block the PAS of glycogen, epithelial gland mucins, thyroid colloid, collagen, and reticulin.

1% HCl in methyl alcohol is required to block nuclear and cytoplasmic staining with azure A, and it destroys the reactivity of argentaffin cells in the methenamine silver procedure.

Geyer (1962) says that methanol–thionyl chloride is more effective than methanol-HCl under some conditions. With 100 ml of absolute methanol in an Erlenmeyer flask, add 4 ml thionyl chloride very slowly down a wall of the flask, held at a slant and away from the face. Use transparent thionyl chloride. Mix immediately before use. Use at 56°C: 30 minutes–2 hours.

Demethylation (Saponification)

Saponification is used to deblock tissue: that is, to return it to the original active form that was blocked by one of the blocking methods. However, it is not always successful in reversing the effect of methylation.

Cover sections with thin nitrocellulose and place in 1% potassium hydroxide in 70% ethyl alcohol, 25°C: 30 minutes. This restores reactivity of methylated sulfonic acid and of carboxyl and phosphoryl groups, but not that of sulfate groups.

Blocking Tryptophane and Cystine

Performic acid oxidation (Pearse 1960) is specific for the indole ring of tryptophane.

98% formic acid	40.0 ml
30% H_2O_2 (must be fresh)	4.0 ml
sulfuric acid	0.5 ml

Use within 8 hours and stir vigorously before using, room temperature: 15–60 minutes.

Peracetic acid reagent, see p. 259.

Blocking Tyrosine

Treat with 1% 2,4-dinitrofluorobenzene (DNFB) in 90% ethyl alcohol containing 0.01 M NaOH, room temperature: 16–20 hours.

Lillie (1968) and Pearse (1968) include detailed discussions of blocking methods.

Chapter 17
Staining Lipids and Carbohydrates

LIPIDS

Lipids (lipoids, fatty substances) include a large number of substances that have been grouped together because of their solubility properties. They are insoluble in water and soluble in fat solvents such as alcohol, ether, chloroform pyridine, benzene, and acetone. Classification of the lipids is not undertaken here, but a few familiar groups can be mentioned: carotenoids (vitamin A), fatty acids, triglycerides (acetone soluble) (neutral fats), phospholipids, cerebrosides, sterols, and lipid pigments (essentially insoluble in acetone) (lipofuscins—p. 245).

For the fixation of fats, formalin is best, particularly if 1% of calcium chloride is added to make the phospholipids insoluble (Gomori 1952). Because of the use of fat solvents, the tissue cannot be embedded in paraffin or nitrocellulose, but it can be embedded in Carbowax (Rinehart and Abul-Haj 1951b), polyvinyl alcohol (Feder 1962; Masek and Birns 1961), polyethylene glycol (Hack 1952), or ester waxes (p. 450). Frozen sections are simpler to make and are the most frequently used. During any processing, alcohol higher than 70% must be avoided. (For the fixation of lipoids, see Elftman 1958.)

The dying of lipids is one of the simplest forms of staining: The coloring agent merely dissolves in a fluid contained within the tissues. In addition, it should be emphasized that a dye solvent that does not dissolve the lipid itself should be used. The dye, therefore, must meet certain requirements:

1. The dye must be strongly colored.
2. It must be soluble in the substance that it is intended to show, but it must not be soluble in water, the major constituent of cells.

3. It must not attach itself to any tissue constituents except by solution.
4. It must be applied to tissues in a solvent that will not dissolve the substance to be dyed, and it must be less soluble in the solvent than in the substance.

Baker (1958) suggests the name *lysochromes* for these dyes that dissolve in the tissue elements to be colored. The name is derived from the Greek *lúsis*, meaning solution. The Sudans were the first synthetic dyes of this sort, followed by the Nile blues and reds.

These dyes are used in saturated solutions and often create the problem of dye precipitate on the tissue. Vlachos (1959) makes the sensible proposal that the precipitate is probably formed by the solution becoming oversaturated and that perhaps a saturated solution is unnecessary. He suggests two alternatives:

1. Make up the concentration below the saturation point; for example, 0.25 g Sudan IV per 100 ml 60% alcohol.
2. Desaturate the solution either by dilution of a saturated solution with equal volumes of 60% alcohol, or by refrigeration. When the solution has been refrigerated long enough to have acquired the refrigerator temperature, filter. Neither method alters the staining quality of the solution, and the amounts of precipitate produced are negligible.

Lipid Staining

Oil Red O for Lipids

FIXATION

10% formalin or other aqueous general fixatives.

SOLUTION

oil red O (C.I. 26125)	0.7 g
absolute isopropanol	200.0 ml

Shake; leave overnight. Filter. Dilute 180 ml of oil red O solution with 120 ml of distilled water. Leave overnight at 4°C. Filter. Let stand 30 minutes and filter. Ready for use. Stable for 6–8 months.

PROCEDURE

1. Mount frozen sections (15μ) on subbed slides. Allow to dry: 5–10 minutes. Loose, unmounted sections also may be stained this way.
2. Rinse in 60% isopropanol: 30 seconds.
3. Stain in oil red O: 10 minutes.
4. Rinse in 60% isopropanol: few seconds.
5. Wash in running water: 2–3 minutes.
6. Stain in Mayer hematoxylin (or the like): 2–3 minutes.
7. Wash in tap water: 3 minutes.
8. Blue in Scott solution (p. 545): 3 minutes.
9. Wash in tap water: 5 minutes. Mount in gum syrup or glycerin jelly. For permanency, ring cover glass.

RESULTS

lipids—orange red or brilliant red
nuclei—blue

COMMENTS

Gurr Aquamont (p. 108) can be used as a mountant for oil red O sections and does not require sealing.

This is by far the best of the oil red O methods that I have used. It never leaves a stain precipitate on the sections.

For a blue coloration, oil blue N (C.I. 61555) can be used, but the color is not as intense as that of oil red O. Use a red nuclear stain.

Sudan IV or Sudan Black B (Chiffelle and Putt 1951)

FIXATION

10% formalin, no alcohol.

SOLUTION

Dissolve 0.7 g Sudan IV (C.I. 26105) or Sudan black B (C.I. 26150) in 100 ml propylene or ethylene glycol. Add small amounts at a time, and heat to 100–110°C, stirring, for a few minutes. Do not exceed 110°C or a gelatinous suspension is formed. Filter hot through Whatman no. 2 paper and cool. Filter again through fritted glass filter with the aid of suction, or through glass wool by vacuum.

PROCEDURE

1. Frozen sections, 15μ, into water.
2. Place in pure propylene or ethylene glycol: 3–5 minutes, 2 changes. Agitate.
3. Stain, 2 changes: 5–7 minutes each, agitate occasionally.
4. Differentiate in glycol and water (85 : 15): 3–5 minutes. Agitate.
5. Wash in distilled water: 3–5 minutes.
6. Counterstain in hematoxylin if desired.
7. Mount in glycerin jelly.

RESULT

lipids—Sudan IV, orange to red; Sudan black B, blue black to black

COMMENTS

Sudan black B is considered the most sensitive of the lipid dyes.

Chiffelle and Putt recommend glycol as a perfect solvent for a fat stain because it does not extract lipids.

Zugibe et al. (1958, 1959) suggest Carbowax 400 as a solvent for oil red O and Sudan IV.

Gomori (1952) questions the use of glycols because they are solvents for so many water-insoluble substances, and he prefers triethylphosphate. It has a low volatility and is harmless to lipids. His method follows.

1. Frozen sections are rinsed in water and transferred into 50% alcohol: few minutes.
2. Stain in a saturated, filtered solution of any of the fat dyes in 60% triethylphosphate: 5–20 minutes.
3. Differentiate in 50% alcohol: 1 minute.
4. Counterstain in hematoxylin or any preferred stain.
5. Mount in glycerin jelly.

Osmic Acid (Mallory 1944)

Oleic acid or triolein react with the osmic acid by oxidation of their double bonds.

FIXATION

10% formalin, no alcohol.

SOLUTION

> osmic acid (osmium tetroxide) ....   1.0 g ampule
> distilled water ....   100.0 ml

With file, score a circle around ampule and drop it into bottle with the distilled water. Several sharp shakes will break the ampule and allow the water to dissolve the crystals. This method eliminates the possibility of breathing the fumes. However, new packaging methods have virtually eliminated this problem.

PROCEDURE

1. Frozen sections, 15μ, into water.
2. Transfer to osmic acid solution: 24 hours.
3. Wash in several changes distilled water: 6–12 hours total.
4. Treat in absolute alcohol: 4–5 hours, for secondary staining of fat.
5. Wash well in distilled water: 5 minutes or more.
6. Mount in glycerin jelly.

RESULTS

lipids—black
background—yellowish brown

Naphthalene Yellow (Sills and Marsh 1959)

Naphthalene yellow in 60% acetic acid was used to restore the yellow color to the fat of formalin-fixed gross specimens. This cannot be used for sections; the color is too pale.

Fluorescent Method (Metcalf and Patton 1944; Peltier 1954)

FIXATION

10% formalin (salts of heavy metals—$HgCl_2$—have a quenching effect on fluorescence), or use fresh tissue.

SOLUTION

phosphine 3R[1]	0.1–1.0 g
distilled water............................	100.0 ml

PROCEDURE

1. Cut frozen sections, 10μ.
2. Wash in distilled water.
3. Stain in phosphine solution: 5 minutes.
4. Rinse quickly in distilled water.
5. Mount in glycerin or examine as a water mount.

RESULT

lipids (except fatty acids, cholesterol, and soaps) will fluoresce brilliant white

Sudan Black B Method for Phospholipids (Elftman 1957*b*)

FIXATION

mercuric chloride, 5% aqueous	100.0 ml
potassium dichromate.....................	2.5 g

Adjust pH to 2.5 with HCl. Mix fresh. Fix for 3 days; oxidizes phospholipids and decreases their solubility in fat solvents such as are used in paraffin embedding. Nonetheless, leave tissues for short periods in all solvents.

SOLUTIONS

Sudan black B, see p. 286.

Ethylene glycol solution

ethylene glycol	85.0 ml
distilled water............................	100.0 ml

---

[1] Roboz Surgical Instrument Co.

PROCEDURE

1. Deparaffinize and transfer slides into absolute alcohol: 2 minutes.
2. Transfer to absolute ethylene glycol.
3. Stain in Sudan black solution: 30 minutes.
4. Differentiate in 85% ethylene glycol: 2–3 minutes.
5. Wash in distilled water.
6. Counterstain, if desired.
7. Wash well, mount in von Apathy's gum syrup or in glycerin jelly.

RESULT

phospholipids—black

COMMENT

Phospholipids are increasingly important in the study of mitochondria and Golgi apparatus, and are an important constituent of myelin.

Roozemond (1969) recommends fixing phospholipids in 4% formalin with 1% of calcium chloride added and then cutting frozen sections: fixation at 0°C reduces the breakdown of lipids. He suggests that a strong interaction between calcium and phospholipids may aid in the retention of the phospholipids when they are surrounded by a calcium-containing medium.

Acid Hematein for Phospholipids (Hori 1963)

FIXATION

Formalin-calcium (10% formalin/1% calcium chloride), formalin-calcium-cadmium (1% cadmium chloride added to above), or glutaraldehyde-formalin-calcium chloride (2.5 : 25 : 1) in 0.1 M sodium cacodylate, pH 7.0, (p. 387): 18 hours.

SOLUTIONS

Acid hematein

hematoxylin	50.0 mg
0.01% sodium iodate	50.0 ml

Heat until it begins to boil, cool, and add:

glacial acetic acid .... 1.0 ml

Prepare fresh.

Borax-ferricyanide

borax ($Na_2B_4O_7 \cdot 10H_2O$) .... 250.0 mg
potassium ferricyanide [$K_3Fe(CN)_6$] .... 250.0 mg
distilled water............................. 100.0 ml

Keeps well in refrigerator.

PROCEDURE

1. Cut frozen sections.
2. Chromatize sections in 5% aqueous potassium dichromate, 60°C: 4 hours. Transfer sections with a glass rod.
3. Wash in several changes of distilled water.
4. Stain in acid hematein, 37°C: 30 minutes.
5. Wash briefly in distilled water.
6. Differentiate in borax-ferricyanide, 20–25°C: 18 hours.
7. Dehydrate, clear, and mount.

RESULT

phospholipids and some proteins—blue, blue gray or blue black.

COMMENTS

This is a simplification of Baker's (1946) original acid hematein test for phospholipids. If it is desired to distinguish between phospholipids and certain proteins that also give a positive reaction in this test, apply Baker's pyridine extraction method:

1. Fix small pieces of tissue in weak Bouin solution: 20 hours.

picric acid, saturated aqueous .... 50.0 ml
formalin, concentrated .... 10.0 ml
glacial acetic acid .... 5.0 ml
distilled water............................. 35.0 ml

2. Transfer to 70% alcohol: 1 hour.
3. Transfer to 50% alcohol: 30 minutes.
4. Wash in running water: 30 minutes.
5. Transfer to pyridine, room temperature, 2 changes: 1 hour each.
6. Transfer to pyridine, 60°C: 24 hours.
7. Wash in running water: 2 hours.
8. Transfer to step 2 of acid hematein method.

RESULTS

phospholipids—unstained
nuclei—stain after extraction, not before extraction
erythrocytes—stain before and after extraction

COMMENTS

Luxol fast blue can be used for phospholipids, see p. 197.

For unsaturated lipids, see Norton et al. (1962); for lipoproteins, "masked" lipids, see Carmichael (1963); Holczinger and Bálint (1961); Serra (1958).

Chromatin makes phospholipids nonextractable; this may be due to the formation of complexes between the phospholipid and chromium ions. Formol-calcium is preferred as the primary fixative when post chromation is used. The calcium aids in preventing leaching of the phospholipids.

Ellender and Lojda (1973) recommend a modification of Weigert hematoxylin for phospholipids.

1. Postfix fresh frozen sections in formol-calcium.
2. Stain in 3 parts A and 1 part B, freshly mixed: 6–8 minutes.

Solution A

distilled water............................	298.0 ml
HCl concentrated	2.0 ml
ferric chloride ($FeCl_3 \cdot 6H_2O$)	2.5 g
ferric sulfate ($FeSO_4 \cdot 7H_2O$)................	4.5 g

Solution B

distilled water............................	100.0 ml
hematoxylin	1.0 g

Dissolve with low heat.

3. Wash, running water: 5 minutes.
4. Differentiate in 0.2% HCl.
5. Wash, running water: 10 minutes.
6. Dehydrate, clear, and mount.

CARBOHYDRATES (SACCHARIDES)

Polysaccharides are a complex group of substances that are found throughout animal tissues. Complete agreement in nomenclature is still lacking, but in general they are classified as follows (Pearse 1968; Zugibe 1970).

Polysaccharides (glycans) include glycogen, starch, and cellulose. Glycogen is the only naturally occurring member of this group that remains in animal tissue after aqueous fixation and embedding.

Polysaccharide-protein complexes include (1) acid mucopolysaccharides—hyaluronic acid (Wharton's jelly), chrondroitins (cartilage), and heparin; (2) neutral mucopolysaccharides—chitin; (3) mucoproteins—lens capsule and vitreous humor, part of epithelial, glandular, and ductal mucins.

Glycoproteins and glycopolypeptides—ovomucoid and salivary gland mucoid.

Glycolipids—cerebrosides and gangliosides.

Glycolipid-protein complexes—ox brain mucolipid.

Certain specific stains and aldehyde reactions are commonly used to demonstrate carbohydrates. The aldehyde groups must first be liberated by some chemical agent, either oxidized (periodic or chromic acid) or hydrolyzed (dilute hydrochloric acid). Then the specific reagent for aldehydes can be applied. See the Feulgen and periodic acid–Schiff reactions, p. 210.

Alcoholic mixtures are usually recommended as fixatives for saccharides. Most conventional fixatives cause glycogen to migrate to one side of the cell. Gomori (1952) theorized that when glycogen is enclosed in a complex mixture of proteins and lipids, a good protein precipitant may coat the glycogen with a protein membrane. This will be impermeable to the large glycogen molecules and keep them in situ. The fixative should act and harden rapidly. Alcohol or acid fixatives form glycogen in coarse droplets, and fixatives containing formalin distribute the glycogen in fine granules. In some tissues, such as liver, unless the tissue and fixative are chilled, enzymes can cause a loss of glycogen. Although in most tissues glycogen is more stable, Gomori recommended placing the tissue in fixative in the refrigerator. Leske and von Mayersbach (1969) have demonstrated that block fixation does not safely preserve glycogen. Sections should be cut fresh frozen, and fixed in Carnoy fixative or 2.5–10.0% aqueous trichloracetic acid: 5–15 minutes. Murgatroyd (1971) reports that cold formal alcohol or cold Rossman fluid preserve the most glycogen.

After proper fixation paraffin embedding is satisfactory, but after deparaffinizing the mounted slides, rinse them in absolute ethyl alcohol and protect the sections with a coat of 1% nitrocellulose to prevent diffusion of the glycogen during staining. Freeze-drying or freeze substitution methods are satisfactory. Lillie (1962) recommends the use of 5% formic acid if a tissue must be decalcified. Check the tissue each day by the oxalate method (p. 30). Stop decalcification as soon as a negative oxalate test is obtained. Avoid any use of heat. Lillie also suggests that 1–3 days of infiltration of the tissue with 1% nitrocellulose before the decalcification can improve the resistance of glycogen to acid extraction. Trott (1961a) uses EDTA, pH 7.5, for decalcification. I find RDO is satisfactory.

For mucoproteins and mucopolysaccharides, any protein precipitant fixative—Zenker, Helly, or Regaud—is satisfactory (Gomori 1952; Hale 1957; Thompson 1966).

Glycogen Staining

Best Carmine

FIXATION

Avoid aqueous media (McManus and Mowry 1958)

absolute alcohol	9 parts
formalin, concentrated	1 part

Use ice cold.

This method starts dehydration for embedding with 95% alcohol, but the loss of glycogen is about 20–30% (Trott 1961a,b; Trott et al. 1962). Double embedding with nitrocellulose and paraffin can be used to reduce the loss of glycogen.

MOUNTING SLIDES

I have achieved better localization of glycogen by mounting paraffin sections with 95% alcohol in place of water. As soon as sections are spread, drain off excess fluid and continue to dry.

SOLUTIONS

Best carmine stock solution

carmine (C.I. 75470)	2.0 g
potassium carbonate	1.0 g
potassium chloride	5.0 g
distilled water.............................	60.0 ml

Boil gently: 5 minutes. Cool. Filter. Add:

ammonium hydroxide	20.0 ml

Lasts 3 months at 0–4°C.

Working solution

carmine stock solution	30.0 ml
ammonium hydroxide	25.0 ml
methyl alcohol	25.0 ml

Lasts 2–3 weeks.

Differentiating fluid

absolute ethyl alcohol......................	16.0 ml
methyl alcohol	8.0 ml
distilled water.............................	20.0 ml

PROCEDURE

1. Deparaffinize, coat with nitrocellulose, and transfer into absolute alcohol: 3 minutes.
2. Coat with 1% celloidin; dry slightly in air.
3. Dip 2 or 3 times in 70% alcohol and then into water.
4. Stain in Mayer hematoxylin: 5 minutes.
5. Wash, blue in Scott, and wash.
6. Stain, Best carmine working solution: 15–30 minutes.
7. Treat with differentiating fluid: 5–15 minutes.
8. Rinse quickly in 80% alcohol.
9. Dehydrate, clear, and mount.

RESULTS

glycogen—red
mast cell granules, mucin, and fibrin—light red
nuclei—blue

COMMENTS

Pearse (1968) uses Lison "Gendre-fluid" method of fixation, claiming that
it shows a localization of glycogen comparing favorably with that in tissue
preserved by freeze-drying.

picric acid, saturated in 96% alcohol	85 parts
formalin	10 parts
glacial acetic acid	5 parts

Cool before use to $-73°C$ in an acetone–CO_2 snow mixture. Use small
pieces; fix for 18 hours.

Trott et al. (1962) fix in 1% periodic acid in 10% formalin, which they
claim shows greater amounts of glycogen than tissues fixed in acetic
acid–formalin–alcohol. Schiff reagent methods are replacing the classical
staining methods for glycogen.

Since it has been suggested that carmine staining is due to the hydrogen
bonding between glycogen and carminic acid (Goldstein 1962), other dyes
that carry groups capable of forming hydrogen bonds also should stain
glycogen (Murgatroyd and Horobin 1969). Acid alizarine blue SWR and
alizarine brilliant blue BS stained glycogen red. Alizarine red S stained it
red; gallein, purple; hematein and hematoxylin, red changing to brown.
The staining solution was made as follows: 1 g dye, 1 g potassium carbonate, and 5 g potassium chloride are added to 60 ml of boiling distilled
water. Let stand 1 hour. To 20 ml of dye solution add 15 ml of ammonia
and 15 ml of absolute methanol in that order. The solution is stable for
only 24 hours. Stain: 2–5 minutes. Dehydrate in 2–3 changes of absolute
methanol, clear, and mount.

Staining of Acid Mucopolysaccharides

For metachromatic methods see p. 318.

Alcian Blue Method, pH 2.8 (Putt 1971)

At this pH the Alcian blue stains both sulfated and nonsulfated muco-substances.

FIXATION

Any general fixative but formalin–99% alcohol (1 : 9) or Rossman fluid are recommended.

SOLUTIONS

Alcian blue, pH 2.5

calcium chloride (CaCl$_2$ · 2H$_2$O)	0.5 g
Alcian blue 8GX (C.I. 74240)	1.0 g
distilled water	100.0 ml

Dissolve CaCl$_2$ first. Filter; add thymol crystal to prevent mold.

Kernechtrot, see p. 247.

PROCEDURE

1. Deparaffinize and hydrate slides to water.
2. Rinse in 3% acetic acid: 3 minutes.
3. Stain in Alcian blue: 30 minutes.
4. Rinse in water and treat with 0.3% sodium bicarbonate aqueous: 30 minutes.
5. Wash in running water: 10 minutes.
6. Counterstain with Kernechtrot: 5 minutes.
7. Rinse in distilled water.
8. Dehydrate, clear, and mount.

RESULTS

acid mucopolysaccharides—blue green
nuclei—red to bluish red

COMMENTS

Alcian blue solutions must be acid to prevent the staining of other tissue elements. Alcian green may be substituted for the blue and preceded by the nuclei stained with the celestine blue–hematoxylin combination (p. 117) or Slidder hematoxylin (p. 115). These stains will not be destained by the acidic Alcian solution.

Can be followed by PAS (p. 210) for demonstration of both acidic groups and 1,2 glycols, or the Feulgen reaction (p. 213).

Lison (1954) counterstains with chlorantine fast red for efficient differentiation of mucin from collagen (mucin—bluish green; collagen—cherry red). Follow steps 1, 2, and 3 above, and then substitute the following steps.

4. Treat with phosphomolybdic acid, 1% aqueous: 10 minutes.
5. Rinse in distilled water.
6. Stain in chlorantine fast red: 10–15 minutes.

   chlorantine fast red 5B[2] ....   0.5 g
   distilled water ....   100.0 ml

7. Rinse in distilled water.
8. Dehydrate, clear, and mount.

Williams and Jackson (1956) note possible diffusion of acid mucopolysaccharides under aqueous conditions. The fixative should form complexes insoluble in water and alcohol. They suggest two possible solutions containing organic chemicals that form such insoluble complexes with acid mucopolysaccharides:

a. 0.5% cetylpyridium chloride (CPC) in 4% aqueous formalin. (See also Conklin 1963 and Wolfe 1964).
b. 0.4% 5-aminoacridine hypochloride in 50% ethyl alcohol.

Čejková et al. (1973) reduced acid mucopolysaccharide leakage to a minimum with unfixed frozen sections followed by alcohol fixation.

Spicer and Meyer (1960) precede staining with Alcian blue by immersing the tissue for 5 minutes in aldehyde-fuchsin (p. 326), transferring it to 70% alcohol, and washing it in running water.

Mowry (1960) and Scott and Mowry (1970) report that lots of Alcian blue have differed in staining efficiency and fastness. Current stocks of Alcian blue 8GX are more soluble, less fast, and more stable; the 8GX prepared by ICI, Ltd., Blackly, Manchester, England, is recommended.

---

[2] Edward Gurr or Ciba, Ltd.

Alcian blue staining is attributed to the presence of carboxyl groups and in some reactions to sulfates. All acidic mucopolysaccharides contain either sulfates alone or sulfate and carboxyl groups, and the stainability should be consistent. Yamada (1963), however, thinks that the basis for staining remains to be determined precisely and that it is better to examine acidic groups by using metachromatic dyes at different pH levels. See below for staining at pH 1.0 for sulfated carbohydrates.

The Alcian green method (Putt and Hukill 1962) substitutes Alcian green[3] for Alcian blue. Since Alcian green stains more rapidly than Alcian blue, the time required for staining (step 2) can be reduced from 30 minutes to 10 minutes. The procedure and results are identical, except that acid mucopolysaccharides stain deep green rather than blue green.

Carlo (1964) identifies sulfated groups with Alcian blue and carboxyl groups with Alcian yellow.

Wolfe (1964) stains extremely water-soluble mucopolysaccharides in a series of toluidine blue dissolved in decreasing concentrations of alcohol. Horn and Spicer (1964a,b) use azure A at a low pH of 4.0–5.0.

For fluorescent staining of mucopolysaccharides, see Saunders (1962).

Alcian Blue Method, pH 1.0 (Luna 1968)

FIXATION

Formalin–95% alcohol (1 : 9) or Rossman fluid.

SOLUTION

Alcian blue 8GX	1.0 g
0.1 N hydrochloric acid	100.0 ml

PROCEDURE

1. Deparaffinize and hydrate sections to distilled water.
2. Stain in Alcian blue: 30 minutes.
3. Do not rinse the sections in water; blot them dry with filter paper.
4. Dehydrate in 95% alcohol, absolute alcohol, and clear and mount.

RESULTS

sulfated mucosubstances—greenish blue
nonsulfated mucosubstances—unstained

---

[3] ESBE Laboratory Supplies.

Iron Diamine for Acid Mucosubstances

N,N-dimethyl-meta-phenyldiamine dihydrochloride and its para-monohydrochloride isomer form a colored cationic oxidation product which becomes bound to the anionic site of acidic mucosubstances. Addition of ferric ions seems to promote oxidation and improves staining. In low concentration the iron permits staining of sulfated mucosubstances and sialomucins; with a content of high iron only the sulfomucins are stained. Both methods can be followed by Alcian blue. Consult Spicer et al. (1967) for more details.

High Iron Diamine Method

FIXATION

Any general fixative.

SOLUTION

N,N-dimethyl-*m*-phenylenediamine-dihydrochloride[4]	0.12 g
N,N-dimethyl-*p*-phenylenediamine-mono-hydrochloride[4]	0.02 g

Add *together* to:

distilled water.............................	50.0 ml

When dissolved pour immediately (within 5 min) into a coplin jar containing:

NF. ferric chloride[5]	1.4 ml

PROCEDURE

1. Deparaffinize and hydrate slides to water.
2. Rinse in distilled water and place slides in coplin jar with diamine solution. Do not crowd slides, one slide only to a slot: overnight.
3. Rinse off excess solution in distilled water, but do this briefly and one slide at a time.

---

[4] Eastman Organic Chemicals 9145 and 492.
[5] Fisher Scientific Co., ferric chloride solution, 40% W/V, purified; use undiluted.

4. Dehydrate in 95% and absolute alcohol (ethyl or isopropyl), clear and mount.

RESULTS

most sulfated mucosubstances—purple black
acid mucopolysaccharides lacking sulfate esters—unstained

COMMENTS

Following step 3 the slides can be stained with Alcian blue 2.8 (p. 297) and the sialomucins and hyaluronic acid will be stained. When I do not use Alcian blue, I counterstain with orange G (p. 118) following step 3. The orange stain gives better contrast to the background than the dull gray of unstained tissue.

The low iron diamine method uses 0.03 g of *m*-diamine and 0.005 g of *p*-diamine with 0.5 ml of ferric chloride. The rest of the procedure is exactly like that of the high iron diamine. The low iron stains sulfated and carboxylic acid mucopolysaccharides gray to purple black.

Warning: The diamine chemicals appear to have a limited life even when stored in the refrigerator. I have had disappointing results with them after 8 months refrigerated storage.

Colloidal Iron Method (Mowry 1958)

SOLUTIONS

Colloidal iron stock solution

Ferric chloride 29% (USP XI)	4.4 ml
distilled water............................	250.0 ml

Bring water to boil. While the water is boiling, pour in the ferric chloride solution. Stir. When solution is dark red, remove from heat and allow to cool. The color must be dark red and clear. It is stable for months.

Working solution

glacial acetic acid	5.0 ml
distilled water............................	15.0 ml
stock colloidal iron solution	20.0 ml

pH should be 1.1–1.3. Nonspecific staining takes place at pH 1.4 or higher. Effective for 1 day.

Hydrochloric acid–ferrocyanide

2% HCl (2ml/98 ml water)	50.0 ml
2% potassium ferrocyanide (2 g/100 ml water) .	50.0 ml

Mix immediately before use.

PROCEDURE

1. Deparaffinize and hydrate slides to water: remove $HgCl_2$.
2. Rinse in 12% acetic acid (12 ml/88 ml water): 30 seconds (to prevent dilution of reagent).
3. Transfer to freshly prepared colloidal iron (working solution): 60 minutes.
4. Rinse in 12% acetic acid, 4 changes: 3 minutes each.
5. Treat with hydrochloric acid–ferrocyanide, room temperature: 20 minutes. (Include a control slide, not treated with colloidal iron.)
6. Wash in running water: 5 minutes.
7. Counterstain if desired: Feulgen, PAS, or hematoxylin.
8. Optional: dip in aqueous picric acid to color cytoplasm and erythrocytes; rinse for few seconds in tap water. (Acetic–orange G can also be used.)
9. Dehydrate, clear, and mount.

RESULTS

acid mucopolysaccharides—bright blue; uncolored in control slide
mucins of connective tissue, epithelium, mast cell granules, capsules of some microbial agents, pneumococci—bright blue

COMMENTS

Hale was the first to suggest the use of colloidal iron for the demonstration of acid polysaccharides, claiming that acidic groups can react with metals and cations. In order that such acidic groups may be demonstrated, the acid moiety should be a firmly bound part of a fixed and insoluble tissue constituent and be free to react. Also the link between the metal and acid should be strong enough that it will not be disrupted when water is used to wash away excess unbound metal. Using a positively charged colloidal solution is the easiest way to prevent this disruption. The metal is held tightly, probably by the multiple valencies of both reacting particles. Since the range of reactivity of the colloidal iron reagent depends on the pH of the solution, it is possible to stain strongly acidic groups selectively by making the solution strongly acid. If strongly acidic groups are blocked by bound iron, it should be possible to stain the weakly acid groups with another colloidal solution. Wolman (1956, 1961) developed his Bi-Col staining technique by using two colloidal metal solutions—colloidal iron and colloidal gold (Wolman 1956).

Mucin

Fluorescent Method (Hicks and Matthaei 1958)

FIXATION

10% formalin preferred.

SOLUTIONS

Iron alum

ferric ammonium sulfate	4.0 g
distilled water.............................	100.0 ml

Acridine orange

acridine orange (C.I. 46005)	0.1 g
distilled water.............................	100.0 ml

PROCEDURE

1. Deparaffinize and hydrate paraffin sections to water, or cut frozen sections and place in distilled water.
2. Treat with iron alum solution: 5–10 minutes.
3. Wash briefly in distilled water.
4. Stain in acridine orange: $1\frac{1}{2}$ minutes.
5. Rinse briefly in distilled water, blot, and mount in Harleco Fluorescent Mountant.[6]

---

[6] Hartman-Leddon Co.

RESULT

mucin—brilliant reddish-orange fluorescence

COMMENTS

This is not a permanent slide. The iron inhibits the production of fluorescence with acridine orange in nearly all tissue components except mucin. Hicks and Matthaei also suggest that the mucins of acid polysaccharides (rather than mucoproteins) fluoresce.

For a fluorescent antibody technique, see Kent (1961).

Lev and Stoward (1969) use eosin as a fluorescent dye to demonstrate mucous cells. The concentration of eosin can be 0.1% to 10%; the pH should be 9.5 to 5.7 but no lower. Staining time is 5–10 minutes. The fluorescence is stronger after Carnoy than after formalin fixation. Isopropyl extracts less stain that ethyl alcohol, thereby producing a stronger fluorescence. Hematoxylin used with the eosin improves the contrast due to the orange-red color it gives to nuclei.

Chapter 18
Staining Cellular Elements

THE ARGENTAFFIN REACTION

The argentaffin reaction should not be confused with silver impregnation (p. 162). In argentaffin reaction some substances (ascorbic acid, aldehydes, uric acid, polyphenols, and others) reduce silver solutions under specific conditions. This reaction with the tissue itself can therefore be used histochemically to identify these substances.

The only source of error in the method is found in calcification areas, but only if these are in large masses. Most silver phosphates and carbonates will be dissolved out during the process. If not, treat the slides with 0.2 to 0.5% nitric acid or hydrochloric acid in absolute alcohol: 2–3 minutes before step 3. Wash off the acid in absolute and 95% alcohol and proceed with step 3 (p. 307).

ENTEROCHROMAFFIN (EC) CELL STAINING

Fontana Method (Culling 1957) **and Methenamine Silver** (Gomori 1952, 1954*b*)

FIXATION

Zenker formol preferred. Do not use alcohol—it dissolves argentaffin granules. During embedding, do not expose to paraffin longer than 30 minutes.

SOLUTIONS

Silver solution A (Fontana method)

silver nitrate, 10% (2.5 g/25 ml water)	25.0 ml
ammonia, 28%	as needed
distilled water.............................	25.0 ml

To silver nitrate add ammonia drop by drop until the precipitate that forms is almost dissolved. Add distilled water. Store for 24 hours in brown bottle. Filter before use. Good for 2 weeks.

Methenamine silver solution B (Gomori method)

Solution I

methenamine, 3% (3 g/100 ml water)	100.0 ml
silver nitrate, 5% (5 g/100 ml water)	5.0 ml

Shake until white precipitate disappears. Keeps for several months.

Solution II

$M/5$ boric acid (12.368 g/1000 ml water)	80.0 ml
$M/5$ borax (19.071 g/1000 ml water)...........	20.0 ml

The pH should be 7.8 to 8.0.

Working solution

Solution I	30.0 ml
Solution II...............................	8.0 ml

Gold chloride

gold chloride stock solution, 1%	10.0 ml
distilled water.............................	100.0 ml

Safranin (or Kernechtrot, p. 247)

safranin O (C.I. 50240).....................	1.0 g
distilled water.............................	100.0 ml

Add a few drops of glacial acetic acid.

PROCEDURE

1. Deparaffinize and hydrate slides to water.
2. Treat with Lugol solution (p. 544): 30 minutes to 1 hour.
3. Wash in running water: 3 minutes.
4. Bleach in 5% aqueous sodium thiosulfate: 3 minutes.
5. Wash in running water: 5 minutes. (Steps 2–4 suppress background staining.)
6. Treat with silver solution A in the dark, room temperature: 18–48 hours. *Alternate method:* Use methenamine silver solution B at 60°C for 3.5 hours; or at 37°C for 12–24 hours—until EC cells stand out black.
7. Rinse in several changes of distilled water.
8. Tone in gold chloride: 10 minutes.
9. Rinse in distilled water.
10. Fix in 5% aqueous sodium thiosulfate: 3 minutes.
11. Wash in running water: 5 minutes.
12. Counterstain in safranin (or Kernechtrot).
13. Rinse in 70% alcohol: few seconds.
14. Dehydrate, clear, and mount.

RESULTS

argentaffin granules—black
melanin—black
other tissue elements—reds and pinks
background grayish to blackish—slide is overstained

COMMENTS

Chromaffin material is found only in the adrenal medulla and the gastrointestinal tract. It received its name because of its reaction with chromium salts and certain other metals to produce a yellowish to brown color, a brownish chromium oxide. To preserve this material, either use a dichromate fixative or postchromate with potassium dichromate (p. 28). Melanin, also brownish, is not a chromaffin substance.

Methenamine silver methods are also used for fungi and bacteria, see p. 367.

Enterochromaffin cells and granules also are argyrophilic and are impregnated by silver solutions when a reducer is present. This method, however, shows other cells as well, so it is not specific.

Diazo-Safranin Method (Lillie et al. 1953)

FIXATION

10% formalin buffered with 2% calcium acetate: 1–3 days.

SOLUTION

Diazotized safranin

The three parts of the diazotized safranin solution can be kept as stock solutions. To prepare the working solution they must be mixed *immediately before use*.

Acid safranin

safranin O (C.I. 50240)	3.6 g
distilled water	60.0 ml
N HCl	30.0 ml

This solution will remain stable for several weeks.
N sodium nitrite

sodium nitrite ($NaNO_2$)	6.9 g
distilled water	100.0 ml

Keep in refrigerator, stable for 3 months.
Disodium phosphate, $M/10$

disodium phosphate (Na_2HPO_4), anhydrous . . .	14.2 g
distilled water	1000.0 ml

To prepare the working solution, add 0.5 ml of sodium nitrite solution to 4.5 ml of ice-cold safranin solution. The resulting mixture turns deep blue and foams. Keep at 0° to 5°C for 15 minutes for diazotization. Dilute 1 ml of the solution with 40 ml of ice-cold disodium phosphate solution and use immediately (pH should be about 7.7).

PROCEDURE

1. Deparaffinize and hydrate slides to distilled water.
2. Place slides in a previously chilled coplin jar and pour over them the freshly prepared diazo-safranin solution: 5 minutes.

3. Decant stain and wash slides in 3 changes of $N/10$ aqueous hydrochloric acid or acid-alcohol (1 ml concentrated HCl/99 ml 70% alcohol): 10–13 seconds total, to remove most of adherent stain. Longer extraction lightens the colors but does not improve contrast.
4. Wash off acid with water (if $N/10$ HCl is used) or 95% alcohol (if acid-alcohol is used).
5. Dehydrate, clear, and mount.

RESULTS

enterochromaffin granules—black
gastric gland, chief cell granules—dark red
Paneth cell granules—red
mucin—unstained
nuclei and cytoplasm—pink to red

Ferric-Ferricyanide Method (Modified Schmorl Technique) (Lasky and Greco 1948; Lillie and Burtner 1953*a*)

FIXATION

10% formalin buffered with 2% of calcium acetate: 1–3 days.

SOLUTION

Ferric-ferricyanide

potassium ferricyanide, 1% aqueous	10.0 ml
ferric chloride, 1% aqueous	75.0 ml
distilled water.............................	15.0 ml

PROCEDURE

1. Deparaffinize and hydrate slides to water.
2. Stain in freshly prepared ferric-ferricyanide: 5 minutes.
3. Rinse in 3 changes of distilled water.
4. Counterstain in safranin, or the like.
5. Dehydrate, clear, and mount.

RESULTS

enterochromaffin granules—dark blue
nuclei—red

COMMENTS

Lillie and his coworkers devoted several years to research on the enterochromaffin granules along with melanins, etc. (Glenner and Lillie 1957*b*; Lillie 1955, 1956*a,b,c,* 1957*a,b,c,* 1960, 1961; Lillie et al. 1953; Lillie et al. 1957; Lillie et al. 1960).

Azo-Coupling Method (Gurr 1958)

FIXATION

Formalin or Bouin.

SOLUTION

Garnet reagent

garnet GBC salt[1]	0.5 g
distilled water.............................	100.0 ml
borax, saturated aqueous...................	2.5 ml

If it is necessary to purify the garnet GBC see comments on p. 411.

PROCEDURE

1. Deparaffinize and hydrate to water.
2. Transfer to garnet solution: 30–60 seconds.
3. Wash in running water: 30 seconds.
4. Stain in Mayer (or other) hematoxylin: 3 minutes.
5. Wash in running water: 5 minutes.
6. Dehydrate, clear, and mount.

RESULTS

argentaffin granules—red
nuclei—blue
background—light yellow

------

[1] Imperial Chemical Industries; Dajac Laboratories; Roboz Surgical Instruments.

AMYLOID STAINING

Congo Red (Puchtler et al. 1962)

FIXATION

10% formalin or alcoholic fixative. Carnoy fixative increases red binding of the Congo red.

SOLUTIONS

Solution A

1% aqueous sodium hydroxide	0.5 ml
80% ethanol saturated with NaCl	50.0 ml

Use within 15 minutes.

Solution B

80% ethanol saturated with Congo red (C.I. 22120) and NaCl	50.0 ml
1% aqueous sodium hydroxide	0.5 ml

Filter. Use within 15 minutes.

PROCEDURE

1. Deparaffinize and hydrate slides to water.
2. Stain in Mayer hematoxylin: 10 minutes.
3. Rinse in 3 changes distilled water: few seconds each.
4. Treat with solution A: 20 minutes.
5. Stain in solution B: 20 minutes.
6. Dehydrate rapidly in 3 changes absolute ethanol, clear, and mount.

RESULTS

amyloid—red
nuclei—blue

COMMENTS

Considerable research is being done on amyloid, a predominantly ex-tracellular deposition of protein-mucopolysaccharide complexes and other substances. The disease occurs principally in aging tissues of the animal species. I have seen it in several of our old marmosets. Staining is not always reliable and may not stain small deposits. Black and Jones (1971) consider frozen sections the best aid in amyloid detection. Freeze immediately without fixation.

The working solutions must be freshly made, but their major components, 80% ethanol saturated with NaCl and 80% alcohol saturated with Congo red and NaCl, will keep for several months. The latter should stand 24 hours before use.

This method is used in preference to Bennhold's or Highman's because in it there is no need for differentiation. Bennhold's method lacks uniformity, and Highman's easily overdifferentiates.

Staining is more intense in alcohol-fixed tissues than in formalin-fixed tissues, but even in the latter small deposits of amyloid are visible.

Puchtler et al. suggest that the selectivity of Congo red for amyloid may be due to the polysaccharide moiety of amyloid. Puchtler et al. also found that certain direct cotton dyes stain amyloid. See Puchtler and Sweat (1965) and Puchtler and Sweat (1966) for a review of the concepts of amyloid substance. See also metachromatic methods, p. 319.

Navagiri and Bubey (1976) use a Leishman stain; it stains normal parenchymal cells faint to deep blue and the amyloid, purple to violet.

Fluorescent Method

Thioflavine T is recommended for amyloid by Vassar and Culling (1959, 1962), but it colors amyloid only faintly. Puchtler and Sweat (1965) use their Congo red method (see above) with excellent fluorescence and pink to red coloring of amyloid, and with other tissues greenish gray. Mount in Harleco Fluorescent Mountant[2] for a black background. Synthetic mountants, such as Permount, have only moderate fluorescence, but slides mounted with them can be used for fluorescent observation; the faintness of the fluorescence does not interfere too seriously with slide study.

MAST CELL STAINING

Mast cells are of common occurrence in connective tissue. Because of their cytoplasmic granules, however, staining methods for these cells have been included here. The specific staining of these granules is the primary means of identification of mast cells. (See also metachromatic methods, p. 319.)

---

[2] Hartman-Leddon Co.

Chrysoidin (Harada 1957)

FIXATION

Any general fixative.

SOLUTION

chrysoidin Y (C.I. 11270)	0.5 g
distilled water............................	100.0 ml

PROCEDURE

1. Deparaffinize and run slides down to 80% alcohol; remove $HgCl_2$ if present.
2. Stain in chrysoidin: 5–10 minutes.
3. Rinse in distilled water.
4. Dehydrate, clear, and mount.

RESULTS

mast cell granules—deep brown to black
other elements—yellowish

COMMENTS

Staining with chrysoidin can be preceded with hematoxylin or PAS.

Toren (1963) combines Pollak (1944) trichrome stain with a Giemsa stain. Barlow (1957) uses a tribasic stain of acridine red, neutral red, and basic fuchsin. Menzies (1962a) combines methylene blue and basic fuchsin. Spatz (1960) uses Bismarck brown.

Fluorescent Method (Jagatic and Weiskopf 1966)

FIXATION

Any general fixative.

SOLUTIONS

Weigert hematoxylin, see p. 113.

Acridine orange

    acridine orange (C.I. 46005) ....   0.1 g

    distilled water............................   100.0 ml

PROCEDURE

1. Deparaffinize and hydrate sections to water.
2. Stain in hematoxylin: 5 minutes.
3. Wash in running water: 3–5 minutes.
4. Stain in acridine orange: 5–6 minutes.
5. Rinse in distilled water: 1 minute.
6. Dehydrate in 95% ethanol or absolute isopropanol.
7. Clear and mount in Harleco Fluorescent Mountant.

RESULT

mast cell granules—bright red orange

COMMENTS

Old H-and-E slides have been restained by this method; also sections from paraffin blocks 10 to 15 years old.

METACHROMASIA

A few tissue elements are stainable by a particular group of cationic dyes, changing in the tissues from blue (the usual orthochromatic form) to purplish red or reddish purple. Such a dye, called metachromatic, is of considerable value in the study of specific elements of connective tissue. Among the metachromatic dyes most commonly used are toluidine blue O, thionin, methylene blue, azures, crystal violet, cresyl violet, methyl violet, safranin O, celestin blue, gallocyanin, and pinacyanole. Some of the tissues identified by this means are mast cells, amyloid, cartilage, and mucus materials (Schubert and Hamerman 1956).

    The methods are tricky, and a technician must learn to distinguish a true metachromasia from a false one. The difficulty in preserving metachromasia lies in the dehydration of the tissue after staining. Increasing strengths of alcohol (ethyl) revert the dye back to the orthochromatic form. Sections can be examined in an aqueous condition, but this produces at best a semipermanent preparation. Some workers use acetone, isopropyl, or tertiary butyl alcohol for dehydration, then follow with one of the clearing agents. Padawer (1959) uses ether 181 (tetramethylene glycol ether),[3] and Levine (1928) uses oil of clove for dehydration. Always stain at room temperature; heat can destroy metachromasia.

Amyloid

Crystal Violet (Lieb 1947)

FIXATION

10% formalin or alcoholic fixative.

SOLUTION

Crystal violet stock solution

crystal violet (C.I. 42555), 14.0–15.0 g saturated in 95% ethyl alcohol	100.0 ml

Working solution

stock solution	10.0 ml
distilled water...........................	300.0 ml
hydrochloric acid, concentrated	1.0 ml

PROCEDURE

1. Run frozen sections or deparaffinized sections down to water.
2. Stain in working solution: 5 minutes to 24 hours.
3. Rinse in water.
4. Mount in Abopon (p. 107).

RESULTS

amyloid—purple
other tissue elements, including hyalin—blue

COMMENTS

Acid in the staining solution makes it self-differentiating, and staining time is flexible. Lieb suggests that if there is only a small amount of amyloid in

---

[3] Ansul Chemical Co.

the tissue, thicker sections should be cut to make the color reaction more clearly visible. Slides mounted in Abopon have retained their color for 2 years. Abopon mounts must be sealed with ringing cement for permanency.

Gomori (1952) outlines a simple method for permanent mounts that preserve the metachromasia of the crystal violet.

1. Float paraffin sections on dye solution: 15–20 minutes.
2. Float sections on distilled water to remove excess dye.
3. Float on 1–2% aqueous acetic acid, to differentiate.
4. Mount on slides in usual fashion and dry.
5. Remove paraffin with xylene and mount.

See Bancroft (1963) for staining with methyl violet and methyl green. See also amyloid staining, p. 311.

Mucin

Toluidine Blue (Lillie 1929)

FIXATION

Any general fixative, but an alcoholic one is preferred. Do not use a chromate fixative.

SOLUTION

> toluidine blue O (C.I. 52040)................   0.2 g
> distilled water............................   100.0 ml

PROCEDURE

1. Deparaffinize and hydrate slides to water; remove $HgCl_2$.
2. Stain in toluidine blue: 1 minute.
3. Wash in water: 2–3 minutes.
4. Dehydrate in acetone, 2 changes: 3–5 minutes each.
5. Clear in xylene and mount.

RESULTS

mucin—reddish violet
nuclei and bacteria—deep blue

cytoplasm, fibrous tissue—bluish green
bone—bluish green
cartilage matrix—bluish violet
muscle—light blue
cell granules—blue violet
hyalin and amyloid—bluish green

COMMENTS

Gomori (1952) uses 0.02–0.05% toluidine blue O in citrate buffer, pH 3.5–4.5: 10–15 minutes, or until the nuclei are blue and the mucin, intense pink.

For nonmetachromatic staining of mucin, see p. 303.

Thionin (Mallory 1944)

FIXATION

Any general fixation, but an alcoholic one is preferred.

SOLUTION

thionin (C.I. 52000)	1.0 g
25% ethyl alcohol	100.0 ml

PROCEDURE

1. Deparaffinize and hydrate slides to water; remove $HgCl_2$.
2. Stain in thionin: 15 minutes to 1 hour.
3. Differentiate in 95% alcohol.
4. Dehydrate in absolute alcohol, clear, and mount.

RESULT

mucin—light to dark red or purple

Acid Mucopolysaccharides

Thionin (Gurr 1958)

FIXATION

10% formalin or other general fixative.

SOLUTIONS

Thionin

thionin (C.I. 52000) saturated aqueous	0.5 ml
distilled water............................	100.0 ml

Molybdate-ferricyanide solution

ammonium molybdate, 5% aqueous	50.0 ml
potassium ferricyanide, 1% aqueous	50.0 ml

Make up solutions fresh each time.

PROCEDURE

1. Deparaffinize and hydrate slides to water.
2. Stain in thionin: 5–15 minutes.
3. Rinse in distilled water.
4. Treat with molybdate-ferricyanide: 2 minutes.
5. Wash in distilled water: 2–3 minutes.
6. Dehydrate, clear, and mount.

RESULTS

acid mucopolysaccharides—purple
other cell elements—bluish

COMMENTS

The molybdate-ferricyanide solution prevents loss of metachromasia.
    See Kuyper (1957) for suggestions concerning fluorescent methods for mucopolysaccharides. See also acid mucopolysaccharides, p. 297.

Mast Cells

For nonmetachromatic staining of mast cells, see p. 313.

Thionin Method (Lillie 1954 or 1965)

FIXATION

Any general fixative.

SOLUTION

thionin (C.I. 52000)	0.5 g
0.01 M acetate buffer (see results)	100.0 ml

PROCEDURE

1. Deparaffinize and hydrate slides to water; remove $HgCl_2$.
2. Stain in thionin: 30 minutes for light stain.
3. Rinse in water.
4. Dehydrate, clear thoroughly in xylene, and mount.

RESULTS

mast cell granules—red purple
nuclei—faint blue violet

COMMENTS

At pH 2, metachromasia stains only mast cells and cartilage; at pH 5, muscle and connective tissue stain light green; at pH 3–2, cytoplasm stains poorly.

Quick Toluidine Blue Method[4]

FIXATION

Any general fixative.

---

[4] Personal communication from Marlies Natzler, U.C.L.A.

SOLUTION

toluidine blue O (C.I. 52040)................	0.2 g
60% ethyl alcohol	100.0 ml

PROCEDURE

1. Deparaffinize and run slides down to 60% alcohol. Remove HgCl₂ if present.
2. Stain in toluidine blue: 1–2 minutes.
3. Rinse quickly in tap water.
4. Dehydrate in acetone, 2 changes: 2–3 minutes each.
5. Clear in xylene and mount.

RESULTS

mast cells—deep reddish purple
background—faint blue

Toluidine Blue (Conroy and Toledo 1976)

FIXATION

Formalin-alcohol preferred, 10% formalin satisfactory.

Donaldson et al. (1973) observed that after mercuric chloride fixation and iodine treatment, mast cell staining was suppressed. Shorten the fixing time to 4–6 hours, wash in 80% alcohol: 40 hours to 6 days, and omit iodine treatment.

formalin	10.0 ml
95% alcohol	90.0 ml

SOLUTION

toluidine blue O (C.I. 52040)................	0.1 g
distilled water............................	100.0 ml

pH should be 6.8–7.2. If this is not possible with distilled water, make up the stain in a buffer, such as McIlvaine (p. 557).

PROCEDURE

1. Deparaffinize and hydrate sections to water.
2. Stain lightly with hematoxylin (1 minute Harris, 3 minutes Mayer, for example), wash, and blue.
3. Stain with eosin and rinse off excess eosin with 95% and absolute ethyl alcohol: 3 dips each.
4. Rinse in tap water: 5 dips.
5. Stain in toluidine blue 0: 7–10 seconds, agitate gently.
6. Rinse in distilled water: 1 dip.
7. Dip once in 95% alcohol.
8. Dehydrate in absolute alcohol, 2 changes: 4 dips and 6 dips. Isopropanol may be used and the dehydration time lengthened.
9. Dehydrate in absolute alcohol, 2 changes, 10 dips each.
10. Xylene, 2 changes: 2 minutes each, and mount.

RESULTS

mast cells—blue to purple
collagen—pink
cytoplasm—pink
nuclei—blue
erythrocytes—red

ENDOCRINE GLAND STAINING

Pituitary Cells

Staining the cytoplasm and its elements is used to characterize and differentiate the cells of the anterior pituitary, but the naming of cell types is still in a state of confusion. Different systems of nomenclature are offered by different investigators. The possibilities include names for cell types based on the hormones they secrete or the morphological characteristics they display during staining. The Greek terminology is still in contention. To add to the confusion—the glands of all species do not react with the same color specificity. It is, therefore, impossible to say "this is it" of any one stain. Some of the more useful stains, each valuable for some particular cell or group of cells, are Mallory types, PAS, aldehyde-fuchsin, aldehyde-thionin, and Alcian blue. When followed with care, Herlant methods rate well among the best methods, particularly for basophilic staining, and I use these. An aldehyde-fuchsin method is included, and the Azan (Mallory type) stain can be found on p. 139. For extensive study of the pituitary, see Harris and Donovan (1966).

In the anterior pituitary, the glandular cells are sometimes classed as either chromophils or chromophobes. 75% of the chromophils are normally acidophilic (acidophils, sometimes called alpha cells) and 25% are basophilic (basophils). The basophilic group is made up of gonadotrophs (delta cells) and thyrotrophs (beta cells). These cells are often listed as 7 types: alpha, beta 1, beta 2, beta 3, delta 1, delta 2, and chromophobe.

Thyrotrophs and gonadotrophs have an affinity for Schiff reagent; thyrotrophs also stain with Gomori aldehyde-fuchsin or aldehyde-thionin. Acidophilic granules differentiate sharply in Elftman's (1960) method. Orange G also stains the acidophilic granules. Paget and Eccleston (1959, 1960) use Luxol fast blue to distinguish them. Separating the chromophobes from the chromophils is not always as simple as it may sound; transitional stages, or the state of secretion of the cells, can make the two types of cells difficult to identify. The cells of the pars tuberalis contain no cytoplasmic granules. Those of the pars intermedia are pale, with a few of the cells containing basophilic granular cytoplasm. The pars nervosa (posterior lobe) does not show well-organized cell structure, but the neurosecretory substance (NSS) can be demonstrated. Gomori chrome hematoxylin is perhaps the most commonly used method for its demonstration, but Bargmann's modification should be used. The substance is easily identified, even though the nuclei, Nissl substance, lipofuscin, and the basophils of the anterior pituitary are also stained. Since the NSS is rich in cystine, the performic acid–Alcian blue method of Adams and Sloper (1955, 1956) and Sloper and Adams (1956) demonstrate the substance.

For additional information, some of the following may be of assistance: Elftman (1957a, 1959a,b, 1960); Glenner and Lillie (1957a); Gomori (1941b); Kerenyi (1959); Kerenyi and Taylor (1961); Landing (1954); Landing and Hall (1955); Lazarus (1958); Paget and Eccleston (1959, 1960); Pearse (1949, 1950, 1960); Shanklin et al. (1959); Thompson, 1966.

Herlant (1960) Pituitary Stain I

FIXATION

Any good general fixative, but Zenker formol preferred.

SOLUTIONS

Erythrosin

erythrosin B (C.I. 45430)	1.0 g
distilled water	100.0 ml

Mallory II

aniline blue WS (C.I. 42755)	0.5 g
orange G (C.I. 16230)	2.0 g
distilled water	100.0 ml
glacial acetic acid	1.0 ml

Acid alizarine blue

acid alizarine blue (C.I. 63015)	0.5 g
aluminum sulfate	10.0 g
distilled water	100.0 ml

Bring to boil, approximately 5 minutes. Cool. Adjust to 100.0 ml with distilled water and filter. Stable, but add thymol.

Phosphomolybdic acid

phosphomolybdic acid	5.0 g
distilled water	100.0 ml

PROCEDURE

1. Deparaffinize and hydrate to water; remove $HgCl_2$ if present.
2. Stain in erythrosin: 5 minutes.
3. Rinse briefly in distilled water.
4. Stain in Mallory II: 5–10 minutes.
5. Rinse briefly in distilled water.
6. Stain in acid alizarine blue: 5–10 minutes.
7. Rinse briefly in distilled water.
8. Treat with phosphomolybdic acid: 5–10 minutes.
9. Rinse briefly in distilled water.
10. Differentiate in 70% ethanol.
11. Clarify stain in 70% ethanol with a few drops of acetic acid added.
12. Dehydrate, clear, and mount.

RESULTS

  alpha cells—yellow
  gamma cells—violet
  beta cells—pale blue
  delta cells—dark blue
  epsilon cells—red to rose
  nucleus—dark blue to violet

Herlant (1960) Pituitary Stain II

FIXATION

Bouin or Hollande Bouin preferred.

SOLUTIONS

  Permanganate solution

2.5% aqueous potassium permanganate	10.0 ml
5% aqueous sulfuric acid	10.0 ml
distilled water............................	60.0 ml

Mix just before use.

  Alcian blue, pH 3.0

Alcian blue 8GX (C.I. 74240)	1.0 g
distilled water............................	100.0 ml
glacial acetic acid	1.0 ml

  Alcian blue, pH 0.2

Alcian blue 8GX (C.I. 74240)	1.0 g
10% sulfuric acid	100.0 ml

The Alcian blue dissolves less easily in the acid solution than in water.
Warm the solution until stain dissolves, cool, and filter. Good for 1 month;
filter occasionally.

  Sodium metabisulfite

sodium metabisulfite	5.0 g
distilled water............................	100.0 ml

Periodic acid

periodic acid..............................	1.0 g
distilled water............................	100.0 ml

Schiff reagent, see p. 210.

PROCEDURE

1. Deparaffinize and hydrate slides to water.
2. Oxidize in permanganate solution: 1–2 minutes.
3. Rinse in distilled water: few seconds.
4. Bleach in metabisulfite solution: 1 minute.
5. Wash in running water: 5 minutes.
6. Rinse in distilled water.
7. Stain in Alcian blue: 15–30 minutes.
8. Wash in running water: 5–10 minutes.
9. Treat with periodic acid: 10 minutes.
10. Rinse in distilled water: 1–2 minutes.
11. Treat with Schiff reagent: 10–20 minutes.
12. Wash in 3 changes of metabisulfite solution: total of 5 minutes.
13. Wash in running water: 15 minutes.
14. Dehydrate, clear, and mount.

RESULTS AND COMMENTS

Step 13 can be followed by staining with hematoxylin if desired.

This method separates the two kinds of gonadotrophic cells. The potassium permanganate oxidation confers on the alpha cells only a faint affinity for the PAS reaction. They will stain a faint rose, much less marked than the color of the gamma cells. If step 13 is followed by staining with 1% orange G, the rose color of the alpha cells is masked, and they become yellow; beta cells, violet; delta cells, dull blue; gamma cells, brick red.

With Alcian blue, increased acidity improves staining specificity. At pH 3, the beta and delta cells show a similar affinity for the stain and react only feebly with PAS. At pH 0.2 the two cells are more easily distinguished; the beta cells are paler than the delta cells and PAS does not change the blue of the latter, but colors the beta cells violet. Apparently the beta cells are so weakly stained by the Alcian blue that they do not mask the PAS.

Ewen (1962) Modification of Cameron and Steele Aldehyde-Fuchsin Method

This modification increases the method's specificity.

FIXATION

Fix in Bouin with 0.5–1.0% trichloracetic acid instead of acetic acid. Helly also is good.

SOLUTION

Aldehyde-fuchsin

Add 1 g basic fuchsin (C.I. 42500) to 200 ml boiling water: boil 1 minute. Cool and filter. Add 2 ml concentrated hydrochloric acid and 2 ml paraldehyde. Leave stoppered at room temperature. When mixture has lost reddish fuchsin color and is deep purple (3–4 days), filter it and discard filtrate. Dry precipitate on filter paper in an oven. Remove and store in bottle. Makes about 1.9 g. To make stock solution, dissolve 0.75 g in 100.0 ml of 70% ethyl alcohol. Keeps 6 months.

Paraldehyde decomposes readily. Freshly opened paraldehyde will give excellent results, but a bottle that has been open for a long time may give negative results (Gairdner 1969).

Working solution

stock solution	0.75 g
70% ethyl alcohol	75.0 ml
glacial acetic acid	1.0 ml

Potassium permanganate, 0.3%

potassium permanganate	0.3 g
distilled water..............................	100.0 ml
sulfuric acid, concentrated	0.3 ml

Sodium bisulfite, 2.5%

sodium bisulfite	2.5 g
distilled water..............................	100.0 ml

Halmi mixture (1952)

distilled water..............................	100.0 ml
light green SF, yellowish (C.I. 42095)	0.4 g
orange G (C.I. 16230)	1.0 g

| chromotrope 2R (C.I. 16570) .... | 0.5 g |
| glacial acetic acid .... | 1.0 ml |

Keeps indefinitely.

PROCEDURE

1. Deparaffinize and hydrate slides to water; remove $HgCl_2$.
2. Oxidize in potassium permanganate: 1 minute.
3. Rinse in distilled water.
4. Bleach in sodium bisulfite until permanganate color is removed.
5. Wash in running water: 5 minutes.
6. Transfer to 70% alcohol: 2 minutes.
7. Stain in aldehyde-fuchsin: 2–10 minutes.
8. Wipe off back of slide and rinse in 95% alcohol.
9. Differentiate in acid alcohol (0.5 ml HCl/100.0 ml absolute ethyl alcohol): 10–30 seconds.
10. Transfer through 70% alcohol and water.
11. Mordant in the following: 10 minutes.

phosphotungstic acid	4.0 g
phosphomolybdic acid	1.0 g
distilled water............................	100.0 ml

12. Rinse in water and stain for 1 hour in a modified Halmi's mixture.
13. Wipe off back of slide, differentiate in 95% alcohol plus 0.2% acetic acid: 2–3 minutes.
14. Rinse in fresh 95% alcohol.
15. Dehydrate in absolute alcohol, clear, and mount.

RESULTS

granulation of beta cells—dark purple
delta cells—green
acidophilic granules—orange
nucleoli—bright red
coagulated contents of cytoplasmic granules—orange

Lead Hematoxylin Method (Solcia et al. 1969)

FIXATION

An aldehyde fixative is best.

SOLUTION

Stock stabilized lead solution

Mix equal volumes of 5% aqueous lead nitrate and saturated aqueous ammonium acetate. Filter. Add 2 ml of 40% formalin for every 100 ml of filtrate. Store at room temperature. Keeps for several weeks.

Working solution

Add 0.2 g hematoxylin dissolved in 1.5 ml 95% ethanol to 10 ml of lead stock solution. Dilute with 10 ml of distilled water. Stir repeatedly. After 30 minutes, filter. Makes up to 75 ml with distilled water. Use immediately.

PROCEDURE

1. Deparaffinize and hydrate to water.
2. Stain in lead hematoxylin, 37°C: 2–3 hours.
3. Wash in running water: 10 minutes.
4. Dehydrate, clear, and mount.

RESULTS

    endocrine cell types, pancreas A and D cells, thyroid C cells, hypophysis MSH (melanocyte-stimulating hormone) and ACTH (corticotrophin hormone) cells—blue black
    nucleoli and nuclear chromatin—dark blue
    calcium deposits, nerve fibers, A and Z bands of striated muscle—blue

Chrome-Hematoxylin, Bargmann Modification for NSS (Pearse 1968)

FIXATION

Bouin preferred. Susa and Stieve satisfactory. If tissue is alcohol-fixed, float sections on Bouin solution instead of on water when mounting sections on slides.

SOLUTIONS

Bouin chrome alum

Bouin solution (p. 14)	100.0 ml
chrome alum	3.0–4.0 g

Potassium permanganate–sulfuric acid

2.5% aqueous potassium permanganate	1 part
5% aqueous sulfuric acid	1 part
distilled water	6–8 parts

Oxalic acid

oxalic acid	1.0 g
distilled water	100.0 ml

Chromium hematoxylin, see p. 331.

Acid alcohol

hydrochloric acid, concentrated	1.0 ml
70% ethyl alcohol	100.0 ml

Phloxin

phloxin B (C.I. 45410)	0.5 g
distilled water	100.0 ml

Phosphotungstic acid

phosphotungstic acid	5.0 g
distilled water	100.0 ml

PROCEDURE

1. Deparaffinize and hydrate sections to water.
2. Mordant in Bouin chrome alum solution, 37°C: 12–24 hours.
3. Wash in running water until sections are colorless.
4. Oxidize in permanganate solution: 2–3 minutes.
5. Wash in distilled water: 1 minute.
6. Bleach in oxalic acid: 1 minute.
7. Wash in running water: 5 minutes.

8. Stain in chrome hematoxylin: 10 minutes.
9. Differentiate in acid alcohol: 30 seconds.
10. Wash in running water: 2–3 minutes.
11. Stain in phloxin: 2–3 minutes.
12. Treat with phosphotungstic acid: 2 minutes.
13. Wash in running water: 5 minutes.
14. Dehydrate, clear, and mount.

RESULTS

NSS—deep purple
nuclei—lighter purple
backgrounds—pinkish red

COMMENTS

McGuire and Opel (1969) oxidize with freshly prepared permanganate–sulfuric acid and oxalic acid (See Bargmann method above), and stain 15 minutes in the following:

resorcin fuchsin[5] (C.I. 11430)	1.0 g
70% ethanol	98.0 ml
HCl concentrated	2.0 ml

Can be used immediately and is stable 20 days if kept tightly closed. Counterstain with light green.
Tan (1973) oxidized with peracetic acid: 10–15 minutes.

acetic acid glacial	72.0 ml
hydrogen peroxide	226.0 ml
sulfuric acid	2.0 ml

Let stand 1–3 days before use. At 4–8°C is stable 5–6 months.
Stain with aldehyde- or resorcin-fuchsin and counterstain with Halmi mixture.
  Shyamasundari and Rao (1975) demonstrate NSS and mucosubstances simultaneously.

---

[5] Chroma Gesellschaff, Roboz Surgical Instruments.

Pancreas Cells

Lane in 1908 established the difference between the zymogen granules and the islet granules, and showed that there were two types of islet cells. He named cells with alcohol-soluble granules, beta cells; those with water-soluble granules, alpha cells. The water-soluble granules are larger than those of the beta cells. Various techniques have been used to bring out the difference in the granules. Mallory's will stain alpha cells red and beta cells blue, but the Azan method is better, and good differentiation depends on proper fixation. Glenner and Lillie (1957*b*) specifically stain alpha cells by a postcoupled benzilidine reaction for indoles. The alpha cells are associated with a strong nonspecific esterase reaction, and the beta cells, with a strong acid phosphatase reaction. Gomori chrome–hematoxylin is a favorite method for staining the islets.

Recently, a D cell was added to the A and B cells. See Epple (1967) for methods of demonstrating D cells and the use of THF as an aid in overcoming difficulties encountered in embedding pancreas blocks; also Solcia et al. (1968, 1969). See Bussolati and Bass (1974) for staining of B cells.

Chromium-Hematoxylin-Phloxin (Gomori 1941*b*)

FIXATION

Bouin preferred. Stieve satisfactory. Zenker, Carnoy, and formalin unsatisfactory.

SOLUTIONS

Bouin solution, see p. 14.

Potassium dichromate–sulfuric acid

This can be made up as separate 0.3% solutions and mixed, or it can be made as follows:

potassium dichromate	0.15 g
distilled water	100.0 ml
sulfuric acid, concentrated	0.15 ml

Hematoxylin solution

hematoxylin	0.5 g
distilled water	50.0 ml

When dissolved, add:

> potassium chromium sulfate (chrome alum), 3%
>   aqueous ....   50.0 ml

Mix well and add:

> potassium dichromate, 5% aqueous ....   2.0 ml
> $N/2$ sulfuric acid (about 2.5 ml/100 ml water) ..   2.0 ml

Allow to ripen for 48 hours. Can be used until a film with a metallic luster does not form on its surface after 1 day's standing. Store at 0–4°C. Filter before use.

Phloxin

> phloxin B (C.I. 45410) ....   0.5 g
> distilled water............................   100.0 ml

PROCEDURE

1. Deparaffinize and hydrate slides to water; remove $HgCl_2$.
2. Refix in Bouin solution: 12–24 hours.
3. Wash in running water: 5 minutes.
4. Treat with potassium dichromate–sulfuric acid: 5 minutes.
5. Decolorize in 5% aqueous sodium bisulfite: 3–5 minutes.
6. Wash in running water: 5 minutes.
7. Stain in hematoxylin solution until beta cells are deep blue (check under microscope): 10–15 minutes.
8. Differentiate in hydrochloric acid, 1% (1 ml/99 ml water): about 1 minute.
9. Wash in running water until clear blue: 5 minutes.
10. Stain in phloxin: 5 minutes.
11. Rinse briefly in distilled water.
12. Treat with 5% aqueous phosphotungstic acid: 1 minute.
13. Wash in running water: 5 minutes. Sections turn red again.
14. Differentiate in 95% alcohol. If the sections are too red and the alpha cells do not stand out clearly, rinse 15–20 seconds in 80% alcohol.
15. Dehydrate in absolute alcohol, clear, and mount.

RESULTS

beta cells—blue
alpha cells—red

delta cells (not present in all animals)—pink to red, actually indistinguishable from alphas

COMMENTS

See also Heidenhain Azan method, p. 139.

If the zymogen granules (acidophilic) are to be preserved in the acinar cells of the pancreatic lobules, avoid fixatives containing acetic acid.

For fluorescent differentiation of the alpha and beta cells, see Hartroft (1951).

Herlant stain I (p. 322) is also excellent for pancreas.

Monroe and Spector (1963) use tannic acid, hematoxylin, Alcian blue, and basic fuchsin.

Kallman (1971) stains for alpha and beta cells: 1 hour in aldehyde fuchsin; wash in 2 changes 70% ethanol; a rinse in distilled water; and stain 4 minutes in 0.05% toluidine blue 0 in 0.2 M McIlvaine phosphate buffer, pH 5. The beta cells have violet-red granules and the alpha cells, light blue. The preferred fixative is Bouin; formalin is good, but Zenker, Helly, and Susa are poor.

Klessen (1971) uses a permanganate-HID technique for zymogen granules.

Thyroid Cells

Outstanding features of the thyroid are the secretory epithelial cells and the extracellular amorphous colloid—the stored secretion. The colloid consists of proteins associated with carbohydrates. Protein reactions can be used; some of the proteins are of the basic type and contain arginine, which can be demonstrated by Sakaguchi modifications. The ferric ferricyanide reduction technique (Lillie 1965) reacts with the colloid. Since the colloid is strongly PAS positive, mucopolysaccharides are probably present, and a positive dialyzed iron reaction would indicate acid mucopolysaccharides. The colloid is basophilic. The thyroid uses iodine to make up its hormone, and since radioactive iodine was one of the first isotopes prepared for biologic purposes, the isotope tracer technique has been used for some time in thyroid studies. Bélanger and Bois (1964) recommended AFA fixation—1 part acetic acid, 5 parts formalin, and 15 parts absolute ethyl alcohol: 24 hours. This gives excellent fixation of the colloid, with less vacuolization than after formalin alone, and well-stained colloid and cell detail with any of the staining techniques. However, I find that preservation of cellular detail and colloid is just as good after fixation in cold (4–10°C) calcium formalin (p. 15). The C cells (assumed to pro-

duce polypeptide hormones) of the thyroid can be demonstrated by Mac-
Conaill lead hematoxylin as modified by Solcia et al. (1969). p. 327.

Ljungberg (1970) uses 0.05% cresyl fast violet aqueous: 1–5 minutes,
rinse in water, dehydrate, and clear.

Hot HCl eliminates diffuse tissue basophilia and increases the
basophilia of certain endocrine cells, particularly thyroid C cells. The
basophilia of secretory granules is increased in part to stored proteins.
Treat for 6 hours in 1 N HCl, 60°C, and wash well. Stain with basic dyes,
such as methylene blue, toluidine blue O, or Azure A (Petkó, 1974; Solcia
et al, 1968).

Cell Membranes

See Puchtler and Leblond (1958).

Chapter 19
Staining Golgi Apparatus, Mitochondria, and Living Cells

GOLGI APPARATUS STAINING

The Golgi apparatus (Golgi bodies, Golgi substance, Golgi complex) is usually lost in routine fixation and requires special treatment. Since the methods are not always predictable under all conditions, it may be necessary to modify the fixing and/or staining time in order to attain precise results. In osmium and silver methods the Golgi appears as either a dark net, a granular mass, a cord, or even a more diffuse condition. These reactions seem to indicate the presence of lipids, principally phospholipids. The Golgi apparatus (or adjacent lysosomes) can have a high level of acid phosphatase activity and lesser levels of alkaline phosphatase and other enzymes. Nucleoside diphosphatase is restricted to the Golgi apparatus.

There is a great deal of controversy about the responses of the Golgi apparatus to different conditions, physiological changes, and techniques. Golgi composition appears to differ from cell to cell, and it can be related to the specific activity of a specific cell. Silver or osmic acid leaves the Golgi in the form of globules that have been described as spheres with an osmophilic cap surrounding an osmophobic center. But a thin lamella also is found adjacent to the spherical elements, indicating a duplex structure. Most of the fixatives recommended for Golgi fixation contain heavy metal salts; yet it has been claimed that these are not necessary and the light metal or organic salts or acids do just as well (Pollister and Pollister 1957; *Proceedings of the Histochemical Society* 1953). (See histochemistry, acid phosphatase, p. 405.)

Much of the current research on the Golgi apparatus is being carried out with electron microscopes. It has demonstrated that the metals are laid down on the membranes of the cisternae and vesicles.

Osmium Tetroxide, Ludford Method (Cowdry 1952; Lillie 1954*b*, 1965)

FIXATION

Mann osmic sublimate: 18 hours.

osmic acid, 1% aqueous	50.0 ml
mercuric chloride, saturate aqueous, plus 0.37 g	
sodium chloride	50.0 ml

PROCEDURE

1. Wash blocks of tissue in distilled water: 30 minutes.
2. Impregnate:
   2% osmic acid: 3 days, 30°C
   2% osmic acid: 1 day, 35°C
   1% osmic acid: 1 day, 35°C
   0.5% osmic acid: 1 day, 35°C.
3. Wash in distilled water: 1 day.
4. Dehydrate, clear, and embed.
5. Section 6–7 μ, mount, and dry.
6. Deparaffinize, clear, and mount.

RESULTS

Golgi apparatus—black
yolk and fat—black (these may be bleached out with turpentine)

COMMENTS

If it is advantageous to have mitochondria stained on the same slide, follow deparaffinization (step 6) by hydrating to water (include cautious treatment with 0.125% potassium permanganate) and stain by the Altmann Method (p. 341). Mitochondria will be crimson.

When using silver and osmic acid techniques, considerable experimentation may be necessary. During the impregnation keep solutions in a dark place and follow instructions for temperature carefully. When the solutions turn dark, renew them.

Nassonov-Kolatchew Method (Nassonov 1923, 1924)

This method does not use the graded osmic acid series.

FIXATION

24 hours in:

3% aqueous potassium dichromate	10.0 ml
1% aqueous chromic acid	10.0 ml
2% aqueous osmic acid	5.0 ml

PROCEDURE

1. Wash in running water: 24 hours.
2. Place in 2% aqueous osmic acid, 40°C: 8 hours, or 35°C: 3–5 days.
3. Wash in running water: overnight.
4. Dehydrate rapidly, clear, embed, and section, 2–4μ.
5. If sections are too dark, turpentine (preferably old and oxidized) will remove most of the excess color.
6. Proceed to deparaffinize, etc., according to desired process. Mounted sections may be counterstained with Altmann's stain, p. 341.

RESULTS

Golgi apparatus—black
background—yellow
mitochondria after Altmann's—red

COMMENT

In interpreting the results, bear in mind that the mitochondria may become impregnated if the period in osmic acid is prolonged.

Saxena Method (1957)

FIXATION

3–6 hours in:

barium chloride	1.0 g
distilled water...........................	85.0 ml

Add just before use:

> formalin ....   15.0 ml

Ramón y Cajal reducing solution

> hydroquinone ....   1.5 g
> formalin ....   15.0 ml
> distilled water..........................   100.0 ml
> sodium sulfite, anhydrous ....   0.5 g

Gold chloride

> gold chloride stock solution (p. 543) ....   1.0 ml
> distilled water...........................   80.0–90.0 ml

PROCEDURE

1. Rinse blocks of tissue in distilled water.
2. Impregnate in 1.5% aqueous silver nitrate, room temperature: 1–2 days. (Use 1% for very small pieces or embryonic tissues, 2% for fatty tissues and spinal cord.)
3. Rinse, 2 changes, in distilled water.
4. Cut blocks into slices thinner than 2 mm. Reduce in developer: 5 hours.
5. Wash thoroughly in distilled water.
6. Dehydrate, infiltrate, and embed.
7. Section, 6–7μ, mount, and dry.
8. Deparaffinize and hydrate to water.
9. Tone in gold chloride: 2 hours.
10. Rinse in distilled water and fix in 5% aqueous sodium thiosulfate: 3 minutes.
11. Wash in running water: 5 minutes.
12. Countestain, if desired, in hematoxylin, thionin, carmalum, etc.
13. Dehydrate, clear, and mount.

RESULTS

Golgi—black
cytoplasm—gray
mitochondria—medium to dark gray or black

COMMENTS

The silver preparations depend on fixation with salts of a heavy metal; barium, in the Saxena method (p. 337). Aoyama (1930) varies the method, using:

cadmium chloride	1.0 g
formalin	15.0 ml
distilled water	85.0 ml

The rest of the procedure is the same as that of Saxena.

Cold-blooded animal tissues require longer fixation and impregnation than do warm-blooded animal tissues (Aoyama 1930).

McDonald (1964) uses nitrocellulose embedding for silver-impregnated Golgi tissues, because the material is very friable and often difficult to handle when embedded with paraffin.

Direct Silver Method (Elftman 1952)

This is a one-step procedure, silver and fixative in one action.

PROCEDURE

1. Immerse small blocks of fresh tissue in silver nitrate in formalin (2 g/100 ml 15% formalin): 2 hours.
2. Rinse briefly in distilled water.
3. Develop for 2 hours in:

hydroquinone	2.0 g
formalin, 15% (15 ml/85 ml water)	100.0 ml

4. Return to 10% formalin to complete fixation: at least overnight.
5. Wash, dehydrate, and embed.
6. Section, 6–7μ; mount, and dry.
7. Deparaffinize, clear, and mount.

RESULT

Golgi—black

COMMENTS

Do not use buffered formalin, which may limit the solubility of the silver salts.

If the silver is too dense, Elftman suggests bleaching it with 0.7% iron alum. Check under the microscope and stop the reaction by washing thoroughly in running water.

Because the silver is readily oxidized, gold toning is usually preferable for a more permanent slide. This can follow deparaffinization; see other Golgi methods. Also counterstaining may be included before dehydrating and clearing slides.

Elftman warns that all the black is not necessarily Golgi.

Sudan Black Method (Baker 1944)

FIXATION

Formalin-calcium (p. 15): 3 days.

EMBEDDING AND SECTIONING

Embed in gelatin (p. 430). Harden the block in formalin-calcium-cadmium:

formalin, concentrated	10.0 ml
10% aqueous calcium chloride	10.0 ml
10% aqueous cadmium chloride	10.0 ml
distilled water............................	70.0 ml

Wash in running water: 3–4 hours. Section on freezing microtome, 15μ.

PROCEDURE

1. Sections can be affixed to slides as for frozen sections, or they can be stained first and then mounted on slides.
2. Transfer through 50% and 70% alcohol: 1–2 minutes each.
3. Place in saturated solution of Sudan black in 70% alcohol: 7 minutes.
4. Remove excess stain in 3 changes of 50% alcohol: 1–2 minutes each.
5. Rinse in distilled water.
6. Counterstain with a light hematoxylin or a red nuclear stain.
7. Mount in glycerin jelly or Von Apathy gum syrup (p. 104).

RESULT

Golgi—black. The vesicles, not the network, are stained. (See also Malhotra 1961).

MITOCHONDRIA STAINING

Mitochondria (from Greek *mittos,* filament; and *kondria,* granule) are tiny complex cell organelles, which are bounded by a double membrane and include inner membranes (cristae), a ground substance (matrix), and occasionally dense granulations. Chemically the mitochondria matrix consists mainly of proteins and lipids, and the membranes are high in insoluble protein, phospholipid, and insoluble enzyme content. The osmiophilic state of the membranes is due to the lipid component. Mitochondria are able to concentrate a large number of substances, such as proteins, lipids, metals, ferments, virus-like particles, and several chemical substances. They are known to contain numerous enzymes— oxidative enzymes, cytochrome oxidase, cytochrome C, succinic dehydrogenase, glutamic dehydrogenase, and adenosine triphosphatase, to name only a few. Mitochondria participate in forming cell organelles; probably of mitochondrial origin are the granules of granulocytes, platelets, neutrophilic myelocytes, zymogen bodies, yolk platelets, secreting granules in neurohypophysis, etc. The electron microscope is responsible for great contributions to the knowledge of the ultrastructure of mitochondria (Novikoff and Podber 1957; Rouiller 1960; *Symposium: The Structure and Biochemistry of Mitochondria,* 1953). (See also histochemistry, succinic dehydrogenase, p. 415).

Altmann Method

FIXATION

Regaud fixative (p. 24): change every day for 4 days; store in refrigerator. Mordant in 3% potassium dichromate: 8 days, change every second day. Wash in running water overnight; dehydrate and embed. Cut sections $2–4\mu$.

SOLUTIONS

Altmann aniline fuchsin

Make a saturated solution of aniline in distilled water by shaking the two together. Filter. Add 10 g acid fuchsin (C.I. 42685) to 100 ml of filtrate. Let stand for 24 hours. Good for only 1 month.

PROCEDURE

1. Deparaffinize and hydrate slides to water.
2. Treat with 1% aqueous potassium permanganate: 30 seconds (see comments below).
3. Rinse briefly in distilled water and bleach in 5% aqueous oxalic acid: 30 seconds.
4. Rinse in several changes distilled water: 1–2 minutes total.
5. Place in steaming Altmann's solution: 5–10 minutes. Remove heat when the slides have been placed in the solution.
6. Differentiate in dilute (0.1%) sodium carbonate until the cytoplasm is pale pink or almost colorless.
7. Stop differentiation and heighten the color by a brief dip in 1% HCl.
8. Wash in distilled water: several dips.
9. Stain in 0.5% methylene blue: 5–6 seconds.
10. Rinse in distilled water and a quick dip in 1% HCl.
11. Wash in distilled water: several seconds.
12. Dehydrate, clear, and mount.

RESULTS

mitochondria—bright red
nuclei—blue to bluish green

COMMENTS

Iron hematoxylin can also be used for mitochondria. Best fixative is Regaud; mordant in 3% potassium dichromate: 7 days. Use the long method of staining: overnight in iron alum and overnight in hematoxylin. The method is not as specific as Altmann; other granules also stain.

Osmic Method (Newcomer 1940)

FIXATION

Zirkle solution: 48 hours.

potassium dichromate......................	1.25 g
ammonium dichromate	1.25 g
copper sulfate.............................	1.0 g
distilled water............................	100.0 ml

PROCEDURE

1. Wash tissue blocks 8 hours to overnight.
2. Impregnate in 2% aqueous osmic acid: 4–6 days. Change solutions on alternate days.
3. Wash: 8 hours or overnight.
4. Dehydrate, clear, and embed. Use benzene, not xylene, for clearing.
5. Cut 5μ sections, mount on slides.
6. Deparaffinize and hydrate slides to water.
7. Bleach in 1% aqueous potassium permanganate: 5 minutes.
8. Rinse in distilled water.
9. Treat with 3% aqueous oxalic acid: 2–3 minutes.
10. Wash in running water: 15 minutes.
11. Dehydrate, clear, and mount.

RESULT

mitochondria—black

COMMENTS

Newcomer used his method on plant cells. A counterstain such as acid fuchsin can be added.

Short Acid-Fuchsin Method (Novelli 1962)

FIXATION

10% formalin or 1% osmic acid, room temperature: 24 hours. Formalin-fixed tissues are easier to process. Wash in running water: 4 hours.

EMBEDDING AND SECTIONING

Paraffin method, $1-4\mu$

SOLUTION

acid fuchsin (C.I. 42685)	0.2 g
methyl blue (C.I. 42780)	0.1 g
N hydrochloric acid	100.0 ml

PROCEDURE

1. Deparaffinize and hydrate slides to water.
2. Stain: 5 minutes.
3. Rinse gently in distilled water.
4. Dehydrate quickly through 95% alcohol and 2 changes of absolute alcohol.
5. Clear and mount.

RESULTS

mitochondria—purple red with peripheral blue wall
chromatin and collagen—blue
nucleoli and erythrocytes—brilliant red

Acid-Hematein Method (Hori and Chang 1963)

SOLUTIONS

Acetone solution

acetone	50.0 ml
uranyl nitrate	0.01–0.02 g
chloral hydrate	0.01–0.02 g

Baker acid hematein. See Hori method, p. 290.

Borax-ferricyanide solution

borax ($Na_2B_4O_7 \cdot 10H_2O$)	250.0 mg
potassium ferricyanide ($K_3Fe(CN)_6$)	250.0 mg
distilled water............................	100.0 ml

Keeps in refrigerator.

PROCEDURE

1. Prepare and section tissues by freeze-substitution method of Chang and Hori, p. 393.
2. Transfer the sections into a screw-cap bottle of acetone mixture. Prechill the solution in crushed ice in a Dewar flask and leave buried in dry ice overnight. The presence of metal in the acetone intensifies the lipid reaction.
3. Rinse briefly in acetone at room temperature and mount on cover glasses.
4. Chromate in 5% aqueous potassium dichromate, 60°C: 5–6 hours.
5. Wash in distilled water: 5 minutes.
6. Stain in Baker acid hematein: 30 minutes.
7. Wash in distilled water.
8. Differentiate in borax-ferricyanide, room temperature: 18 hours.
9. Dehydrate, clear, and mount.

RESULTS

mitochondria—sharply stained blue black
lipids in other cellular elements—may also stain

COMMENTS

Caulfield (1957) reports that tonicity has a direct effect on the appearance of the mitochondria; low tonicity produced swollen mitochondria, and high tonicity shrank them. He recommends fixing 1 hour at 0–4°C in the following.

Caulfield solution

stock buffer (see below)	5.0 ml
0.1 N hydrochloric acid	5.0 ml
distilled water.............................	2.5 ml
2% osmic acid	12.5 ml

Adjust pH to 7.4, if necessary, by adding a few drops of 0.1 N HCl or stock buffer. Then add 0.045 g of sucrose per ml of solution.

Stock buffer

sodium veronal............................	14.714 g
sodium acetate (3H₂O)	9.714 g

Dilute to 500.0 ml.

See Chang (1956) for method with frozen dried tissues, Benés (1960) for method using amidoblack 10B, Avers (1963) for fixing and staining for the electron microscope, and p. 415 of this book for demonstrating enzymes in mitochondria.

Takaya (1967) uses the Luxol fast blue method (p. 197) but counterstains with 0.5% aqueous phloxin instead of PAS.

SUPRAVITAL STAINING

Certain dyes will penetrate living cells and stain specific organelles within the cells. Neutral red reacts on certain granulations, products of the cytoplasm, secretory granules, vacuoles of digestion, and others. These vacuoles or granules appear to be heterogeneous and contain a phospholipoprotein complex, acid phosphatase, lipase, and some alkaline phosphatase, and they apparently correspond to lysosomes (Koenig 1963; Ogawa and Okamoto 1961). Neutral red is commonly used in combination with Janus green B for the vital staining of Golgi and mitochondria in a single preparation. Neutral red, however, does not stain a reticular Golgi apparatus, but does stain vesicles (lysosomes) in the region of the apparatus.

Janus green B does vitally stain the mitochondria and is dependent upon the enzymatic activites of the cell. The staining will take place under partial anaerobic conditions, but it will decolorize when all oxygen is removed, and it is reversibly inhibited by cyanide. Janus green B can be reduced in both mitochondrial and nonmitochondrial portions of the cell, but the localization of a cytochrome oxidase system in the mitochondria slows down the rate of reduction in that organelle, while reduction proceeds more rapidly in the other portions of the cell. It is erroneous to consider all isolated structures taking up Janus green B as mitochondria; secretion granules of the islets of Langerhans and of enterochromaffin and others will stain as well.

Neutral Red–Janus Green Method

The dyes should be certified for vital staining so they are not toxic to living cells. The solutions used for staining are so weak that it is convenient to make them up as stock solutions of greater strength. They will keep indefinitely if stored in glass-stoppered bottles. If using a pipette with a rubber bulb, do not allow the dye to run up into the bulb. Make the diluted solutions in small quantities, and if they are mixed together, use immediately. The mixture is not stable. Slides must be clean. They should be cleaned in dichromate solution, washed well in running water and then in distilled water, and stored in 95% alcohol. Rub dry just before spreading with stain.

SOLUTIONS

Neutral red stock solution

neutral red (C.I. 50040)	0.5 g
neutral absolute ethyl alcohol	100.0 ml

Janus green B stock solution

Janus green B (C.I. 11050)	0.5 g
neutral absolute ethyl alcohol	100.0 ml

The alcohol must be neutral and should be distilled over calcium.

PROCEDURE

1. Dilute neutral red stock solution, 1 ml to 10 ml of neutral absolute ethyl alcohol.
2. Mix 0.4 ml of Janus green B stock with 3 ml of dilute neutral red solution.
3. Flood clean slides with mixed dyes and touch edge of slide to absorbent material to draw off excess solution. Allow to dry horizontally in warm air, but keep them protected from dust. Be sure to mark the stained surface. The stain should be distributed thinly and evenly. When dry, the slides can be stored in a dustfree box. Sometimes successful smears can be made by placing a drop of solution at the end of the slide and smearing it like a blood film.

   Another method is to make enough solution to fill a coplin jar or bottle of comparable size. Gently flame a clean slide and plunge it while hot into the stain. Withdraw the slide and drain it upright on absorbent paper. Wave it rapidly in the air to quick-dry it and to preserve even distribution on the slide. If the humidity is high, dry the smears in hot air by using a hair dryer or other drying equipment. The smear may not be as uniform as one following slower drying, but it is necessary to speed up the drying because humidity allows the alcohol to take up moisture and ruin the dye film before the evaporation is complete.
4. Place a drop of live cells or organisms on a clean cover glass and carefully lower onto a stained slide. Seal with vaseline.

Blood, bone marrow, and other tissue fluids or cells may be diluted with an equal amount of autogenous heparinized plasma and examined on a stained slide. After a few trials and errors, the proper size of drop will be realized. With blood, the red cells should spread in a single layer. If the preparations are too thick, the leukocytes will round up instead of flatten. If air bubbles are trapped under the cover glass, the cells do not spread properly. If the cover glass is dropped on the stain or if there is too little of the suspension, the cell membranes may rupture.

5. If a warm stage is not available, place slides of tissue cells in a 37°C incubator for 20 minutes. Then examine immediately. Cells begin to round up as the slide cools.

6. When the mitochondria or neutral red bodies are sufficiently stained, if it is desirable to make the slides semipermanent, clean off the vaseline and slip the cover glass toward the edge of the slide, so that the blood or other fluid is spread evenly on both the slide and the cover glass. Slip the cover glass off and rapidly dry both the slide and the cover glass. (Both carry cell preparations.) If dried too slowly, the cells will be distorted. The drying may be completed in a vacuum desiccator (2–4 hours) or by shaking in several changes of anhydrous ether. The anhydrous ether is essential to remove fat from bone marrow. Clear in xylene and mount.

RESULTS

mitochondria—green
neutral red bodies—red

Schwind Method (1950)

Schwind replaces Janus green B with pinacyanole, which does not interfere with the staining of vacuolar systems by neutral red, does not fade from the mitochondria as rapidly as Janus green B, and does not interfere with the cell motility.

SOLUTIONS

Neutral red stock solution

| neutral red (C.I. 50040) .... | 0.2 g |
| neutral absolute ethyl alcohol .... | 50.0 ml |

Keeps indefinitely.

Pinacyanole stock solution

pinacyanole[1]	0.05 g
neutral absolute ethyl alcohol	50.0 ml

Keeps for 6 months.

Working solution

neutral red stock	30	drops
pinacyanole stock	9	drops
neutral absolute ethyl alcohol	5.0 ml	

Mix and make films.

RESULTS

mitochondria—deep blue
chromatin—purplish blue

Jackson Fluorescent Method (1961) for Blood

SOLUTIONS

Acridine orange stock solution

acridine orange	0.1 g
95% ethyl alcohol	50.0 ml

Allow excess dye to settle on bottom of bottle.

Working solution

acridine orange stock	5	drops
95% ethyl alcohol	5.0 ml	

Do not draw up any of excess dye from stock solution; use from top of solution.

---

[1] Eastman Chemicals.

PROCEDURE

Flood clean slides and drain off excess stain. Dry in vertical position. Store in dustfree box. Apply cover glass and cells as above. Examine with ultraviolet light (McClung 1939; Pienaar 1962; Schwind 1950; Scott 1928).

RESULTS

neutrophils—bright green nuclei, orange granules
eosinophils—same as neutrophils, but with larger granules
basophils—yellow granules
lymphocytes—green nucleus, usually one large red granule, occasionally more than one granule
monocytes—typically folded bright green nucleus, dense green cytoplasm
mature red cells—no fluorescence
reticulocytes—bright orange reticulum
platelets—pale green with pale orange granules
LE bodies—pale green

Vital–Nonvital Stain (DeRenzis and Schechtman 1973)

Trypan blue alone does not accurately distinguish vital from nonvital cells. Dead cells are stained blue but live cells do not take up color and are difficult to identify or can be overlooked. This two-dye method corrects the problem.

Add 0.5 ml of cell suspension to 0.5 ml of 0.04% neutral red in balanced salt solution. Incubate at 37°C: 10 minutes. Add 0.5 ml of 0.5% trypan blue in balanced salt solution for 2–3 minutes. Mix well and place in a hemocytometer. The cells are now ready to be counted and have good color contrast. Red cells are viable; blue ones are not. If a cell contains both colors, it was damaged in processing. The method can be used on cells growing on glass. Remove culture medium and add stain.

Chapter 20
Staining Microorganisms

In this chapter, to simplify the specificity of staining methods, the parasitic microorganisms will be broken down into the following groups: bacteria, spirochetes, fungi, rickettsia, and viruses (Burrows 1954).

BACTERIA STAINING

Bacteria are customarily studied by direct microscopic observation and differentiated by shape, grouping of cells, presence or absence of certain structures, and the reaction of their cells to differential stains. Bacteria may be stained with aniline dyes—in a single dye, in mixed dyes, in polychromed dyes, or by differential methods. One of the most universally used stains was developed by the histologist Gram, while he was trying to differentiate the bacteria in tissue. His method separates bacteria into two groups: (1) those that retain crystal violet and are said to be Gram positive; (2) those that decolorize to be stained by a counterstain and are said to be Gram negative.

Some bacteria of high lipid content cannot be stained by the usual methods but require heat or long exposure to the stain. They are also difficult to decolorize. Because they resist acid alcohol, they have been given the name acid fast. The spiral forms will stain only faintly if at all and must be colored by silver methods.

Many bacteria form a capsule from the outer layer of the cell membrane, and the capsule appears like a halo around the organism, or over a chain of cells. This capsule will not stain in the customary stains; Hiss stain (p. 359), however, is simple and usually effective in this situation.

Some bacteria are able to form spores that can be extremely resistant to injurious conditions (heat, chemicals). Boiling will destroy some of these spores, but many are more resistant. Some bacteria have flagella—filamentous appendages for locomotion.

Bacteria can be classified according to shape: coccus, spherical; bacillus, rod shaped; spiral, a curved rod.

Gram Staining

When a Gram-staining procedure has been applied, a Gram-positive cell or organism retains a particular primary dye. This process includes mordanting with iodine to form (with the dye) a precipitate that is insoluble in water and is neither too soluble nor insoluble in alcohol, the differentiator. According to Bartholomew and Mittwer (1950, 1951), the mechanism of stainability can be explained by differences in the permeability of the cell membrane. The bacteria stain by linkage between acidic groups of the bacteria and alkaline groups of dye. Iodine forms a complex with the dye, and this complex is dissociated by alcohol. If alcohol passes easily through the membrane, decolorization is rapid and the reaction is Gram negative. If the membrane is hardly or not at all penetrated by alcohol, the reaction is Gram positive. The condition of the membrane, therefore, is important; it must be intact.

Pearse (1960) suggests that the initial acidic protein–crystal violet complex is broken by the iodine, which then combines with the basic groups at the ends of the triphenylmethane molecule to form a relatively insoluble crystal violet–iodine precipitate. This precipitate is not easily removed if the following conditions exist: (1) if more of the crystal violet–iodine complex has formed where originally there had been more crystal violet in combination with protein; (2) if there are physical barriers, consisting of lipid or lipo-protein membranes, that resist extraction of the crystal violet–iodine.

For more details about Gram staining, see Bartholomew and Finkelstein (1959) and Mittwer et al. (1950).

Gram-Weigert Method (Krajian and Gradwohl 1952)

FIXATION

10% formalin or Zenker formol.

SOLUTIONS

Eosin

eosin Y (C.I. 45380)	1.0 g
distilled water............................	100.0 ml

Sterling gentian violet

crystal violet (C.I. 42555) (gentian violet)	5.0 g
95% ethyl alcohol	10.0 ml
aniline oil	2.0 ml
distilled water............................	88.0 ml

Mix aniline oil with water and filter. Add the crystal violet dissolved in alcohol. Keeps for several weeks to months.

Gram iodine solution, see p. 544.

PROCEDURE

1. Deparaffinize and hydrate slides to water. Remove $HgCl_2$.
2. Stain in eosin: 5 minutes.
3. Rinse in water.
4. Stain in Sterling gentian violet solution: 3 minutes for frozen sections; 10 minutes for paraffin sections.
5. Wash off with Gram iodine and then flood with more of same solution: 3 minutes.
6. Blot with filter paper.
7. Flood with equal parts of aniline oil and xylene; reflood until color ceases to rinse out of sections.
8. Clear in xylene and mount.

RESULTS

Gram-positive bacteria, and fungi—violet
Gram-negative organisms—not usually stained
fibrin—blue black

Leaver et al. (1977) Substitute for Brown and Brenn

FIXATION

10% formalin preferred.

SOLUTIONS

Crystal violet

crystal violet (C.I. 42555) (gentian violet)	1.0 g
distilled water.............................	100.0 ml

Gram iodine, see p. 544.

Sandiford stain

malachite green (C.I. 42000)	0.05 g
pyronin Y (C.I. 45005)	0.15 g
distilled water.............................	100.0 ml

PROCEDURE

1. Deparaffinize and hydrate sections to water.
2. Stain in crystal violet: 3 minutes.
3. Rinse in tap water.
4. Treat with Gram iodine: 3 minutes.
5. Rinse in tap water and blot almost dry.
6. Differentiate in equal parts of acetone and absolute alcohol until no more blue color comes off.
7. Wash in running tap water: 2–3 minutes.
8. Counterstain with Sandiford stain: 2 minutes.
9. Rinse in tap water and blot almost dry.
10. Differentiate for a few seconds in 95% alcohol and dehydrate in absolute alcohol, clear, and mount.

RESULTS

Gram-positive organisms—purple black
Gram-negative organisms—red
background—blue green

COMMENTS

I have found this method to show the best gram-stained organisms of any of the Brown and Brenn type methods.

Acid-Fast Staining

In acid-fast staining, the presence of phenol in the stain solution appears to be essential; it must, in some way, influence the dye or substance that imparts the acid-fast character. Lartique and Fite (1962) propose that phenol decreases the solubility of fuchsin in water and increases its solubility in the lipids of the bacillus. The acid-fastness can be enhanced by artificially coating the bacilli with oil. Actually, no specific chemical role can be assigned to phenol; it does not combine with the fuchsin, nor does it cause capsular disruption or protein denaturation. Harada's (1976) work indicated that adding phenol made water-soluble dyes, such as basic fuchsin, more lipid soluble. He also concluded it was possible that the presence of hydroxyl and free carboxyl groups are necessary in a mycolic acid type of long chain fatty acids to assure acid-fastness.

Staining of lepra bacilli is improved if, when removing paraffin from sections, 15% of mineral oil is added to the xylene (5 minutes). Wash 30 seconds in detergent solution (Haemosol), wash in water, and proceed to stain.

Harada Method (1973) a Ziehl-Neelsen type.

FIXATION

Any general fixative, but 10% formalin is best.

SOLUTIONS

Carbol-fuchsin

basic fuchsin (C.I. 42500), saturated alcoholic .	10.0 ml
phenol, 5% aqueous	90.0 ml

Methylene blue

methylene blue (C.I. 52105)	3.0 g
absolute ethyl alcohol	30.0 ml
potassium hydroxide, 0.01% aqueous	100.0 ml

PROCEDURE

1. Deparaffinize and hydrate sections to water.
2. Oxidize in 1% potassium permanganate, aqueous: 1 hour.

3. Rinse in tap water and bleach in 1% oxalic acid, aqueous: 3 minutes.
4. Stain in carbol-fuchsin, warmed to steaming: 5 minutes.
5. Rinse in distilled water.
6. Decolorize in 1% HCl in 70% alcohol: approximately 20 seconds.
7. Wash in running water: 2–3 minutes.
8. Counterstain in Loeffler methylene blue diluted 1 : 9 with distilled water: 15 seconds.
9. Rinse in tap water.
10. Dry smears and mount. Dehydrate sections, clear, and mount.

RESULTS

acid-fast bacteria—red
red blood corpuscles—pink
mast cell granules—deep blue
other bacteria—blue
nuclei—blue

COMMENTS

Marti and Johnson (1951) add 1 drop Turgitol 7 to every 30 ml of carbol-fuchsin solution and counterstain with malachite green, 1% aqueous.

Pottz et al. (1964) also use a wetting agent, dimethylsulfoxide (DMSO). Their solution is:

basic fuchsin	4.0 g
95% ethyl alcohol	25.0 ml
phenol, liquified	12.0 g
glycerol	25.0 ml
DMSO	25.0 ml

Bring solution to 160 ml with distilled water.

See Harada (1976) for a methenamine-silver method.

Fite-Formaldehyde Method (Wade Modification, 1957)

FIXATION

Zenker preferred; removal of $HgCl_2$ not necessary, since it disappears during processing.

SOLUTIONS

Phenol new fuchsin

new fuchsin (C.I. 42520) (magneta III)	0.5 g
phenol (carbolic acid)	5.0 g
ethyl or methyl alcohol	10.0 ml
distilled water to make	100.0 ml

Van Gieson, modified

acid fuchsin (C.I. 42685)	0.01 g
picric acid	0.1 g
distilled water............................	100.0 ml

PROCEDURE

1. Deparaffinize in turpentine–paraffin oil (2 : 1) 2 changes: 5 minutes total.
2. Drain, wipe off excess fluid, blot to opacity, place in water.
3. Stain overnight in phenol-fuchsin, room temperature.
4. Rinse off excess stain in tap water.
5. Treat with formalin (full strength): 5 minutes. (The tissue will turn blue.)
6. Wash in running water: 3–5 minutes.
7. Treat with sulfuric acid, 5% (5 ml/95 ml water): 5 minutes.
8. Wash in running water: 5–10 minutes.
9. Treat with potassium permanganate, 1% (1 g/100 ml water): 3 minutes.
10. Wash in running water: 3 minutes.
11. Treat with oxalic acid, 2–5% (2–5 g/100 ml water) individually with agitation: not more than 30 seconds. Slides can remain in water while others are being treated.
12. Stain with modified Van Gieson: 3 minutes. *Do not wash*.
13. Rinse for few seconds in 95% alcohol.
14. Dehydrate, clear, and mount.

RESULTS

acid-fast bacteria—deep blue or blue black
connective tissue—red
other tissue elements—yellowish

COMMENTS

Beamer and Firminger (1955) emphasize care in the use of formalin. Old formalin yields poor results; a redistilled form produces the most brilliant staining. The reagent grade in 16 oz. brown bottles, if kept tightly closed, gives a good stain.

Tilden and Tanaka (1945) outline a method for frozen sections (essentially the same as the Fite-formaldehyde method), after the sections have been mounted with a celloidin protective coat (or try subbed slides). They also emphasize the need for good formalin; if the sections fail to turn blue, try a different batch.

Fluorescent Method (Bogen 1941; Richards 1941; Richards, et al. 1941)

FIXATION

10% formalin for sections; smears by heat.

SOLUTIONS

Auramin stain

auramin O (C.I. 41000).....................	0.3 g
distilled water............................	97.0 ml
melted phenol (carbolic acid)	3.0 ml

Shake to dissolve dye or use gentle heat. Solution becomes cloudy on cooling, but is satisfactory. Shake before using.

Decolorizing solution

hydrochloric acid, concentrated	0.5 ml
70% ethyl alcohol	100.0 ml
sodium chloride	0.5 g

PROCEDURE

1. Deparaffinize and hydrate sections to water.
2. Stain in auramin, room temperature: 2–3 minutes.
3. Wash in running water: 2–3 minutes.
4. Decolorize: 1 minute.

5. Transfer to fresh decolorizer: 2–5 minutes.
6. Wash in running water: 2–3 minutes.
7. Dry smear and examine. Mount sections in fluorescent mountant (p. 108) or Harleco Fluorescent Mountant for examining.

RESULT

bacilli—golden yellow

COMMENTS

Bogen counterstains with Loeffler alkaline methylene blue solution before mounting (p. 355).

See also Braunstein and Adriano (1960); Moody et al. (1958); Yamaguchi and Braunstein (1965).

Carter and Leise (1958) use a single fluorescent antiglobulin for various bacteria.

Capsule Staining

Hiss Method (Burrows 1954)

FIXATION

Any good general fixative; 10% formalin is satisfactory.

SOLUTIONS

Stain solution

Either of the following may be used:

A.
basic fuchsin (C.I. 42500) .... 0.15–0.3 g
distilled water............................. 100.0 ml

B.
crystal violet (C.I. 42555) .... 0.05–0.1 g
distilled water............................. 100.0 ml

Copper sulfate solution

copper sulfate crystals .... 20.0 g
distilled water............................. 100.0 ml

PROCEDURE

1. Deparaffinize and hydrate slides to water.
2. Flood with either staining solution and heat gently until the stain steams.
3. Wash off the stain with copper sulfate.
4. Blot, but do not dry.
5. Dehydrate in 99% isopropyl alcohol, 2 changes: 1 minute each.
6. Clear and mount.

RESULTS

capsules—light pink (basic fuchsin) or blue (crystal violet)
bacterial cells—dark purple surrounded by the capsule color

SPIROCHETE STAINING

These are spirally shaped organisms, which multiply by transverse fission. They cause relapsing fever, syphilis, and yaws.

Dieterle Method (Beamer and Firminger Modification 1955)

FIXATION

10% formalin.

SOLUTIONS

Dilute gum mastic

gum mastic, saturated in absolute alcohol	30	drops
95% ethyl alcohol	40.0 ml	

Developing solution

distilled water............................	20.0	ml
hydroquinone	0.5	g
sodium sulfite	0.06	g

While stirring add:

formalin	4.0 ml
glycerin	5.0 ml

When thoroughly mixed add, drop by drop, with constant stirring:

gum mastic saturated in absolute alcohol	4.0 ml
absolute alcohol	4.0 ml
pyridine	2.0 ml

PROCEDURE

1. Deparaffinize and hydrate slides to water.
2. Remove any formalin pigment in 2% ammonium hydroxide (2 ml/98 ml water): few minutes.
3. Transfer to 80% alcohol: 2–3 minutes.
4. Wash in distilled water: 5 minutes.
5. Treat with uranium nitrate 2–3% (2–3 g/100 ml water), previously warmed to 60°C: 10 minutes.
6. Wash in distilled water: 1–2 minutes.
7. Transfer to 95% alcohol: 1–2 minutes.
8. Treat with dilute gum mastic: 5 minutes.
9. Wash in distilled water, 3–4 changes, until rinse is clear.
10. Impregnate with silver nitrate, 2% (2 g/100 ml water), previously warmed to 60°C: 30–40 minutes.
11. Warm developing solution to 60°C. Dip slide up and down in solution until sections turn light tan or pale brown.
12. Rinse in 95% alcohol, then in distilled water.
13. Treat with silver nitrate, 2% (2 g/100 ml water): 1–2 minutes.
14. Wash in distilled water: 1–2 minutes.
15. Dehydrate, clear, and mount.

RESULTS

spirochetes—black
background—yellow

COMMENTS

Use thin sections, 4–5μ.

The uranium nitrate prevents the impregnation of nerve fibers and reticulum.

Warthin-Starry Silver Method (Bridges and Luna 1957; Faulkner and Lillie 1945a; Kerr 1938)

Caution: All glassware must be cleaned with potassium dichromate–sulfuric acid. Avoid contamination; coat forceps with paraffin. Solutions must be fresh (no more than 1 week old) and made from triple-distilled water. Carry a known positive control slide with test slide through the process or, preferably, a control section on the same slide.

FIXATION

10% formalin.

SOLUTIONS

Acidified water

triple-distilled water	1000.0 ml
citric acid	10.0 g

The pH can range from 3.8 to 4.4.

2% silver nitrate

silver nitrate	2.0 g
acidified water	100.0 ml

1% silver nitrate

Dilute a portion of the 2% silver nitrate solution with equal volume of acidified water.

0.15% hydroquinone

hydroquinone	0.15 g
acidified water	100.0 ml

5% gelatin

sheet or granulated gelatin of high degree of purity	10.0 g
acidified water	200.0 ml

Developer

2% silver nitrate	1.5 ml
5% gelatin	3.75 ml
0.15% hydroquinone	2.0 ml

Warm solutions to 55–60°C and mix in order given, with stirring. Use immediately.

PROCEDURE

1. Deparaffinize and hydrate slides to acidified water.
2. Impregnate in 1% silver nitrate, 55–60°C: 30 minutes.
3. Place slides on glass rods, pour on warm developer (55–60°C). When sections become golden brown or yellow, and the developer brownish black (3–5 minutes), pour off. The known positive can be checked under microscope for black organisms.
4. Rinse with warm (55–60°C) tap water, then distilled water.
5. Dehydrate, clear, and mount.

RESULTS

spirochetes—black
background—yellow to light brown
melanin and hematogenous pigments—may darken
Underdevelopment will result in pale background, very slender and pale spirochetes.
Overdevelopment will result in dense background, heavily impregnated spirochetes, obstructed detail, sometimes precipitate.

COMMENTS

Faulkner and Lillie (1945*a*) use water buffered to pH 3.6–3.8 with Walpole *M*/5 sodium acetate–*M*/5 acetic acid buffer (p. 553).

Levaditi Method for Block Staining (Mallory 1944)

FIXATION

10% formalin.

SOLUTIONS

Silver nitrate

silver nitrate	1.5–3.0 g
distilled water	100.0 ml

Reducing solution

pyrogallic acid	4.0 g
formalin	5.0 ml
distilled water	100.0 ml

PROCEDURE

1. Rinse blocks of tissue in tap water.
2. Transfer to 95% ethyl alcohol: 24 hours.
3. Place in distilled water until tissue sinks.
4. Impregnate with silver nitrate, 37°C, in dark: 3–5 days.
5. Wash in distilled water.
6. Reduce at room temperature in dark: 24–72 hours.
7. Wash in distilled water.
8. Dehydrate, clear in cedarwood oil, and infiltrate with paraffin.
9. Embed, section at 5μ, mount on slides, and dry.
10. Remove paraffin with xylene, 2 changes, and mount.

RESULTS

spirochetes—black
background—brownish yellow

For a fluorescent method, see Kellogg and Deacon (1964).

FUNGI STAINING

Open or draining lesions are difficult to examine for fungi because of heavy bacterial contamination, but dermatophytes are easily demonstrated. Scrapings from horny layers or nail plate or hair can be mounted in 10–20% hot sodium hydroxide. This dissolves or makes transparent the tissue elements and then the preparation can be examined as a wet mount. Fungi in tissue sections are readily stained.

Hotchkiss-McManus Method (McManus 1948)

FIXATION

10% formalin or any general fixative.

SOLUTIONS

Periodic acid

periodic acid	1.0 g
distilled water	100.0 ml

Schiff reagent, see p. 210.

Differentiator

potassium metabisulfite, 10% aqueous	5.0 ml
N HCl (p. 542)	5.0 ml
distilled water	100.0 ml

Light green

light green SF, yellowish (C.I. 42095)	0.2 g
distilled water	100.0 ml
glacial acetic acid	0.2 ml

PROCEDURE

1. Deparaffinize and hydrate slides to water; remove $HgCl_2$ if present.
2. Oxidize in periodic acid: 5 minutes.
3. Wash in running water: 15 minutes.
4. Treat with Schiff reagent: 10–15 minutes.
5. Differentiate, 2 changes: total 5 minutes.
6. Wash in running water: 10 minutes.
7. Stain in light green: 3–5 minutes. If too dark, rinse in running water.
8. Dehydrate, clear, and mount.

RESULTS

fungi—red. Not specific, however; glycogen, mucin, amyloid, colloid, and others may show rose to purplish red
background—light green

COMMENTS

To remove glycogen, starch, mucin, or RNA, see p. 211.

DePalma and Young (1963) use 0.1% basic fuchsin for 2 minutes with agitation instead of Schiff reagent. Differentiate only 1–2 minutes with agitation.

Gridley Method (1953)

FIXATION

Any good general fixative.

SOLUTIONS

Chromic acid

chromic acid..............................	4.0 g
distilled water............................	100.0 ml

Stable for 2 months.

Schiff reagent, see p. 210.

Sulfurous rinse

sodium metabisulfite, 10% (10 g/100 ml water) .	6.0 ml
N hydrochloric acid	5.0 ml
distilled water............................	100.0 ml

Aldehyde-fuchsin, see p. 326.

Metanil yellow

metanil yellow	0.25 g
distilled water............................	100.0 ml
glacial acetic acid	2 drops

PROCEDURE

1. Deparaffinize and hydrate slides to water; remove $HgCl_2$.
2. Oxidize in chromic acid: 1 hour.

3. Wash in running water: 5 minutes.
4. Place in Schiff reagent: 15 minutes.
5. Rinse in sulfurous acid, 3 changes: 1.5 minutes each.
6. Wash in running water: 15 minutes.
7. Stain in aldehyde-fuchsin: 15–30 minutes.
8. Rinse off excess stain in 95% alcohol.
9. Rinse in water.
10. Counterstain lightly in metanil yellow: 1 minute.
11. Rinse in water.
12. Dehydrate, clear, and mount.

RESULTS

mycelia—deep blue
conidia—deep rose to purple
background—yellow
elastic tissue, mucin—deep blue

Gomori Methenamine—Silver Nitrate Method (Grocott Adaptation, 1955; Mowry Modification, 1959)

FIXATION

10% formalin or any good general fixative.

SOLUTIONS

Methenamine–silver nitrate stock solution

silver nitrate, 5% aqueous	5.0 ml
methenamine, 3% aqueous	100.0 ml

Silver nitrate is considered to be stable for 2 weeks in refrigerator, methenamine for 1 month, but I recommend freshly prepared solutions for superior results.

Working solution

borax, 5% aqueous	2.0 ml
distilled water............................	25.0 ml
methenamine–silver nitrate stock solution	25.0 ml

Use within 24 hours.

Light green stock solution

light green SF, yellowish (C.I. 42095)	0.2–0.5 g
distilled water............................	100.0 ml
glacial acetic acid	0.2 ml

Working solution

light green stock solution	10.0 ml
distilled water............................	100.0 ml

PROCEDURE

1. Deparaffinize and hydrate slides to water; remove $HgCl_2$.
2. Oxidize in periodic acid, 0.5% (0.5 g/100 ml water): 10 minutes.
3. Wash in running water: 3 minutes.
4. Oxidize in chromic acid, 5% (5 g/100 ml water): 45 minutes.
5. Wash in running water: 2 minutes.
6. Treat with sodium bisulfite, 2% (2 g/100 ml water): 1 minute to remove chromic acid.
7. Wash in running water: 5 minutes.
8. Rinse in distilled water, 2–3 changes: 5 minutes total.
9. Place in methenamine–silver nitrate, 58°C: 30 minutes. Do not use metal forceps. Sections appear yellowish brown.
10. Wash thoroughly, several changes distilled water.
11. Tone in gold chloride (10 ml stock solution/90 ml water) until sections turn purplish gray, fungi are black.
12. Rinse in distilled water.
13. Fix in sodium thiosulfate, 5% (5 g/100 ml water): 3 minutes.
14. Wash in running water: 5 minutes.
15. Counterstain in light green: 30 seconds.
16. Dehydrate, clear, and mount.

RESULTS

fungi—black
background—light green

COMMENTS

The Mowry modification uses oxidation with both periodic and chromic acid (former methods use only the latter) with the result that the final staining is stronger and more consistent than that of the original Gomori method.

Gold toning may be omitted and the methenamine stain followed by 8 minutes in each of 3 changes of 10% thiosulfate to give a brown instead of a black stain. An aldehyde-fuchsin stain may also be used for fungi.

Metachromatic Method (Kelly et al. 1962)

FIXATION

10% formalin.

SOLUTIONS

Sulfation reagent

Slowly add cold concentrated sulfuric acid to an equal volume of iced diethyl ether.

Toluidine blue

toluidine blue O (C.I. 52040)................	0.01 g
3% aqueous acetic acid	100.0 ml

PROCEDURE

1. Deparaffinize sections and transfer to absolute ethyl alcohol.
2. Dry sections in air: 5–10 minutes.
3. Sulfate: 5 minutes.
4. Wash in 3% aqueous acetic acid: 1 minute. Change solution after 10 slides.
5. Stain in toluidine blue: 5 minutes.
6. Wash in 3% aqueous acetic acid: 1 minute.
7. Dehydrate in absolute ethyl alcohol: 1 minute.
8. Clear and mount.

RESULTS

fungi—red
background—pale blue or colorless

COMMENTS

Some fungal forms are detected by this method only; sulfation reveals previously undetected mycelia of *Candida*. The 1,2 glycol groups (—CHOH—CHOH—) in fungi apparently are esterified by sulfuric acid to form ester sulfate groups (—OSO_3), which react metachromatically with toluidine blue.

Schneider (1963) simplifies a Gram stain for fungi.

Fluorescent Method (Pickett et al. 1960)

FIXATION

10% formalin, Zenker, alcohol, and other general fixatives.

SOLUTIONS

Acridine orange

acridine orange	0.1 g
distilled water	100.0 ml

Weigert hematoxylin, see p. 113.

PROCEDURE

1. Deparaffinize and hydrate sections to water.
2. Stain in Weigert hematoxylin: 5 minutes.
3. Wash in running water: 5 minutes.
4. Stain in acridine orange: 2 minutes.
5. Rinse in tap or distilled water: 30 seconds.
6. Dehydrate in 95% alcohol: 1 minute.
7. Dehydrate in absolute alcohol, 2 changes: 2–3 minutes.
8. Clear in xylene, 2 changes: 2–3 minutes.
9. Mount in a nonfluorescing medium.

RESULTS

All fungi fluoresce except *Nocardia* and *Rhizopus;* colors of the fluorescing genera appears as follows:

Coccidioides, Rhinosporidum—red
Aspergillus—green

Actinomyces, Histoplasma—red to yellow
Candida, Blastomyces dermatitids, Monosporium—yellow green
Blastomyces brasiliensis—yellow

COMMENTS

Old hematoxylin and eosin slides may be decolorized and restained as above.

Weigert hematoxylin staining is necessary as a quenching agent because some fungi are difficult to see against a background which also fluoresces. With the hematoxylin, the fungi fluoresce brightly against a dark setting.

For a fluorescent method, see Clark and Hench (1962), Mote et al. (1975).

For fluorescent antibody techniques, see Batty and Walker (1963); Procknow et al. (1962).

STAINING OF RICKETTSIAE AND INCLUSION BODIES

Rickettsiae are very small, Gram-negative coccobacillary-type microorganisms associated with typhus and spotted fever and related diseases. They may appear as cocci or short bacilli, and may occur singly, in pairs, or in dense masses. Most of them are intracellular. Some species are found only in the cytoplasm; others prefer the nucleus. Rickettsiae stain best in a Giemsa-type stain or by the Ordway-Machiavello method.

Viruses are microorganisms too small to be visible under the microscope, and they are capable of passing through filters. Viruses are responsible for many diseases, including yellow fever, poxes, poliomyelitis, influenza, measles, mumps, shingles, rabies, colds, infectious hepatitis, infectious mononucleosis, trachoma, psittacosis, and foot-and-mouth disease. In tissue sections and smears, viruses are characterized by elementary bodies and inclusion bodies. Elementary bodies are infectious particles, and inclusion bodies are composed of numerous elementary bodies. Both types of bodies vary in size and appearance; some, such as rabies, psittacosis, and trachoma, are located in the cytoplasm of infected cells; and some, such as poliomyelitis, are intranuclear. Special staining methods can demonstrate them effectively.

Modified Pappenheim Stain (Castañeda 1939)

FIXATION

Any general fixative; Regaud recommended.

SOLUTIONS

Stock Jenner stain

Jenner stain	1.0 g
methyl alcohol, absolute	400.0 ml

Stock Giemsa stain

Giemsa stain	1.0 g
glycerin	66.0 ml

Mix and place in oven 2 hours, 60°C. Add:

methyl alcohol, absolute	66.0 ml

Working solution A

distilled water	100.0 ml
glacial acetic acid	1 drop
Jenner stock solution	20.0 ml

Working solution B

distilled water	100.0 ml
glacial acetic acid	1 drop
Giemsa stock solution	5.0 ml

PROCEDURE

1. Deparaffinize and hydrate slides to water.
2. Stain in solution A, 37°C: 15 minutes.
3. Transfer directly to solution B, 37°C: 30–60 minutes.
4. Dehydrate quickly, 2 changes absolute alcohol.
5. Clear and mount.

RESULT

rickettsiae—blue to purplish blue

Castañeda Method (Gradwohl 1956)

FIXATION

Any general fixative; Regaud recommended.

SOLUTIONS

Buffer solution

Solution A

dibasic sodium phosphate ($Na_2HPO_4 \cdot 12H_4O$) .	23.86 g
distilled water.............................	1000.0 ml

Solution B

monobasic sodium phosphate, anhydrous	
(NaH_2PO_4)	11.34 g
distilled water.............................	1000.0 ml

Working solution

solution A	88.0 ml
solution B	12.0 ml
formalin	0.2 ml

Methylene blue solution

Dissolve methylene blue (C.I. 52015)	2.10 g
in 95% alcohol	30.00 ml
Dissolve potassium hydroxide	.01 g
in distilled water	100.00 ml

Mix the two solutions and let stand 24 hours.

Staining solution

Mix buffer working solution with 1.0 ml methylene blue solution.

PROCEDURE

1. Deparaffinize and hydrate slides as far as 50% alcohol.
2. Stain in methylene blue: 2–3 minutes.
3. Wash in running water: 30 seconds.
4. Counterstain in 1% aqueous safranin: 1–2 minutes.
5. Dip briefly in 95% alcohol.
6. Dehydrate in 2 changes absolute alcohol, clear, and mount.

RESULT

rickettsiae and inclusion bodies—light blue

COMMENTS

.Burrows (1954) recommends this as one of the best methods for rickettsiae. The Giménez (1964) is a good one, too.

For a fluorescent method, see Anderson and Grieff (1964).

Ordway-Machiavello Method (Gradwohl 1956)

FIXATION

Regaud recommended.

SOLUTION

Poirier blue (C.C.), 1% aqueous	20.0 ml
eosin bluish (C.I. 45400), 0.45% aqueous	15.0 ml

Mix just before use. Add slowly with constant shaking:

distilled water...........................	25.0 ml

Use within 24 hours.

PROCEDURE

1. Deparaffinize and hydrate slides to water.
2. Stain: 6–8 minutes.
3. Decolorize in 95% ethyl alcohol until slides appear pale bluish pink.
4. Dehydrate in absolute alcohol: 1 minute.
5. Clear and mount.

RESULTS

rickettsiae and inclusion bodies—bright red
nuclei and cytoplasm—sky blue

Giemsa Method (American Public Health Association 1956)

FIXATION

Any general fixative; Regaud recommended.

SOLUTIONS

Giemsa stock solution, see p. 372.

Buffer solutions

Solution A

dibasic sodium phosphate, anhydrous (Na_2HPO_4)	9.5 g
distilled water	1000.0 ml

Solution B

monobasic sodium phosphate ($NaH_2PO_4 \cdot H_2O$)	9.2 g
distilled water	1000.0 ml

Working solution, pH 7.2

solution A	72.0 ml
solution B	28.0 ml
distilled water	900.0 ml

Giemsa working solution

Dilute 1 drop Giemsa stock solution with 5 ml of buffer working solution.

PROCEDURE

1. Deparaffinize and hydrate slides to water. (Smears can be carried directly to water: wash well in running water: 5–10 minutes.)
2. Leave slides in Giemsa working solution overnight, 37°C.
3. Rinse thoroughly in distilled water. Dry between blotters.
4. Dip rapidly in absolute alcohol. If overstained, use 95% alcohol to decolorize them, dip in absolute alcohol.
5. Clear and mount. (Smears, after treatment with absolute alcohol, can be washed in distilled water: 1–2 seconds, and blotted dry. Examine with oil or mount with a cover glass.)

RESULT

rickettsiae and inclusion bodies (psittacosis)—blue to purplish blue

Modified Gomori Method for Trachoma

SOLUTION

Dilute 1 drop of stock Giemsa with 2 ml of buffer working solution above.

PROCEDURE

1. Hydrate slides to water (these will be smears).
2. Stain in working solution, 37°C: 1 hour.
3. Rinse rapidly, 2 changes, 95% ethyl alcohol.
4. Dehydrate in absolute alcohol, clear, and mount.

RESULT

inclusion bodies—blue to purplish blue

Schleifstein Method for Negri Bodies (1937)

FIXATION

Zenker preferred.

SOLUTION

basic fuchsin (C.I. 42500)	1.8 g
methylene blue (C.I. 52015)	1.0 g
glycerin	100.0 ml
methyl alcohol	100.0 ml

For use add about 10 drops to 15–20 ml of dilute potassium hydroxide (1 g/40,000 ml water). Alkaline tap water may be used. Keeps indefinitely.

PROCEDURE

1. Deparaffinize and hydrate slides to water; remove $HgCl_2$.
2. Rinse in distilled water and place slides on warm electric hot plate.
3. Flood amply with stain and steam for 5 minutes. Do not allow stain to boil.
4. Cool and rinse in tap water.

5. Decolorize and differentiate each slide by agitating in 90% ethyl alcohol until the sections assume a pale violet color. This is important.
6. Dehydrate, clear, and mount.

RESULTS

Negri bodies—deep magenta red
granular inclusions—dark blue
nucleoli—bluish black
cytoplasm—blue violet
erythrocytes—copper

COMMENTS

Schleifstein outlines a rapid method for fixing and embedding so the entire process can be handled in 8 hours, including fixing, embedding, and staining.

Massignani and Malferrari Method for Negri Bodies (1961)

FIXATION

10% formalin or saturated aqueous mercuric chloride–absolute alcohol (1:2).

EMBEDDING AND SECTIONING

Paraffin method, sections 4μ.

SOLUTIONS

Harris hematoxylin, see p. 113.

Dilute hydrochloric acid

hydrochloric acid, concentrated	1.0 ml
distilled water.............................	200.0 ml

Dilute lithium carbonate

lithium carbonate, saturated aqueous	1.0 ml
distilled water.............................	200.0 ml

Phosphotungstic acid–eosin stain

Grind together 1 g eosin Y (C.I. 45380) and 0.7 g phosphotungstic acid. Mix thoroughly into 10.0 ml of distilled water and then bring volume up to 200.0 ml with distilled water. Centrifuge at 1500 rpm: 40 minutes. Pour off supernatant solution but do not throw it away. Dissolve the precipitate in 50 ml of absolute alcohol. When dissolved, add to the supernatant solution. The solution is ready to use. If the eosin and phosphotungstic acid are not ground together and then dissolved, it requires 24 hours for the dye-mordant combination to form.

PROCEDURE

1. Deparaffinize and hydrate sections to water; remove $HgCl_2$.
2. Stain in hematoxylin: 2 minutes.
3. Wash in running water: 5 minutes.
4. Dip 8 times in dilute hydrochloric acid.
5. Wash in running water: 5 minutes.
6. Blue in dilute lithium carbonate: 1 minute.
7. Wash in running water: 5 minutes.
8. Dehydrate to absolute alcohol.
9. Stain in phosphotungstic acid–eosin: 8 minutes.
10. Rinse in distilled water.
11. Dehydrate by quick dips in 50%, 70%, 80%, and 90% alcohol; then follow with 95% alcohol: 1 second.
12. Complete dehydration in absolute alcohol, 2–3 changes: 4 minutes each.
13. Clear and mount.

RESULT

Negri bodies—deep red

COMMENTS

Massignani and Refinetti (1958) adapted the Papanicolaou stain (p. 463) for Negri bodies. Because further study led to the discovery that eosin combined with phosphotungstic acid is responsible for Negri body staining, the Massignani and Malferrari stain above was developed specifically for these bodies.

ANTIGENS AND ANTIBODIES

The subject of immunity is complex and has filled several large textbooks. It is mentioned briefly here to bring to mind certain principles used in this field, and to highlight one of the most important techniques developed for locating antigents and antibodies in tissues.

If foreign materials, living or nonliving—bacteria and viruses, for instance—invade the body, certain substances are formed in body fluids to combat these foreign materials and to "neutralize" them. The defense substances are proteins called antibodies, which have the power to combine specifically with the invading foreign materials (antigens), which induced the formation of the antibodies. Eventual immunity to an infection can be induced by entrance of the antigen, either naturally or artificially (by "shots," for instance). For artificial immunity, serum, called antiserum (immune serum), from an artificially immunized animal can be injected into a nonimmune animal to induce immunity. Antiserum can be used in this way to combat an infection already present (mumps, measles, anthrax, tetanus); or it can be used to prevent infection (measles, poliomyelitis, tetanus, diphtheria).

If the fluid portion of blood plasma is allowed to clot (fibrinogen precipitates out), a relatively stable serum remains. This is made up of two protein fractions: albumin and globulin. The globulins consist of two alpha globulins, one beta globulin, and one gamma globulin. It is the gamma globulin that is associated with immunity. Antibodies have been classically identified with gamma globulin (to a lesser extent with the other globulins) and are considered modified serum globulin. It has become a common practice to use for immune serum the gamma globulin fraction, which can be isolated from the other serum proteins and used in a concentrated form. The difference, therefore, between immune and normal serum globulin lies in the ability of the former to combine with an antigen.

If antigens can be labeled, sources of antibody production in tissues can be seen microscopically. Coons and his associates have coupled antibodies with fluorescein isocyanate and isothiocyanate to form fluorescent carbamide-proteins. The conjugates thus formed from dye and antibodies retain the specific reaction of the antibody for the antigen that causes its formation. Labeled antibodies are poured over the tissues or cells and will leave in the tissues a labeled protein that can be observed under a fluorescent microscope.

Some of the solutions, antisera (immune serum), and conjugates have to be prepared in the laboratory; some can be purchased,[1] but it is recom-

---

[1] Arnel Products Co. (Sylvania Chemicals); Baltimore Biological Laboratory.

mended that all conjugates be purified by absorption against tissue powders. For the reasons and means of doing this, see Coons (1958); Coons et al. (1955); and Mellors (1959). The method is complicated and should not be undertaken without a thorough study of the source material.

With a few exceptions (some polysaccharides), chemical fixatives must be avoided in order to retain the specific activity of the antigen. Smears, touch preparations, tissue cultures, and cell suspension can be used. If sections are preferred, freeze-dried or unfixed frozen sections cut in a cryostat (p. 393) are mandatory. Tissues may be quick-frozen in petroleum ether previously chilled to −65°C with a dry ice–alcohol mixture in a Dewar flask. When the tissues are completely frozen, blot and store in tightly stoppered test tube at −20°C to −70°C until used. Long storage is to be avoided because the tissues dehydrate (Tobie 1958). Coons et al. (1955) place small pieces of tissue against the wall of a small test tube and plunge it into alcohol cooled to −70°C with solid carbon dioxide. Store at −20°C until used. Saint-Marie (1962) details a paraffin-embedding method.

After the tissues have been sectioned by either method, sections must be fixed before the antigen-antibody solution is applied. Fixation depends on the antigen.

Coons Fluorescent Antibody Technique

FIXATION

Proteins: 95% ethyl alcohol, sometimes absolute methyl alcohol. Polysaccharides: can be fixed in the block by picric acid-alcohol-formalin (Rossman fluid, p. 24) and paraffin-embedded. Lipids: 10% formalin. Viruses: acetone.

DIRECT METHOD

The direct method consists of staining directly with fluorescein-labeled antibody against a specific antigen. Slides are dried after fixation, rinsed in buffered saline (0.8% sodium chloride with 0.01 M phosphate, pH 7.0), and the excess saline wiped off except over the sections. Labeled antibody is pipetted over the section; the slides, covered with a petri dish with moist cotton or filter paper attached to its undersurface, are allowed to stand at room temperature: 20–30 minutes. The slide is wiped dry except for the section and mounted in buffered glycerol (9 parts anhydrous glycerol, 1 part buffered saline), then covered with a cover glass. Examination is made under the fluorescent microscope.

INDIRECT METHOD

A tissue section containing antigen is covered with unlabeled specific antiserum to allow the antibody molecules to react with it and be fixed in situ (humid environment): 20 minutes. The slide is rinsed off in buffered saline: 10 minutes, then wiped dry except for the section, and fluorescein-labeled antiglobulin serum (prepared against the species that furnished the specific serum) added: 20 minutes. The slide is washed in buffered saline and mounted in buffered glycerol. Cells that reacted with the antiserum will be covered with antibody (globulin) and will have reacted with the labeled antiglobulin. This has been termed layering because the bottom layer is antibody, the middle layer is the antigen, and the labeled antibody lies on top.

COMMENTS

Coons (1958) recommends control slides and uses a blocking technique. Unlabeled specific antiserum is added to a slide and allowed to react for 20 minutes. Then, if the section is washed with buffer solution and treated with a drop of labeled antibody, only a faint fluorescence, if any, should be exhibited.

Silverstein (1957) described a means of applying contrasting fluorescent labels for two antibodies, using rhodamine B with fluorescein.

The Coons methods have been applied for the detection of bacteria, viruses, rickettsiae, epidemic typhus, mumps, influenza, canine hepatitis, chicken pox, canine distemper, measles, psittacosis, and poliomyelitis.

This description is intended as only a brief résumé of Coons techniques in order to acquaint the technician with their potentialities. The following references contain extensive and pertinent discussions of the procedures: Coons (1956 and 1958); Goldman (1968); Mellors (1959); Nairn (1962).

A tremendous bibliography is developing in this field; the following references will lead to many others: Buckley et al. (1955); Cohen et al. (1955); Coons and Kaplan (1950); Coons et al. (1942, 1950, 1951, 1955); Kaplan et al. (1950); Lacey and Davis (1959); Leduc et al. (1955); Marshall et al. (1958, 1961); Mellors et al. (1955); Noyes (1955); Noyes and Watson (1955); Tobie (1958); Watson (1952); White et al. (1955); Zwaan and Van Dam (1961).

Peroxidase-Labeled Antibody Technique (Nakane and Pierce 1967)

An improvement over the fluorescent antibody technique has been developed by conjugating enzymes instead of the fluorescent compound to antibodies. An antibody (to a particular hormone) in rabbit serum attaches to that hormone when applied to sections on microscope slides. Then an anti–rabbit gamma globulin, to which a peroxidase is coupled, is applied to the sections and becomes attached to the antigen-antibody complex. By allowing the peroxidase to oxidate a substrate (3,3'-diamino-benzidine), an insoluble brown precipitate is left at the site of the reaction. The enzyme reaction products are visible both for light and electron microscopy.

This method overcomes some of the problems of Coons method. Dark-field ultraviolet is not required; the preparations are permanent; the peroxidase-antibody conjugates are stable and can be stored indefinitely in a frozen state for several months at 4°C.

Part III
HISTOCHEMISTRY AND MISCELLANEOUS SPECIAL PROCEDURES

Chapter 21
Histochemistry

By definition the field of histochemistry is concerned with the localization and identification of a chemical substance in a tissue. In a broad sense this might include staining, combining chemical and physical reactions, and using chemical methods to demonstrate basic and acidic properties of tissue. But strictly speaking, histochemistry applies only to chemical methods that immobilize a chemical or enzyme at the site it occupies in living tissue. These methods can apply to inorganic substances, such as calcium, iron, barium, copper, zinc, lead, mercury, and others, as well as to organic substances, such as saccharides, lipids, proteins, amino acids, nucleic acid, enterochromaffin substance, and some pigments. Some substances are soluble and react directly; others are insoluble and must be converted into soluble substances before a reaction takes place. Occult or masked materials are part of a complex organic molecule, which has to be destroyed by an unmasking agent before the chemical can react. Some chemicals may be fixed in place; others that are soluble or diffusible have to be frozen quickly and prepared by the freeze-drying method, without a liquid phase. Sometimes it is advisable to make control slides, thereby preventing confusion between a genuine reaction and a nonspecific one that gives a similar effect.

It is difficult and somewhat impractical to make a sharp distinction between staining and histochemical methods. Some will disagree with the arrangement of methods in this book, but a sequence according to similar tissue or cell types seems most adaptable to general laboratory application—the primary intent of the book. This section on histochemistry will, therefore, cover mainly procedures used to identify enzymatic activity.

The field is a tremendous and exciting one of increasingly extensive activity. It is impossible and impractical to include all histochemical methods and their variations in this type of manual. Because there are available many excellent books and journals on histochemistry that describe the newest developments, only methods most familiar (at least to the author) and practical are incorporated in this text. The procedures can be adapted to a minimum of glassware, equipment, chemicals, and time, and yet demonstrate some of the more cooperative of the enzymes. Students who enter research in any aspect of histochemistry or enzymology will have to consult the more specialized and the most recent literature as their interest continues to progress.

For greater details start with the following: Burstone (1962); Cassellman (1959); Danielli (1953); Davenport (1960); Glick (1949); Gomori (1952); Gurr (1958); Lillie (1965); Pearse (1968, 1972); Thompson (1966).

FIXATION

The use of fixed tissue for enzyme demonstration is debatable, but cutting frozen unfixed tissues may result in a greater loss of enzyme into the incubating medium and more cell damage than if fixed tissue is used. Many enzymes will tolerate some exposure to fixatives and should be so treated, if possible, for ease of processing and reduction of diffusion. Even succinic dehydrogenase can tolerate calcium-formalin fixation at 5–10°C for 8–16 minutes, if the pieces are kept small. The addition of sucrose (10–15%) to fixatives will often improve the preservation of some enzymes and should be tried.

Aldehydes

Calcium formalin (Baker 1944) is used cold, 4°C, and recommended for the preservation of many enzymes (acid phosphatase, esterase, and lipase, but not for alkaline phosphatase or aminopeptidase).

calcium chloride, anhydrous	1.0 g
distilled water	60.0 ml
formaldehyde, concentrated	10.0 ml

Adjust pH to 7.0–7.2 with 1 N NaOH. Add distilled water to make a total of 100.0 ml. Store in refrigerator. Check pH frequently, or just before use, and adjust if necessary.

Glutaraldehyde

Sabatini et al. (1964) discovered that 2–6% glutaraldehyde in a phosphate buffer can be used for some enzymes (alkaline phosphatase and partial preservation of acid phosphatase and esterase). Sea water can be used as diluent for marine tissues and tyrode solution for tissue culture cells.

50% solution of glutaraldehyde[1]	4.0–12.0 ml
0.1–0.2 M phosphate buffer	100.0 ml

The pH should be 7.4. If shrinkage occurs in the tissue, use the dilution at 2% (4 ml of glutaraldehyde per 100 ml of buffer, not a higher concentration). One percent of sucrose can be added to the glutaraldehyde solutions for better preservation of tissue constituents. Store in the refrigerator.

Cacodylate buffer, pH 7.2

1 M cacodylic acid (137.99 g/1000 ml)	54.6 ml
1 N NaOH (40 g/1000 ml)	50.0 ml
distilled water	895.4 ml

Glutaraldehyde-fixed tissue frequently is postfixed in 1% osmic acid buffered with 0.1 M phosphate buffer and glucose or veronal-acetate-sucrose after cacodylate-buffered glutaraldehyde.

Physiologic buffer

Solution A. 2.26% $NaH_2PO_4 \cdot H_2O$
Solution B. 2.52% NaOH
Solution C. 5.4% glucose
Solution D. Isotonic disodium phosphate

solution A	41.5 ml
solution B	8.5 ml

Adjust pH to 7.4–7.6 with solution B.

Phosphate-buffered fixative

solution D	45.0 ml
osmic acid	0.5 g
solution C	5.0 ml

Stable for several weeks at 4°C.

---

[1] Biological Grade, G–151, Fisher Scientific Co.

Wash tissues in buffer several hours or overnight. The maximal stability of glutaraldehyde is at concentrations between 2% and 10%. Low temperatures and storage in the dark also are advisable. See Anderson (1967) about the purification of glutaraldehyde.

Paraformaldehyde

0.2 M s-collidine buffer

Solution A

s-collidine (2,4,6 trimethylpyridine)[2]	2.64 ml
distilled water.............................	47.36 ml

Solution B

1.0 N HCl	9.0 ml
distilled water.............................	41.0 ml

Working solution

Equal parts of A and B, pH 7.4 to 7.45

Paraformaldehyde solution

paraformaldehyde	4.0 g
distilled water to make	66.0 ml

Add distilled water to the paraformaldehyde in a graduated cylinder until combined total is 66.0 ml. Warm in flask to 60°C. Depolymerize by adding 0.1 N NaOH, drop by drop; shake until solution clears.

Working solution

paraformaldehyde solution	66.0 ml
0.2 M s-collidine buffer	33.0 ml
0.5 M CaCl$_2$ (5.55 g/100 ml)	1.0 ml

Adjust pH to 7.4 if necessary.

---

[2] Eastman Organic Chemicals.

Acetone (absolute)

Acetone, 4°C or colder (refrigerated or kept in the freezer), is used for some enzymes, such as aminopeptidase and alkaline phosphatase.

Buffered acetone (Kaplow and Burstone 1963) is recommended for several enzymes, such as alkaline phosphatase, esterase, peroxidase, and dopa oxidase.

acetone	300.0 ml

Add, with stirring, to a mixture of

0.03 M sodium citrate (dihydrate)	32.0 ml
0.03 M citric acid (monohydrate)	168.0 ml

The pH should be 4.2. Store in freezer.

Preparation for Fixation and Storage of Tissues

For the best results always fix very small pieces of tissue, no larger than 1–2 mm thick. They can be left in calcium formalin or glutaraldehyde in the refrigerator overnight. Then proceed to sectioning or store for several days or weeks in a gum sucrose solution or glycerin.

Gum Sucrose (Holt et al. 1960)

Store tissue at 0° to −2°C.

sucrose	30.2 g
distilled water.............................	100.0 ml
gum acacia	1.0 g

Prepare only enough for immediate use. If prepared in excess, stock mixtures should be kept frozen to prevent growth of mold. This can be done in small vials, then only single vials are melted as needed.

The gum sucrose preserves the enzymes, facilitates cutting, and produces good morphological detail.

Glycerin (Turchini and Malet 1965)

Store tissue at −20°C.

glycerin	50.0 ml
distilled water............................	50.0 ml

Place fresh tissue immediately into the glycerin solution. It can be stored in this condition for at least nine months. For processing, remove the block of tissue and fix for routine histological staining or in calcium formalin. Nuclear and cytoplasmic alterations may occur, but enzymatic activities will be well preserved.

See specific enzymes for additional fixation and preservation recommendations.

DEHYDRATING AND EMBEDDING

Freeze-Drying Method

If the apparatus is available and controllable, freeze-drying is considered the proper tool for the preparation of tissues for enzyme study. There are many types of freeze-drying apparatus on the market; many have been expensive, but cheaper ones are now available. Temperatures sometimes are difficult to control and the drying time may be long. The principle of freeze-drying is that frozen tissue is kept under high vacuum until all water molecules are removed and condensed onto a cold surface or collected by a desiccant. Small pieces (approximately 1 mm) are frozen solid instantly with isopentane or Freon 12 (obtainable in a can) cooled by liquid nitrogen to a temperature of −150°C. The tissue must be frozen rapidly to prevent large crystal formation, which would disrupt the cells. The initial freezing is commonly called quenching; it stops all chemical reactions in the tissues. Immediately after freezing and while still frozen, the tissue is dehydrated in a drying apparatus in vacuo at a temperature of −30° to −40°C, and the ice is sublimed into water vapor and removed. There is no liquid phase, and therefore no diffusion of enzymes.

When the material is dry, allow it to rise to room temperature, infiltrate with degassed paraffin or Carbowax, and embed.

Because of sensitivity to water, the sections cannot be floated on water, but must be applied directly to warm albumenized slides (Mendelow and Hamilton 1950). Embedded blocks must be kept stored in a desiccator.

Many types of freeze-dry apparatus are offered by the manufacturers. Thieme (1965) gives directions for the construction of an inexpensive and convenient model that is easy to operate and dries tissue overnight. Stumpf and Roth's (1967) model is available as a kit.[3] Wijffels (1971) describes a model that is easy to construct and operate.

Freeze-Substitution Method

A simpler and cheaper method that produces excellent results is the freeze-substitution procedure. The tissue is rapidly frozen and then the ice formed within the tissue is slowly dissolved in a fluid solvent such as ethyl or methyl alcohol. Other polar solvents also have been tried (Patten and Brown 1958). When the tissue is free of ice and completely permeated by the cold solvent, it is brought to room temperature and embedded.

Advantages of this method are these: there are no disruptive streaming movements in the tissue—the ice simply dissolves—and the substitution solution can sometimes be chosen to contain a specific precipitant for a specific substance; and the method can be used for radioautography, for fluorescent preparations, for studying many enzymes, and for other cytohistochemical preparations. There can, however, be cases in which the freeze-drying procedure is preferred—for instance, when an enzyme might be damaged by the denaturing action of the alcohol as it warms up.

Freeze-Substitution Technique (Feder and Sidman 1958; Hancox 1957; Patten and Brown 1958)

PROCEDURE

1. Cool a beaker of liquid propane-isopentane (3 : 1) or Freon 12 to $-170°$ to $-175°C$ with liquid nitrogen.
2. Place tissue specimen (small piece, 1.5–3 mm) on thin strip of aluminum foil and plunge into the cold propane-isopentane. Stir vigorously: 30 seconds.
3. Transfer frozen specimen to ice solvent, previously chilled to $-60°$ to $-70°C$ (see comments below). After a few minutes the tissue can be shaken loose from the foil. Store in dry-ice chest for 1 or 2 weeks: has been stored for 52 days without harm (Patten and Brown).
4. Remove fluid and tissue to refrigerator and wash 12 hours or longer in 3 changes of absolute alcohol in refrigerator.

---

[3] Delmar Scientific Glass Products, catalog no. 7126.

5. Transfer to chloroform: 6–12 hours in refrigerator, to remove any traces of water. (This step may be omitted in most cases.)
6. Transfer to cold xylene (or similar agent), −20°C; and remove to room temperature for 10 minutes.
7. Transfer to fresh xylene, room temperature: 10 minutes.
8. Transfer to xylene–Tissue Prep (50 : 50), 56°C, vacuum: 15 minutes.
9. Infiltrate, Tissue Prep no. 1, vacuum: 15 minutes.
10. Transfer to Tissue Prep no. 2, vacuum: 15 minutes.
11. Transfer to Tissue Prep no. 3, vacuum: 15 minutes.
12. Embed.

COMMENTS

Ice solvents that can be used are absolute methyl, absolute ethyl alcohol, and *n*-butyl alcohol. Feder and Sidman include complete directions for obtaining, storing, and disposing of liquid nitrogen, isopentane, and propane; also details about equipment. *Warning:* There are potential explosion hazards.

See Moline and Glenner (1964) for a method that involves coating the tissue with powder to permit faster cooling of the tissue. It has the advantage of safety, simplicity, and speed over methods using hydrocarbon baths cooled in liquid nitrogen.

See Bullivant (1965) concerning the use of glycerin preservation prior to freeze-substitution.

Blank and Delamater (1951) use glycols if fixation is to be avoided, alcohol or acetone if fixation is permissible. This method is as follows:

1. Place slices (2–3 mm) of tissue in cold propylene glycol in deep freeze, below −20°C: 1–2 hours.
2. Transfer to mixture of Carbowaxes, 55°C: 2–3 hours.

Carbowax 4000	9 parts
Carbowax 1500	1 part

3. Prepare blocks (p. 445).
4. Cut sections and float on slides according to Blank and McCarthy method (p. 446).
5. Stain or use for autoradiographs. Water-soluble substances will not have leached out. For autoradiographs, before each cut coat the block face with thin molten wax. Press wax-reinforced sections against emulsion for exposure. See also Feder and Sidman (1958); Freed (1955); Masek and Birns (1961); Meryman (1959, 1960); Woods and Pollister (1955).

Freeze-Substitution of Sections (Chang and Hori 1961; Patten and Brown 1958)

PROCEDURE

1. Quench 1–2 mm slices in liquid nitrogen or isopentane prechilled by liquid nitrogen.
2. Section at −15°C in cryostat.
3. Sweep sections into screw-cap vial of acetone prechilled in dry ice.
4. Return vial to dry ice and complete dehydration.

Chang and Yokoyama Method (1970)

PROCEDURE

1. Cut small pieces of fresh tissue and section in cryostat. Transfer sections to cold cover glasses and immediately thaw by placing a finger on the glass under the section.
2. Immediately immerse in acetone at dry-ice temperature for freeze substitution: 2 hours to overnight.

This method permits the cutting of many tissue samples to be preserved at dry-ice temperature until they are processed or incubated. A number of enzymes can be preserved in this way.

CRYOSTAT SECTIONING

Carbowax or paraffin can be cut at room temperature, but many enzymes and other tissue constituents must be sectioned at low temperatures of −25° to −20°C or they are lost. Since embedding is impractical at these temperatures, tissue is frozen in isopentane or Freon 12 (−160°C) or in dry ice–ethyl alcohol mixtures (−70°C). The tissue must be maintained at least at dry-ice temperature (−20°C) and sectioned and mounted at this temperature. This is performed inside a cold-chamber cryostat[4] (−25° to −20°C). The object holder is precooled (object holders can be stored in the cold chamber) and the tissue is frozen on it with a few drops of water. Sectioning is facilitated by using O.C.T. compound (p. 427) to support and protect the tissue. I usually use O.C.T.; it melts away against the slide and

---

[4] American Optical; Fisher Scientific Co.; Harris Refrigeration Co.; International Equipment Co.; Lab-Tek; Lipshaw Manufacturing Co.; Scientific Products.

does not interfere with enzymatic reactions. *Caution:* if the slides are to be used for autoradiography, remove the O.C.T. by washing slides in distilled water before dipping in emulsion.

The knife used in sectioning must be kept sharp and dry. Carefully wiping it with a finger tip before each section keeps it dry and clean and reduces compression in the section. If O.C.T. compound is used and the sections are to be mounted on slides, cut through the compound and through the tissue just until the knife edge reaches the O.C.T. along the opposite edge of the tissue. This means that the cut is complete through the tissue but not through the O.C.T. (Fig. 21-1). Stop sectioning. (As the cut is being made, lightly hold the section against the knife with a small cold brush. This prevents curling.) Flatten the section with the brush and finish cutting through the O.C.T. With the cold brush (sometimes two brushes) lift the section to a cold slide. Thaw the section by placing the undersurface of the glass against a finger or the cushiony part of the palm.

When thawed, the sections are ready for processing. If the sections are to go directly into a substrate or fixative, lift them, curled or uncurled, with the cold brush and drop them into the solution. Usually a slight tap of the brush against the side of the container will shake them loose. Frozen tissues can be mounted on cover glasses in the same way as on slides and in some cases the section can be picked up by touching it with the cover glass. If the tissues have been suspended in gum sucrose, $6-10$ μ sections are easily cut and then dropped into the proper substrate for incubation or mounted on cover glasses or slides. Gelatin embedding (Burkholder et al. 1961; Taylor 1965) can be tried for better tissue support; it does permit easier sectioning and handling.

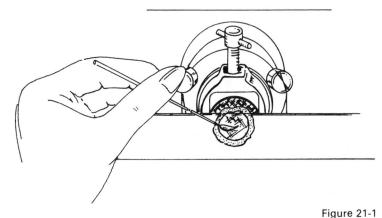

Figure 21-1
Flattening and correcting folds in section: the cut through O.C.T. compound (strippled area) is not yet complete.

Fitz-William et al. (1960) recommend that immediately before each section is cut, the tissue block be coated with 20% polystyrene dissolved in methylene chloride (volatile at cryostat temperature). Wait until the highly reflecting surface that forms disappears and then cut section. Curling of sections is minimal and they can be picked up with fine forceps. The polystyrene solution should be stored in the cryostat. These authors also include complete directions on the care of the microtome used in the cryostat.

Consult the following references for more about cryostat sectioning: Blank and Delamater (1951); Feder and Sidman (1958); Freed (1955); Gersh (1932); Glick and Malstrom (1952); Hancox (1957); Hanson and Hermodsson (1960); Ibanez et al. (1960); Jennings (1951); Mellors (1959); Meryman (1959, 1960); Patten and Brown (1958); Woods and Pollister (1955); Zlotnik (1960).

SECTIONING WITHOUT A CRYOSTAT

If a cryostat is not part of the laboratory equipment, frozen sections of many of the enzymes can be made on the clinical freezing microtome, or by the method developed by Adamstone-Taylor.

Adamstone-Taylor (1948) Cold Knife Technique

PROCEDURE

1. Quickly freeze fresh tissue (2 mm blocks) on freezing block of microtome or in liquid nitrogen, isopentane or Freon 12 chilled with nitrogen, or petroleum ether containing solid carbon dioxide.
2. Store in dry ice ($-75°C$) or in deep freeze ($-25°C$).
3. Chill knife by fastening solid blocks of dry ice on each end. Use Scotch tape or thin sheets of metal foil.
4. When the knife is chilled, freeze the frozen block of tissue in place on freezing head.
5. Cut sections and hold them flat on the knife with a camel's hair brush. Do not allow them to thaw.
6. Remove section to a small metal scoop, which is kept cold with a few bits of dry ice. Transfer the section to a slide. If the slide is cold, speed of transfer is not as essential as it is if the slide is at room temperature.
7. Press the section in place, and as it begins to soften but before it completely thaws, immerse slide in fixative.

COMMENTS

Unthawed sections can be removed rapidly to a beaker of cold acetone surrounded by dry ice. With needles chilled in the dry ice, mount the sections in the acetone onto chilled cover glasses. Remove, allow to dry, then fix.

This method is difficult in warm and humid atmospheres; the sectioning preferably should be done in a cold room.

ACETONE FIXATION AND EMBEDDING

Some enzymes can be carried through rapid paraffin embedding with good results, but great care must be taken to keep the duration of time in solutions to a minimum and temperatures as low as possible. Even with extreme care, results can occasionally be disappointing.

Gomori Method (1952)

PROCEDURE

1. If possible, chill tissue for a short time before immersing in acetone. It may prevent some shrinkage artifacts. But fix within 10 minutes of death for best results. Cut very thin slices, 1–2 mm, no larger.
2. Fix in cold acetone, 2–3 changes: 12 hours each.
3. Transfer to ether–absolute alcohol (1 : 1): 2 hours in refrigerator.
4. Infiltrate with dilute celloidin or nitrocellulose (approximately 10%): overnight in refrigerator.
5. Drain on cleansing tissue. Place in benzene or chloroform: 30 minutes–1 hour. Stirring on a magnetic mixer aids penetration of fluid.
6. Infiltrate with paraffin 52–56°C (52°C is best): 15–20 minutes in oven without vacuum; 30 minutes with vacuum; and 10–15 minutes without vacuum.
7. Embed. Store in refrigerator.
8. Section. Float on lukewarm water, briefly (10 minutes maximum).
9. Drain off water and dry.
10. If to be kept for some time, put in oven 5–10 minutes to melt paraffin, forming a protective coating against atmospheric influence. Store in cold room or refrigerator. The safest way to store is in uncut paraffin block in refrigerator.

COMMENTS

Step 4 can be reduced to 2–3 hours in nitrocellulose, and the entire procedure following fixation can thus be completed in one day with good results.

Alternate Short Gomori Method

1. Fix in cold acetone overnight.
2. Transfer through 2 changes of cold petroleum ether: 1 hour each.
3. Remove petroleum ether and tissues to room temperature: 15 minutes.
4. Infiltrate in 45°C paraffin in vacuum: 30 minutes.

Novikoff et al. (1960) find acetone fixation is good for some cryostat preparations.

Frozen Sections Fixed in Acetone (Burstone 1962)

PROCEDURE

1. Cut unfixed frozen sections. Mount on slides.
2. Immerse in acetone: 1–5 minutes.
3. Transfer to 90–95% acetone-water: 1–3 minutes.
4. Transfer to 80–85% acetone-water: 30 seconds–1 minute.
5. Transfer to 70% acetone-water: 30 seconds.
6. Rinse in water: 10–20 seconds.
7. Incubate.

COMMENTS

The quality of fixation is not the best but the method is simple and useful when cytological detail is not of great importance.

GENERAL SUGGESTIONS

Stock Solutions

The following solutions can be kept in storage and ready for immediate use: in the refrigerator—calcium-formalin and glutaraldehyde; in the freezer—buffered acetone, absolute acetone, absolute methyl alcohol, and gum sucrose. (See p. 402 concerning methyl alcohol.)

Read labels on all substrates and diazonium salts. Some must be kept in the refrigerator or freezer. If in doubt, store such materials in the refrigerator for longer life.

Control Slides

Most enzyme reactions should be accompanied by a control slide in order to help detect false positives. An enzyme can be inactivated by any one of the following methods:

Place slide in distilled water and bring to boil: 10 minutes.

Immerse in 1 N HCl: 15 minutes; wash in tap water: 0.5 hour.

Incubate slide in substrate medium containing an inhibitor (NaCN or KCN).

Incubate untreated slide in a medium lacking substrate.

For additional information, see each specific enzyme method.

Preservation of Incubation Media

Klionsky and Marcoux (1960) prepared twelve different media, froze them in a mixture of dry ice and acetone, and stored them in a dry-ice chest. For at least 6 months, the solutions remained as good as freshly prepared solutions. The types of media tested were:

Azo dye and metal precipitate methods for acid and alkaline phosphatases.

Alpha-naphthyl and indoxyl acetate for esterase.

5-nucleotidase.

Nitro-BT for succinic dehydrogenase.

Postcoupling technique for beta-glucuronidase.

Triphosphopyridine nucleotide and diphosphopyridine nucleotide diaphorase.

Storage of Slides and Tissues

Storage for long periods of time is not generally recommended for enzymes; even at −40°C they can be kept only about 6 hours. If, however, storage is unavoidable, quenching at −160°C and storage at −85°C will preserve many enzymes in the tissue. Some enzymes on slides can be kept for 3 months at −25°C, but −85°C is even better and good for a year.

ALKALINE PHOSPHATASE

Sections containing active phosphatase are incubated in a mixture of calcium salt and phosphate ester, usually sodium glycerophosphate, so that calcium phosphate is precipitated at the sites. The enzyme liberates the phosphate ions, which are held on the spot by salts of metals whose phosphates are insoluble. (Magnesium ions are often required as an activator and are added in the form of magnesium sulfate or chloride. The mechanism of activation is unknown.) The insoluble phosphates are made visible by cobalt nitrate and ammonium sulfide, which produces cobalt phosphate and finally black cobalt sulfide.

Phosphatase is dissolved or destroyed by ordinary fixatives, but alcohol or acetone will preserve it. Celloidin or paraffin embedding can be used safely, but paraffin sections must be covered with a film of celloidin to protect against dissolution of enzyme. Freeze-substitution preserves it well. See Burstone (1960*b*) for fluorescent demonstration of alkaline phosphatase.

Gomori Method (Frankel and Peters Modification 1964)

FIXATION

Thin slices (1–2 mm) are fixed at once in chilled acetone, 3 changes: 24–48 hours. Other fixatives may be 80%, 95%, and absolute alcohol, or cold formalin (4°C), followed by freezing method.

INFILTRATION AND EMBEDDING

For paraffin embedding use chloroform or petroleum ether and paraffin at no higher than 56°C melting point, preferably 52°C. The short methods are recommended. Cut sections 6–10μ. Do not dry slides at higher than 37°C. Store all blocks at 6°C until needed.

INCUBATING SOLUTION

0.8% paranitrophenyl phosphate[5]............	2.0 ml
2% sodium barbitol (a drug)	2.0 ml
distilled water............................	1.0 ml
2% calcium chloride	4.0 ml
5% magnesium sulfate	0.2 ml

PROCEDURE

1. Deparaffinize in light petroleum or chloroform: 1 minute.
2. Hydrate with 3 or 4 dips in each of absolute acetone, 80% acetone–water, 50% acetone–water.
3. Rinse with several dips in water, or until surface of slide appears homogeneous.
4. Incubate in incubating solution, 37°C: 30–45 minutes.
5. Wash well in distilled water.
6. Treat with 2% cobalt nitrate: 3–5 minutes.
7. Rinse well in distilled water.
8. Treat with 2% yellow ammonium sulfide: 1–2 minutes.
9. Wash well in distilled water.
10. Counterstain with hematoxylin, safranin, Kernechtrot, or the like.
11. Dehydrate, clear, and mount.

RESULTS

sites—black, sharp and clear, nondiffuse
nucleus and nucleolus—show a minimum of gray deposit of cobalt
   sulfide

COMMENTS

Paranitrophenyl phosphate is split by alkaline phosphatase at a faster rate than other phosphates (phenyl phosphate, glycerol phosphate, and phenolphthalein phosphate) that have been used for this reaction. Since the paranitrophenyl phosphate is so rapidly hydrolyzed, the time of the reaction is reduced to 30 minutes and diffusion artifact formation is minimized. The sites correspond to those demonstrated by the azocoupling method below. The color in the coupling method tends to fade within a year, but both methods should be tried.

Freeze-substituted and cryostat-cut sections can be used.

---

[5] Sigma Chemical Co.'s 104 phosphatase substrate.

The pH optimum is 8.5–10.0. The specific activator of alkaline phosphatase is Mg^{++}; other divalent cations, such as Mn^{++}, Zn^{++}, and Co^{++}, may activate in very low concentrations. Inhibitors include EDTA, zinc chloride, formaldehyde, oxidizing agents such as potassium permanganate and iodine, cysteine, and sodium arsenate.

False positives can be due to hemosiderin or melanin, and the sites should be checked for the presence of these pigments.

For additional reading about the Gomori method, see Ackerman (1958); Gomori (1941a, 1946); Kabat and Furth (1941); Moffat (1958); and Pearse (1968).

Azo-Coupling Method

One of the very important naphthols is 3-hydroxy-2-naphthoic acid. It is made when sodium naphtholate is treated with carbon dioxide and is used to produce the so-called naphthol AS (naphthol anilid–säure) compounds. Naphthol AS phosphates are phosphate esters of these complex compounds and release highly insoluble naphthols upon enzymatic hydrolysis. These naphthols couple immediately with diazonium salts to form insoluble azo dyes, thereby demonstrating microscopic localization of enzymes in properly fixed tissues. Several of these naphthol phosphate compounds are suitable for demonstrating alkaline phosphatase, but probably the most popular is AS-MX (others are AS-KB, AS-TR, AS-BS, AS-AN, AS-E, and AS-BI).

The azo-coupling method can be used on tissues prepared in the same manner as those for the modified Gomori technique. After deparaffinizing and hydrating, transfer the sections into the incubation solution below.

For the beginner in enzyme methodology, smears and touch preparations are simple to prepare and can be processed in conjunction with sections of the tissue to verify the presence or absence of an enzyme. For example, if tissue blocks are made of liver, also make and fix touch preparations from the same organ. Stain sections and touch preparations in the same incubating medium. If the sections are negative, but the touch preparations are positive, the enzyme is being lost somewhere in the processing for sectioning.

The occurrence of alkaline phosphatase in the neutrophils of the peripheral blood of most animals (not mice or chickens) makes these cells excellent and easily prepared subjects for the demonstration of this enzyme. Spin whole blood in a small tube (hematocrit or bacteriological culture tube, 11 mm in diameter) 2500–3000 rpm for 10 minutes. With a Pasteur pipette collect the buffy coat of white cells lying on top of the packed red cells and place a *small* drop on a slide. Immediately smear like a blood smear (p. 220), or place another slide on top of the drop. It should be placed so the drop is close to one end of the slide, as shown in Figure 21–2. Wait until drop stops spreading; then quickly pull off the top slide. When preparing smears, do not allow the cells to be overcrowded; the cells must be well flattened to demonstrate precisely the alkaline phosphatase granules. Allow the smears to dry thoroughly; slight heat, such as that from a slide warmer, for a few seconds, causes no damage to the enzyme and helps flatten the cells.

When the smears are dry, place them in cold fixative (0° to −10°C) consisting of 5.0 ml of concentrated formaldehyde and 45.0 ml of absolute methyl alcohol, and leave them overnight. The alcohol must be fresh and kept tightly closed. Alcohol from the final third of a gallon bottle may not preserve the enzyme adequately: I purchase only one-pound bottles, and store them in a refrigerator or freezer. Mixed fixative can be kept in a freezer, but not longer than one week.

Temperature is very important; if it is too high or too low, the fixation of

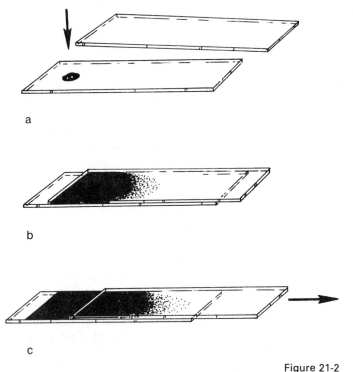

a

b

c

Figure 21-2
Type of smear useful for enzyme preparations: (a) Place a slide on top of drop of white cells; (b) allow cells to spread between slides; (c) when spreading stops, pull top slide length of and off bottom slide.

the enzyme will be imperfect and either diffuse or reduced activity will result. Incubation of smears and touch preparations can be undertaken on the same day after 30 minutes of fixation. Such preparations can be kept in the cold fixative for 2 nights and then be processed with sections prepared from tissue blocks that have been fixed overnight, rapidly embedded, and sectioned. Do not leave smears and touch preparations in the fixative more than 2 days; to do so reduces the activity. Storing in gum-glucose does not preserve alkaline phosphatase activity.

When drawing blood for buffy coats, use heparin, not EDTA, as an anticoagulant. The EDTA is an inhibitor of alkaline phosphatase.

Although a phosphatase score will drop very slightly within the first couple of hours after blood withdrawal, good scores can be obtained within 6 hours after withdrawal. Keep the blood at room temperature or 37°C. Do not refrigerate it; low temperatures inhibit the activity. If a centrifuge is not handy, allow a buffy coat to form by sedimentation for 1–2 hours. Then collect the white cells.

Kaplow Method, Modified (1963)

SOLUTIONS

Propanediol stock solution

2 amino-2-methyl-1,3-propanediol	10.5 g
distilled water...........................	500.0 ml

Buffer working solution, pH 9.7

propanediol solution	25.0 ml
0.1 N HCl	5.0 ml
distilled water to make	100.0 ml

Incubating solution

naphthol AS-MX phosphate[6]................	5.0 mg

Dissolve in:

N-N dimethyl formamide (DMF)[7]	0.2–0.3 ml

---

[6] Nutritional Biochemicals Corp.
[7] Fisher Scientific Company.

Add:

buffer working solution .... 60.0 ml
fast red violet salt LB[8] .... 30.0–40.0 mg

Shake for 30 seconds and filter onto slides.

PROCEDURE

1. Deparaffinize sections in chloroform: 1 minute. Hydrate through acetone: few dips; through 70% acetone–water: few dips; through 50% acetone–water: few dips; then wash in distilled water: 30–60 seconds. Wash smears and touch preparation in distilled water: 20–30 seconds, and blot.
2. Incubate, room temperature: 20–25 minutes.
3. Wash in running water: 1 minute.
4. Stain in Mayer hematoxylin: 3 minutes.
5. Wash in running water: 1–2 minutes.
6. Blue in Scott solution: several dips.
7. Wash in running water: 1 minute.
8. Stain in 0.5% aqueous wool green S: $\frac{1}{2}$ minute.
9. Rinse in water (tap or distilled): 3–4 seconds.
10. Dehydrate in 2 changes of 95% ethyl alcohol: few seconds in each.
11. Dehydrate in 2 changes of absolute ethyl alcohol: 2–3 minutes in each.
12. Clear in 2 changes of xylene and mount.

RESULTS

sites—red
nuclei—blue
cytoplasmic staining—pale green
eosinophilic granules—brilliant green

COMMENTS

Fast green FCF or light green can be substituted for wool green, but the stain is paler and less interesting. Kaplow does not use the green, but addition of it provides background cytoplasmic staining and distinctive green color to the eosinophilic granules.

Kaplow blots and dries the slides after step 5, and mounts in glycerol

---

[8] Verona Dyestuffs.

jelly or Permount. Bluing in Scott solution and counterstaining does not alter the score of the phosphatase in the neutrophils.

Gelva (p. 423) is satisfactory as a mounting medium. It may prevent some fading, but the resolution is not quite as precise as when a synthetic resin is used. The problem of fading is not critical; 8 months after slide preparation and mounting in synthetic resin, I obtained a score only a few points below the score obtained on the day of preparation.

For scoring the amount of alkaline phosphatase in the neutrophils, see Kaplow (1963).

Coupling can be done with other salts, such as fast blue RR (Sigma), which gives blue sites.

For references about alkaline phosphatase, see Burstone (1962); Gomori (1951); Manheimer and Seligman (1949); and Pearse (1968).

ACID PHOSPHATASE

This enzyme exhibits optimal activity at pH below 7.0, usually between 3.8 and 6.0. The best tissues for its demonstration are spleen, liver, kidney, and blood leukocytes. In many cells, acid phosphatase appears to be associated almost exclusively with the lysosomes, so much so that acid phosphatase is coming to be considered a "marker" for them. Lysosomes are rich in hydrolytic enzymes—ten or more of them, all showing an acid pH optimum. Esterase also can be demonstrated in lysosomes. The small granules of concentrated acid phosphatase found in the Golgi region are tentatively considered to be lysosomes.

Care must be taken in the preparation of tissues for acid phosphatase. The fact that the enzyme is readily made soluble and released from the tissue is a probable cause of failures in past methods. The simplest and most dependable fixation is to place small pieces (2 mm) in cold calcium-formalin (p. 386), 4–10°C, overnight. Large organs from large animals often require 1–2 minutes of perfusion with fixative before a small block of tissue is removed. The tissue should then be stored in gum sucrose, 4°C, for 24 hours.

Section on freezing microtome and drop sections into incubation medium. Mount sections on slides and allow to air dry: 2 hours. Proceed to substrate or store overnight in refrigerator. If sectioning cannot be done immediately after overnight fixation, transfer the blocks of tissue into gum sucrose (p. 389) and refrigerate. Krus et al. (1961) say it can be preserved for 40 days. (Freeze-substitution preserves 75–80% of the enzyme.)

Tissues can be acetone-fixed and paraffin-embedded (short method), but deparaffinizing the sections and mounting them on slides may inactivate the enzyme.

Cold 2% glutaraldehyde can be used for good preservation of vacuoles and other structures, but it should be accompanied by calcium-formalin fixed blocks. There are disadvantages to using glutaraldehyde. It penetrates slowly; even after overnight fixation the center of a block may remain unfixed, and diffusion of the enzyme into the outer fixed area may have taken place. (Never cut blocks thicker than 1–2 mm for enzyme studies.) Glutaraldehyde leaves a yellow background color, which may be considered undesirable. See Janigan (1965) for studies on aldehyde fixation for acid phosphatase.

DMSO (dimethylsulfoxide) is being used in enzyme research because it protects against freezing damage and increases cell permeability. But it must be used with caution; a more than desirable diffuse reaction can take place. Brunk and Ericsson (1972) use it and sucrose in their solutions and thereby reduce the incubation time. Sucrose is added to a total of $0.1 M$ to the fixative and the wash contains $0.25 M$ sucrose and 10% DMSO. The incubation medium is Gomori's made $0.22 M$ with sucrose plus 1% of DMSO.

As recommended for alkaline phosphatase, touch preparations and smears are useful adjuncts to sections. Two methods for fixing such preparations are suggested. Prepare duplicate slides; leave one dry and unfixed and fix the other for 20–30 seconds (no longer; do not store in fixative) in cold buffered acetone (p. 389) or cold calcium-formalin. Drain against filter paper or cleansing tissue and allow to dry. Incubate both slides with tissue sections. Postfix the unfixed slide before counterstaining (see step 4, azo-coupling method).

Protozoans make fine subjects for lysosome studies and for other enzymes. Prepare some on subbed slides for better adherence but always make control slides that have no adhesive on them, since it is possible that "sub" material or albumen may inhibit an enzyme.

Gomori Method (1952)

SOLUTIONS

Sodium acetate buffer, pH 5.0

0.6% acetic acid	300.0 ml
sodium acetate (27 g/1000 ml water)	700.0 ml

Substrate

acetate buffer	25.0	ml
distilled water	100.0	ml
lead nitrate	0.12	g
3% sodium glycerophosphate (freshly		
prepared)	10.0	ml

Keep at 37°C for 24 hours. Filter. Add 2–5 ml of distilled water to prevent precipitation on evaporation. Keep in refrigerator. If the solution becomes turbid, discard. It is best when used within 3–4 days.

PROCEDURE

1. Deparaffinize sections as for alkaline phosphatase and hydrate to water. Smears and frozen sections can be placed directly into substrate.
2. Incubate in substrate, 37°C: 1–24 hours; average is 4 hours.
3. Rinse in distilled water: few seconds.
4. Wash in 1–2% aqueous acetic acid: 1 minute (see comments below).
5. Wash in distilled water: 1 minute.
6. Transfer to 1% aqueous ammonium sulfide: 2 minutes.
7. Wash in distilled water: several minutes.
8. Counterstain in hematoxylin, Kernechtrot, or the like.
9. Wash well in tap water to remove excess stain.
10. Mount in glycerin jelly; or dehydrate, clear, and mount in synthetic resin.

RESULT

sites—black

COMMENTS

Wachstein et al. (1962) use pH 6.0.

Gomori followed step 3 by a wash in dilute acetic acid to remove nonspecific bound lead, but Goldfischer et al. (1964) imply that the rinse can remove all the reaction product from an enzyme site of low activity. One slide should therefore be processed as a test slide with the acetic acid rinse omitted.

Gomori method is analogous to the alkaline phosphatase technique in that phosphate is released by enzymatic activity and is precipitated as lead phosphate, which is finally converted to a black lead sulfide.

Controls: since sodium fluoride inhibits acid phosphatase, add the fluoride to the incubating solution to a final concentration of $0.005\,M$. The sodium glycerophosphate can then be left out of the incubating solution as a control.

Thin blocks of tissue can be incubated, embedded in paraffin or some other medium, thin $2–3\mu$ sections cut, mounted on slides, cleared, and cover slipped.

For unfixed frozen sections, see Bitensky (1963) for modification of Gomori method.

Azo-Coupling Method (Kaplow and Burstone 1964)

This also is analogous to the alkaline phosphatase method, but employs a buffer at pH 5.0.

SOLUTION

Substrate

naphthol AS-MS phosphate..................	5.0 mg

Dissolve in:

N-N dimethyl formamide (DMF).............	0.2–0.3 ml

Add:

distilled water............................	25.0 ml
0.2 M acetate buffer (see Gomori method above)	25.0 ml
fast red violet salt LB	30.0–40.0 mg

Shake for 30 seconds and filter onto slides. One may substitute 60 ml of a 0.1 M citrate buffer, pH 5.2, for the water and acetate buffer.

PROCEDURE

1. Deparaffinize sections in petroleum ether or chloroform, and hydrate quickly through acetone solutions to water. Frozen sections, smears, and touch preparations go directly into substrate.

2. Incubate, 37°C: 2 hours.
3. Wash in running water: 1 minute.
4. Fix unfixed smears in 10% formalin: 30 seconds, and wash in running water: 2–3 minutes. This step may be omitted, but better nuclear staining is obtained after brief fixation.
5. Stain in Mayer hematoxylin: 3 minutes.
6. Wash in running water: 5 minutes.
7. Mount in glycerin jelly or polyvinyl alcohol (p. 423).

RESULTS

sites—red
nuclei—blue

COMMENTS

Gelva preserves acid phosphatase, but bubbles develop in the mounting medium after a few weeks, even when the cover glass is sealed. PVP/VA does not preserve it adequately. See p. 103 for plastic mountant.

Since the lysosomal membrane is less permeable to glycerophosphate than to naphthol AS phosphates, the reaction is slower in the glycerophosphate technique (Maggi and Riddle 1965).

Schajowicz and Cabrini (1959) say that if decalcification is necessary, 2% formic acid and 20% sodium citrate (1:1), pH 5, is satisfactory up to 15 days. There is loss of enzyme if Versene is used.

For additional reading about acid phosphatase, see Burstone (1959b, 1962); De Duve (1959, 1963a,b); Gomori (1950a, 1956); Novikoff (1960, 1961b); Pearse (1968); Weissmann (1964); Zugibe (1970).

AMINOPEPTIDASE (PROTEOLYTIC ENZYME)

The proteolytic enzymes hydrolyze proteins and amino acid compounds. The aminopeptidase (and exopeptidase) reaction depends on the hydrolysis of a peptide bond adjacent to a terminal alpha amino group, resulting in the liberation of a chromogenic moiety. The enzyme hydrolyzes L-leucinamide at a highly preferential rate, which makes leucyl naphthylamide a favored substrate. It can be activated by bound metal ions such as Mg^{++}, Mn^{++}, and Co^{++}. It can be destroyed by a one-hour immersion in water at 70°C; it can be inhibited by heavy metal ions, Zn^{++}, Cu^{++}, Hg^{++}, Cd^{++}, Pb^{++}, and Fe^{++}, or by fixing in absolute methanol and 40% formaldehyde.

High activity will be found in the kidney, parathyroid, small intestine, and uterus; some activity in salivary glands, thyroid, liver, and thymus. Buffy coats make excellent preparations; fix in cold acetone: 1 hour. Fix tissue blocks in cold acetone ($-10°$ to $0°C$) overnight and embed by the short method (p. 397). Xylene, 1–2 hours, can be substituted for petroleum ether. Freeze-dried cryostat sections or fresh tissue cut frozen can be used.

A large number of diazotates are stabilized as double zinc salts, and since aminopeptidase is sensitive to metallic ions like zinc, it is important to use the proper diazonium salt in the coupling reaction. Garnet GBC is available as a stabilized diazonium sulfate.

Burstone and Folk Method (1956)

SOLUTIONS

Substrate stock solution

L-leucyl-b-naphthylamide	1.0 g
distilled water............................	100.0 ml

Tris buffer, pH 7.19

0.2 M tris (hydroxymethyl) aminomethane	25.0 ml
0.1 N HCl	45.0 ml
dilute with distilled water to	100.0 ml

Working solution

stock substrate	1.0 ml
distilled water............................	40.0 ml
tris buffer	10.0 ml
garnet GBC[9] (diazotized o-aminoazotoluene) ..	30.0 mg

Shake and filter.

For substrate, the alanyl compound may be substituted: 10 mg in 40 ml hot distilled water (90°C). Cool and add 10 ml of tris buffer and 30 mg garnet GBC.

PROCEDURE

1. Chloroform or petroleum ether, 2 changes: 4 minutes each.
2. Absolute acetone: 1 minute.

---

[9] Roboz Surgical Instruments, catalog no. 10597.

3. 95% acetone: 1 minute.
4. 85% acetone, several dips.
5. Distilled water, few dips.
6. Substrate: 15 minutes to 1 hour, room temperature, or 37°C.
7. Tap water: 5 minutes.
8. Hematoxylin: 3 minutes.
9. Tap water: 10 minutes.
10. Mount in glycerin jelly.

RESULT

sites—red

COMMENTS

If necessary to purify GBC salt (not necessary with Imperial Chemical
Industries product[10]):

1. Dissolve 10 g salt in 50 ml methyl alcohol. Let stand 30 minutes.
   Filter.
2. Repeat with residue.
3. Dissolve second residue in 500 ml ether. Let stand 2–3 hours.
4. Filter. Wash residue 2–3 times with ether. Store dry powder in refrigerator.

Aminopeptidase preparations are only semipermanent. Bubbles develop in glycerin jelly mounts within a few weeks, and the reaction fades
in PVP.

For further information about aminopeptidase, see Burstone (1962);
Pearson et al. (1963); Zugibe (1970).

See Lojda and Havrankova (1975) for a method using bromoindolyl
leucinamide.

ESTERASES (NONSPECIFIC) AND LIPASES

These terms are used in reference to enzymes that hydrolyze esters of
carboxylic acids. There is an overlapping between enzyme types. Some
seem to occupy intermediate positions between lipases and esterases, and
react like both forms (Burstone 1962; Chessick 1953; Gomori 1952). In

---

[10] Roboz Surgical Instruments, catalog no. 10597.

general, esters of short-chained fatty acids are acted on by esterases; long-chained esters, by lipases. A large number of presumably specific enzymes have been reported and demonstrated in lysosomes.

Esterases or lipases are not destroyed by acetone fixation, are resistant to heat, and can be embedded in paraffin if not used excessively (3 hours maximum, 58°C). Sections fixed overnight in calcium-formalin or 10% formalin and frozen sections can be used for esterase and lipase. Infiltration with gum sucrose, 12–24 hours or longer, facilitates sectioning. Freeze-drying gives good preservation.

Fix buffy coats in formalin vapor (1–2 ml of concentrated formaldehyde in a coplin jar) add dried smears, and cover tightly: 30 minutes. Store dry until use. Unfixed smears can accompany fixed smears through incubation; then, before counterstaining, postfix the unfixed smears in 10% formalin: 60 seconds; or methyl alcohol: 30 seconds.

Lipases are incubated with the long-chained fatty acid esters or sorbitan and mannitan, water-soluble commercial products by the name of Tween. On hydrolysis, the liberated fatty acids form insoluble calcium soaps deposited at the sites. These are converted into lead soaps and then into brown sulfide of lead.

Esters of naphthols also are used, and naphthol is liberated and demonstrated by azo-coupling.

Different esterases react to different inhibitors, but for nonspecific esterases, use E600 (diethyl-p-nitrophenyl phosphate), 10^6 M to 10^{-3} M, as an inhibitor, or leave the substrate out of the solution.

Esterase (Moloney et al. Modification 1960)

SOLUTIONS

Buffer, pH 7.4

Sodium barbiturate (1.03 g/50.0 ml water a drug)	50.0 ml
0.1 N HCl	31.0 ml
distilled water to make	100.0 ml

Substrate

naphthol AS–D chloroacetate[11]	10.0 mg

---

[11] Verona Dyestuffs.

Dissolve in:

>   acetone ....   0.5 ml

Add:

>   distilled water............................   25.0 ml
>   buffer ....   25.0 ml
>   fast blue RR[12] or garnet GBC salt ....   30.0 mg

Shake 30 seconds and filter onto slides.

PROCEDURE

1. Deparaffinize and quickly hydrate paraffin sections. Wash fixative out of fixed smears or frozen sections.
2. Incubate in substrate, room temperature: 30 minutes.
3. Rinse in running water: 1 minute. (2–3 changes of water for loose sections.)
4. Counterstain in Mayer hematoxylin: 3 minutes.
5. Wash in running water: 3 minutes. (2–3 changes of water for loose sections.) Do not blue; the hematoxylin will remain purplish and present fair contrast to the dark blue of fast blue RR.
6. Mount in glycerin jelly.

COMMENTS

Postfixing in 10% formalin improves cellular detail and nuclear staining of both fixed and unfixed smears.

Kernechtrot can be substituted for hematoxylin, but it fades in glycerol jelly. I prefer the fast blue RR salt; the garnet GBC produces an undesirable yellowish background. Glycerin jelly mounts keep well; PVP and Gelva mounts show leaching and fading after standing for a few weeks.

Burstone (1962) uses a tris buffer at pH 7.1.

>   0.4 M tris (24.2 g/500 ml water) ....   500.0 ml
> N HCl ....   189.0 ml
>   distilled water to make.....................   1000.0 ml

See also Holt (1956) concerning the indoxyl methods.

----

[12] Sigma Chemical Co.

Lipase (George and Ambadkar 1963; George and Iype 1960)

SOLUTION

5% Tween 85 (see comment below)	2.0 ml
barbiturate buffer, pH 7.4 (p. 412) (a drug)	5.0 ml
10% calcium chloride	2.0 ml
distilled water.............................	40.0 ml

Incubate mixture, 40°C for 8–10 hours before use, in order to precipitate free fatty acids. Filter. Add crystal of thymol.

PROCEDURE

1. Quickly freeze fresh tissue and cut 40μ sections. Drop into cold (4°C) 6% neutral formalin.
2. Wash in cold distilled water: 1 hour.
3. Mount on albumenized slides and dry.
4. Coat slides with a thin layer of 5–10% aqueous gelatin. Preserve gelatin with thymol; phenol inactivates the enzyme. (See comment below.)
5. Place in freezer to solidify gelatin.
6. Fix in 6% cold formalin: 1 hour.
7. Wash in running water: 2 hours. Rinse in distilled water.
8. Incubate, 37°C: 8–12 hours.
9. Wash well in warm (40°C) distilled water.
10. Treat with 1% lead nitrate: 20 minutes.
11. Wash in warm distilled water.
12. Wash in running water: 10 minutes. Rinse in distilled water.
13. Treat with 1% aqueous yellow ammonium sulfide: 1–2 minutes.
14. Wash in 1% aqueous acetic acid to remove yellow color: few minutes.
15. Wash in distilled water and mount in glycerin jelly or polyvinyl alcohol (p. 423).

RESULT

sites—brownish black

COMMENTS

Eapen (1960) proved that lipase is activated by gelatin; he coated sections to give better results.

If Tween 80 is used, it is acted on by nonspecific esterases as well as lipase.

Counterstaining with hematoxylin can be done.

Bokdawala and George (1964) refine the method by using alizarine red S stain for the calcium soap that is formed.

Additional pieces of tissue can be fixed in calcium-formalin and embedded in gelatin, and the frozen sections can be cut and stained for lipids.

To inactivate the lipase, use heat or Lugol iodine solution.

Sources of error: Brownish pigment or calcareous deposits, pigments visible in unincubated section. These may be hemosiderin (perform Prussian blue reaction) or calcium deposits (remove before incubation in citrate buffer of pH ±4.5: 10–20 minutes).

SUCCINIC DEHYDROGENASE

Oxidation-reduction reactions, which take place in biological systems, are characterized by hydrogen atoms or electrons being transferred from molecule to molecule. Oxidation refers to the loss of an electron. It follows that the substance donating hydrogen or electrons is oxidized and the one accepting is reduced. These reactions are catalyzed by enzymes, and the living cell contains a series of electron donors and acceptors, which act in a chain. Molecular oxygen is the final electron acceptor in most reactions, and hydrogen combines with the oxygen to form water. Some enzymes are unable to use oxygen directly and the hydrogen atoms must be transferred by intermediate compounds. Dehydrogenases are the first compounds of the chain of enzymes, which transfer the hydrogen atoms from the substrate to the molecular oxygen. Succinic dehydrogenase acts as an intermediate carrier.

Among the techniques for the demonstration of dehydrogenases, the tetrazolium technique has become the preferred one. The tetrazolium salts, characterized by a heterocyclic ring structure containing one carbon and four nitrogen atoms, are colorless or pale and are readily reduced to form intensely colored water-insoluble formazans. The tetrazolium salts, however, do not accept the electrons from the dehydrogenases, but from subsequent links in the chain.

The tetrazolium salts are:

TPT—2,3,5-triphenyl tetrazolium chloride: simplest, but with limited sensitivity.

BT—2,2′,5,5′-tetraphenyl-3,3′ (3,3′-dimethoxy-4,4′-biphenylene) ditetrazolium chloride: better pigment qualities.

Nitro   BT—2,2'-di-*p*-nitrophenyl-5,5'-diphenyl-3,3'-(3,3'-dimethoxy-4,4'-bi-phenylene) ditetrazolium chloride.

Most methods require anaerobic (absence of oxygen) conditions to prevent atmospheric oxygen from competing with the tetrazolium for the electrons of the dye and thus to decrease reduction. Sodium succinate is the substrate.

Mitochondria are a complete biochemical unit involved in biological oxidations and have the succinic oxidase (also cytochrome oxidase) system built within their structure. If mitochrondria are well preserved and the substrate is not oxidized by nonmitochondrial components, the staining could be interpreted as mitochondrial. But not all sites are necessarily mitochondria; formazan can form crystals on the surface of lipid droplets, lipofuscin granules, fat vacuoles, etc.

For further details concerning dehydrogenases and tetrazolium salts, see Burnstone (1962).

Most sections are fresh frozen, are cut at 30μ (not less than 20μ), and are improved by cutting in a cryostat. They may be floated on a beaker of the incubating fluid inside the cryostat or mounted on slides and air dried in the cryostat. For anaerobic conditions, remove air from the solution with vacuum, or boil the substrate and then cool to 37°C. In either method, keep the solution in full, tightly closed bottles (small weighing bottles are excellent). Drop the sections in and close the bottle immediately.

Warning: The freezing and thawing of sections may disrupt the morphological structure of the tissue and may result in small formazan crystal artifacts, which can be confused with mitochondria. Cardiac muscle is the best tissue to use for an unfixed frozen preparation.

Novikoff (1961*a*) relates that localization is not possible unless the cell structure is better preserved than is that of cryostat-cut unfixed sections. Try fixing small (1 mm) slices of tissue in cold calcium-formalin, 0° to 5°C: 15 minutes. Quench in isopentane, −70°C: 1–2 minutes, and cut in cryostat at −25°C. Transfer to cover glasses and incubate or transfer directly into incubation solution.

Touch preparations and smears are always easy to prepare, and they make beautiful demonstrations of succinic dehydrogenase. Some cells may be broken and produce some diffusion, but those with unbroken membranes can always be recognized. Both fixed and unfixed slides can be incubated together. Fix the touch preparations and smears in cold calcium formalin: 15–20 seconds.

Nitro BT Method

SOLUTIONS

Tris buffer, pH 7.4

N HCl	17.0 ml
0.4 M tris (24.2 g/500 ml water)	83.0 ml

Substrate

Nitro BT[13]	5.0 mg

Dissolve in:

N-N dimethyl formamide (DMF)	0.25 ml

Add:

sodium succinate	500.0 mg
distilled water...........................	10.0 ml

Add:

tris buffer	20.0 ml

Shake the solution after each addition. Filter onto slides.

PROCEDURE

1. Incubate fixed or unfixed smears, frozen or cryostat-cut sections, 37°C: 30–45 minutes.
2. Wash in running water: 1 minute. Following wash, unfixed preparations may be fixed in 10% formalin or calcium formalin, room temperature: 10–30 seconds.
3. Counterstain in Mayer hematoxylin: 2–3 minutes.
4. Wash in running water: 3 minutes. Do not blue the hematoxylin.
5. Mount in glycerin jelly or polyvinyl alcohol; or dehydrate, clear, and mount.

RESULTS

sites—blue
nuclei—purple blue

---

[13] Dajac Laboratories.

COMMENTS

Control slides: Replace succinate with distilled water and incubate as above.

Aronson and Pharmakis (1962) observed irregular staining in hearts and fatty degeneration. By adding 25.0 ml of 0.05% potassium cyanide to 75.0 ml of substrate, they improved the staining.

A 0.1 M to 0.06 M phosphate buffer, pH 7.5, may be substituted for the tris buffer.

For more information about succinic dehydrogenase, see: Burstone (1962); Farber and Bueling (1956); Farber and Louviere (1956); Morrison and Kronheim (1962); Nachlas et al. (1957); Novikoff (1961a); Novikoff et al. (1960); Pearson (1958); Rutenburg et al. (1950, 1953); and Zugibe (1970).

THE OXIDASES

The oxidases are enzymes that catalyze the transfer of electrons from a donor substrate to oxygen. Most oxidases contain iron or copper. Tyrosinase contains copper and catalyzes the aerobic oxidation of tyrosine and of some phenols. When tyrosine is acted upon by tyrosinase, dihydroxyphenylalanine (dopa) is formed. Dopa is converted to O-quinone, and is then condensed to form hallochrome, a red pigment, which is spontaneously transformed into a black pigment, melanin. The enzyme tyrosinase, therefore, is responsible for the conversion of tyrosine into melanin, but the so-called tyrosinase system is very complicated, and care should be taken in interpreting its reaction (Burstone, 1962; Pearse, 1968).

Laidlaw Dopa Oxidase Method (Glick 1949)

FIXATION

Only fresh tissue with no fixation (or 5% formalin for no longer than 3 hours, but results are inferior to those from fresh tissue).

SOLUTIONS

Dopa stock solution

DL *b* (3.4 dioxyphenylalanine) phenyl tyrosine[14]	0.3 g
distilled water.............................	300.0 ml

Store in refrigerator. Discard when it turns red.

Buffer solution A

disodium hydrogen phosphate	11.0 g
distilled water.............................	1000.0 ml

Buffer solution B

potassium dihydrogen phosphate	9.0 g
distilled water.............................	1000.0 ml

Buffered dopa solution

stock dopa solution	25.0 ml
buffer solution A	6.0 ml
buffer solution B	2.0 ml

Filter through fine filter paper. Should be pH 7.4.

Buffered control solution

distilled water.............................	2.0 ml
buffer solution A	6.0 ml
buffer solution B	2.0 ml

Cresyl violet

cresyl violet acetate	0.5 g
distilled water.............................	100.0 ml

---

[14] Eastman Kodak Co., no. 4915.

PROCEDURE

1. Cut frozen sections, 10μ.
2. Wash in distilled water: few seconds.
3. Treat with buffered dopa solution, 30–37°C. The solution becomes red in about 2 hours, and gradually turns sepia brown in 3–4 hours. Do not allow sections to remain in a sepia-colored solution; they may overstain. After first 30 minutes, change to a new solution.
4. Wash in distilled water.
5. Counterstain in cresyl violet.
6. Dehydrate, clear, and mount.

RESULTS

dopa oxidase—black
leukocytes, melanoblasts—gray to black
melanin—yellow brown

COMMENTS

Freeze-dried and embedded tissues give a good dopa reaction.

Tyrosinase is inhibited by compounds which form a stable complex with copper; potassium cyanide and hydrogen peroxide are good examples.

PEROXIDASE METHODS

Peroxidases are hemoproteins that catalyze the transfer of two electrons from substrates to hydrogen peroxide to form water and oxidized dyes. Benzidine was introduced in 1904 as a reagent for peroxidase and is used as an acceptor in the transfer of oxygen from hydrogen peroxide in the reaction catalyzed by this enzyme. This produces a blue product which in a short time fades into brown and diffuses. Treatment with nitroprusside stabilizes this blue reaction product by forming a more stable salt than is formed by the benzidine and peroxidase alone. Peroxidase activity occurs in the mitochondria of striated muscle and heart and the granules of myeloid and mast cells. It takes place also in the acinar cells of thyroid and salivary glands, the medulla of the kidney, the Kupffer cells of the liver, and the hair follicles.

Wachstein and Meisel (1964) use fresh frozen sections (cut 10–20μ), which they spread on lukewarm water and then put directly into the incubation mixture.

Smears, air-dried, can be fixed in formalin–95% alcohol (1:9) or acetone: 30 seconds.

Straus Method (1964)

FIXATION

Fix thin slices of tissue in 10% formalin plus 30% of sucrose, 0–4°C: 18 hours, not less than 15 hours.

SOLUTIONS

Benzidine solution

50–70% ethanol	50.0	ml
benzidine[15]	0.1–0.2	g
hydrogen peroxide	0.015	ml

Warning: benzidine is considered a carcinogen.

Nitroprusside solution A

sodium nitroprusside......................	4.5 g
70% ethanol	50.0 ml

Nitroprusside solution B

30% aqueous sodium nitroprusside	15.0 ml
absolute methanol	35.0 ml
0.2 M acetate buffer, pH 5.0 (p. 552)	2.5 ml

PROCEDURE

1. Cut frozen sections, 4–6μ, attach to slides, and dry 1 hour.
2. Treat in ice-cold benzidine solution until peroxidase is distinctly blue. Stop reaction before blue pigment crystals form.
3. Transfer into nitroprusside solution A, ice-cold: few seconds.
4. Transfer into nitroprusside solution B, ice-cold in refrigerator: 1–2 hours.

---

[15] Dajac Laboratories.

5. Dehydrate through 3 changes 70% alcohol, 2 changes 95% alcohol, absolute alcohol, absolute alcohol–xylene, and 2 changes xylene: 2–3 minutes each. Mount.

RESULT

sites—gray green to blue black.

COMMENTS

For blood pictures the smears or sections can be counterstained in Giemsa: 10 minutes.

Giemsa stock solution (p. 231)	4.0 ml
acetone	3.0 ml
0.1 M phosphate buffer, pH 6.5	2.0 ml
distilled water	31.0 ml

Then dehydrate through isopropanol. Do not use a buffered fixative or a calcium fixative; the use of salts appears to increase the formation of crystals in the blue reaction product. Washing the fixed sections for 24 hours in 30% sucrose also reduces this tendency. Storage in glucose at 0–4°C for several weeks preserves the enzyme.

If catalase is added to the incubation mixture, no reaction occurs because the catalase destroys the peroxide. Boiling the tissue also destroys activity. If the slides are kept in the refrigerator, they are usable for several months. If kept at room temperature, they fade in a few days.

Wachstein and Meisel (1964) have found that frozen sections are satisfactory, but the red blood cells hemolyze in the aqueous medium and do not react. Calcium-formalin fixation, 4°C: 30 minutes, preserves a positive stain in the hemogloblin and improves activity in muscle. Longer fixation, however, suppresses the activity.

A peroxidase method can be used to demonstrate microbodies (cytoplasmic bodies Type II, peroxisomes) in hepatocytes of livers and renal proximal tubules of kidneys. See Essner (1970); Roels et al. (1970, with bibliography); and Straus (1967).

Cytochrome Oxidase

Cytochrome oxidase is an iron-porphyrin enzyme that catalyzes the reduction of oxygen to water and accounts for the reduction of most of the oxygen used by tissues. It is closely associated with cytochrome C, a hemoprotein capable of undergoing a reversible oxidation reduction. Cytochrome oxidase is demonstrated by the indophenol blue or "nadi" reaction. It was named nadi from the first two letters of *na*phthol and *di*amine (alpha-naphthol and N,N-dimethyl-p-phenylenediamine), the two chemicals that are used to produce the reaction (Burstone 1962).

For information about oxidases, see Burstone (1962); Lillie and Burtner (1953); and Wachstein and Meisel (1964).

MOUNTANTS

Gelva

Burstone (1957*a*) recommends the following procedure for some azo-dye methods:

1. Mount slides directly from 75–80% alcohol in 20% solution of polyvinyl acetate (Gelva 2.5)[16] in 80% alcohol. It is soluble also in acetic acid, *n*-butanol (90%), aniline, pyridine, and ethyl Cellosolve.
2. It preserves esterase, acid and alkaline phosphatase, PAS, hematoxylin, celestin blue, aldehyde fuchsin, eosin.
3. Cover glass is immovable after 1 hour. No bubble formation. Index of refraction is 1.3865, but it increases after evaporation of alcohol and water.

Polyvinyl Alcohol (McCurdy and Burstone 1966)

Add 20 g AIRCO[17] to 100 ml cold or tepid water with stirring. When all particles are wet, heat the slurry to the boiling point for about 15 minutes or until all particles are dissolved. A pale yellow color forms. A thin film forms on the surface but can be removed when using the solution.

Polyvinyl alcohol is recommended for various enzyme systems— alkaline and acid phosphatase, esterase, succinic dehydrogenase, and cytochrome oxidase in particular.

McCurdy and Burstone also mention two acrylic resins.

---

[16] Shawinigan Resins Corp.
[17] Air Reduction Chemical and Carbide Co.

PVP (Vinyl Pyrrolidone–Vinyl Acetate Copolymers)

PVP/VA, E-735 (Plastone C),[18] is also recommended by Burstone (1962) for some of the enzymes. Mix the PVP with an equal volume of water, stirring well until the plastic is thoroughly mixed into the water. Mounting can be done from water or alcoholic solutions. In my experience esterase fades in PVP, but acid and alkaline phosphatase are reasonably well preserved, although with some diffusion. If smears or mounted sections are taken from the water and dipped 1 or 2 times in 95% alcohol, mounting in PVP is expedited.

Other Mounting Media

Other mounting media are Nyssco-Plastic (Diacumakos et al. 1960) and prepolymerized butyl methacrylate (Izard 1964).

SUBSTRATE FILM METHODS

In Daoust's method, a film of gelatin containing a substrate is placed in contact with tissue sections. During the exposure, the tissue areas that contain the appropriate enzyme attack the substrate in the film. After exposure, either the unaltered substrate or the products of the reaction are visualized by staining. An "autograph" is made in the film and can be compared with the corresponding tissue sections to reveal the sites of enzyme activity. For example, if checking for DNAse activity, add DNA to the gelatin. After exposure the DNA remaining in the film is stained with toluidine blue. Unstained regions correspond to areas of tissue in which DNAse activity occurs (Daoust 1957, 1961; Daoust and Amano 1960; Mayner and Ackerman 1962, 1963).

In another method, a blackened photographic plate is used, and the gelatine film is digested by protease enzymes, thereby leaving an autograph (Adams and Tuqan, 1961).

OSMIUM BLACK METHODS

Seligman and Hanker and their coworkers have developed substrates which use groups capable of reacting with osmic acid and can be used for demonstrating enzymes, such as esterase, phosphatases, dehydrogenase,

---

[18] Antara Chemicals, General Aniline and Film Corp.

and other substances. These osmiophilic reagents can be used for stable preparations that are nonvolatile and electron-opaque for the electron microscope and have good pigment qualities for the light microscope. The end product is called osmium black. A new principle of osmium bridging has been developed which uses thiocarbohydrizide (TCH) as a ligand to bond osmium black to other metals. The application of this principle has been described in numerous reports; start with Hanker et al. (1972); Seligman et al. (1968); and their references.

Chapter 22
Special Procedures I

FREEZING TECHNIQUES

The freezing technique for preparing specimens is unsurpassed in certain situations. It is rapid, making it the best choice for diagnostic purposes where speed is essential. It can be used for the preparation of sections containing such elements as fats, enzymes, and some radioisotopes that would be lost in alcohol, in paraffin solvents, or by heat. One disadvantage is that some distortion may be caused by the freezing and cutting.

Freezing methods formerly used ether, ethyl chloride, or other volatile liquids. Present methods employ compressed carbon dioxide gas blown into a chamber. There the gas expands, cooling the chamber, and freezing the tissue mounted on it. For this procedure most laboratories use the clinical microtome with a freezing attachment. An attachment can also be used on sliding microtomes, but not efficiently on rotary models. However, Strike (1962) and Stagg (1963) have adapted rotary microtomes for the freezing technique. There are also dry ice holders available, such as the Dry Ice Freezing Chamber.[1] Dry ice is held inside the chamber, and the specimen is frozen simply by pushing the dry ice up against the metal top holding the specimen. Nickerson (1944) fitted the object clamp of a rotary microtome with a small metal box holding dry ice. A new freezing system, which uses Freon, is on the market—the Histo-Freeze,[2] a portable freezing unit, 12×18 inches, with Freon 12 circulating through the unit. Model 67027A fits American Optical, 67027B fits Bausch and Lomb,

---

[1] Fisher Scientific Co.
[2] Scientific Products.

and 67027C fits the Sartorius freezing microtomes. Lipshaw has a tissue-freeze refrigerating unit, counter or desk height, fitting American Optical, Bausch and Lomb, Sartorius, or Reichert microtomes. Tissu-Freez,[3] Bailey Freezing Stage,[4] and Cryo-Histomat[5] are solid state transistorized cooling devices using the Peltier effect to convert electricity into cold; they can be used with any freezing microtome. National Tissue Freeze[6] is an aerosol-type coolant for quick freezing of tissue. It uses a pistol grip with fingertip control and an extended nozzle.

FIXATION, BLOCKING, AND SECTIONING

Fresh tissues may be used; but it is preferable to fix the material, usually in formalin, which gives the tissue an ideal consistency for the freezing technique. Formalin is a water solution and requires no washing. Hartz (1945), however, recommends Bouin as better than formalin, but says that the tissue must be washed for a short time before freezing. If the fixative contains mercuric chloride, the crystals may be removed immediately after sectioning with an iodine solution, such as Lugol's, which contains no alcohol. If the tissues are extremely friable, it may be advantageous to embed them in gelatin (p. 430), or in alginate gel (Lewis and Shute 1963), or to immerse them in a thick aqueous syrup, such as gum arabic. O.C.T. Compound[7] or Cryoform[8] facilitates cutting and handling of frozen sections. It is applied directly on the tissue immediately before freezing is started. O.C.T. Compound comes in three types for three ranges of temperature, $-10°$ to $-20°$, $-20°$ to $-35°$, and $-35°$ to $-50°$.

Purists among technicians may object to the following method of fixation, and it is recommended only for very limited use. It does, however, preserve sufficient cell structure for immediate diagnosis on surgical biopsies from patients being held under anesthesia. A small piece of the tissue can be boiled for a couple of minutes in 10% formalin. Rinse in water and freeze. Reiner (1953) adds tissue to undiluted formalin at room temperature, then heats the solution and tissue to 56–60°C in a water bath for 2 minutes. Agitate the tissue during the heating. Reiner claims inferior results without agitation and says that placing tissues in preheated formalin is disadvantageous. Additional pieces of tissue can be fixed in a more conventional manner for later study.

---

[3] Eric Sobotka Co.
[4] Bailey Instruments Co.
[5] Hacker Instruments.
[6] Allied Chemical Corp.
[7] Lab-Tek Instruments Co.
[8] Fisher Scientific Co.

Friedland Method, Using Agar (1951)

1. Boil tissue in formalin.
2. Replace formalin with 2% sterile agar (previously heated to its melting point) in water. Boil gently 1 minute.
3. Pour off and freeze tissue immediately.

The agar can be kept ready in Pyrex tubes and melted quickly over an open flame.

Lev and Thomas Method (1955)

This method fixes tissue at 80°C and produces excellent results. The following solution is placed in a baby's bottle warmer and comes to a boil in 1 minute. Add the tissue and allow it to boil for 1 more minute.

formalin	10.0 ml
glacial acetic acid	1.0 ml
95% ethyl alcohol	80.0 ml

The tissue must be washed to remove the alcohol before freezing can be attempted. This can be a disadvantage if speed is important.

The pieces to be cut should be no more than 2–4 mm thick for rapid and uniform freezing. One surface should be flat to provide sufficient contact with the tissue carrier. If, during freezing, there is a tendency for the tissue to break loose from the carrier, cut a piece of filter paper to fit the top of the carrier and place it, wet, under the tissue to hold the tissue in place while freezing and sectioning. Young (1962) suggests, for extremely small tissues, the use of a small piece of sponge as a base to raise them high enough to prevent the knife from striking the metal of the object holder.

Consider the shape and type of tissue when it is oriented on the carrier. If one side is narrower than the other, place that side, or any straight side, parallel to the knife. If there is a tough membrane on one side of the tissue, place that side toward the knife to prevent the membrane from breaking away from the rest of the block during sectioning. Usually the amount of water carried over with the tissue is sufficient for firm attachment to the carrier. (Use normal saline for fresh tissues.) Avoid too much water and do not allow it to settle around the tissue in a wall of ice that can deflect the knife or tear through the tissue during sectioning. Uneven or torn sections result. If, however, gum or O.C.T. Compound is being used, it is necessary to build the compound around the tissue while freezing it. In this case the tissue must be completely encased by the medium.

With a finger, gently press the tissue onto the tissue carrier. Slowly freeze the tissue by turning on CO_2 for a moment or two, and then turn it off. A series of successive gas jets freezes better and wastes less gas than does a continuous stream. Other freezing methods, such as Histo-Freeze and Tissu-Freez, use a continuous action. When the tissue is frozen fast and the finger can be removed, move the knife over the block so it too is in the path of the cooling gas. This aids in uniform freezing of the tissue by deflecting some of the gas onto the top of it, and at the same time cools the knife.

Good sections depend on proper tissue temperature. Material frozen too hard forms white brittle fragments on the knife; if too soft it forms a mushy mass. In both cases the sections break up when placed in water. These difficulties can be avoided by slightly overfreezing the block and then, as it warms and reaches the correct temperature, a number of sections can be cut in close succession. When these sections fall on the tissue carrier, allow several to accumulate and pick up the group with one sweep of the finger. Or, as soon as the section is cut, remove it from the knife with a finger tip and place it in a dish of water. Do not allow water to collect on the knife. Keep both knife and fingers dry. Some tissues, such as adipose, tend to stick to the finger and will not shake loose in water. Transfer them to 70% or 80% alcohol.

Occasionally, when there is no time for gelatin embedding, one encounters a tissue difficult to section because it tends to fall apart. Bush and Hewitt (1952) support the tissue with Kodak Frozen Section Film. The film is softened in water, excess moisture is sponged off, and the film is pressed down on the freshly cut surface of the frozen tissue. The film freezes and adheres to the tissue while a section is cut; this film is carefully raised as the knife passes under it. The film with the section is laid face down on a slide. As the section thaws, it will cling to the slide, and a thin layer of moisture develops by capillarity between it and the film. When this has dried, the section can be stained. (See also use of cellulose tape, p. 526.)

MOUNTING

If the sections are collected in a glass dish, place the dish on a black surface, such as a tabletop or a piece of black paper. The sections are more easily manipulated over the black background, and the best ones can be selected. Dip one end of an albumenized slide into the water. With a needle, or preferably a small round-tipped glass rod, gently bring a section against the slide. Hold the section at one corner; by maneuvering under water, creases can be unfolded and the section unrolled as the slide is drawn out of the solution. If the wrinkles refuse to straighten, drop the section on 70% alcohol and then remove it to a slide. Drain off excess water or pull it off with filter paper. Press section in place with filter paper moistened with 50% alcohol. With a pipette add absolute alcohol directly on top of the section; let stand for approximately 30 seconds. Replace with more absolute alcohol (must be ethyl alcohol). Drain thoroughly, and flood slide with 1% nitrocellulose (dissolved in absolute ethyl alcohol–ether, 1:1) for 1 second. Drain and wave slide in air until almost dry. Immerse in 70% or 80% alcohol, 5–10 minutes or longer; it can be left here for several hours or days. (*Caution:* Use an old solution of alcohol, either one from the staining series, or one which has been mixed for at least a day. A freshly prepared solution forms bubbles between section and slide.)

Fresh tissue sections may be removed from normal saline to the slide and thoroughly drained. Add 95% alcohol to remove the water; let stand 30–60 seconds. Drain off and allow the section to almost dry. Place in 95% alcohol for 1–2 minutes and proceed to stain.

A gelatin fixative can be used to affix the sections. Spread fixative on the slides and dry them in an incubator. (They can be stored for some time in a dustfree box.) Float the sections on the gelatin, drain off excess water, and place the slides in a covered dish over formalin. This converts the gelatin into an irreversible gel, holding the sections in place. After the slides have been over the formalin for 30 minutes, wash them in running water for 10 minutes and proceed to stain. Subbed slides can be used; I prefer them over other methods. Mount the sections out of the water onto the subbed slides, tilt the slides to drain off excess water, and allow to dry in a vertical position.

Snodgrass and Dorsey (1963) use an egg albumen method for embedding. It appears to be simple and practical.

Gelatin Embedding

This is recommended for tissues as difficult to handle as lung and bone marrow.

Pearse Method (1953)

1. Fix in cold formalin (15%) at 4°C if enzymes are to be preserved: 10–16 hours.

2. Wash in running water: 30 minutes.
3. Infiltrate with gelatin, 37°C: 1 hour.

gelatin	15.0 g
glycerin	15.0 ml
distilled water..........................	70.0 ml
small crystal of thymol	

4. Cool and harden in formalin (40%), 17–22°C: 1 hour. Wash.
5. Store at 4°C or below until sectioned.

STAINING

Nitrocellulose-mounted sections are transferred from the 70% alcohol to water and then into any desired stain. After staining they probably will require special care. If after dehydration the sections are placed in absolute ethyl alcohol and show a tendency to fall off because the nitrocellulose dissolves, follow the 95% alcohol with isopropyl alcohol, 2 changes, and then xylene. Another method is to follow the 95% alcohol with carbol-xylol (p. 543) or terpineol-xylol (1:9), several changes, until the sections look clear. Fogginess or colored droplets on the slides indicate traces of water. If carbol-xylol is used, rinse it off by dipping the slide for 2–3 seconds in xylene. Carbol-xylol must be removed; any of it left in the sections will fade stains in a short time, almost overnight. (*Caution:* Keep fingers out of carbol-xylol; it contains carbolic acid. Wash it off immediately if any spills on the skin, and use lanolin for skin irritation.)

Irregular staining usually is caused by an uneven coat of nitrocellulose; this can be prevented by tilting the slide immediately after the solution is applied. If the nitrocellulose is thicker than 1%, uneven staining may result. A colored background indicates a thick coat of nitrocellulose. Sometimes irregular staining is due to loosening of parts of the section. Sections will adhere to subbed slides better than to nitrocellulose-coated ones, and can be carried directly into water followed by stain.

Detec, the section clarifier, is no longer available, but Coolidge (1972) has come up with a substitute. This dissolves the protein coagulated around cellular elements and makes them as clear as paraffin sections.

1. Rinse frozen sections in 3 changes of distilled water.
2. Treat in Prestone-Tergitol, 19:1 for 5 seconds only.
3. Rinse briefly in 3 changes of distilled water.
4. Mount on albuminized slides (I use subbed). Dry 10 minutes, 60°C. Stain.

See Albrecht (1956); Jacobson (1963); Peters (1961*a*); and Thompson (1957) for staining in quantity.

Rapid Staining Methods

Pinacyanole Method (Humason and Lushbaugh 1961; Proescher 1933)

Staining and mounting time: 3–4 minutes.

PROCEDURE

1. Mount frozen section (fresh or fixed) on subbed slide; drain off excess water and blot with two sheets of filter paper.
2. Cover section with several drops of stain: 3–5 seconds. (pinacyanole,[9] 0.5% in either 70% methyl or ethyl alcohol; keep stock solution in refrigerator.)
3. Wash gently in tap water: 5–10 seconds or until free of excess stain (longer washing will not alter intensity). Blot.
4. Dehydrate in isopropyl alcohol, 2 changes: 1–2 minutes.
5. Clear in 2 changes xylene and mount.

Because differential staining is lost in paraffin sections, this method is recommended only for fresh or frozen sections. In order to preserve cytoplasmic staining, do not dehydrate the sections in ethyl or methyl alcohol; they decolorize cytoplasmic structures, and only the nucleus will retain appreciable amounts of stain. Steps 4 and 5 can be omitted and the sections mounted with an aqueous mountant or glycerine.

RESULTS

chromatin—well-differentiated blue to reddish blue
connective tissue—pink
elastic tissue—dark violet
muscle—violet to purple
plasma cells—red granuloplasm
hemosiderin—orange
hemoglobin; neutrophilic and eosinophilic granules—unstained
neutral fat—colorless to faint bluish violet
lipoids—bluish violet to purple
amyloid—carmine red

---

[9] Eastman Organic Chemicals.

Humphrey Method (1936)

This is perhaps the fastest method.

1. Place section on a slide and wipe off excess water.
2. Add 1 drop of 0.5% brilliant cresyl blue in saline.
3. Cover with cover glass and examine.

This is never a permanent mount.

Thionin or Toluidine Blue Method

1. Remove section from water to a solution of either 0.5% thionin or toluidine blue O in 20% ethyl alcohol plus a few drops of acetic acid: 30 seconds.
2. Rinse in water and float on slide.
3. Drain off excess water, blot around edges of section, add drop of glycerine, and cover glass.

This is not a permanent slide.

The above two methods stain the tissue elements in shades of blue and purple.

NITROCELLULOSE METHOD

This form of embedding is often classified as the celloidin technique; celloidin has become something of a generic term including the various cellulose compounds, such as nitrocellulose and soluble guncotton or collodion. These are solutions of pyroxylin consisting chiefly of cellulose tetranitrate. Obviously a purified nonexplosive form of pyroxylin is necessary, and there are several: Parloidin, Celloidin, and Photoxylin. LVN is an excellent embedding medium:[10] a solution of 30% low-viscosity nitrocellulose in 35% ether and 35% absolute ethyl alcohol. The special low-viscosity nitrocellulose allows more rapid infiltration with solutions of higher concentration than is possible with many of the other so-called celloidins. This permits formation of harder blocks in a shorter time, and cutting of thinner sections.

---

[10] Tissue Embedding Solution no. 4700, Randolph Products Co.

The chief advantages in using the nitrocellulose method are that larger and harder pieces can be cut in nitrocellulose than in paraffin, and the consistency of nitrocellulose allows mixed (hard and soft) tissues to be cut. This quality is useful for organs such as eyes, teeth, and bones and their surrounding tissues, and for problems of shrinkage and the formation of artificial spaces. But a slow dehydration and prolonged infiltration are preferable in these cases, and the method is costly.

In this method toluene cannot be used as a solvent, but there are many other solvents for nitrocellulose. One of the most common is a combination of two so-called latent solvents, not in themselves efficient solvents, but possessing excellent solvent qualities when mixed with other compounds: diethyl ether (generally referred to as ether) and ethyl alcohol. The solution is made up of equal parts of ether and absolute ethyl alcohol.

DEHYDRATING AND INFILTRATING

The tissue is fixed, washed, and can be stored in 70% alcohol as usual. The next steps are as follows:

1. 95% ethyl alcohol: 4–24 hours, or longer.
2. absolute ethyl alcohol, 2 changes: 4–24 hours or longer in each.
3. ether-alcohol (equal parts of absolute ethyl alcohol and anhydrous ether): 4–24 hours or longer.
4. 10% nitrocellulose (dissolved in absolute ethyl alcohol–ether, 1 : 1): 2 days or longer, in a screw-cap jar to keep evaporation to a minimum.
5. 33–35% nitrocellulose (in absolute ethyl alcohol–ether, 1 : 1): 2 days or longer, in tight jar.

Nitrocellulose solutions should be stored in the dark; light causes the solution to deteriorate. Ferreira and Combs (1951) warn that old light-affected solutions cause fading of nervous-tissue blocks.

EMBEDDING

Slow Method

Place the tissue in a small glass-covered dish (stender dish), and over it pour 33–35% nitrocellulose until it is $\frac{1}{4}$ to $\frac{1}{2}$ inch above the tissue. Over this pour a thin layer of ether-alcohol (1 : 1). Cover the dish tightly, and allow the solvent to evaporate slowly until the embedding medium reaches a proper consistenty. If bubbles appear in the medium, evaporation is proceeding too rapidly. A little Vaseline on the ground-glass edge of the cover will help to seal it more tightly. In dry warm weather it may be necessary to enclose the evaporation dish in another dish or jar to slow down the evaporation of solvent. The process should take several days to a week or more, the slower the better. Do not allow the medium to become too hard; it should reach the consistency of hard rubber and should feel dry and not show fingerprints. If a tough film forms quickly on the surface of the solution and sticks to the side of the dish, carefully loosen the film from the glass to allow more efficient evaporation.

When the nitrocellulose is properly formed, trim it down to cutting size, and mount it on a fiber or wooden block (Fig. 22–1). Wrap a band of hard bond paper tightly around the block, and secure the band with string or stick it down with a paper label. (Do not use a rubber band.) The paper must project high enough above the mounting block to enclose the tissue block. Moisten the mounting block and paper with ether-alcohol, and then place a small amount of 33–35% nitrocellulose in the container formed by the two. Roughen the underside of the tissue block with a needle, and cover it with a drop or two of ether-alcohol. Press the roughened surface tightly into the nitrocellulose on the paper-wrapped block. After a few minutes in the air, place the block in a closely capped jar with a small amount of chloroform; leave it there 20–30 minutes. Add more chloroform to immerse the entire block, and if possible allow it to "set" in chloroform overnight. There are other and more haphazard ways of mounting nitrocellulose blocks, but the above method is relatively sure. Instant setting glues can be used, but they have to be scraped off the blocks following use.

Romeis (1948) claims that if wooden blocks are used they should be preseasoned with nitrocellulose. Cook the wooden blocks in distilled water and allow them to dry for several days. Extract with equal parts of 70% alcohol and glycerin for a day and wash the blocks in distilled water until shaking no longer causes a foam. Cover the surface with 8% nitrocellulose in ether-alcohol, dry, and put the blocks in a glass bottle.

Rapid Method

This method will not form as perfect a block as the slow method; the nitrocellulose is not always uniform and may section unevenly.

Wrap paper around the fiber block as for the slow method. Fill the cavity with some nitrocellulose from the infiltrating jar and transfer the tissue into the mass of nitrocellulose. Orient the piece of tissue so that it can be sectioned along the upper surface. Set the fiber block, nitrocellulose, and tissue in a tightly capped jar with a small amount of

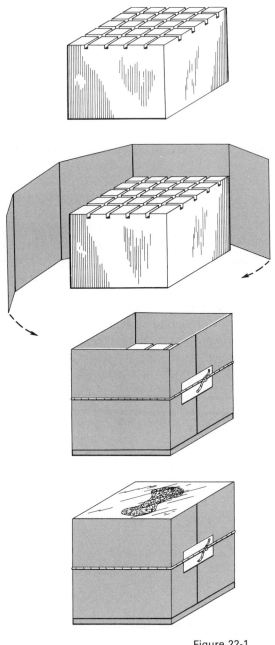

Figure 22-1
Preparation of a nitrocellulose tissue block for sectioning.

chloroform. Leave them in this situation until the block is firm, but check frequently to see that the chloroform does not evaporate completely. When the nitrocellulose has hardened, add more chloroform to completely immerse the block and tissue until it is to be sectioned.

Mounted or unmounted tissue blocks can be stored in either chloroform, or 70%, 80%, or 95% alcohol, but for long storage add a little glycerin to alcohol solutions. (See Sallmen and Sherman 1961 for another rapid method.)

Vacuum can be used to facilitate infiltration.

Hot Celloidin Technique

Koneff and Lyons Method (1937)

This is a rapid method and can be used when materials are not injured by heat. The transfers are handled in screw-cap jars in a paraffin oven, about 56°C.

PROCEDURE

1. 70% alcohol, 2 changes: $\frac{1}{2}$ hour each.
2. 80% alcohol, 2 changes: $\frac{1}{2}$ hour each.
3. 95% alcohol, 2 changes: $\frac{1}{2}$ hour each.
4. Absolute alcohol, 2 changes: $\frac{1}{2}$ hour each.
5. Absolute alcohol–ether (equal parts): 1 hour.
6. 10% nitrocellulose: 1 hour.
7. 25–33% nitrocellulose: overnight.
8. Embed as for rapid cold method.

COMMENTS

Walls (1932) uses a bottle with thick lips and corks and attaches the corks with a wire wound under the lip and over the cork. High pressure builds up in the bottle and, together with the high viscosity of the heated celloidin, produces rapid penetration. He maintains that heat is of no concern in this method; in the paraffin method it is the hot paraffin, not the heat itself, which produces a brittle block if used too long.

Do not hurry the cooling process. If the warm bottles are left on a wooden tabletop until they have reached room temperature, they will not break.

SECTIONING

Clamp the jaws of the tissue carrier tightly against the lower two-thirds of the mounting block. If the jaws are clamped against the upper portion, the resulting compression in the block can loosen the nitrocellulose.

If the knife has a concave surface, clamp that surface uppermost in the knife holder. The knife holder must be adjusted so the knife is pushed back on the instrument far enough to clear the tissue before the tissue is elevated in preparation for the next section. An automatic lever on a small slide (sliding surface) on the back of the knife block can be moved until the correct knife position is determined. If the knife does not clear the tissue before the tissue is elevated, the pressing of the knife can injure the tissue and may alter the thickness of the next section.

The maximum thickness of a section as controlled by the automatic feed lever will be only 30 or 40μ on most microtomes. Thicker sections can be cut by moving the hand feed counterclockwise and then releasing it. The hand feed is a large round knob (American Optical sliding microtome) mounted at the base of the tissue carrier and near the micrometer screw. Each movement of the hand feed will equal the number of microns indicated by the automatic feed.

Keep the surface of the slide for the knife block well oiled with the oil provided with the microtome (or Bear Brand Household Oil). A dry slide produces a jerky knife movement and rough, irregular sections.

Always maintain a pool of 70% alcohol on the edge of the knife and the top of the tissue block. Otherwise the sections may shrivel instead of slicing in a smooth sheet. The following methods are suggested for sectioning.

Method I

Wet knife and block with alcohol. Draw the knife through the block until a section is almost, but not quite, cut to completion, making certain that the cut has cleared the tissue in the block. If the cut does not clear the tissue, a striation will appear in the tissue at the point where the knife stopped or hesitated. The section rolls as it is cut. While the section is still attached at one corner, use the thumb or a camel's hair brush wet with 70% alcohol to unroll the section against the knife, and then finish the cut. Squares of paper cut to a size a little larger than the section are moistened with 70% alcohol. Hold one of these against the undersurface of the knife and slip the section off onto the paper (Fig. 22–2). Invert the section and paper in a dish of 70% alcohol and press them down against the bottom of the dish. The sections can be stacked one on another in this fashion and stored until

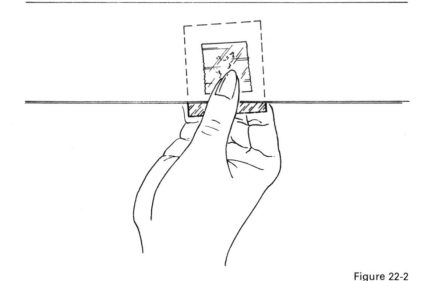

Figure 22-2
A suggested method for removing sections from knife; thumb on top of section and other fingers holding paper under knife.

they are stained. (*Caution:* Use plain white paper, filter paper, or paper towels. Lined paper may stain the nitrocellulose and produce poor staining.)

Method II

Flood the knife with alcohol and moisten a camel's hair brush. Hold the brush close to the nitrocellulose block, so the alcohol is held between the block and brush by capillary action. Do not rest the brush too firmly on the block, lest it be cut by the knife. Pull the knife across the block; raise the brush slightly, and it will guide the section onto the knife, smoothly and without rolling. Remove as above (Walls 1932).

Method III

Walls (1936) developed a dry method that eliminates the constant soaking with alcohol during sectioning. After the block is hardened, soak it in a mixture of cedarwood oil and chloroform (50:50) for 24 hours. Blot the block and allow it to stand in the air for 10–20 minutes. Transfer to cedarwood oil; this lubricates the block and knife during sectioning. The sections, as removed from the knife, are placed in the same kind of oil, then rinsed in 95% alcohol, then 80% alcohol, and transferred to water for staining. This type of oil-soaked block can be sectioned on a rotary microtome as well as a sliding one.

Difficulties in Nitrocellulose Sectioning[11]

1. Scratches in sections. Caused by:
   a. Nick in knife.
   b. Hard material in tissue.
   c. Position where knife was stopped while unrolling section.
2. Tissue soft and mushy; it crumbles or falls out. Caused by:
   a. Imperfect infiltration due to either incomplete dehydration or too short a time in nitrocellulose; reinfiltrate.
   b. Embedding too quickly, as in rapid method.
3. Sections vary in thickness. Caused by:
   a. Worn parts in microtome.
   b. Pressing too hard on knife block.
   c. Insufficient hardening of nitrocellulose, which is too soft and compresses under knife.
   d. Tissue block rising before return stroke of knife has cleared it.
   e. Insufficient tilting of knife; shoulder of facets compresses block instead of cutting it and next section is thick.
   f. Embedding too quickly, as in rapid method.

STAINING AND MOUNTING

Staining Before Mounting

In the methods that follow sections are stained before mounting. Staining may be handled in syracuse watch glasses or similar flat dishes. With forceps or spatula, carry the sections through successive solution in watch glasses and follow each change by draining the section against a paper towel or filter paper. In this way, contamination can be held to a minimum. If, however, the solutions do pick up considerable stain, change them as frequently as seems necessary, or even more often. Also, if they have remained uncovered for some time it is probable that the alcoholic content has been reduced by evaporation. Never, at any time

---

[11] Modified from Richards (1949).

during the transferring of sections, allow them to become dry. They may be carried directly from the 70% alcohol to the stain, if it is an alcoholic stain; or first into water, then into the stain, if it is an aqueous stain. (Remember to remove undesirable pigments or crystals before staining.)

When ready for dehydration, transfer through 50% and 70% alcohol into 95% alcohol. Absolute ethyl alcohol cannot be used with nitrocellulose sections unless about 5–10% of chloroform is added to it to nullify the softening effect of the alcohol. Absolute isopropyl alcohol, however, is prefectly safe to use and can be recommended before clearing action is applied. Do not attempt to dissolve or remove nitrocellulose from the sections; they will become too difficult to manipulate.

Method I

Transfer sections into carbol-xylol (p. 543) or terpineol-xylol (1:9): 2–3 minutes. Keep carbol-xylol covered at all times to prevent evaporation of xylene. Dip the edge of a slide into the carbol-xylol dish and slip the section in place with a needle or forceps. Drain thoroughly and flush off excess carbol-xylol with toluene or xylene, 2 or 3 times. Add several drops of mountant and cover glass. Press cover firmly in place. Several hours later, after the resin has thickened, press down again. If the cover glass still tilts or lifts, place a small weight on it or clamp it with a clothespin overnight. Betram (1958) describes a weight that will not stick to the mountant and cover glass. *Caution:* When using carbol-xylol, keep fingers out of the solution. Wash off any that does get on the fingers and rub lanolin on the skin.

Method II

Transfer the sections to absolute alcohol containing 5% chloroform to prevent the alcohol from dissolving the nitrocellulose. Clear with benzene or toluene and mount as above, omitting the carbol-xylol.

A method particularly good for transferring thin sections to slides is as follows: slip a piece of cigarette paper under a section and spread it smoothly on the paper with a brush or needle. Place the paper plus the section on a slide with the section between the slide and paper. Blot firmly with a piece of filter paper. Peel off the cigarette paper and rinse the section with toluene or xylene. Add mountant and cover glass.

Mounting Before Staining

In the methods below sections are affixed to slides for staining.

Method I

1. Albumenize slide or use subbed slides.
2. Press section on with filter paper.
3. Pour clove oil over it: 5–20 minutes.
4. 95% alcohol, 3 changes: 5–10 minutes total.
5. Absolute alcohol–ether, 2 changes, to dissolve nitrocellulose: 2 minutes total.
6. 70% alcohol, etc., to stain.

Method II (Lewis 1945)

1. Rub one drop Haupt fixative (p. 549) on slide.
2. Transfer section on cigarette paper to slide.
3. Blot and firmly press with filter paper.
4. Roll off cigarette paper rapidly; section must not dry.
5. Place immediately in absolute alcohol–ether until nitrocellulose is dissolved: 10–20 minutes.
6. Remove to 70% alcohol, etc., to stain.

Method III

1. Float sections onto subbed slide and blot firmly with filter paper.
2. Dip slide into 0.5% nitrocellulose for few seconds; drain and wipe back of slide clean.
3. Harden film in chloroform: 5–10 minutes.
4. 95%, 70% alcohol, etc., to stain.

Method IV (Culling 1957)

1. Transfer section to 95% alcohol: 1–2 minutes.
2. Float onto slide, drain for few minutes, blot lightly to flatten section, but do not dry.
3. Pour ether vapor over sections—only vapor, not liquid. This partially dissolves nitrocellulose and causes it to adhere to slide.

4. Place slide in 80% alcohol: 5 minutes, to harden nitrocellulose.
5. Running water: 10 minutes; stain.

Method V—Serial Sections

1. Arrange sections on knife from left to right.
2. Lay cigarette paper on sections and saturate with 70% alcohol.
3. With quick sweeping movement, pull paper off knife and carry to slide; smooth with brush.
4. Place filter paper on top and press flat, until much of alcohol is absorbed.
5. Jerk off paper and cover with clove oil: 3–5 minutes.
6. Submerge in 95% alcohol.
7. 70% alcohol, etc., to stain.

Method VI—Serial Sections (Williams 1957)

1. Coat clean slides by dipping in warm solution of 1% gelatin; allow to drain, and dry in vertical position.
2. Lay sections on blotting paper moistened with 70% alcohol; keep covered to prevent drying.
3. Pick up sections, one at a time, dip in 70% alcohol, and arrange in order on coated slides.
4. Blot sections with dry filter paper, flattening and pressing them into contact with gelatin.
5. Place in coplin jar containing 2–3 ml formalin; do not allow fluid to come in contact with sections because it may cause them to float off slides.
6. Allow slides to remain tightly covered: 2–3 hours, room temperature.
7. 70% alcohol, etc., to stain.

Peters (1961*b*), after firmly pressing the sections onto albumenized slides, allows them to air dry overnight before staining.

I find the staining of unmounted nitrocellulose sections less traumatic than working with mounted ones. For preparing mounted sections, adopt the method that is most consistently effective.

WATER-SOLUBLE WAX EMBEDDING AND SECTIONING

The so-called Carbowax compounds and water-soluble polyethylene glycol waxes are used when it is necessary to go directly from fixative or water to the embedding medium, and no alcohols or clearing agents are required. These waxes are polymers, compounds composed of the same kind of atoms in the same percentage composition but in different numbers and therefore having different molecular weights; a polymer is designated by its average molecular weight.

The Carbowaxes are these: Compound 4000, which is hard and dry and in crystalline flakes; Compound 1540, which is not so firm, and liquefies within a week; Compound 1000, which is in slippery lumps and liquefies within 24 hours; Compound 1500, a blend of equal parts of polyethylene glycol 300 (a fluid) and wax 1540.[12]

Other Carbowax compounds are the polyglycols E9000, E6000, E4000, E2000, and E1000,[13] and HEM (Harleco Embedding Media).[14]

Wade (1952) finds 4000 too hard for his use and suggests a mixture of 1:9 or 2:8 of 1540 and 4000, depending upon the weather; he recommends the 2:8 mixture except in the hottest and most humid climate.

Fixation

Any fixative may be used, but after a potassium dichromate fixative wash tissue for 12 hours before embedding.

Infiltration and Embedding

Use Carbowax 1–3 hours, 50–56°C; agitate occasionally. Rinehart and Abul-Haj (1951b) preceded infiltration with pure Carbowax by placing the tissue for 30 minutes in 70% aqueous Carbowax and 45 minutes in 90% Carbowax. It is wise to include these steps if a tissue shows any tendency to infiltrate poorly under the short method.

Prepare the Carbowax ahead of time by placing the mixture in a beaker and allowing it to melt slowly in an oven. If it is melted rapidly over a flame, a block formed from it will not cut well. The mixture will keep for an indefinite time in the oven.

---

[12] Fisher Scientific Co.; Union Carbide.
[13] Dow Chemical Co.
[14] Hartman-Leddon Co.

Embed by placing the tissue in a second batch of Carbowax in a small container or paper box in the refrigerator until it is hard, approximately 30 minutes. Blocks solidifed at room temperature will not section as well as those made in the refrigerator. Do not chill blocks in water, which dissolves Carbowax. When completely hard, the blocks may be removed from the refrigerator. They will turn opaque as they warm to room temperature. This is of no disadvantage, but keep them in polyethylene bags or containers if storing them for some time. They must not pick up water, even from the atmosphere.

Reid and Sarantakos (1966) have improved their water-soluble embedding by adding a water-insoluble polyvinylacetate resin to the embedding compound. Their method is as follows:

1. Infiltrate with Carbowax 200,[15] 2 changes: 4 hours each.
2. Infiltrate with Carbowax 1540[15] and 4000[15] (3 : 1) mixed with an equal amount of Carbowax 200, 2 changes: 6 hours each.
3. Infiltrate with Carbowax 1540 and 4000 (3 : 1) plus 7.5 g polyvinylacetate resin AYAF[15] to 100 ml of Carbowax mixture, 2 changes: 12 hours each.
4. Cast blocks; keep in a desiccator until used. Section at 24°C or less. Sections can be mounted on water.

Sectioning

A cool, dry room is recommended. The edges of the block must be parallel in order to obtain ribbons. Wade (1952) suggests that if the sections do not adhere, break up on handling, or do not ribbon at all, the block should be exposed to air for a day or two. For immediate use, paint a 25% solution of beeswax in chloroform on the upper and lower surfaces of the block. Make the layer uniform and allow it to become dull dry. Such surfaces should help the sections adhere to each other. Plain water may do as well in some cases.

Hale (1952) found that variable and high humidity produces erratic results, because with the absorption of water on the surfaces of the block, sectioning becomes impossible. He found that at 24–25°C sections ceased to ribbon at 40% humidity; at 17–18°C humidity had no effect. He therefore concluded that as the temperature drops, greater humidity is permissible. At higher temperatures a higher percentage of hard wax may counteract the effect of humidity, but care must be taken that the block does not become too brittle for good sectioning.

---

[15] Fisher Scientific Co.; Union Carbide Co.

Mounting Sections

Water mounting dissolves Carbowax, and will result in distorted sections and surface tension problems when trying to affix sections to slides. Various solutions for this problem follow.

Blank and McCarthy Method (1950)

potassium dichromate......................	0.2 g
gelatin	0.2 g
distilled water.............................	1000.0 ml

Boil for 5 minutes. Filter.

The sections are picked up on a slide while they are floating on this solution and then allowed to dry.

Wade Method (1952)

Wade prefers to albumenize slides with a thin coat of Mayer egg albumen (p. 548). Air dry overnight, or three hours or longer in an oven. To float ribbons that are intact and without wrinkles remains a problem. He suggests the addition of a wetting agent, turgitol 7 (0.005% in distilled water), to reduce surface tension effects. Add 10% of Carbowax 1540 to reduce shrinkage while sections are drying.

Giovacchini Method (1958)

Giovacchini smears the slides with a thin covering of the following solution: Dissolve 15 g gelatin in 55.0 ml distilled water by heating. Add 50.0 ml glycerin and 0.5 g phenol. Place sections on slides and place on warming plate at 58–60°C for 15 minutes. Transfer to drying oven, 58°C, for 24 hours. The slide is then ready for staining.

Jones et al. Method (1959)

Jones and his coworkers float their sections on the following solution:

diethylene glycol	100.0 ml
formalin	7.0 ml

Carbowax ....   1.0 ml
distilled water............................   400.0 ml

In this method more spreading results from increasing the proportion of water. If the tissues overexpand, add a small amount of Zephiran chloride concentrate (5–10 drops to 500.0 ml flotation solution); this reduces surface tension and prevents air bubbles from being trapped between tissue and solution surface. Mount on albumenized slides and dry.

Pearse Method (1953)

This method combines features of the Giovacchini and Jones methods. Pearse smears with Giovacchini's fluid and mounts the sections with the diethylene glycol mixture.

Zugibe et al. Method (1958)

In this method, a section is cut from the ribbon with a razor blade. One edge of the section adheres to the blade. Touch the loose section edge against a slide coated with a flotation fluid and draw the rest of the section onto the slide.

Goland et al. Method (1954)

Goland and coworkers follow Carbowax infiltration with the following:

1. Xylene, 61°C: 10 minutes.
2. Paraffin, 61°C: 30 minutes.
3. Embed.

Chill in refrigerator; do not apply ice directly. This method overcomes some of the disadvantages of Carbowax and produces less shrinkage and distortion than a regular paraffin method. After sectioning, the tissues may be handled as is customary for paraffin sections.

Either albumen or Giovacchini's fluid smeared on slides are successful. It is easier to mount sections out of a water bath (of floating solution, Jones et al. 1959) onto gelatin-smeared slides than to lay the sections directly onto the floating solution on the slide. Albumen slides mount readily either way. The razor blade method of moving the sections, as described above, is tricky, but works.

Other Methods

Reid and Taylor (1964) double embed with nitrocellulose and Carbowax.

Ashley and Feder (1966) combine hydroxyethyl methacrylate with Carbowax 200. Ruddell (1967a,b) combines methacrylate with Carbowax 400.

Riopel (1962) describes a method for botanical material that can be adapted to animal tissues. Before placing the tissue in pure Carbowax, he places it in aqueous dilutions of Carbowax and outlines a method for mounting serial sections.

(*A Final Warning:* Humidity and temperature are not always readily controlled in these techniques. As previously noted, at the usual room temperature, 21–22°C, humidity must not rise above 40% and preferably should be much lower.)

DOUBLE EMBEDDING

Fragile, small, hard objects often crumble when processed by the paraffin method, and some difficulties can be eliminated by using a double-embedding technique.

1. Fix and wash tissue as usual.
2. Dehydrate in 50%, 70%, and 90% ethyl alcohol: 2 hours each.
3. Absolute ethyl alcohol: 2–16 hours.
4. Methyl benzoate–celloidin solution (see below): 24 hours. Pour off and replace with fresh solution: 48 hours. If tissue is not clear, repeat for 72 hours.
5. Pure benzene, 3 changes: 4 hours, 8 hours, and 12 hours, respectively.
6. Mixture of equal parts of paraffin and benzene, in embedding oven: 1 hour.
7. Paraffin, 2 changes: $\frac{1}{2}$ to 6 hours each, depending upon thickness and nature of tissue (3 hours for 5 mm thickness).
8. Embed and proceed as with ordinary paraffin sections.

Methyl Benzoate–Celloidin Solution

To prepare, add 1 g air-dried celloidin flakes to 100.0 ml methyl benzoate. Shake well and allow bottle to stand upright for an hour or longer. Invert for an hour. Lay bottle on side for an hour, and then turn it upright again. Repeat this process until solution is completed, probably the following day.

Mounting Sections and Preparing to Stain

If the sections do not flatten well on slide, but instead curl and separate from it, float them on 95% alcohol to soften the celloidin. If there still is a curling problem, soften the sections with ether vapor, blot with filter paper, and soak in 0.5–1% celloidin or nitrocellulose (Lillie 1954 or 1965).

If albumen continues to fail as an adhesive agent, use a fresh solution of Masson gelatin fixative (p. 549) under the sections and warm slides on warming plate. As soon as the sections have spread, remove the slides and allow them to cool for a few minutes. Drain off excess gelatin solution, blot sections with filter paper, and place in formalin vapor overnight (40–50°C).

When ready to stain, place the slides in chloroform before proceeding to 95% alcohol. The chloroform will remove the paraffin and harden the celloidin simultaneously.

Molnar Double Embedding (1974)

1. Dehydrate blocks to 95% alcohol.
2. Infiltrate with 0.25, 0.5, 1.0 and 1.5% nitrocellulose in methyl benzoate.
3. Harden the nitrocellulose in toluene. This is better than chloroform; the latter hardens the tissue.
4. Infiltrate with paraffin, 2 changes.
5. Place in vacuum oven, 60°C: 1–3 hours. (Not over 70°, the tissue will harden.)

Timing in the solutions is controlled by the size of the tissue. Small soft pieces: 2 hours per solution; 20–25 mm size: 3 hours; large pieces: 6 hours.

Engen (1974) also uses a nitrocellulose-paraffin combination. Phenol is added to the dehydrating alcohol and glycerin to the nitrocellulose to improve the quality of sectioning.

Buzzell (1975) embeds in agar followed by either ester wax 1960 or paraffin. Dioxane or amyl acetate have to be used as the clearing agent. Toluene or xylene dissolve the agar.

Dioxane in Double Embedding

1. Fix, decalcify if necessary, and wash.
2. Dioxane: 2 hours.
3. Dioxane-nitrocellulose: 3 days.

dioxane	70.0 ml
2% nitrocellulose	30.0 ml

4. Dioxane: 2 hours.
5. Dioxane-paraffin (50 : 50): 2 hours.
6. Paraffin, 3 changes: 20 minutes, 30 minutes, 1 hour.
7. Embed and section immediately.

At UCLA, this method was used successfully on hard fish with scales intact. (Also see Brown 1948.)

Crabb (1949) combines rosin with celloidin, and then double embeds with paraffin. Salthouse (1958) double embeds with tetrahydrofuran, particularly for insects.

ESTER WAX EMBEDDING

The ester waxes were introduced by Steedman (1947) as embedding media that combine the advantages of both paraffin and celloidin, but reduce their disadvantages. Tissue structure is well preserved, and structural lipids are retained perhaps better than when either paraffin or celloidin are used. The ester wax method is based on diethylene glycol distearate, whose hardness compares favorably with that of paraffin. Ester wax melts at a lower temperature and does not require the use of hardening hydrocarbons. The wax is soluble in the usual paraffin solvents and is also miscible with n-butyl alcohol, isopropyl alcohol, Cellosolve, glycol monobutyl ether, acetone, and aniline. It supports soft tissues adjacent to hard ones and adheres to smooth surfaces, and it can be sectioned on a rotary microtome.

Steedman Method (1960)

SOLUTION

diethylene glycol stearate	60.0 g
glyceryl monostearate	30.0 g
300 polyethylene glycol distearate	10.0 g

Melt the diethylene glycol distearate and heat until clear. Add glyceryl monostearate. When this is dissolved, add the 300 polyethylene glycol distearate. Filter.

PROCEDURE

1. Transfer fixed tissues to 70% alcohol: 1–2 hours. Time depends on size of tissues.
2. Transfer to 70% alcohol–Cellosolve (ethylene glycol–monoethyl ether) (50 : 50): 1–2 hours.
3. Transfer to pure Cellosolve: 1–3 hours.
4. Transfer to Cellosolve–ester wax (50 : 50): 1–3 hours.
5. Infiltrate in pure ester wax, 3 baths: 24 hours total.
6. Block.

ALTERNATE PROCEDURE

1. Transfer fixed tissues to 95% alcohol: 1–3 hours.
2. Transfer to 95% alcohol–ester wax (50 : 50) in oven (35–40°C): 1–3 hours.
3. Transfer to pure wax: 24 hours.
4. Block.

Have the melted wax about 10°C above the melting point (48°C) for blocking. Use a cold mold or metal L's. Cool block in ice water but do not submerge it. The more quickly it cools, the better the final texture of the wax.

SECTIONING

Ester waxes must be cut very slowly with a sharp knife. The wax is so hard that the knife must be held firmly in the clamps, and all parts of the microtome must be in good repair and tight, or the sections will be of uneven thickness. Thick sections may be cut the same day that the tissue is blocked, but thin sections will cut more easily on the second or third day following the blocking.

Mount as for paraffin sections on albumenized slides. Drain and dry at room temperature.

Sidman et al. modification (1961)

After receiving some poor batches of wax, Sidman and his coworkers tested a number of products and found that polyethylene glycol 400 distearate (Kessler Chemical Co.) gave the most consistent results. This wax can be purchased in 8 lb. pails and is used unmixed. It melts at 35°C and begins to solidify at 30°C. Melt it overnight (no longer) at 56°C and then transfer it to a 37°C oven for use.

1. Fix and dehydrate the tissue as desired.
2. Infiltrate in 3 changes of wax, 37°C: 24 hours at least.
3. Embed. Pour wax in mold and set it on a cube of ice until a solid base of wax forms. Remove from ice and add tissue. Orient it and place mold and tissue in refrigerator: 1 hour.
4. Allow to warm to room temperature: $\frac{1}{2}$ hour. Trim and mount on object disc as for paraffin blocks. Section under cool conditions. If the room is warmer than 25°C, solid CO_2 held in a kitchen strainer on a ring stand about 6 inches above the knife will provide adequate cooling. Cut slowly.
5. Float sections on 0.1% gelatin in water (25°C) and onto clean slides. If the sections are thicker than 10μ, use 30–32°C water. Drain. Dry at room temperature: several hours or overnight. Sections should not float too long on water, because the wax is soluble and loses its cohesiveness. A wet brush is easier to use than needles, because of the stickiness of the wax. Blocks should be stored in a cool place, and unused sections in the refrigerator.
6. For staining, remove the wax with absolute alcohol, hydrate, and stain.

See also Kuhn and Lutz (1958) and Ueckert (1960) for methods of polyester embedding and sectioning. Menzies (1962) recommends a paraffin-beeswax-stearic acid mixture for thin sections. See Wigglesworth (1959) for an excellent double-embedding method using agar and ester wax for sections that are 0.5 to 1.0μ thick.

Distearate 300 or 400 are not satisfactory for warmer climates, so Sage (1972) uses the longer chain polyethylene glycol, distearate 600[16] with the addition of 1-hexadecanol wax (cetyl alcohol). Melt only the amount to be used at one time; storing in a melted condition lowers the melting point. Sections can be cut at 2μ at 21° with stearate 600 and 1% hexadecanol wax added, and at 24° with 10% hexadecanol added. The sections are mounted on albumenized slides with 10% formalin. Excess formalin is blotted off and the sections allowed to dry for 2 days.

---

[16] Curtin Matheson Scientific Co.

METHACRYLATE PROCESSING FOR THIN SECTIONS

Methacrylate-Plexiglas Processing for Thin Sections for Light Microscopy (Zambernard et al. 1969)

FIXATION

Any routine fixation followed by appropriate washing. Formalin must be thoroughly removed.

SOLUTIONS

Solution I

methyl methacrylate[17]	27.0 ml
polyethylene glycol distearate 1540[18]	6.0 ml
dibutylphthalate	4.0 ml

Heat to 56°C to dissolve distearate. Cool.

Solution II

solution I	90.0 ml
benzoyl peroxide[17]	0.6 ml
wetting agent, Igepal 630[19]	3 drops
Plexiglas molding powder A-1[20]	30.0 g

Add Plexiglas gradually while stirring with magnetic stirrer. Store in refrigerator. Bring to room temperature before use.

PROCEDURE

1. Wash out fixative and dehydrate through 70%, 95% and absolute ethanol: 20–40 minutes each.
2. Transfer into anhydrous acetone (anhydrous sodium sulfate: acetone, 2 g/500 ml): 10–15 minutes.
3. Transfer into methyl methacrylate, 2 changes: 20 minutes each.
4. Transfer into equal parts of methyl methacrylate and solution II: 60 minutes.

---

[17] Curtin Matheson Scientific Co.
[18] Ruger Chemical Co.
[19] Chemical Sales Co.
[20] Rohm & Haas Co.

5. Place in solution II: 60 minutes or more, may remain up to 48 hours (see comment 1).
6. Fill Peel-A-Way plstic embedding molds[21] with solution II and add tissue (see comment 2).
7. Polymerize plastic under ultraviolet light[22] at room temperature: 4 hours. Final cure under ultraviolet light, 56°C: 4 hours (see comment 3).
8. Peel off mold, moisten block with 50% alcohol, and trim with a copping saw or sharp knife.
9. Section on sliding or rotary microtome (see comment 4). Keep block moistened with 50% alcohol during sectioning.
10. Transfer moist sections to albumenized or subbed slides. Flatten with a couple of drops of 95% alcohol and dry on 45°C warm plate.
11. For staining, flood dried slides with clove oil: 10–20 minutes; 95% alcohol, 3 changes; and proceed as with deparaffinized slides.

COMMENTS

1. Infiltration of solution II may be increased by placing the solution and tissue in a desiccator and applying negative pressure of 500–600 mm, or rotating the vials at $\frac{1}{3}$ rpm.
2. To keep the tissue from settling too close to the bottom of the mold, a previously polymerized layer (1–2 mm thick) of solution II may be formed in the molds; I have not found this necessary.
3. The final curing described in step 7 can be done overnight in an oven. The molds can be placed in a small box and the ultraviolet light allowed to rest on the box edges, 5–6 cm above the molds; or blocks can be used to support the light.
4. Zambernard and his coworkers section on a sliding microtome, but I have had problems with a lightweight sliding microtome. The hardness of the plastic block lifted the knife and its support and cut a wedge-shaped section, thicker at the point of the knife's entry and thinner at the place of its exit. These problems were overcome by sectioning on a rotary microtome. The block maintains adequate moisture in the vertical position, and the section stays uniform in thickness. With a brush moistened in 50% alcohol, slip the section on the wet knife to one side of the tissue block (raised or lowered, whichever facilitates the action). Carefully place a slide behind the knife (between the knife and the microtome chuck) and with the wet brush slip the section onto the slide (Fig. 22–3). Several sections can be placed on a slide.

---

[21] Lipshaw Manufacturing Co.
[22] Long Wave, UVL-21, Blak-Ray, Ultra-Violet Products, Inc., also Fisher Scientific Co.

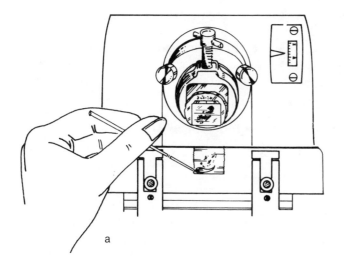

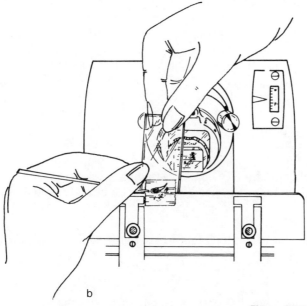

Figure 22-3
Removing a plastic section from a microtome: (a) The section is straightened on the knife with a moist (50% alcohol) brush and is ready to be moved to the left; (b) the section is slipped onto the slide.

5. Clamping the tissue block tightly in the microtome chuck can cause the block to compress and make the top of the block bulge outward. This can be corrected by mounting the tissue block on a wooden block with epoxy cement or quick-setting glues like Eastman 910. Then clamp the chuck on the wooden block instead of on the tissue block.

See Cole and Sykes (1974) for another methacrylate method.

EPOXY RESIN PROCESSING

The epoxy resins, popular for electron microscopy, can be used as well for light microscopy. They produce uniformly hard blocks with very little shrinkage of the tissue. The resin is cured (polymerized) with anhydrides; DDSA (dodecenyl succinic anhydride) produces a soft block, while NMA (nadic methyl anhydride) forms a very hard one. Varying proportions of the anhydrides, therefore, can be used to adjust the resin block to the desired hardness. Dibutyl phthalate, a plasticizer, sometimes is added to impart flexibility to the casting and to reduce viscosity. At final embedding, an accelerator, DMP [2,4,6-tri (dimethylaminomethyl)-phenol] is added. Propylene oxide is commonly used as the clearing agent following dehydration through alcohol. Epon 812 has been used following washing in water (see Craig et al. 1962; Idelman 1964). Other water-miscible embedding media are Aquon, Durcupan, 2-hydroxy-propyl methacrylate, and glycol methacrylate.

There are many recommendations on the use of these resins: Luft (1961) uses Epon 812, highly favored, particularly for electron microscopy; Grimley et al. (1965) prefer Araldite 502 over Epon 812 and use it for light microscopy; I prefer Mollenhauer's mixture, which follows.

Epoxy Resin Processing for Light Microscopy (Mollenhauer 1964)

FIXATION

Any routine fixative, but preferably one of the aldehydes.

SOLUTIONS

Resin mixture

Epon 812[23]	62.0 ml
DDSA[23]	100.0 ml
Araldite 502[23]	81.0 ml
dibutyl phthalate[23]	4.0–7.0 ml

Mix thoroughly; a Teflon stirrer is recommended. To eliminate the difficulties of cleaning resins from glassware, use disposable beakers for mixing and disposable syringes for measuring the reagents. Store for short periods in the refrigerator, but it can be kept for several months in a freezer.

Catalyzed resin

resin mixture	5.0 ml
DMP-30[23]	3 drops

Mix just before use.

PROCEDURE

1. After washing out fixative, dehydrate through 70% and 95% alcohol, and 4 changes absolute alcohol: 30 minutes (or more) each. (Tissue blocks should be no larger than 4–5 mm in thickness.)
2. Transfer to propylene oxide,[23] 2 changes: 30 minutes (or more) each.
3. Transfer to propylene oxide and catalyzed resin (2:1): 1 hour (or longer).
4. Transfer to propylene oxide and catalyzed resin (1:2): 1 hour (or longer).
5. Transfer to catalyzed resin in Peel-A-Way molds and polymerize overnight at 60°C, or leave overnight at 35°C, the next day at 45°C, and then overnight at 60°C. The latter forms a more easily sectioned block.
6. Remove mold and section at 1–3μ with a very sharp knife. The sections can be removed from the knife with fine forceps and placed in a blackened box as in paraffin sectioning (p. 59).
7. With fine forceps and a dissecting needle, pull out any folds in the sections, straighten them as much as possible, and place them on

---

[23] Polysciences, Inc. Shell Oil Co. is discontinuing Epon 812. Araldite 506 can be substituted.

albumen-covered slides. The albumen is not essential, but it does make the mounting solution puddle more uniformly.

8. Place on 45°C warm plate and immediately add 10% aqueous acetone under the sections. Allow the solution to evaporate and continue to straighten the sections with needles, if necessary. As soon as the slide is dry, it can be stained.

COMMENTS

As with the Zambernard method, I prefer to mount the tissue blocks on wooden blocks so that the microtome clamps on the wood rather than on the plastic. Sliding or rotary microtomes can be used, but the sliding microtome must be a heavy-duty model to prevent the same problem as described for the Zambernard blocks. A hard-grade steel or tungsten–carbide steel knife with a 55° cutting bevel edge is recommended. Plastic dulls the average microtome knife; only three or four sections can be cut successfully in a single area of the knife.

If, after curing, the block feels "sticky" soft and will not cut smoothly, return it to the 60°C oven overnight. Quickly peel off the mold and trim off the sloping sides of the block. Do this immediately after removal from the oven and while the block is still warm. It trims more easily in this condition than after it has cooled. Cut the first sections with an old knife, as recommended for paraffin sectioning, until the best part of the tissue is reached. Change to the good knife and collect the desired sections. Keep the edge of the knife clean; wipe it off after every section.

Plastic sections usually adhere to the slides through all solutions, but should loose sections develop, run the slides through a Bunsen burner flame three or four times. Allow them to cool before staining.

A student considering plastic embedding and sectioning should consult the article by Bennett et al. (1976).

Staining Epoxy Sections

Tissues embedded in plastic will not respond to all staining methods; most stains cannot penetrate the plastic. Generally speaking, these sections will stain satisfactorily in alcoholic solutions and in basic dyes in alkaline solutions in the pH 11.0 range. An alcoholic and a buffered aqueous method are described below. Other aids to staining are pretreatment with mordants, hydroxide, peroxide, or iodine. I have had excellent staining results with the hydroxide method below. So far, among the stains I have tried, only PAS has given less than satisfactory returns, but see comments below.

Pinacyanole Stain (Humason and Lushbaugh 1961)

1. Cover sections with several drops of pinacyanole[24] (0.5 g/100 ml 70% ethanol): 10–15 minutes. Cover the slides to prevent evaporation of alcohol.
2. Wash off excess stain in running water: 2–3 minutes.
3. Blot and allow to dry or dehydrate in isopropanol: 2 minutes in each of 3 changes.
4. Clear in xylene and mount.

Toluidine Blue Stain (Trump et al. 1961)

1. Flood slide with freshly prepared and filtered toluidine blue 0, C.I. 52040 (0.1 g/100 ml 2.5% aqueous sodium carbonate, pH 11.1): 30–120 minutes, or until cells are dark reddish purple.
2. Tip slide and draw off extra stain with filter paper.
3. Flood slide gently with tap water from a medicine dropper and draw off water with filter paper.
4. Wash briefly with 90% ethanol and then absolute ethanol. This differentiates the stain, so check it under microscope.
5. Clear in xylene and mount.

Sodium Hydroxide Treatment (Lane and Europa 1965)

1. Prepare a saturated solution of sodium hydroxide in absolute ethanol. Allow to stand: 2–3 days.
2. Immerse slides in solution: 1 hour. Keep container tightly closed to prevent evaporation and deposit of sodium hydroxide (NaOH) on slides.
3. Drain well; do not blot.
4. Transfer through absolute ethanol, 4 changes: 5 minutes each. If the first change becomes milky with accumulated NaOH, discard it, move numbers 2, 3, and 4 up into 1, 2 and 3 positions, and add a new number 4.
5. Treat with phosphate buffer, pH 7.0 (p. 557): 5 minutes.
6. Wash in 3 changes of distilled water: few seconds in each.
7. Treat with phosphate buffer, pH 4.0 (p. 557): 5 minutes.
8. Wash in running water: 5 minutes.
9. Stain.

---

[24] Eastman Kodak.

COMMENTS

Berkowitz et al. (1968) use a dilution of saturated NaOH 1 : 5 with absolute ethanol. I have obtained the best results with the method described above.

Snodgrass et al. (1972) describe a method involving treatment with acetone, benzene, and picric acid subsequent to the NaOH–alcohol.

Litwin et al. (1975) and Litwin and Kasprzyk (1976) made a precise study of PAS reaction on epon-embedded tissues treated with NaOH saturated in absolute alcohol. Treatment beyond 10 minutes produces a diffuse staining, and it must be reduced to 1–5 minutes depending on each batch of NaOH–alcohol. Mixtures vary due to water absorbed and ripening time of the solution. They use 10 minutes in PA and 30 minutes in Schiff reagent.

Pool (1973) uses hydrogen peroxide oxidation for PAS. His solution is 15 ml of hydrogen peroxide (H_2O_2) and 30 ml distilled water. Adjust the pH to 3.2 with approximately 0.18 ml of 0.1 N H_2SO_4. Make fresh; pH must be 2.9–3.2. Oxidize for 1–2 minutes, rinse in distilled water, stain in Schiff 1–3 minutes, wash, drain, dry, and mount.

For the peroxide method, see Aparicio and Marsden (1969) or Pool (1969). For the iodine treatment, see Yensen (1968), and for permanganate mordanting, see Grimley et al. (1965) and Shires et al. (1969).

Alsop (1974) stains epoxy sections by heating a polychrome stain dissolved in PEG 200. Chang (1972) heats his hematoxylin and eosin stains on plastic sections. Sato and Shamoto (1973) use a warmed polychrome stain, an alkaline solution of basic fuchsin and methylene blue. Jha (1976) modifies their method.

For descriptions of other stains, see Flax and Caulfield (1962); Grimley (1964); Grimley et al. (1965); Hendrickson et al. (1968); Hoefert (1968); Martin et al. (1966); Richardson, Jarett, and Finke (1960); Schantz and Schechter (1965); Spurlock et al. (1966); Trump et al. (1961).

Chapter 23
Special Procedures II

EXFOLIATIVE CYTOLOGY

Exfoliative cytology (study of cells, as opposed to histology, study of tissues) concerns the preparation and examination of desquamated cells. These are cells that have been shed or pulled off from a superficial epithelium, mucus membranes, renal tubules, or the like. For example, the horny layer of epidermis is shed normally, but in disease or inflammation the process may become exaggerated and form abnormal-size flakes or scales. These may be placed on slides, stained, and examined.

Methods in this field are especially useful in diagnostic pathology. By screening slides made from smears (vaginal, cervical, prostatic, etc.) or from body fluids (peritoneal, pleural, gastric, urine, spinal, etc.), rapid diagnosis of malignancies is possible. I have found them extremely useful in animal research.

The cells in question degenerate rapidly, and smears should be made and fixed immediately, in equal parts of 95% alcohol and ether, for a minimum of 15 minutes. The slides may remain in this fluid as long as a week before staining.

Add to body fluids an equal volume of 95% alcohol and centrifuge as soon as possible, 2000 rpm: 3 minutes. Remove supernatant fluid. (If smears cannot be made immediately, cover the sediment with absolute alcohol and place in refrigerator.) Drop some of the sediment on an albumen-coated slide and, using another slide, smear the sediment as evenly as possible over the albumenized area. Allow the smear to begin to dry around the edge but to remain still moist in the center. Fix in ether-alcohol: 1 hour.

After fixation, the slides should not be allowed to dry at any time before or during the staining procedure. In emergencies, they can be shipped dry, but some protection for the smear is advisable if possible. Ehrenrich and Kerpe (1959) add 5% of polyethylene glycol to the fixative. Following complete fixation, allow the slides to dry 5–10 minutes, and then prepare them for shipping. Papanicolaou (1957) protects smears for shipping by covering the fixed smears with Diaphane[1] solution: 2 parts of Diaphane to 3 parts of 95% alcohol. Allow to dry, 20–30 minutes, and prepare for shipping. The polyethylene glycol or the Diaphane form a protective coating over the smear.

Richardson (1960) uses the following solution for shipping purposes; it is economical, noncombustible, and nonvolatile. Biopsy pieces or slides can be shipped in it.

monoethylene glycol	344.6 ml
diethylene glycol	18.1 ml
borax	3.6 g
water	577.6 ml
glacial acetic acid	50.0 ml

Brandi et al. (1962) fix in 1% nitrocellulose in 95% ethyl alcohol–ether (1:1): 15 minutes. Drain until dry.

Spray-Cyte[2] can be sprayed on the slides from an aerosol can and allowed to dry.

Membrane filters are used for screening large volumes of fluids. The suspended cells are concentrated on the filter and processed there with relative ease. The Gelman membrane filters[3] were developed for this purpose and do not retain stain or clear more readily than do other cytological membranes. If possible, fluids should be filtered first and the cells washed and fixed after they are collected on the membrane. Directions and suggestions for the processing of various types of fluids can be obtained with the filters. For more about filter techniques, see Bernhardt et al. (1961); Hulton (1958); McCormick and Coleman (1962); Nedelkoff and Christopherson (1962); Thomison (1961).

Papanicolaou Method (1942, 1947, 1954, 1957)

This can be used as a routine method for paraffin sections. Excellent preparations of the following solutions can be purchased from the Ortho

---

[1] Will Corp.
[2] Clay Adams.
[3] Gelman Instrument Co.

Pharmaceutical Corp.: Harris hematoxylin, EA 65, EA 50, and OF 6. Otherwise the stains can be prepared as follows:

SOLUTIONS

Harris hematoxylin, see p. 113 (omit acetic acid).

Orange G 6 (OG 6)

orange G 6, 0.5% (0.5 g/100 ml 95% alcohol) ..	100.0	ml
phosphotungstic acid	0.015 g	

Eosin-azure 36 (EA 36 or EA 50, the commercial designation)

light green SF yellowish (C.I. 42095), 0.5% (0.5 g/100 ml 95% alcohol)	45.0 ml
Bismarck brown (C.I. 21000), 0.5% (0.5 g/100 ml 95% alcohol)	10.0 ml
eosin Y (C.I. 45380), 0.5% (0.5 g/100 ml 95% alcohol)	45.0 ml
phosphotungstic acid	0.2 g
lithium carbonate, saturated aqueous	1 drop

EA 65 is the same, except for the light-green content; 0.25% in 95% alcohol is used. This gives a lighter, more transparent stain, which is desirable in smears containing a lot of mucus. Differentiation between acidophilic and basophilic cells is better, however, with EA 36 (EA 50), and therefore it is preferable for vaginal, endocervical, and endometrial smears.

PROCEDURE

1. Out of fixative, hydrate slides to water. Deparaffinize sections before hydration. Rinse in distilled water. A few seconds in each solution is adequate, or dip slides up and down until their surface has a homogeneous appearance.
2. Stain in Harris hematoxylin, either of following methods:
   a. Papanicolaou dilutes Harris with an equal amount of distilled water: 8 minutes.
   b. Harris hematoxylin without acetic acid: 4 minutes.
3. Wash in tap water, running water if flowing only slightly; do not wash off or loosen parts of smear: 3–5 minutes.
4. Differentiate nuclei in 0.5% hydrochloric acid in 70% alcohol (0.5 ml/100 ml) until nuclei are sharp against a pale blue cytoplasm.

5. Wash in slightly running tap water: 5 minutes, or until nuclei are a clear blue.
6. Rinse in distilled water.
7. Transfer through 70%, 80%, and 95% alcohol: few seconds in each or until surface appears homogeneous.
8. Stain in OG 6: 1–2 minutes.
9. Rinse in 95% alcohol, 3 changes: few seconds each.
10. Stain in EA 36 (EA 50): 2–3 minutes.
11. Rinse in 95% alcohol, 3 changes: few seconds each.
12. Dehydrate in absolute alcohol, 2 changes: 1 minute each.
13. Clear and mount.

RESULTS

nuclei—blue
acidophilic cells—red to orange
basophilic cells—green or blue green
cells or fragments of tissue penetrated by blood—orange or orange green
blood vessels—orange red

COMMENTS

Papanicolaou cautions against agitating slides excessively, or crowding them while staining. If parts should float off from a positive slide onto a negative one, a false positive can occur.

A modification of Shorr's stain may be substituted for above.

50% ethanol	100.0	ml
Biebrich scarlet (C.I. 26905)	0.5	g
orange G (C.I. 16230)	0.25	g
fast green FCF (C.I. 42053)	0.075	g
aniline blue	0.04	g
phosphomolybdic acid	0.5	g
acetic acid glacial	1.0	ml

Replace steps 8, 9, 10 with this stain and proceed to step 11.

These stains can be applied to paraffin sections with excellent results.

Fluorescent Method (von Bertalanffy and Bickis 1956; von Bertalanffy et al. 1958)

If proper equipment is available, the fluorescent procedure requires less time for preparation and less skill on the part of the examiner than the Papanicolaou method. The polychrome staining is so brilliant that cells are seen in sharp contrast against a black background, rendering mass screening simple and quick. Under low power atypical cells show increased amounts of fluorescence; malignant cells are a brilliant, flaming, reddish orange (they seem to glow) in contrast to normal cells, which are usually greenish gray with whitish-yellow nuclei. The method is based on the differentiation of RNA and DNA by acridine orange NO (AO). The RNA in the cytoplasm and nucleolus fluoresces red, and the DNA of the nucleus fluoresces green.

SOLUTIONS

Acridine orange stock solution

acridine orange (C.I. 46005)	0.1 g
distilled water.............................	100.0 ml

Keep in dark bottle in refrigerator.

M/15 phosphate buffer, pH 6

$M/15$ sodium phosphate, dibasic ($Na_2H_2PO_4$) (11.876 g/1000 ml water)	7.0 ml
$M/15$ potassium phosphate (KH_2PO_4) (9.078 g/1000 ml water)	43.0 ml

Keep in refrigerator.

Working solution

acridine orange stock solution	5.0 ml
buffer solution	45.0 ml

Calcium chloride, M/10:

calcium chloride	11.0 g
distilled water.............................	1000.0 ml

PROCEDURE

1. Smears are fixed by Papanicolaou method, ether–95% alcohol: at least 15 minutes.
2. Hydrate to water: few seconds in each solution.
3. Rinse briefly in acetic acid, 1% aqueous (1 ml/99 ml water), followed by wash in distilled water: few minutes.
4. Stain in acridine orange: 3 minutes.
5. Destain in phosphate buffer: 1 minute.
6. Differentiate in calcium chloride: 30 seconds. Wash with buffer.
7. Repeat step 6. If smear is very thick, a longer differentiation may be necessary. Nuclei should be clearly defined.
8. Wash thoroughly with phosphate buffer. Calcium chloride must be removed.
9. Blot carefully and mount with a drop of buffer.

Frozen Sections, Method A

1. Wash out ether-alcohol, freeze, and cut.
2. Omit step 2 and proceed as usual. Use small, shallow dishes; mount sections on slide after staining.

Frozen Sections, Method B

1. Section immediately after biopsy or wrap tissue in saline-soaked gauze and wax paper surrounded by ice. Keep in refrigerator, 4°C, until processed. Cut, transfer to slides, and fix in alcohol or Carnoy's fluid.
2. Hydrate and stain.

RESULTS

DNA of chromatin of nucleus—green, from whitish to yellowish hues
connective tissue fibrils, cornified epithelia; structures—greenish
mucus—dull green
leukocytes—bright green
RNA of basophilic structures of cytoplasm, main portion of nucleoli—
    red and orange fluorescence
sites of strongly acidic groups—deep carmine
malignant cells under low power—brilliant flaming red orange

proliferating malignant cells under high power—intense fluorescence
   cytoplasm—flaming red orange containing reddish granules, patches,
      or fine granules
   nuclei—yellowish hues, yellow orange to yellow green
   nucleoli—brilliant orange red
degenerating and necrotic malignant cells—characterized by a gradual
   loss of cytoplasmic RNA and therefore of the flaming red orange
   cytoplasm—faint orange or brick red, RNA often concentrated at
      periphery in a rim of dark brick red
   nuclei—brilliant green, green yellow or pale yellow gray
   nucleoli—often enlarged and multiple, orange red

COMMENTS

Paraffin sections can also be stained by this method.

The buffer mount is a temporary mount, even when ringed. If Elvanol 51–50[4] is dissolved in the buffer until the solution is of thin, syrupy consistency, a clearer and more lasting mount results. Seal the cover slip, however, with one of the ringing cements (p. 109). Many scientific supply houses are now advertising media for fluorescent mounting.

After the smear has been used for fluorescent examination, it can be stained by the Papanicolaou technique.

1. Remove cover glass in distilled water.
2. Transfer to 50% alcohol: approximately 1 minute total, 2 changes.
3. Proceed to Papanicolaou method.

Bacteria and parasites will fluoresce and can confuse the results; for example: bacteria—bright orange; *Trichomonas,* cytoplasm—bright orange, and nucleus—whitish yellow. These are easily recognized in vaginal smears. See Umiker and Pickle (1960) for lung cancer diagnosis.

See Taft and Lojananond (1962) for evaluation of the fluorescent method; see Sani et al. (1964) for an excellent article on the cytodiagnosis of cancer.

---

[4] PVA, polyvinyl alcohol, E. I. du Pont de Nemours and Co.

Dart and Turner Method (1959)

SOLUTIONS

McIlvaine buffer

sodium phosphate (Na_2HPO_4)	10.81 g
citric acid, monohydrate	13.554 g
distilled water.............................	1000.0 ml

Keep in refrigerator. Good for 1 week.

Buffered acridine orange

Stock solution

acridine orange (C.I. 46005)	0.1 g
distilled water.............................	100.0 ml

Add 2 ml of Tween 80 to 1000 ml of stock.

Working solution

acridine orange stock solution	1 part
McIlvaine buffer	9 parts

PROCEDURE

1. Hydrate slides to water, 5 dips in each of 80%, 70%, and 50% alcohol.
2. Dip 4 times in acetic acid, 1% aqueous (1 ml/99 ml water).
3. Rinse in distilled water: 2 minutes.
4. Transfer to McIlvaine buffer: 3 minutes.
5. Stain in buffered acridine orange: 3 minutes.
6. Differentiate in McIlvaine buffer: 4 minutes.
7. Wipe excess material from ends of slides, gently blot with coarse blotting paper.
8. Apply cover glass with buffer. Let stand 2 minutes before examining.

COMMENT

Can be restained, as above, with Papanicolaou. After this, can be decolorized and restained with acridine orange.

SEX CHROMATIN

Barr and Bertram's (1949) demonstration of sex differences in cat neurological tissue created interest in the cells of all parts of the body, and sex chromatin has been identified in 50–80% of the cells in thin (5μ) sections of various tissues from females. In males a similar smaller chromocenter occurs in an average of 5% of the cells, but may not be homologous with the sex chromatin of the female (Hamerton 1961). The female chromatin is a dark-staining mass (Barr body) usually lying against the nuclear membrane.

A most interesting and easily observed form of this phenomenon is the drumstick of the neutrophil leukocyte. The drumstick is a round hyperchromatic appendage attached to one lobe of the nucleus by a stem of chromatin. The drumstick should not be confused with smaller clubs, sessile nodules, tags, or "racket" formations. Although the chromatinrich sessile nodules and tags are more frequent in females, the small clubs and racket structures are perhaps more common to the male (Davidson and Smith 1954). Before a sample is considered chromatin-negative, 500 neutrophils should be examined. If sessile nodules are seen, it is advisable to search beyond the 500 cells for sex chromatin. There is wide variation in the number of cells displaying drumsticks among women: some may show them in as many as 6% of the neutrophils; others may have as few as 1 or 2 per 1000 neutrophils (Briggs 1958). In the case of a low count, the examination should be repeated several times (Caratzali et al. 1957; Kosenow 1962).

Since the sex chromatin is composed largely of DNA and histochemically resembles chromosomes, it can be stained by any stain specific for DNA or for chromatin. With a culture method, Riis (1957) unmasked differences in chromatin structure between male and female lymphocytes. Moldovanu (1961) and Seman (1961) demonstrated sex differences in lymphocytes and monocytes.

Klinger and Hammond Method (1971)

FIXATION

Peripheral blood smears in methyl alcohol: 1–2 minutes. See comment below for "buffy coat" suggestion. Oral mucosal and vaginal smears in absolute alcohol–ether (1 : 1): 2–24 hours. Fix blocks of tissue in:

Davidson fixative

95% ethyl alcohol	30.0 ml
formalin	20.0 ml
glacial acetic acid	10.0 ml
distilled water.............................	30.0 ml

SOLUTIONS

Pinacyanole

pinacyanole[5]	0.25 g
70% ethyl or methyl alcohol	100.0 ml

Wright buffer

monobasic potassium phosphate	6.63 g
dibasic sodium phosphate	3.20 g
distilled water.............................	1000.0 ml

PROCEDURE

1. Deparaffinize sections and hydrate to water.
2. Extract slides in 5 *N* HCl, room temperature; smears: 2 minutes, sections: 3–6 minutes.
3. Wash in running water: 2 minutes.
4. Stain with pinacyanole: 45 seconds.
5. Differentiate in buffer: 45 seconds.
6. Wash in running water: 5 seconds.
7. Dehydrate in 2 changes absolute tertiary butanol or isopropanol: 1 minute each.
8. Clear in xylene and mount.

RESULTS

Staining: chromatin—blue
Counting: Oral mucosal and vaginal smears—6 cells out of 100 counted must be positive in order for female sex chromatin to be considered female; if fewer than 4 per 100 are positive, the specimen is male.

Blood smears—at least 6 rich-staining "drumsticks" must be counted in 500 polymorph cells for a female count; no more than 2 drumsticks

---

[5] Eastman Chemicals.

should appear in a male count. (See excellent pictures in Greenblatt and Manautou, 1957.)

COMMENT

Stain intensity can be controlled by the dye concentration or staining time. Change the buffer solution if it becomes discolored. Old faded slides can be restained by this method. The acid extraction has been added to this method because it removes basophilic substances in the cyoplasm and clarifies the chromatin staining.

For sex chromatin counts, the preparation of a buffy coat is recommended. It gives the examiner a fighting chance at a relatively easy cell count. Place the collected blood, plus an anticoagulant, in a glass tube and allow it to stand for several hours, or use mild centrifugation, 2500–3000 rpm. The cellular elements will settle out and leave a straw-colored solution on top. The leukocytes are not as dense as the erythrocytes and will settle out last, remaining as the so-called buffy coat between the red cell layer and the plasma on top. Smear this buffy coat with its concentration of leukocytes on the slide and fix.

Klinger and Ludwig Method (1957)

FIXATION

95% alcohol for smears and thin embryonic membranes: 30 minutes to 4 hours. Tissue blocks in Davidson fixative (see above): 3–24 hours, and transfer to 95% alcohol. If material was fixed in some other fixative, transfer immediately to 95% alcohol, but do not leave in first fixative for longer than 24 hours for good results.

SOLUTIONS

Buffered thionin

Solution A

thionin (C.I. 52000), saturated solution in 50% alcohol. Filter.

Solution B

sodium acetate	9.714 g
sodium barbiturate	14.714 g
distilled water, CO_2-free	500.0 ml

Solution C

hydrochloric acid, sp gr 1.19	8.50 ml
distilled water...........................	991.5 ml

Working solution, pH 5.7

solution A	40.0 ml
solution B	28.0 ml
solution C	32.0 ml

PROCEDURE

1. Deparaffinize and hydrate slides to water.
2. Hydrolyze in 5 N HCl (p. 542), 20–25°C: 20 minutes.
3. Rinse thoroughly in distilled water, several changes; no acid should be carried over into thionin.
4. Stain in buffered thionin: 15–60 minutes.
5. Rinse in distilled water, and in 50% alcohol.
6. Rinse in 70% alcohol until clouds of stain cease to appear.
7. Dehydrate, clear, and mount.

RESULTS

sex chromatin—deep blue violet
nuclear chromatin—lightly colored
cytoplasm—unstained
fibrin and related structures will show metachromasia

COMMENTS

The Feulgen technique may be used for sex chromatin body identification, but the Klinger and Ludwig method is simpler and quicker. The principle is the same in both methods: the sex chromatin body differentiates from nonspecific nuclear chromatin because of a higher content of DNA, which takes longer to extract. If the basophil shell of the nucleolus is not visible, no attempt at sexing should be made (Klinger and Ludwig 1957). In the

Feulgen method, the sex chromatin stains more deeply than any other Feulgen-positive material present.

Skin biopsies are reliable if the sections are of good quality, but the simplest specimens are oral scrapings, easy to obtain and equally easy to stain. Scrape the mucosa of the cheek with a tongue depressor and smear scrapings on an albumen-coated slide. Fix immediately in Davidson's or alcohol. Papanicolaou fixative (equal parts of ether and 95% alcohol) also may be used.

For other methods, see Guard (1959) and Beckert & Garner (1966), who use Biebrich scarlet and fast green; and Lennox (1956), who uses gallocyanine. Romanovsky stains, like Wright, stain the drumstick intensely.

Makowski et al. (1956) identified sex chromatin in the cellular debris of amniotic fluid, spreading the centrifuged sediment on albumenized slides. Vernino and Laskin (1960) identified it in bone decalcified in EDTA. Barr et al. (1950) describe it in nerve cells. See also: Klinger (1958); Marberger et al. (1955); Marwah and Weinmann (1955); Moore and Barr (1955); Moore et al. (1953); Taylor (1963).

CHROMOSOMES

This is one of the most difficult sections on techniques to cover adequately. Present-day literature abounds in new methods and modifications of old ones. Forgive me if I have failed to include a favorite; I have included only those I feel I can recommend after having worked with them.

Chromosome Squashes

Giant chromosomes are found in some of the somatic tissues of two-winged flies and reach the largest size in the salivary glands of the larvae. In well-made squashes, the cross striations of the chromosomes stain in varying degrees of intensity. *Drosophila* larvae are the familiar organisms used for demonstrating these striations. But Barley (1964) found the larvae of black flies (*Simulium vittatum*) even easier to work with. Martin (1966) uses the larvae of *Chironomus* ("bloodworms") found in decaying vegetation along edges of ponds.

Acetocarmine or Aceto-Orcein Staining

SOLUTIONS

Acetocarmine stock solution

Boil an excess (approximately 0.5 g/100 ml) of carmine in 45% acetic acid, aqueous: 2–4 minutes. Cool and filter.

Working solution

Dilute stock solution with 45% acetic acid, aqueous, 1:2. An iron-mordanted stain is often favored because of darker, bluish-tinged red. Belling (1926) adds a few drops of ferric hydrate in 50% acetic acid, but only a few drops: too much iron produces a precipitate in a short time. Moree (1944) determined quantitatively the amount of ferric chloride to add and includes tables of various normalities and volumes. 1% by volume of lactic acid can be added to the solution; this intensifies the staining of pachytene chromosomes (Yerganian 1963).

Aceto-orcein solution

Add 1–2 g orcein[6] to 45 ml of hot acetic acid. When cool, add 55 ml of distilled water. LaCour (1941) used 2% orcein in 70% acetic acid.

Alternate solution

Mix 85% lactic acid and acetic acid in equal amounts. Heat to boiling before adding stain (1–2 g/100 ml solution). Cool and filter. This stain does not precipitate (Yerganian 1963).

PROCEDURE

1. Choose large, sluggish, *Drosophila* larvae, which have crawled up the side of the jar. Hold the posterior end of the larva (in cold-blood saline) with forceps, and with a dissecting needle pull. away the mouth parts. The salivary glands will be attached to the mouth parts.
2. Trim off the fat bodies and place the glands in Carnoy fixative or directly into a small amount of stain on an albumenized clean slide.

---

[6] G. T. Gurr's synthetic orcein is expensive, but excellent.

3. Place a cover glass on top of preparation and cover with bibulous paper. Apply pressure.
4. Seal edges with paraffin, vaseline, or dentists' sticky wax, and allow to stand one day.

RESULTS

acetocarmine—red
iron acetocarmine—deep bluish red
aceto-orcein—dark purple

COMMENTS

Sometimes it is advantageous to allow material to remain in stain 5–10 minutes. Then remove to a fresh batch of stain before squashing.

Fragments of animal tissue (ovaries, testes, biopsies) can be put into a tube with a large excess of stain for 2–7 days. Remove to fresh solution on slide and squash.

Barley (1964) stores in Carnoy's. Before making squashes, he treats the glands with acid alcohol (hydrochloric acid–absolute alcohol, 1 : 1) for 2–3 minutes, returns glands to Carnoy's for 5 minutes, and proceeds to stain. This method facilitates rupture of the cells.

Properly applying the pressure on the cover glass is very important for good squashes. Press the thumb in the middle of the cover glass (covered with bibulous paper). Gently roll the thumb toward an edge of the cover glass, so a small amount of stain oozes out and is absorbed by the paper. Continue to roll the thumb toward the edges of the cover glass until no fluid squeezes out. Always roll from the center toward the edges; the fluid must not roll back toward the center and disrupt the cells.

Moorman (1971) stains with a modification of iron hematoxylin.

Jona (1963) squashes under Scotch tape no. 665; Murín (1960) squashes under cellophane.

Lactopropionic Orcein Method (Dyer 1963)

SOLUTIONS

Stock solution

natural orcein	2.0 g
lactic acid	50.0 ml
propionic acid	50.0 ml

Working solution

stock solution	45.0 ml
distilled water	55.0 ml

PROCEDURE

1. Fix 5 minutes in the following:

absolute ethyl alcohol	10 parts
glacial acetic acid	2 parts
chloroform	2 parts
formalin	1 part

2. Macerate in 1 N HCl, 60°C: 5 minutes.
3. Trap out into stain: 2 minutes.
4. Squash.

RESULT

chromosomes—deep reddish purple

Permanent Mounts

If a slide is to be made into a permanent mount and there is a chance of losing the specimens, use subbed slides (p. 550).

Smith Method (1947)

1. Remove paraffin or vaseline seal with razor blade and xylene.
2. Soak in equal parts of acetic acid and 95% alcohol until cover glass comes off.
3. Place in equal parts of 95% ethyl alcohol and tertiary butyl alcohol: 2 minutes.
4. Transfer to tertiary butyl alcohol, 1–2 changes: 2 minutes each.
5. Briefly drain slide and cover glass against absorbent paper or blotter.
6. Add thin resin mountant to slide and return cover glass to same position as before.

Step 2 can be followed by 2 changes of 95% ethyl alcohol; mount in Euparal or Diaphane.

Nolte Method (1948)

1. Place slide in covered dish so edge of cover glass dips into 95% alcohol: 6–12 hours.
2. Immerse in 95% alcohol: 1–2 hours.
3. With sharp needle, gently pry cover glass free while immersed.
4. Drain off excess alcohol and add drop of Euparal or Diaphane and replace cover glass.

Lamination Method

See p. 527.

Other Permanent Methods

Peary (1955) uses a frozen method of dehydration, also counterstains with fast green. See also Bradley (1948*b*); Celarier (1956); and McClintock (1929).

Delamater (1951) describes a freezing method.

Conger and Fairchild (1953) freeze smears while they are in stain on a block of dry ice: 30 seconds. Pry off cover glass with razor blade. Place before thawing in 95% or absolute alcohol, 2 changes: 5 minutes. Mount in Euparal or Diaphane.

Freeze slides in a mixture of absolute methyl or ethyl alcohol and dry ice. Pry off cover slip. If liquid nitrogen is available, freeze slides in it and pry off cover glass.

Hrushovitz and Harder (1962) make permanent mounts in Abopon. See Gardner and Punnett's (1964) method with Hoyer medium.

Elston and Sheehan (1967) spray the underside of the slide with Freon-aerosol mixture.

Restoration of Deteriorated Slides (Persidsky 1954)

If dye has precipitated as dark crystals because the preparation has dried:

1. Remove sealing material.
2. Place 1 drop of 2 N HCl at one edge of cover glass. Apply blotter to opposite edge to help draw the acid under: 3–5 minutes.
3. Replace HCl with acetocarmine by same method.

4. Heat gently; do not boil. Crystals are redissolved and specimen restained.
5. Reseal.

Pachytene Chromosomes

Pachytene chromosomes, the prophase stage of meiosis, are filaments somewhat irregular in outline, but they are clearly double, forming two parallel strands (chromatids) and are united at one point (the centromere). Squash preparations must be processed so that the chromosomes are well spread and all chromosomes can be identified. See Clendenin (1969); Gardner and Punnett (1964); Ohno (1965); Schultz and St. Lawrence (1949); Welshons et al. (1962); Williams et al. (1970); and Yerganian (1957).

Somatic Chromosomes: Culturing and Spreading Techniques

Progress in cytogenetics has advanced rapidly since 1956 with refinements in techniques for establishing the correct number of chromosomes in animals, including human beings (Ford and Hamerton 1956; Tijo and Levan 1956). Squash techniques have been applied for years, particularly to the chromosomes of plants and invertebrates, but tissue culture methods have now been simplified and spreading techniques have also been developed. Cultures now can be made from bits of tissue, bone marrow, or peripheral blood cells.

The bone marrow method is disadvantageous in that it causes some discomfort to the subject, if not its death. Cultures from the skin cause less discomfort, but the final preparations are the result of many divisions in vitro. The karyotype may possibly alter. Peripheral blood cultures are relatively simple and require only a short-term duration of culture, 2–3 days. The cells are likely to be in first mitosis, or perhaps at most in the second or third mitosis. Apparently, the lymphocytes are the cells that undergo mitosis and have to be separated out from the red cells. Between 5 and 10 ml of blood is ample for several cultures. Other methods have been developed for small samples: Brown and Fleming (1965); Edwards and Young (1961); Hungerford (1965); and Shelley (1963).

At metaphase chromosomes are more concentrated and can be most easily recognized individually and be counted; in their natural state they lie closely packed and must be separated for identification. The mucoprotein plant extract, phytohemagglutinin, was used originally to agglutinate the erythrocytes as a means of separating the leukocytes from whole blood. Then it was found to be a specific initiator of mitotic activity as well; lymphocytes appeared to be altered to a state in which they were capable of division (Nowell 1960). This stimulation of mitoses is probably an antigenic effect (Goh 1965). Phytohemagglutinin now is used as a stimulant of mitosis as well as a precipitant of red cells. Two forms of phytohemagglutinin (PHA) can now be purchased, M and P. PHA-P is approximately 50 times more potent than PHA-M in its hemagglutinizing capacity and should be used with caution. The desiccated mucoprotein is rehydrated by adding 5 ml of bacto-hemagglutinin buffer or sterile triple-distilled water. Keep the PHA in the freezer and never let the stock reach room temperature. Discard rehydrated stock after one month. PHA-M is recommended for the beginner; if P is used, reduce the amount by at least 50%.

After a period of incubation, a spindle inhibitor is added to the culture to interfere with the formation of the mitotic spindle; arrest the dividing cells and allow them to accumulate at metaphase. Either colchicine or its analogue, Colcemid (CIBA), are used for this purpose. The concentration of colchicine or colcemid and the length of treatment should be carefully controlled. (Minimum time: 90 minutes; maximum time: 14–18 hours if cells have a generation time of 12–24 hours.) The degree of chromatin condensation varies according to duration of treatment: the longer the action, the more contracted the chromosomes; the longer the chromosomes, the more affected they are by condensation action. Highly contracted chromosomes tend to have thin centromeres, which are located more medially than are those in less contracted chromosomes. Thus the degree of condensation can affect their relative lengths when considered in terms of percentage of the whole complement (Sasaki 1961).

The use of hypotonic solutions (or water) swells the cells and aids in the dispersal of the chromosomes. Squashing techniques have been improved but are surpassed by methods for spreading the chromosomes.

Identification and analysis for karyotypes are made from photographs of suitably spread metaphases (Figs. 23–1, 23–3). The chromosomes are cut individually from the photographic print and arranged in decreasing size, measured, and paired. Human chromosomes are identified numerically and alphabetically according to a combination of the Denver System ("Denver Report" 1960) and that of Patau (1960, 1961); this system of identification is called the London System ("London Conference" 1964) (Figs. 23–2, 23–4).

Culture methods vary greatly but space does not permit inclusion of all of them. Only a few methods used successfully by myself and my colleagues at Los Alamos and the Oak Ridge Associated Universities are given here, although some of the other methods will be cited. Variations on techniques should be tried. Failures must be expected but should not

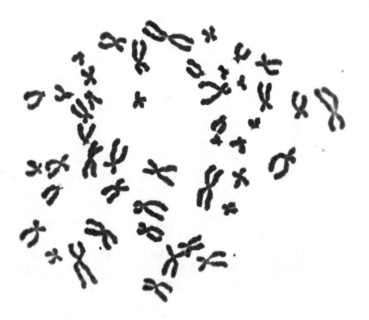

Figure 23-1
Chromosome spread of human male.

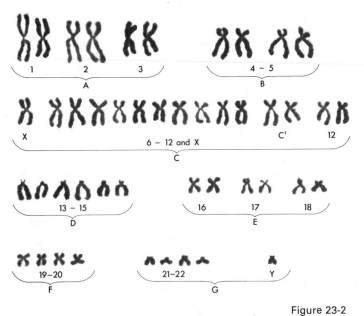

Figure 23-2
Karyotype of human male chromosomes, identified by the London system.

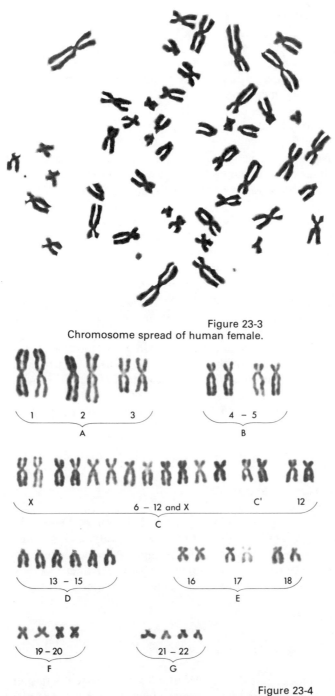

Figure 23-3
Chromosome spread of human female.

Figure 23-4
Karyotype of human female chromosomes, identified by the London system.

discourage the beginner. When preparing cultures from abnormal blood, try two types of cell suspensions, one with and one without PHA. Experiment with cultures: vary the concentration and combination of the constituents of the medium; try different incubation temperatures; or sample the culture for metaphases at the end of 48, 72, and 96 hours, particularly if abnormal cells are present.

Leukocytes suspended in plasma can be kept 96 hours at 5°C, but do not leave them at room temperature (Hungerford and Nowell, 1962). Blood in a sterile container can be transported on ice for about 5 hours, or kept in the refrigerator at 5°C for 24 hours, without detriment to future culturing (Patau et al., 1961).

Since mammalian cells are sensitive to all conditions, a number of precautions must be taken when culturing them. Use only glassware or plasticware designed for tissue culture work. For washing use nontoxic cleaners sold specially for such containers. Rinse in distilled water. All containers must be sterilized and maintained in a sterile condition. Use stainless steel hypodermic needles; white, silicone, and most amber pure gum rubber stoppers are satisfactory. Keep a constant pH control; make certain the containers are gastight by using screw caps with rubber liners or tight rubber stoppers, or maintain cultures in a controlled CO_2 atmosphere.

Always flame (run through flame of a Bunsen burner) all instruments and openings of bottles used in culturing, before adding cells and after adding cells and before stoppering. When in doubt—flame. Wipe off stopper of heparin bottle with 70% alcohol before inserting syringe needle.

Culturing Peripheral Blood Leukocytes (Goh, 1965)

PREPARATION OF BLOOD

1. Using sterile techniques, draw 5–10 ml of venous blood into a syringe containing 0.1 ml of heparin (Lilly, USP, 100 units per ml)
2. Gently mix by inverting the syringe. With sterile technique, cap the needle or transfer the blood to a 15 ml centrifuge or other small tube, and stopper it with a sterile silicone stopper.

SEDIMENTATION

Must be done under sterile conditions.

1. Let syringe or tube stand vertically at room temperature: 1–2 hours, or until red cells are separated from leukocytes and plasma.

2. Inject from syringe, or aspirate from tube, approximately 1.0 ml of plasma and suspended leukocytes and place in sterile culture bottle (medicine bottles, milk dilution bottles, or others) containing this medium:

TC 199[7]	7.0 ml
PHA-M	0.1 ml
autologous plasma	2.0 ml

3. Add 100 units penicillin and 100 mg streptomycin for each ml of final concentration.
4. Gently shake bottle.

INCUBATION

1. Incubate in closed bottles, 37.5°C: 68–72 hours.
2. Add 0.1 ml of $10^{-6} M$ colchicine per 1 ml of culture: $1\frac{1}{2}$–2 hours (up to 4 hours is satisfactory). Leave at 37.5°C. (Sterile conditions no longer required.)
3. Centrifuge, 600 rpm: 6–10 minutes. Discard supernate.

HYPOTONIC TREATMENT

1. Add slowly 1.0 ml of 0.37% sodium citrate (see comments). Shake well after first few drops of citrate have been added and again after each addition of more citrate. Cells must be unclumped and resuspended. Let stand at room temperature: 20 minutes. At 10 minutes, bottle may be shaken again.
2. Centrifuge, 300 rpm: 60 minutes. Pipette off and discard supernate. Resuspend cells.

FIXATION

1. Slowly add, with shaking, 1 ml of freshly mixed methanol–acetic acid–chloroform (6:1:3). Merck's glacial acetic acid is recommended. Ethyl alcohol may be substituted for methyl alcohol, but for best results with either, the alcohol must be 100 proof. Keep bottles tightly closed so contents cannot absorb moisture and keep only one-pound bottles as stock. Resuspend cells thoroughly after

---

[7] Baltimore Biological Laboratory; Difco Laboratory; Hyland Laboratory; Microbiological Associates.

first few drops of fixative. Add remaining solution slowly, with shaking, and leave at room temperature: 10 minutes.

2. Centrifuge, 800 rpm: 5–10 minutes. Remove supernate and add fresh fixative.
3. Repeat step 2 at least once more. Several washings with fresh fixative improve the final spreads. Use about 0.5 ml of fixative for final wash.

SLIDE PREPARATION

1. Place clean slides in ice water. Remove a chilled slide, shake off excess moisture, and immediately add a drop or two of cell suspension. Tilt slide against blotter or cleansing tissue to draw off excess fluid and to help spread cells.
2. Dry rapidly over a flame or hot plate. Dry thoroughly before staining.

COMMENTS

Hungerford (1965) uses 0.075 M potassium chloride warmed to 37°C in place of sodium citrate and adds 16 USP units of heparin per ml of the hypotonic KCl to reduce the surface adhesion among suspended cells. The clumps then break up more readily.

Additional suggestions for slide preparation: Refrigerate slides in distilled water for several hours; drain them and drop the cell suspension from a height of 8–12 inches. Do not use too thick a suspension. Help smear the cells by blowing across them. Dry flat on a 60° hot plate: 1–2 minutes. Keep dry at least one hour, but better overnight. Stain within 2–5 days.

Alternate methods are given by Ford and Woollam (1963) and Ford and Zeiss (1961). For squashing methods see p. 475.

Staining Somatic Chromosomes

Aceto-Orcein Method

1. Dissolve 2 g orcein in 45 ml of acetic acid. Boil with stirring. Cool to 50°C and add 55 ml distilled water. Cool to room temperature and filter. Refilter before use to eliminate precipitate.
2. Stain: 30 minutes.
3. Rinse off excess stain with tertiary butyl alcohol (isopropyl or ethyl alcohol may also be used): 3 washes.

4. Mount while still moist in Euparal, or clear in xylene and mount in synthetic resin.

Giemsa Method for Specific Banding (Drets and Shaw 1971)

Chromosome banding has revolutionized cytogenetics. The chromosomes appear to be made up of an arrangement of clusters of DNA containing different base compositions. This permits identification of individual chromosomes and parts of chromosomes and detecting structural changes in them.

SOLUTIONS

Alkaline solution, pH 12.0

sodium hydroxide	2.8 g
sodium chloride	6.2 g
distilled water...........................	1000.0 ml

Saline-citrate solution, pH 7.0

sodium chloride	105.2 g
trisodium citrate	52.9 g
distilled water...........................	1000.0 ml

Adjust pH to 7.0 with $0.1 N$ HCl (p. 542) if necessary.

Buffered Giemsa

Giemsa stock solution (p. 231, or purchased solution)	5.0 ml
absolute methanol	3.0 ml
$0.1 M$ citric acid (pp. 551, 557)..............	3.0 ml
distilled water...........................	89.0 ml

Adjust pH to 6.6 with $0.2 M$ Na_2HPO_4 (p. 552 or 555) if necessary.

PROCEDURE

1. Treat fixed and flamed slides in alkaline solution, room temperature: 30 seconds. Note: if the fixing (step 1, p. 483) is extended to 24 hours, banding may improve.
2. Rinse in saline-citrate solution, 3 changes: 5–10 minutes in each.

3. Incubate in saline-citrate solution, 65°C: 60–72 hours (try 24 hours).
4. Treat with 3 changes of 70% ethanol and 3 changes of 95% ethanol: 3 minutes each.
5. Air dry.
6. Stain in buffered Giemsa: 5 minutes.
7. Rinse briefly in distilled water.
8. Air dry and mount.

COMMENTS

Gurr's[8] Giemsa powder and solution gives better results than other commercial solutions. The dried slides, before staining, may be subjected to trypsin treatment to swell the chromatids and increase the size of the chromosomes (Seabright 1971). Use 0.025–0.1% trypsin[9] in a Hanks balanced salt solution at pH 7: 10–15 minutes, 25–32°C. Trypsin also may be added to the staining solution:

phosphate buffer, pH 6.8	36.5 ml
methanol	12.5 ml
Giemsa solution (p. 231, or purchased solution)	1.0 ml
0.1% trypsin in buffer......................	0.25 ml

Scheres (1977) stains with basic fuchsin.

Comings (1975) has a precise analysis of banding and good references.

Chresman (1976) modifies the banding technique for mouse embryo chromosomes, and Klášterská and Natarajan (1976) modify it for diffuse diplotene chromosomes.

The fluorescent and quinacrine mustard methods have more or less been superceded by the Giemsa banding technique. In the latter method, no special microscopic equipment is required, and the preparations are permanent. The following are a few references to the quinacrine mustard method: Caspersson et al. (1971); George (1971); Khudr and Benirschke (1973).

Stains for Squashes and Spreads

Aceto–basic fuchsin: Tanaka (1961).
Aceto–iron hematoxylin: Lowry (1963); Wittman (1962, 1963, 1965).
Acridine orange: Schiffer and Vaharu (1962).

---

[8] Biomedical Specialties.
[9] Difco Bacto-trypsin #0153.

Carbol-fuchsin: Carr and Walker (1961).
Feulgen: Rafalko (1946) and Sachs (1953).
Gomori's hematoxylin: Melander and Wingstrand (1953).
Inverted Feulgen: Koulischer and Mulnard (1962).
Lactic-acetic orcein: Welshons et al. (1962).
Lacto-propionic orcein: Dyer (1963).
Iron hematoxylin: Chen (1944a); Griffin and McQuarrie (1942).
Iron hematoxylin with acetocarmine: Austin (1959).
Pinacyanole: Klinger and Hammond (1971).
Silver-hexamethylentetramine: de Martino et al. (1965).
Sudan black B: Cohen (1949).
Sudan black with acetocarmine: Bradley (1957).

Methods of Culturing for Animals

Birds: Krishan (1962).
Dogs: Ford (1965).
Domestic fowl: Newcomber and Donnelly (1963).
Drosophila: Horikawa and Kuroda (1959).
Macaques: Egozcue and Egozcue (1966); Sanders and Humason (1964).
Opossums: Shaver (1962).
Snakes and other cold bloods: Beçak et al. (1962, 1964).
Spider monkeys: Bender and Eide (1962).
Insects: Crozier (1968).
Pigs: Srivastava and Lasley (1968).
Mouse embryos: Wroblewska (1969).
Rana pipiens: Picciano and McKinnell (1977).

Hunsaker Method for Lizards[10]

SOLUTIONS

Medium

TC 199	100.0 ml
chicken serum	25.0 ml
PHA-M	2.5 ml

---

[10] Personal communication.

Antibiotic solution

penicillin G	350.000 units
streptomycin..............................	350.0 mg
Hank solution (sterile)	100.0 ml

Incubate at 30°C, swirling bottles twice daily.

PROCEDURE

Draw 1.0 ml blood from postcaval or mesenteric vein or heart puncture and sediment it 30 minutes. Plant 0.1 ml of plasma and leukocytes in 5.0 ml of medium and 0.1 ml of antibiotics.

Techniques for culturing biopsies of tissues such as skin will not be included here. The methods are not as simple as those for peripheral blood and require longer periods of culturing. See Berton and Phillips (1961); Ford and Woollam (1963); Harnden (1959); Harnden and Brunton (1965); Hsu and Kellogg (1960); Puck et al. (1958); Swanson and McKee (1964); Tÿo and Puck (1958).

For an in vivo method of cell culture see Berman and Kaplan (1959); Berman and Newby (1963); and Gengozian et al. (1964). Tiny chambers are made of millipore filters and lucite rings. Various kinds of cell suspensions can be placed in these chambers, which are then planted in the peritoneal cavities of mice. The cells will grow in vivo, with the mice serving as a tissue culture medium.

New methods are continually appearing in the literature.

Bone Marrow Preparations

Bone marrow provides the simplest of all methods for preparing chromosome spreads. Only a small sample is needed, and sterile conditions can be forgotten. The following method[11] has been modified at Oak Ridge Associated Universities[12] to give excellent well-spread and well-defined chromosomes.

---

[11] Courtesy of Ruby I. Thompson Zuelzer, Childrens Hospital of Michigan.

[12] Dr. Littlefield and staff, Cytogenetics, Medical and Health Sciences Division.

Oak Ridge Modification of Zuelger Method

PROCEDURE

1. Collect 1–5 drops of fresh marrow; place in 10 ml of warm TC 199 (37°C) containing 10 units of heparin and 0.002% colchimide/1 ml.
2. Incubate at 37°C: 1 hour.
3. Centrifuge, 600 rpm: 6–10 minutes. Discard supernate.
4. Add 2–3 ml 0.075% potassium chloride with shaking to break up the cell button. Make certain the cells are well dispersed in solution. Let stand 7 minutes.
5. Centrifuge and discard supernate.
6. Add fresh methanol–acetic acid (3 : 1) with continuous shaking. The cell button must be broken and the cells dispersed in the fixative.
7. Repeat step 6 at least 2–3 times. In the last fixing solution, stir the cells into suspension (the solution should appear cloudy) and scatter several large drops on a clean slide. Immediately turn the slide upside down and run it through the flame of a Bunsen burner. When the solution ignites, turn the slide right side up and set it aside to cool.
8. Stain with cresyl violet, orcein, buffered Giemsa, or any desired stain.

COMMENTS

Do not use "bloody" marrow; a high content of red cells reduces the chances of a successful preparation. In step 1, break up any large clumps of cells; very small groups of cells and, ideally, individual cells make the best spreads.

I have used this method on marmoset marrow with beautiful results.

Chromosome Analysis

Human chromosomes are classified in 7 groups on the basis of total length and arm ratio. Certain indices are relied on for analysis: (1) the length of each chromosome relative to total length of a normal X-containing haploid set (the sum of the lengths of 22 autosomes and the X chromosome expressed per 1000); (2) arm ratio of the chromosomes expressed as the length of the longer arm relative to the shorter one; (3) centromeric index (used sometimes), expressed as the ratio of the length of the shorter arm to the length of the whole chromosome; and (4) the presence of additional distinguishing features such as satellites and secondary constrictions (Fig. 23–5).

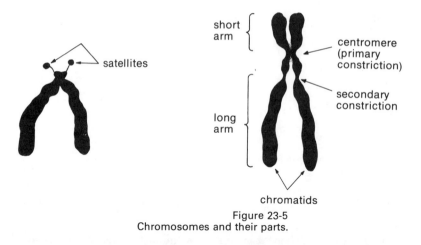

Figure 23-5
Chromosomes and their parts.

All is not as simple and reliable as it may sound. Considerable variation does exist even between the homologs in the same nucleus. One may be more contracted than the other, and chromosomes lying near the periphery of a spread tend to be larger than those in the center. Thus the value of measurements is questionable. Each arm should be measured along its axis, but the choice of exact termination of the arm can be arbitrary. In the best of photographs, the ends of the arms are always fuzzy. However, if one person makes all measurements on a set of chromosomes, some constancy in results can be achieved. Measurements may have to be relied on when other distinguishing features are absent. Labeling with tritiated thymidine will identify some chromosomes. In many chromosome analyses, two or more methods may be required to corroborate the final results.

The members of each of the 7 groups are easily distinguished from those of the other 6, but it is not always easy to distinguish those within a group from one another. Often it becomes necessary to speak of a chromosome as a C chromosome, for example, rather than to designate it by number. They should be arranged according to length, decreasing in size. This form of alphabetical-numerical arrangement is being adopted for other animals.

Exaggerating the constrictions has been tried by Kaback et al. (1964); Patau (1965); Saksela and Moorhead (1962); Sasaki and Makeno (1963); and Upadhya (1963).

For the measurements of normal human chromosomes, see Burdette (1962).

Preparation of Karyotypes

A karyotype is the systematized arrangement of the chromosomes of a single cell prepared by a camera lucida drawing or a photograph (see Figs. 23–2, 23–4). An idiogram is a diagrammatic representation of a karyotype and may be based on the measurements of the chromosomes of several or many cells.

Contrast process panchromatic film can be used for photographing the spreads. Enlargement prints can be made on Velour black A1, A2, A3, or A4 paper, depending on the contrast required for the film. See Christenson (1965) for photographic suggestions.

Two enlargements are developed and printed, one for determining arm measurements and one for cut-ups for the karyotype. The chromosomes can be cut from the enlarged photograph and be arranged and pasted on Bristol board or the like. Photographic mounting tissue (Kodak) can be lightly attached to the back of the photograph with a mounting iron. Then cut out the chromosomes and apply them to the Bristol board with a mounting iron. Cut-out chromosomes can be lightly mounted by one edge with double-coated Scotch tape on Bristol board (sprayed with plastic) or on a heavy sheet of plastic. Then if necessary, they can be pulled loose with fine forceps and rearranged.

If precise arm measurements are required, they can be made according to Puck.[13] Apply a strip of Scotch linerless no. 665 double-coated tape over each arm of each chromosome, so it follows the central axis of the arm. Using surgical silk[14] follow the axis of each arm, including all curves from proximal to distal end. Cut the silk with a sharp razor, pick up the silk with fine forceps, and stretch it on a section of white Scotch tape no. 666. Measure it to tenths of a millimeter with a millimeter ruler. From these measurements, total lengths, arm ratios, centromeric index, percentage total, and complement lengths can be determined. Select metaphases that show a minimum of contraction; slender chromosomes with well-defined centromeres are easier to measure and have more constant arm ratios than short fat chromosomes.

Measuring and calculating by the above method is slow and time consuming. La Bauve et al. (1965) have developed a digitized comparator to obtain the initial linear measurements on punchcards. Final matching, however, still requires some visual evaluation, as in the above method.

---

[13] Personal communication.
[14] 3–0, Edward Weck and Co.

Figure 23-6
Measuring a chromosome arm with calipers.

If chromosomes are well spread with reasonably straight arms and with a minimum of curls and wiggles, and precise measurements are unnecessary, calipers with fine pointed tips can check the arm lengths (Fig. 23–6) or total chromosome length of one chromosome against that of another.

Computer analysis of chromosome spreads is also possible (Neurath 1968).

Scientists at the City of Hope Medical Center and Caltech's Jet Propulsion Laboratory, California, are developing a karyotype maker that turns out pictorial karytypes from chromosome spreads.

Chapter 24
Special Procedures III

PREPARATION OF INVERTEBRATES FOR WHOLE MOUNTS AND SECTIONS

It is impossible to cover specifically all the members of this huge group of organisms, but certain generalizations can be made that will be applicable to the common forms and perhaps be adaptable to the less common ones. Fixatives can be applied directly to many of the invertebrates, but not to others that, because of their propensity to contract or ball up, pull in their tentacles or other appendages, and thereby make whole mounts or sections practically worthless. Careful anesthetizing or narcotizing of these invertebrates must precede killing and fixation. As soon as the narcotization is complete and before death, if possible, fixation can be successful.

Anesthetizing and Narcotizing Agents (See Laboratory Safety, p. 574, concerning drugs.)

Magnesium chloride or magnesium sulfate

Either is widely and successfully used on sea anemones, corals, annelids, tunicates, and nudibranchs, to name a few. Crystalline magnesium sulfate can be tied in a bag suspended above and just touching the water surface, or a 33% aqueous solution siphoned slowly in, controlled by a screw clamp. When the organisms are anesthetized (no reaction to the touch of a needle), siphon off the water until the animals are barely covered and carefully add fixative. Disturb the animals as little as possible. When partially hardened, transfer to fresh fixative.

Cocaine (a drug)

This can be used for cilitates, rotifers, bryozoans, hydras, some worms, and nudibranchs. A 1% aqueous solution is added to the water in proportions of about 1.0 ml to 100.0 ml of the water containing the animals. Eucaine hydrochloride can be used in the same manner. Check for contraction and fix.

Menthol

Sprinkle on the water surface and leave overnight. Good for difficult-to-narcotize sessile marine animals, coelenterates, some bryozoans, hydroids, also flukes. It is more efficient when combined with chloral hydrate (a drug, may be unobtainable) in proportions of 45.0 g menthol and 55.0 g chloral hydrate. Grind together in a mortar with a little water. Drop on surface of the water. Fix animals when they no longer contract. Large marine forms probably will require overnight treatment. Chloral hydrate can be used alone, sprinkled on water surface for annelids, molluscs, tunicates, bryozoans, and turbellarians.

Chloretone

A 0.33 to 1% solution can be used, but it is slow acting.

Chloroform

This can be dropped on the water surface for many aquatic forms and is used in special bottles for insects and arachnids (see below).

Ether and alcohol

These can be used by dropping on the water, or alcohol can be added gradually to the water by a tube controlled with a screw clamp until the proportion of alcohol to water is approximately 10%. It is particularly good for fresh water forms and earthworms. Ether can be used like chloroform for insects.

Asphyxiation

Boil water to remove the air and seal it in a jar. Particularly good for gastropods (snails); place them in the boiled water after it has cooled.

Cold

Partially freeze organisms in salt and ice mixture or in freezing compart-
.ment of refrigerator, or in icewater until they are relaxed. Good for
tapeworms. Transfer to lukewarm water and then fix.

Hanley solution (Gray 1954)

water	90.0 ml
ethyl Cellosolve	10.0 ml
eucaine hydrochloride	0.3 g

Add 1 drop per 10 ml of water in which animals are living. Good for
rotifers and bryozoans.

Propylene phenoxetol[1] (Owen 1955; Owen and Steedman 1956, 1958;
    Rosewater 1963)

Introduce a large globule of the compound into the water containing the
animals in an amount equally 1% of water volume, or shake vigorously 5
ml of the compound with 15–20 ml of sea water. Add to water containing
animals. Good for molluscs.

Tips on Special Handling

Porifera

Small forms can be dropped directly into osmic-mercuric chloride (water,
250.0 ml; osmic acid, 2.5 g; mercuric chloride, 9.0 g). Large forms fix
better in alcoholic sublimate (Gilson, Carnoy).

Calcareous sponges can be decalcified in 70–80% alcohol plus 3% of
hydrochloric or nitric acid. Siliceous sponges can be desilicified in 80%
alcohol plus 5% hydrofluoric acid added gradually. Perform the latter in a
glass dish coated inside with paraffin. After a few hours, transfer to 80%
alcohol. Do not breathe hydrofluoric acid fumes. Small spicules will section easily without desilification.

Coelenterates

Hydrae

Place in a small amount of water (few drops) in a shallow dish. When
animals are extended, rapidly pipette warm mercuric chloride–acetic acid

---

[1] Goldschmidt Chemical Corp.

(95.0 ml saturated aqueous mercuric chloride, 5.0 ml acetic acid) at them. Work the fixative from base toward oral region, thereby preventing tentacles from contracting or curling.

Sea anemones

Anesthetize overnight with menthol; or 30% magnesium chloride (50 to 100 ml) can be added gradually over a period of one hour: leave in the solution until there is no more contraction of tentacles. Siphon off the solution until anemones are just barely covered. Add fixative (Susa, Bouin, mercuric chloride combinations, 10% formalin–sea water) slowly down the side of the container. Also pipette some directly into the throats of the anemones. When the anemones are partially hardened, transfer into fresh undiluted fixative.

Jellyfish

Place in sea water; while stirring, add fixative (10% formalin) down side of container until proportions are approximately 10 ml or more to 100.0 ml of sea water. Stir for 3 or 4 minutes. After 2–3 hours, change to fresh formalin.

Medusae

These must be anesthetized; magnesium solution or cocaine (a drug) is best. Allow them to expand in a small dish of water; treat with magnesium solution or add 3–4 ml of 1% cocaine per 100.0 ml of water. When they no longer contract, add fixative (10% formalin) and stir.

Corals with extended polyps

Narcotize with magnesium sulfate and fix in hot saturated mercuric chloride plus 5% acetic acid.

Coelenterates

These are sometimes frozen.

Platyhelminths

Planaria

Starve for a few days before killing. Place in a small amount of water on a glass plate. When the worm is extended, add a drop of 2% aqueous nitric acid directly on it. Follow by pipetting fixative (Gilson or saturated mercuric chloride in saline) on it and then transfer into a dish of fixative.

Combined relaxing and fixing agent (Dawar 1973). The animals will contract and roll at first, but soon will relax and become flat.

distilled water.............................	200.0 ml
nitric acid, concentrated	2.0 ml
formalin	4.5 ml
magnesium sulfate ($MgSO_4$)	2.5 g

Use at room temperature: 24 hours; temperatures lower than 20–30°C cause mucus formation.

Trematodes

Anesthetize large trematodes with menthol: 30 minutes. Then flatten by the following method: saturate filter paper with Gilson and lay on a plate of glass. Lay worms on paper and quickly cover with another sheet of saturated paper and a second plate of glass. Add a weight, but not such a heavy one that the worms are crushed: 8–12 hours. Remove glass and paper and transfer worms to fresh fixative. A single specimen can be flattened between slides, tied together, and dropped into fixative. After 2–3 hours remove the string, but do not disturb the slides. Leave overnight. Remove slides and transfer animal to fresh fixative. Steaming (not boiling) 5% formalin can be used for trematodes.

Small trematodes can be shaken in a small amount of 0.5–1.0% salt solution: 3 minutes. Add saturated mercuric chloride plus 2% acetic acid. Shake for several minutes. Change to fresh fixative: 6–12 hours. If they should be flattened, they can be laid on a slide and covered with another slide or, if they are very delicate, with a cover glass. Pipette fixative carefully along the side of the cover and then lower carefully into fixative.

Cestodes

Place a large tapeworm in icewater until relaxed (overnight). Then flatten between glass plates for microscope-slide whole mounts (as for trematodes; see above). The worm can be wound spirally around slides. Then place a slide against both sides, press, and tie together (Demke 1952). For museum mounts, wind the worm around a tall bottle or 100 ml graduated cylinder and pour fixative over it. Their powers of contraction make them difficult to handle. Sometimes (after chilling) they can be relaxed in alcohol, 70%, or the steaming 5% formalin recommended above.

Small cestodes can be stretched from an applicator stick supported over a tall container and fixative poured down the length until they hang straight. Immerse them completely; then they can be flattened between microscope slides.

Nemertines

Drop into saturated mercuric chloride–acetic acid (95 : 5).

Rotiferae

Narcotize by adding 3–5 drops of the following solution for every ml of culture:

2% benzamine hydrochloride	3 parts
distilled water............................	6 parts
Cellosolve	1 part

When the animals are narcotized so they do not contract when touched, but the cilia are still beating, add 10% of formalin to the culture.

Nemathelminths

Nematodes (hookworms)

Shake 3 minutes in physiological saline. Pour off and drop worms into hot glycerin alcohol (70–80% alcohol plus 10% of glycerin). A sublimate fixative can be used. (This procedure may be used also for ascarids.)
3% formalin at 70°C penetrates and fixes within 3 minutes.

Very small nematodes can be relaxed and killed in a depression slide or watch glass by gentle heat. An incubator regulated to 50–52°C is most reliable. If using a flame or hot plate, be careful not to boil the worms. Transfer to fixative.

Formalin-acetic-alcohol is an excellent fixative for small nematodes:

95% ethyl alcohol	12.0–20.0 ml
glacial acetic acid	1.0 ml
formalin	6.0 ml
distilled water............................	40.0 ml

If a few drops of saturated aqueous picric acid are added, the worms take up a little color.

Slide whole mounts are easy to prepare. Place the worms in the alcohol-glycerine mixture in an incubator (35–37°C) or on top of an oven to permit evaporation of the alcohol. When the solution is almost pure glycerin, mount in glycerin jelly (p. 104). Helminth ova can also be mounted in this way. For other methods, see whole mounts, p. 506.

Yetwin's (p. 107) is an excellent mountant.

Friend (1963) makes a microstrainer for handling small ova. Electron microscope grids are attached to the ends of 3.5 mm (internal diameter) glass tubing with an insoluble resin. This method can also be used for processing small larvae. Helminth ova can be fixed in the steaming 3% formalin mentioned above.

Microfilaria

Blood infected with microfilaria is smeared and dried as for any blood smear. Then fix in any mercuric chloride fixative and stain with Delafield hematoxylin. If it is preferable to have the slides dehemoglobinized, after fixation treat them with 2% formalin plus 1% of acetic acid: 5 minutes. Wash and stain. Alternate method: smear and dry the blood; dehemoglobinize in 5% acetic acid and air dry; fix in methyl alcohol and stain with Giemsa (p. 231).

Bryozoa

Anesthetize with menthol. Fix salt water forms in chromo-acetic solution (7–10 ml 10% chromic acid; 10 ml 10% acetic acid; water to make 100 ml). Fix fresh water forms in 10% formalin.

Brachiopoda

The shells of brachiopods are so fragile the animals must be treated to remain open without forcing the shells. Leave them just covered with sea water and gradually add 95% alcohol until the alcohol concentration is approximately 5%. Let stand for 30–60 minutes. When the shell is easily opened, prop it open with a piece of glass rod and transfer to fixative.

Mollusca

Snails

Place in boiled water or add propylene phenoxetol to the water until snails are limp. Fix.

For decalcifying snail shells, Anderson (1971) uses a mixture of decalcifier-fixative (2 : 1), and the concentration can be as high as 7 : 1 for 6 hours. RDO (p. 50) is his choice for best cellular detail and staining properties. This can be used for other molluscs with calcareous exoskeletons.

Mussels

Sections of undecalcified shell can be made by grinding. Decalcification can be undertaken with 3–4% nitric acid. If the soft parts are to be fixed, wedge the valves apart with a small length of glass rod and place entire animal in fixative. Dissect out after it is fixed. Gills can be removed and fixed flat in a mercuric chloride fixative.

Nudibranchs

Add 1% formalin, a few drops at a time every 15 minutes, to the sea water containing them.

Freshwater molluscs

Gradually warming the water will make them extend their feet. Fix when they no longer respond to the prick of a needle. Bouin, Zenker, or Gilson are satisfactory.

Annelida

If sections of worms (aquatic or terrestrial) are desired, the intestines must be freed of grit and other tough particles.

Cocke (1938) feeds earthworms on cornmeal and agar (1 : 1) and some chopped lettuce for 3 days, changing the food every day. Becker and Roudabush (1935) recommend a container with the bottom covered with agar. Wash off the agar twice a day for 3–4 days. Moistened blotting paper can be used. When the animals are free from grit, place them in a flat dish with just enough water to cover them. Slowly siphon in 50% alcohol until the strength of the solution is about 10% alcohol. Chloroform can also be used for narcotizing. Fix in Bouin's or mercuric chloride saturated in 80% alcohol plus 5% acetic acid. The worms may be dipped up and down in the fixative and then supported by wire through a posterior segment, hanging down in the fixative, or placed in short lengths of glass tubing in fixative to keep them straight. This is necessary if perfect sagittal sections are to be cut. Embedding is probably most successful by the butyl alcohol method. After removal of mercuric chloride in iodized 80% alcohol, transfer to *n*- or tertiary butyl alcohol: 24 hours (change once). Transfer to butyl alcohol saturated with paraffin (in 50–60°C oven): 24 hours; pure paraffin: 24 hours; and embed.

Sea worms can be kept in a container of clean sea water, changed every day for 2 or 3 days, then anesthetized with chloroform and fixed, using fast-penetrating fixative (Bouin or a mercuric chloride fixative).

Arthropoda

Insects and arachnids

Ether, chloroform or potassium cyanide are used for killing. Simplest method: Place a wad of cotton in the bottom of a wide-mouthed jar and cover the cotton with a piece of wire screen. Dampen the cotton with ether or chloroform or lay a few lumps of potassium cyanide on it before adding the screen wire. Keep tightly closed with a cork or screw cap. A piece of rubber tubing soaked in chloroform until it swells and placed under the screen wire will hold chloroform for several days. If the appendages should be spread when fixed, as soon as the insect is dead place it on a glass slide with another slide on top of it. Run fixative in between slides.

For whole organisms, rapidly penetrating fixatives should be used: picro-sulfuric, sublimate fixatives, mixtures containing nitric acid (Carnoy), alcoholic and ordinary Bouin, or Sinha (1953) fixative.

For whole mounts, clearing of exoskeleton is sometimes difficult. Body contents have to be made transparent or have to be removed. Lactophenol (p. 107) mounting will serve the purpose in the first case, but heavily pigmented arthropods probably will have to be bleached in hydrogen peroxide for 12 hours or longer. Fleas, ticks, and the like make better demonstrations if not engorged; in any case they should be treated with 10% potassium hydroxide, 8–12 hours, to swell and dissolve the soft tissues, thus clearing out the body contents. Wash well to remove the potassium hydroxide. Acid corrosives, such as that which follows, are preferred by some because they do not soften the integument as much as alkaline corrosives.

glacial acetic acid	1.0 ml
chloral hydrate (a drug)	1.0 g
water	1.0 mg

Do not use a fixative containing alcohol or formalin if a corrosive is necessary. The organism will not clear.

Vyas (1972) preserves the soft parts as well as the exoskeleton with the following:

glycerin	12.0 ml
formalin	2.0 ml
distilled water	100.0 ml
Add a few crystals of thymol.	

This can be injected into the body cavity of large specimens.

Avoid the use of potassium hydroxide for very small or delicate insects. Transfer them into equal parts of chloral hydrate and phenol for about 2 weeks. If they are not cleared at the end of this period, place insects and solution in a 40°C oven for 2 days. This should complete the clearing. Transfer to absolute alcohol and mount in Berlese mountant or Turtox mountant (p. 107).

For methods of mounting whole mounts, see p. 506.

Because of chitin, sectioning insects can be difficult. Avoid the high-ethyl alcohols. Dioxane methods (p. 450), butyl alcohol (above), or double embedding (p. 448) must be used and provide excellent results. Soaking the tissue blocks overnight or for several days in water (p. 61) simplifies sectioning.

A modified Carnoy is recommended for fixation (Bilstad[2]):

absolute isopropyl alcohol	6 parts
chloroform	3 parts
formic acid	1 part

Fixation, dehydration, and infiltration with Paraplast are all performed under reduced pressure (De Giusti and Ezman 1955).

For mounting heavily chitinized sections, if a protective cover of a nitrocellulose is undesirable, Bilstad suggests Land mounting medium (p. 551). Subbed slides (p. 550) may be most efficient for mounts.

See Beckel (1959) for a sectioning method for heavily sclerotized insects. Barrós-Pita (1971) is able to section exoskeleton as tough as praying mantis. Roden (1975) uses a nitrocellulose method. For other methods see Nelson (1974) and Kimmel and Jee (1975). Farnsworth (1963) uses Scotch tape to transport insect eggs through processing.

Echinodermata

Narcotize echinoderms by placing them in fresh water or sprinkle menthol on the water surface. Magnesium chloride or magnesium sulfate can be gradually added to the water.

Inject fixative into the tip of the rays of starfish. This will extend the feet. Then drop the animal in fixative; mercuric chloride–acetic is good.

See Moore (1962) for collecting and preserving starfish, ova, and larvae.

---

[2] Personal communication.

Staining İnvertebrates

Sections can be stained in any manner, depending on the fixatives and the desired results.

For beautiful, transparent, whole mounts, the carmine stains (p. 508) are the usual choice for most of the invertebrates: obelias, hydras and hydroids, daphnids, Bryozoans, medusae, flukes, tapeworms, small annelids, tunicates, and ammocoetes (larvae), to name a few. But Kornhauser hematein (p. 510) is an excellent substitute, particularly for flukes and tapeworms.

If the cuticle and muscle layers of flukes and tapeworms tend to remain opaque and obscure the details of internal anatomy, either of the two methods can be tried to correct the condition. Dehydrate and clear in an oil, such as cedarwood oil or terpinol. Place in a flat dish under a dissecting microscope and carefully scrape away some of the tissue from both surfaces. With care, this can free the animal of some of the dense tissue material and not harm the internal structures. An alternate method is to finish staining and then wash in water. Transfer to a potassium permanganate solution made from a few drops of 0.5% solution added to water until only a pink color develops. When the worm begins to show a greenish brown sheen, remove it immediately to distilled water: 5 minutes. Transfer the worm to 2–3% aqueous oxalic acid until it is bleached and the sheen is lost. Wash thoroughly in running water at least one hour. Dehydrate, clear, and mount.

References for perparing invertebrates are Becker and Roudabush (1935); Galigher and Kozloff (1964); Gatenby (1950); Gray (1952); Guyer (1953); Hegner, Cort, and Root (1927); Mahoney (1968); and Pantin (1946).

PREPARATION OF CHICK EMBRYOS

The following procedure is a standard and simple one for removing, fixing, and staining chick embryos of any size, from primitive streak to age 96 hours.

With the handle of the scissors, break the shell at the air space end, turn scissors, and cut the shell around the long axis, being careful to keep the tip of the scissors pressed against the inside of the shell while cutting. This will prevent penetration of the yolk or the embryo. With egg submerged in physiological saline at 37°C in a finger bowl, remove top half of the shell. (With rapidity developed from experience, warm water serves just as well as saline, with no ultimate harm to the chick if it is to be fixed immediately.) The chick will be found floating on the upper surface of the yolk. It can be a waste of time to remove the lower half of the shell, but many workers prefer to do so. This is a matter of individual preference. With small scissors, cut quickly around the outside of the vascular area. Except for large embryos, do not cut through the vascular area. Keep one edge gripped by forceps, and slip a syracuse watch glass under the chick. Withdraw the watch glass and embryo with a little of the saline. Bring as little yolk as possible with the chick.

The vitelline membrane most likely will come with the embryo; occasionally it will float free. If not, wave the embryo gently back and forth in the saline until the membrane begins to float loose. Release grip on embryo and remove the membrane. Make certain that the embryo is wrong side up, that is, the side that was against the yolk is now uppermost. Straighten out the chick, taking care that there are no folds. Sometimes washing with saline in a pipette across the top helps to clean and flatten it. Pull off remaining saline and yolk, carefully, not just from one edge but slowly and with gentle pressure on the pipette bulb around the vascular area. Do not get so close that the embryo is drawn up into the pipette.

Have a circle of filter paper cut with the inside hole approximately the size of the vascular area if it is to be retained, or a little larger than the embryo if it only is to be used. Being careful not to disturb the embryo, drop the circle of paper around it. Press gently around the paper to make it adhere to the blastoderm. With a pipette apply fixative (Gilson or Bouin), carefully dropping it first directly on the embryo; ease it on (do not squirt it on) and then work outward toward and over the paper circle, finally adding enough fixative to completely immerse embryo and paper. Leave overnight. The embryo should remain adhering to the paper, and it can be transported in this fashion from solution to solution. Follow fixation by proper washing, depending upon choice of fixative. Stain the embryo in a carmine, hematoxylin, or hematein stain (see whole mounts, p. 508), and dehydrate. Remove the chick from the circle, clear, and mount. The cover glass may have to be supported (p. 510) to protect the delicate embryo from destructive pressure. After washing, the embryos can be dehydrated, cleared, and embedded in paraffin.

Incubation time and determination of age can be puzzling. A perfectly aged chick sometimes is difficult to obtain, because the age is determined not only by the number of hours in, and the temperature of, the incubator, but the length of time the egg was in the hen and the temperature at which the egg was stored. The temperature of the incubator should be 37.5°C, and about 4 hours are usually required to warm up the egg once it is placed in the incubator. Thus a 36 hour chick would require about 40 hours in the incubator for proper development. The accompanying chart (Fig. 24–1) indicates how counting the number of pairs of *somites* in the chick will determine its age; thus 10 pairs indicates an age of 30 hours. This, rather

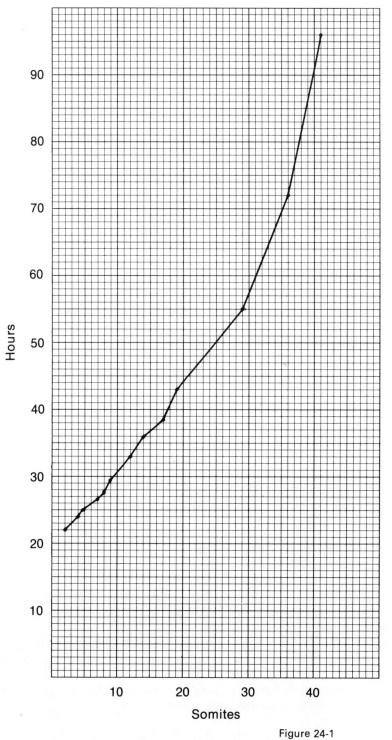

Hours

Somites

Figure 24-1
Development of a chick embryo. Data from Patten (1952).

than the number of hours in the incubator, is the best criterion for development.

Any bird or reptilian embryo can be fixed as outlined for the chick. Amphibian larvae can be dropped directly into fixative, after any surrounding yolk or jelly has been dissected off.

Mammalian embryos should be dissected from the uteri and fixed in Bouin. If the uterine site also is to be preserved, cut through the uterus on either side of the embryo and fix. A little fixative can be injected into the uterine cavity to quicken the fixing action.

Ascarid uteri are fixed in alcoholic Bouin; sometimes Carnoy will give desired results.

WHOLE MOUNTS

Whole mounts are slide mounts of entire specimens small enough to be studied in toto and mounted in resins in toluene or water, in gums or glycerin as aqueous mounts, or as dry mounts. Some examples from the innumerable methods will follow, with the hope that a technician can adapt them to almost any whole-mount venture.

General Preparation

FIXATION

95% alcohol, followed by bleaching if necessary. From 70% alcohol into bleaching fluid (3–4 drops of Clorox in 10 ml of 70% alcohol, or water plus H_2O_2 (1 : 1) with a trace of ammonia): 24 hours.

PROCEDURE

1. Wash with 70% alcohol, 3 changes.
2. Dehydrate in 95% and absolute alcohol.
3. Clearing is often difficult. Try mixture of xylene and beechwood creosote (or aniline or phenol) (prevents whole mounts from becoming brittle), 2 changes. If opaque patches persist, return them to absolute alcohol.
4. Impregnating with mountant can be difficult. Specimen may collapse and turn black or opaque when taken directly from alcohol or clearer to mountant. Impregnate gradually by adding each day a drop or two of mountant to the clearer containing the specimen. Carefully stir the mountant into the solution.

Galigher (1934) uses a cone of filter paper containing the mountant, and allows it to enter the solution gradually and continuously.

When the clearer has obviously thickened, allow concentration to continue by evaporation.

5. Mounting sometimes requires considerable experience (and patience) in judging the correct method and amount of mountant. Large specimens should have supports under the cover glass to prevent tilting of the cover glass; delicate specimens should have supports under the cover glass for protection against smashing. Various materials may be used to support the cover glass: bits of cover glass or slide, thin glass rods, circles, squares, or strip stamped or cut out of heavy aluminum foil, bits of glass wool.

Place a drop of mountant on slide and place specimen in it. Lower cover glass carefully; keep it flat (not tilted); have specimen in center of mount. If several specimens are to be mounted under one cover glass, chance of their slipping out of place is lessened by allowing the mountant to air dry for a few hours. Place a bit more mountant over them and add cover glass. Warming the cover glass aids in bubble prevention.

COMMENTS

Demke (1952) embeds helminths in celloidin to support them. Dehydrate through absolute alcohol, through alcohol-ether, and into thin celloidin. Pour celloidin and specimens into flat dish (petri dish); allow solvent to evaporate slowly. Cut out squares of celloidin containing specimens, dehydrate, clear, and mount. No other support of cover glass required.

Rubin (1951) uses PVA mounting medium (p. 105). See other mounting media in that section. Courtright (1966) uses polyester resins.

Glycerin Jelly Mounts

Many materials, including frozen sections, can be mounted directly from water into glycerin jelly. If there is danger of an object's collapsing, transfer it from 70% alcohol or water into a mixture of 10–15% glycerin in alcohol or water. Leave the dish uncovered until most of the alcohol or water has evaporated. Mount in glycerin jelly. Sections are mounted without cover glass support. Thicker specimens probably will require a supported cover glass. With a turntable, spin a ring of gold size on the slide. Allow it to dry. If it is not high enough, add more layers to the height desired.

Melt glycerin jelly in water bath or in oven. Add a drop or two of jelly onto specimen inside ring. Use just enough to fill ring. (Warm the slide briefly, and the glycerin jelly will spread readily under the cover glass.) Warm cover glass and ease horizontally into place. Carefully wipe off any glycerin jelly that works out from cover glass. After a few tries, it becomes relatively easy to estimate the correct amount of jelly to make a clean mount. Seal cover glass, supported or unsupported, with gold size or other cover glass sealer. (See page 109; also for a method using 2 cover glasses.)

Stained Whole Mounts: Hydras, Embryos, and Flukes

Many fixatives are suitable: Carnoy or Gilson for worms, Bouin or Zenker for embryos, formol acetic or saturated $HgCl_2$ acetic (95/5) for flukes. Follow by proper washing and extraction of any undesirable pigments or crystals. But before fixing consult the previous section on invertebrates or an authoritative source concerning the problem at hand.

Choice of stain and method also depends upon the type of material. A few of wide usage are included here.

Grenacher Borax Carmine (Galigher 1934)

SOLUTION

carmine (C.I. 75470)	3.0 g
borax	4.0 g
distilled water.............................	100.0 ml

Boil approximately 30 minutes or until carmine is dissolved. Mixture may be allowed to stand until this occurs. Add:

70% ethyl alcohol	100.0 ml

Allow to stand 1–2 days. Filter.

PROCEDURE

1. Transfer from 50% alcohol to borax carmine: 3–4 hours or overnight.
2. Add concentrated HCl, slowly, a drop at a time, stirring, until carmine has precipitated and is brick red. Let stand 6–8 hours or overnight.

3. Add equal volume of 3% HCl in 70% alcohol and thoroughly mix. Let stand 2–3 minutes until specimens have settled. Draw off precipitated carmine with pipette. Add more acid alcohol, mix, allow to settle, and draw off fluid. Repeat until most of carmine is removed.
4. Add fresh acid alcohol and allow tissue to destain, checking at intervals under microscope: may require 2 hours or more. If destaining is exceedingly slow, increase percentage of acid in alcohol.
5. When destained, replace acid alcohol with 80% alcohol, several changes, to remove acid: over a period of 1 hour.
6. Dehydrate in 90% alcohol: 30 minutes or more, depending on size of specimen; absolute alcohol: 30 minutes or more.
7. Clear. For delicate objects clearing probably should be done gradually through the following concentration of absolute alcohol and creosote (or aniline or carbol) xylene, and then into pure creosote xylene.

Absolute alcohol	Creosote (aniline or carbol) xylene
80 parts	20 parts
60	40
50	50
40	60
20	80
0	100

8. Mount. For exceedingly delicate objects and roundworms, heed the warning above concerning slow impregnation with the mounting medium. Judgment concerning its use rests with experience, but slowness is invariably necessary for roundworms.

Mayer Carmalum (Cowdry 1952)

SOLUTIONS

Stock solution

carminic acid (C.P. carmine) (C.I. 75470)	1.0 g
ammonium alum	10.0 g
distilled water...........................	200.0 ml

When dissolved, filter and add:

formalin	1.0 ml

Working solution

carmalum stock	5.0 ml
glacial acetic acid	0.4 ml
distilled water............................	100.0 ml

PROCEDURE

Stain for 48 hours, no destaining. Dehydrate as above. Carmalum is good and requires no destaining.

Hematein (Kornhauser 1930)

SOLUTION

Stock solution

hematein	0.5 g
95% ethyl alcohol	10.0 ml
potassium aluminum sulfate, 5% aqueous	500.0 ml

Grind hematein with alcohol and add to the aqueous sulfate.

Working solution

Dilute above 1 : 10 with distilled water.

PROCEDURE

1. Stain overnight.
2. Transfer to 70% alcohol.
3. Destain in acid alcohol, 5% HCl in 70% alcohol.
4. Blue in alkaline alcohol, ammonia, or sodium bicarbonate in 70% alcohol.
5. Dehydrate and mount as above.

Hematein is good for flatworms. Alum cochineal used to be popular but has been largely replaced by carmine stains. Alum hematoxylin may be used for small organisms if they are not too dense. The celestine blue method (Demke 1952) and trichrome stain (Chubb 1963) give good results.

Cochineal-Hematoxylin

SOLUTION

alum cochineal		3 parts
potassium alum.	30.0 g	
cochineal	30.0 g	
distilled water	100.0 ml	
Delafield hematoxylin		1 part
distilled water		25 parts

Can be used immediately, but is better after a few hours. Filter before use.

PROCEDURE

1. After fixation wash in 50% alcohol: 1 hour.
2. Wash in distilled water: 10 minutes.
3. Stain: 4 hours or overnight.
4. Transfer to 70% alcohol: 2 minutes.
5. Differentiate in acid alcohol (70% alcohol : HCl, 99 : 1) until internal structures are visible.
6. Place in 70% alcohol: 1–2 minutes.
7. Blue in 70% alcohol plus a few drops of saturated aqueous lithium bicarbonate: 1–2 minutes.
8. Dehydrate in 95% alcohol: 1 hour, 3 changes of absolute alcohol: 1 hour, 30 minutes, 1 hour.
9. Clear in absolute alcohol–cedarwood oil (1 : 1): $\frac{1}{2}$–1 hour.
10. Finish clearing in cedarwood oil: 1 hour, and mount in one of the synthetic mountants (Permount or like).

Protozoa

An easy method for handling protozoa is the following; it gives excellent results and rarely fails.

Chen (1944a) Cover-Glass Method

FIXATION

Hot Schaudinn, 50°C: 5–15 minutes. Other fixatives that can be used are Bouin, Champy, Flemming, or Worcester. See Merton (1932) for fixing stentors, spirostomids, and vorticellas.

Organisms such as paramecia should be concentrated by centrifuging. Quickly pour off some of the solution and add fixative. Amoebae will settle on the bottom of a clean culture dish. Decant off most of the culture media, leaving the bottom barely covered. Quickly pour hot Schaudinn or Bouin (50–60°C) over the organisms. After a few minutes add an equal amount of 85% alcohol. Carefully loosen any amoebae clinging to the bottom and collect the solution in a centrifuge tube. For both paramecia and amoebae, allow to settle or centrifuge down at low speed. Pour off fixative and wash several times with 70% alcohol, centrifuging after each wash. If a mercuric chloride fixative was used, one of the washes must contain some iodine.

PROCEDURE

1. Follow the 70% washing with 80–85% alcohol: 10 minutes.
2. Smear cover glass with albumen fixative. Cover glasses can be subbed like slides and used here. Place it albumen-side up on slides with a bit of edge projecting beyond the slide. The projecting edge can be easily grasped with forceps when it becomes necessary to move the cover glass. With a pipette pick up a few drops of alcohol containing organisms. Drop them in the center of the cover glass, where they will spread over its surface. The alcohol will begin to evaporate and bring the specimens in contact with the albumen. Avoid complete drying; the edges may dry, but the center will remain slightly moist. Carefully add a couple of drops of 95% alcohol onto the specimens, and then transfer the cover glass to a petri dish of 95% alcohol. Slides lying in the bottom of the dish will help to handle the cover glass, as above.
3. Carefully remove the cover glass from the 95% alcohol, dip it gently in absolute alcohol, and flood it with 1% celloidin or nitrocellulose. Drain off excess celloidin against filter paper. Wave it back and forth a few seconds until it begins to turn dull and place it in 70–80% alcohol. It can remain in this solution until ready for staining.
4. Any stain can be applied—hematoxylin, carmine, Feulgen— depending upon study to be undertaken.
5. The cover glasses are finally dehydrated, cleared, and mounted with the specimens albumen-side down on a drop of mounting resin.

COMMENTS

Subbed slides (p. 550) are excellent. Cells adhere well and rarely rupture. Apply a drop of cells, drain off excess fluid, and allow to dry.

Smyth (1944) uses a quicker method and directly on slides. After fixation he carries the organisms through graded alcohols into absolute alcohol. This is dropped on a film of albumen on a slide. Place the slide in absolute alcohol and continue from there to the stain.

Agrell (1958) places suspensions of minute embryos on albumen-coated slides. Allowed to almost, but not quite, dry, they become flattened and in close contact with albumen. Dip into absolute alcohol, and then into fixative; place them horizontally in 95% alcohol vapor: 1 minute. Then fix. This coagulates the embryos and attaches them.

Paramecia can be fixed to preserve their normal shape without contraction by first adding copper sulfate or acetate to the cell suspension.

Merton Method (Kirby 1947)

1. Add a drop of paramecia to an albumenized or subbed slide: 30 seconds.
2. Add an equal-sized drop of 1% copper sulfate: 7–8 minutes, or 3% copper acetate: 45 seconds.
3. Draw off part of fluid and suspend slide over 2% osmic acid: 45 seconds.
4. Add saturated aqueous mercuric chloride to the drop of organisms: 10 minutes.
5. Transfer to 70% alcohol plus small amount of iodine: 10 minutes.
6. Wash with distilled water.
7. Stain.

Prescott and Carrier Method (1964)

Amoebae, well flattened for tritium or other special studies, can be prepared by this method.

1. Place a small drop of some amoebae on a subbed slide (p. 550).
2. Add a small drop of fixative (70% alcohol or acetic alcohol) to a cover glass and place over the amoebas.
3. Immediately freeze in liquid nitrogen: 15 seconds. Flip off cover glass, rinse slide in 95% alcohol, and air dry. If liquid nitrogen is not available, fix in 50% aqueous acetic acid, and freeze in dry ice. (Alcohol fixative will not freeze in dry ice.)

Staining Protozoa

There are many special and routine stains adapted to protozoa. Trichrome and hematoxylin stains are always good. To stain fibrillar elements with iron-hematoxylin, use a fixative containing chromium, warm the stain to 50°C, and destain with 10% commercial H_2O_2 (Kidder 1933). Rothenbacher and Hitchcock (1962) use a Giemsa method for flagella. Loeffler's stain also can be used (Kirby 1947). In ciliates, the surface pattern, cilia, basal granules, and connecting filaments are clearly shown by the Gelei osmium–toluidine blue method (Kirby 1947 and Pitelka 1945). For the silver line system, the classic techniques of Chatton and Lwoff (1930, 1935, and 1936); Frankel and Hechtman (1968); Gelei (1932, 1935); and Klein (1926), are outstanding. Also see Corliss (1953) for a silver method and Bodian (1937) for an activated protein silver method. Protozoans can be tagged for autoradiographs and are excellent subjects for vital staining and enzyme techniques. Schiff et al. (1967) use a safranin–fast green stain to differentiate the micro- and macronuclei of protozoa. Nigrosin demonstrates various organelles.

Borror (1968) Nigrosin Stain

SOLUTIONS

Solution A

saturated aqueous $HgCl_2$	10.0 ml
glacial acetic acid	2.0 ml
formalin	2.0 ml
tertiary butanol	10.0 ml

Solution B

formalin	20.0 ml
nigrosin (C.I. 50420), water soluble	4.0 g
distilled water	100.0 ml

Working solution

solution A	12 parts
solution B	1 part

PROCEDURE

1. Place a drop of concentrated organisms on slide.
2. Add, from a height of 2–3 cm, a drop of working solution. Leave 3–4 seconds.
3. Gently wash culture and stain to one end of slide with additional drops of working solution. Stained fixed organisms will remain attached to slide. Leave 15 seconds.
4. Dehydrate, clear, and mount.

RESULTS

ciliary organelles—black against gray background

COMMENTS

A drop of 10% aqueous nigrosin can be mixed with a culture of living organisms and the staining of the various organs observed under the microscope. After allowing the slide to dry, the dye will adhere to some of the structures and the slide can be mounted with resin and cover glass. (Repak and Levine 1967)

The food vacuoles of paramecia are demonstrated by adding a small amount of carmine to the culture. After about 5 minutes the food vacuoles are filled with carmine. Mix a couple of drops of the culture with an equal amount of saturated aqueous solution of nigrosin. Air dry and cover. Do not use too much fluid; it slows down the drying (Wilhelm and Smoot 1966).

Sectioning Protozoa

Stone and Cameron (1964) **Agar Method** (Kimball and Perdue Modification 1962)

1. Pour a small amount of melted agar (p. 544) into a short length of glass tubing sealed at one end. Chill until agar is hardened.
2. Pipette a concentration of organisms on top of agar and add fixative.
3. When cells are fixed and settled on the agar, draw off the fixative.
4. Add more agar mixed with a little eosin or other counterstain and chill to solidify agar.
5. With a pipette, force water down side of agar and under it, thereby forcing the block loose and out of the tubing.

6. Dehydrate, clear, and embed block of agar like a piece of tissue.
7. When the tissue is being sectioned, the cells will be found at the junction of colored agar with the uncolored layer.

Alternate method

Form a block of agar and hollow out a small cavity in it. Pipette a drop of concentrated fixed protozoa into the cavity and allow the organisms to settle. Remove as much fluid as possible and seal the cavity with warm agar. Embed in paraffin. Eosin can be added to the organisms to make them more visible in the block. Eosin in the added agar also helps to locate them.

Dry Mounts of Foraminifera and Radiolaria

An opaque type of slide mounting is used for dry objects that are to be examined by reflected light. Glue the shells on a black background, place a cover glass and supporting ring around them, and cement the cover glass and ring in place. Gray (1964) and Kirby (1947) suggest a simpler and perhaps more practical method. Cut two pieces of cardboard to slide size. Punch a $\frac{5}{8}$ inch hole in the center of one piece. Paint a black $\frac{7}{8}$ inch square in the center of the other piece or paste a $\frac{7}{8}$ inch square of black paper on it. Stick the two pieces of cardboard together with dry mounting tissue (use in photography) and a flatiron. The cavity thus formed can be covered with a cover glass glued on with household cement, or with a slide laid over it, and the parts can be held together with a specially designed aluminum holder.[3] Gray recommends gum tragacanth as the best adhesive for sticking the shells to the background. Place little dabs of the gum on the black background, and with a moistened fine brush (red sable) pick up a good specimen and place it in the center of a drop of gum. Breathe on it; moisture is necessary to make it adhere. When all dabs of gum have a shell applied, breathe again several times on the preparation. Then let the slide dry. Turn it over and tap it several times. If some of the shells are not adhering, it may be necessary to add a drop of water to the gum to make the shells stick. But make certain the slide is dry before covering it. Special opaque slides can be purchased for making dry mounts (Ward's), or regular depression slides can be used if the depression is painted black.

---

[3] Ward's Natural Science Establishment #C5054.

ANIMAL PARASITES

Animal organisms parasitic in or on humans include protozoans, platyhelminths, nemathelminths, and arthropods. Clinical parasitology is extensive and therefore only a few methods are incorporated here. Since the Romanovsky-type stains are used on blood parasites (malaria, trypanosomes, filaria, etc.) the preparation of smears is included in the chapter on hematologic elements (p. 220). Tissue sections can be stained in a similar manner. Parasitic roundworms (pinworms, *Trichuris*, *Ascaris*, hookworms); flatworms (lung, intestinal, bile duct, and blood flukes, tapeworms); and arthropods (ticks, lice, mites) are discussed in the sections on invertebrates and whole mounts in this chapter. Methods given there can be adapted for most parasitic helminths and arthropods. Sections of tissue parasitized by protozoa or helminths are effectively stained by hematoxylin methods, also by periodic acid–Schiff; protozoa and worms are strongly PAS positive, because of stored glycogen in both forms, and PAS-positive cuticle in the helminths. A methenamine silver method is excellent on protozoans, particularly flagellates. The scolices hooks in hydatid disease (Echinococcus) do not show with PAS and are better demonstrated against a hematoxylin background. Kenney et al. (1971) use acid-fast staining to demonstrate the hooklets. Ova and larvae can be handled according to directions given on p. 109 or 506. Intestinal protozoa (amoebae, flagellates, ciliates, and coccidia) require the following special methods, both for smears and tissue sections.

Only permanent slide mounts are described. Consult clinical laboratory manuals for temporary and rapid examination methods for immediate diagnosis. Gradwohl (1963) and Lynch et al. (1969) are comprehensive.

Intestinal Protozoa: Smear Techniques

Preparing Concentrate Smears (Arensburger and Markell 1960)

1. Add 1 ml of feces to 10–15 times its volume of tap water. Mix well and strain through 2 layers of wet gauze in a funnel. Collect in a small centrifuge tube. Add 1–2 ml of ether. Using a cork or thumb for a stopper, cautiously shake the tube. Fill with water to 1 cm from top.
2. Centrifuge 45 seconds, 2500 rpm. Break up any plug at top and decant supernatant fluid.
3. Add 2–3 ml of normal saline and shake to resuspend the sediment. Fill tube with normal saline to within 1 cm of top and centrifuge.

4. Decant supernatant fluid. Take a small quantity of the original fecal specimen on the end of an applicator stick and mix well with sediment at bottom of tube.
5. With an applicator stick, transfer as much as possible of the material to a clean slide. Smear as for a conventionally made slide and fix immediately in Schaudinn fluid.
6. Stain as preferred, or with one of the following stains:

Kohn Stain, Combined Fixation and Stain (Faust and Russell 1970)

SOLUTION

Basic solution

90% ethyl alcohol	170.0 ml
methanol	160.0 ml
glacial acetic acid	20.0 ml
phenol liquified	20.0 ml
1% phosphotungstic acid, aqueous	12.0 ml
distilled water	618.0 ml

Grind 5 g chlorazol black E in a mortar at least 3 minutes, add a small amount of basic solution, and grind until a smooth paste. Add more solution and grind 5 minutes. Allow to settle a few minutes and pour off supernatant into a separate container. Add more solution to mortar and continue grinding and mixing until all dye is in solution. Add remaining basic solution and allow to ripen 4–6 weeks. Filter through #2 Whatman paper before use. Keep in tightly stoppered bottle. The prepared solution may be obtained from Harleco.

PROCEDURE

1. Stain according to following chart. Tissue sections require twice as long as smears, and water may be used for dilutions for the sections.

Stain	Basic solution	Hours
undiluted	— — —	2–3
1 part	1 part	2–4 to overnight
2 parts	1 part	2–4
1 part	2 parts	2 to overnight
1 part	3 parts	4 to overnight

2. Dehydrate in 95% alcohol: 10–15 seconds.
3. Dehydrate in absolute alcohol, 2 changes: 5 minutes each.
4. Clear in xylene and mount.

RESULTS

protozoa—gray green, gray, or black
cysts—gray green to dark gray
nuclei, chromatid bodies, karyosomes, cell membranes—dark green to
   black
ingested red cells—pink to dark red

Kessel (1925) **and Chen** (1944*a*) **Smears** (Modified)

FIXATION

Schaudinn (p. 25), 40°C: 10–15 minutes or more.

SOLUTIONS

Iron alum

ferric ammonium sulfate	4.0 g
distilled water............................	100.0 ml

Hematoxylin stock solution

hematoxylin	1.0 g
absolute ethyl alcohol	10.0 ml

Allow to ripen several months or hasten process (p. 112).

Hematoxylin working solution

hematoxylin stock solution	0.5 ml
distilled water............................	99.5 ml

Add 3 drops of saturated aqueous lithium carbonate. If the hematoxylin
working solution looks rusty or muddy brown, it is unsatisfactory and will
not stain efficiently.

PROCEDURE

1. Transfer slides into 70% alcohol (from fixative): 2–3 minutes.
2. Treat with Lugol solution: 2–3 minutes.
3. Wash in running water: 3 minutes.
4. Decolorize with 5% sodium thiosulfate: 2 minutes.
5. Wash in running water: 3–5 minutes.
6. Mordant in iron alum, 40°C: 15 minutes.
7. Wash in running water: 5 minutes.
8. Stain in hematoxylin, 40°C: 15 minutes.
9. Wash in running water: 5 minutes.
10. Destain in iron alum 2% (dilute stock 4% with distilled water, 1 : 1) until nuclei and chromatoidal bodies are sharp against colorless cytoplasm. Check under high dry objective.
11. Wash thoroughly in running water: 15–30 minutes.
12. Dehydrate, clear, and mount.

RESULT

nuclei, chromatoidal bodies—sharp blue black

COMMENTS

Saturated aqueous picric acid may be used for destaining. Follow it with a rinse in dilute ammonia (2–3 drops/100 ml water) and thorough washing in water.

Diamond (1945) added 1 drop of Turgitol 7[4] to the diluted hematoxylin solution just before use (1 drop per 30–40 ml solution). He substituted the Turgitol treatment for heat; it reduces surface tension and increases cell penetration. Staining time was reduced to 5 minutes.

This method may be used for amoebae in tissue as well as in smears.

Lawless Rapid Method (1953)

SOLUTIONS

Schaudinn-PVA fixative (Burrows 1967)

Dissolve 4.5 g $HgCl_2$ in 31 ml of 95% ethanol in stoppered flask. Shake at intervals. Add 5 ml acetic acid and mix. Set aside until needed. Add 5 g

---

[4] Carbide and Carbon Chemical Corporation.

PVA powder[5] to 1.5 ml glycerin and mix thoroughly until all particles of PVA appear to be coated with glycerin. Transfer to 125 ml flask, add 62.5 ml distilled water. Stopper and leave at room temperature: 3 hours to overnight. Occasionally swirl the mixture. Place flask in 70–75°C water bath: 10 minutes, swirling frequently. When PVA is mostly dissolved add fixative mixture. Continue to mix in bath: 2–3 minutes, or until PVA is dissolved and solution clears. Cool at room temperature. If solution gels during storage, warm in 56°C water bath. Some protozoans fix better in warm PVA fixative.

Stain

chromotrope 2R (C.I. 16570)	0.6 g
light green SF, yellowish (C.I. 42095)	0.15 g
fast green FCF (C.I. 42053)	0.15 g
phosphotungstic acid	0.7 g
glacial acetic acid	1.0 ml
distilled water............................	100.0 ml

Add acetic acid to dyes and phosphotungstic acid; let stand 15–30 minutes. Add water.

PROCEDURE

A small portion of stool is fixed in PVA fixative (1 part stool to 3 parts fixative): 15 minutes to 1 hour or more. In this form it can be shipped in a vial. When ready to make slides, decant off excess PVA solution. Replace cap of vial and shake the emulsion. Remove cap and cover vial opening with gauze. Place 3 or 4 drops of strained material on cleansing tissue or blotter. Allow absorption of PVA for about 5 minutes. Scrape up moist residue and spread with an applicator stick on slide or cover glass. Dry thoroughly and drop into iodine-alcohol (p. 13). If smears wash off, too much PVA was carried over; leave material for longer time on cleansing tissue. (In the dry condition, smears can be stored for several months.)

Smears may be fixed directly on slides. Take 1 drop of fecal material to 3 drops of PVA fixative. Smear over a large area of the slide and dry in oven overnight. Immerse in iodine-alcohol.

1. Leave in iodine-alcohol: 1 minute.
2. Decolorize in 70% alcohol, 2 changes: 1–2 minutes each.
3. Stain: 5–10 minutes.

---

[5] Delkote: specify the grade as pretested for use in Brooke and Goldman fixative.

4. Differentiate in acidified 90% alcohol (1 drop acetic acid/10 ml alcohol): 10–20 seconds or until stain no longer runs from smear.
5. Dehydrate in absolute alcohol; rinse twice, dipping up and down.
6. Dehydrate in second change of absolute alcohol: 1 minute.
7. Clear and mount.

Warning: Press the cover glass in place very gently; the smear is somewhat brittle and is easily loosened. Too much pressure on the cover glass may cause parts of the smear to float around on the slide.

RESULTS

background—predominantly green
cysts—bluish-green cytoplasm, purplish-red nuclei; green is more intense than background stain
engulfed red cells—vary: green, red, or black
    chromatoidal bodies—intense green, but may be reddish
karyosomes—ruby red
helminth eggs and larvae—usually red

COMMENTS

If cysts do not stain, or do not stain predominantly red, fixation is incomplete. Warming the fixative may help, although cold fixation yields more critical staining. A newly prepared stain is predominantly red in color and reaction; older stains show more violets and greens. The stain is more transparent than hematoxylin stain, but it fades after a few years.

Hajian (1961) fixes in Bouin for better staining of karyosomes.

Tomlinson and Grocott (1944) describes a phloxine–toluidine blue stain for malaria, leishmania, microfilaria, and intestinal protozoa in tissue.

Vetterling and Thompson (1972) use hematoxylin and a polychromatic stain.

Yang and Scholten (1976) use celestine blue B:

Solution A

celestine blue B (C.I. 51050)	0.6 g
glacial acetic acid	20.0 ml
distilled water	80.0 ml

Keeps indefinitely in a brown bottle.

Solution B

ferric alum	4.0 g
distilled water............................	100.0 ml

Keeps indefinitely in refrigerator.

Mix equal parts. Deteriorates in 2 weeks. From water transfer to stain: 5 minutes, but is not overstained if left an hour. Rinse in tap water, dehydrate, clear, and mount. It can be followed by a trichrome stain.

For immunofluorescent techniques see Hoffmann and Miller (1976) and Nayebi (1971).

Chapter 25
Special Procedures IV

SPECIAL MOUNTS

Sections Mounted on Film (Pickett and Sommer 1960;
Sommer and Pickett 1961)

Pickett and Sommer use film in place of glass slides for supporting tissue
sections. The flexibility of film and the low cost of material and time can
be advantageous. Many sections mounted on a long strip of film are
stained in one process by using a developing reel and then are sprayed
with plastic as a unit instead of having to be individually coverslipped. A
length of film can be stored in a small space, and single sections or groups
of sections can be clipped or stapled to records or reports and filed or
shipped in envelopes. Film with section also can be mounted in film
holders. If desired, one section of a series can be mounted on film and a
successive section mounted on glass. Sections already mounted on film
can be cut out of it and mounted on glass slides in the conventional
manner. Section details are clear under low and medium power. For
oil-immersion or high-power examination, a cover glass should be
mounted on the section with a nondrying immersion oil (cedarwood oil
dissolves plastic). Only low-power projection can be used with this film.
Pickett and Sommer include the design of a holder to keep the film flat
while it is being scanned under the microscope.

REQUIRED MATERIALS

35 mm film, Cronar, P-40B leader film[1]
Krylon Plastic Spray, Crystal Clear Spray Coating no. 1302[2]
Nikor 35 mm developing reel, stainless steel

PROCEDURE

1. Cut strips of film of desired length; clean by dipping in acid alcohol; dry with soft cloth (not gauze).
2. Coat film with albumen fixative.
3. Place tissue sections on water bath, maneuver film under ribbon, and pick up sections. Dry in 56°C oven. Film can be cut in 3×1 inch strips; lay on glass slide; smear with albumen. Place section on film; pipette water under it, and dry on slide warmer as usual for slides.
4. After drying, load lengths of film in reel and stain as usual. Short pieces can be handled individually. To guard against loose sections, protect with coating of 0.5% celloidin or nitrocellulose. Proceed as follows:
   a. Xylene: 1–2 minutes.
   b. Absolute alcohol: 1–2 minutes.
   c. 0.5% celloidin: 5 minutes.
   d. Air dry: 3–4 minutes.
   e. 80% alcohol: 5 minutes.
   f. Wash and stain.
   g. Dehydrate through 95% alcohol.
   h. Mixture of $\frac{1}{3}$ chloroform and $\frac{2}{3}$ absolute alcohol, 2 changes.
   i. Xylene.
5. Place face up on a blotter; do not allow to dry. The film must lie flat; buckling results in uneven plastic coat. Use weights along edges. Apply spray quickly from end to end, 2–3 times until surface appears smooth. Hold can about 6 inches above film. Allow to harden: 5–10 minutes. Repeat 2–3 times. Allow to dry.
6. Label with India ink if desired.

COMMENTS

Mylar polyester film (Du Pont), 0.25 mm thick, also has been recommended (Johnston 1960).

---

[1] E. I. du Pont de Nemours and Co.
[2] Krylon.

Burstone and Flemming (1959) recommend film for smears, touch preparations, and sprayed-on suspensions. Adherence is excellent, and the preparations can be floated on incubation media, staining solutions, etc. The film can be mounted in glycerin jelly or water-soluble plastics.

Berton and Phillips (1961) use film for tissue cultures. The live cells attach and adhere firmly to film.

Shanklin and Laite (1963) immerse the reel of film (after clearing) in Krylon crystal clear 150 base plastic material: 10–15 minutes. Drain: 5–10 minutes. Dip again: 1–2 minutes. Several more short dips may be added, 5–10 minutes between dips. Dry on blotting paper but strip the film from the reel before film dries completely. Stand film on edge to dry.

Cellulose Tape Mounting (Palmgren 1954)

This method has been devised for difficult sections, such as tissues containing a large amount of yolk, chitinous material, or hard tissue that breaks away from soft parts and falls out of sections. Cellulose tape is pressed firmly on the section surface of the paraffin block or the tissue frozen on the freezing microtome. The sticky surface of the tape attaches to all parts of the section and prevents its wrinkling or shattering during sectioning; 1μ to 100μ cuts will perform equally well. Using this method, Palmgren has cut whole adult mice for radiography on the freezing microtome without loss of parts. Since the sticky material is soluble in xylene, the paraffin cannot be dissolved while on the tape, and it is preferable to transfer the sections to a glass slide or plate.

PROCEDURE

1. Mount section and tape with section-side up on glass. Hold it in place with small strips of tape, but do not cover the section.
2. Spread a thin layer of 0.5% nitrocellulose over the section, but not over the tape. Let dry.
3. Immerse in xylene in a petri dish, and leave until the nitrocellulose film and section can be loosened easily from the tape. Leave in xylene a little longer to remove all sticky material.
4. Smear clean slide liberally with albumen fixative or use subbed slides (p. 550). Float section on it and press in place. Blot dry.
5. Transfer immediately to absolute alcohol–ether (1 : 1) and leave until nitrocellulose is dissolved. Section should be adhering to slide.
6. Transfer to 95% alcohol, hydrate to water, and stain.

Frozen sections must be either dried (for autoradiographs) or carried from water up to absolute alcohol before the nitrocellulose is applied for histological staining.

Serial sections can be made by adding sections in sequence to continuous tape, rolled at each end of tissue block.

Laminated Thermoplastic Mounts (LaCroix and Preiss 1960)

LaCroix and Preiss describe a practical and simple plastic mounting method for chromosome smears, whole mounts, and cross and sagittal sections. The cellular detail is good, and the plastic mounting makes these preparations practical for class study and demonstration. There is no danger of breakage, and if the plastic gets marred, the scratches can be polished away. See Tonna and Love (1961) for a high-pressure method.

AUTORADIOGRAPHY

Autoradiography will be retained here in preference to the alternate form, radioautography (Tauxe et al. 1954). *Auto-* is added to the noun *radiography* to describe such techniques as gamma radiography and X-radiography, in which a radioactive substance is placed in direct contact with a photographic plate. Autoradiography can be broken down further into micro-, macro-, or even histo-autoradiography.

Radioactive isotopes of stable compounds can be incorporated in some tissue compounds, and autoradiographs (ARGs) will reveal the location of the compound that was metabolically active during the time of the application of the labeled material. Autoradiographs have become one of the most useful methods for studying biochemical reactions in cells. Many of the most dependable tracer isotopes are those of elements normally found in the body, and the selection of isotope will depend on the study to be undertaken: for example, strontium or calcium for bone, iodine for thyroid, thymidine for DNA.

After proper preparation, the specimen containing the radioactive material is placed in contact with a photographic emulsion. Ionizing radiations are emitted during decay of the isotope and change the emulsion to produce blackening during its development. The number of developed grains will depend on the emulsion and the kind of ionizing particles emitted from the specimen. Alpha (α) particle tracks are large and dense, are not easily deflected, and move in a straight path only a few micra in length. Beta (β) particles are small and produce longer tracks, but they are easily deflected and produce grains far apart and more scattered. Gamma (γ) rays are useless with long random paths, but many isotopes that give off γ rays also produce low-energy β particles, and these will give the picture on film. High-energy β particles (^{32}P) may travel a millimeter or so before producing a grain, but low-energy β particles (^{131}I and ^{14}C) have short paths and give very satisfactory resolution on photographic emulsion.

Many varieties of material can be used—tissue sections, chromosome squashes and spreads, microorganisms, whole cells, smears, cultures, sections of whole animals, etc. The specimen and film are placed in contact for a certain length of exposure time. The radioactive atoms are decaying, and the emitted radiation strikes the photographic emulsion.

Monochrome film emulsions consist of mixtures of gelatin and crystalline silver halides, primarily bromides. The halide crystals are imperfect and provide electron traps to produce a concentration of electrons. The positively charged silver atoms are attracted by the negatively charged electrons and form a latent image. Development reduces the silver halide to metallic silver, and the unreduced grains of silver are dissolved away in the hypo solution. Also, the emulsion is hardened by the hypo. Ordinary photographic emulsions cannot be used because they contain only about 30% silver halide and have large grains. Special emulsions have been developed to intercept the maximum number of rays. The nuclear emulsions contain up to 90% and 95% silver halide and have small grains. American Kodak nuclear emulsions are NTA, NTB, NTB2, and NTB3, arranged according to ascending sensitivity. The last two are the most commonly used, particularly for tritium and ^{14}C. Emulsions of Ilford Ltd., England, are also popular for use because of their sensitivity and grain size.

As mentioned above, the grains are not uniform; some carry a silver speck, and others are relatively insensitive. Some grains develop in an unexposed emulsion, and some become accidentally developable by darkroom light, cosmic rays, or other sources of ionizing radiation. When evaluating autoradiographs, such "background" values have to be considered and checked. Take care to avoid unnecessary background accumulation.

The following are good texts on autoradiography:

Baserga, Renato, and Malamud, Daniel. *Autoradiography. Techniques and Application.* New York: Harper & Row, 1969.
Gude, William D. *Autoradiographic Techniques.* Biological Techniques Series. Englewood Cliffs, N.J.: Prentice-Hall, 1968.

Rogers, Andrew W. *Techniques of Autoradiography.* New York: Elsevier, 1967.

Roth, Lloyd J., and Stump, Walter E. *Autoradiography of Diffusible Substances.* New York: Academic Press, 1969.

PROCEDURES FOR AUTORADIOGRAPHS

Fixation and Slide Preparation

Knowing the half-life of the isotope that will be used for autoradiographs helps to determine the approximate time of exposure required for ARGs and the length of time that the tissue can be stored before processing. The half-life is the time required for disintegration of one-half of the atoms present. For example, ^{32}P has a half-life of 14.2 days and must be processed immediately. Within 42 days only 12% of the original radioactivity will remain. ^{3}H has a half-life of more than 12 years and loses only 0.5% of its activity each month. ^{14}C has a half-life of 5568 years, so you have your lifetime plus for processing. ^{125}I has 60 days, ^{131}I, only 8.07 days. For the latter, fix in ethanol and process in a hurry.

The aldehydes are usually satisfactory for fixation, but avoid solutions containing mercuric chloride, which will produce artifacts on the emulsion. If a fixative such as Zenker has been used, treat the sections with iodine and thiosulfate, followed by 0.2 *M* cysteine for 10 minutes. This will reduce desensitization of the emulsion. Carnoy, methanol, or acetic-alcohol (1 : 3) can be used for smears and tissue cultures. Allow smears to dry and flatten the cells before fixation. In all cases the fixative must be washed out of the tissue; some fixatives cause air bubbles. This applies as well to smears; wash them in several changes of 70% alcohol and distilled water.

Solubilities of isotopes must be checked carefully; some leach out in water or paraffin solvents and must be prepared by freeze-drying methods (Holt and Warren 1953). In some cases Carbowax can be used, but it will dissolve water-soluble isotopes (Holt et al. 1949, 1952; Holt and Warren 1950, 1953). Mounting must be handled with care to prevent leaching (Gallimore et al. 1954).

Witten and Holmstrom (1953) freeze the tissue at the microtome immediately after excision. The knife is kept cold with dry ice, keeping the section frozen after it is cut. Then the section is carried still frozen to the photographic emulsion. As the section thaws it produces a bit of moisture and thereby adheres to the emulsion.

Bone must *not* be decalcified for this process. In some of the best techniques for bone, methacrylate or epoxy embedding is used before sectioning (Arnold and Jee 1954*a,b;* Norris and Jenkins 1960; and Woodruff and Norris 1955).

If the isotope leaches out slowly and quantitative results are not required, paraffin embedding can be undertaken, but keep the periods in fixative and alcohol solutions to a minimum. Butyl alcohol extracts less of some isotopes (^{125}I and ^{131}I) than does isopropyl or ethyl alcohol. Use xylene, not toluene, for clearing before paraffin infiltration. A brownish discoloration of the sections indicates improper infiltration with paraffin. Vacuum infiltration will correct this, but if vacuum is not available when the sections are mounted on slides later, leave them on a hot plate or run them through a Bunsen burner just long enough to melt the paraffin (p. 66). Check for leaching by counting the reagents after use in an appropriate counter. If a significant count is given off, avoid the use of that reagent. When the tissue is embedded in paraffin, leaching is no longer a problem.

Slides used for mounting must be clean; dust can cause air bubbles and distortion of the emulsion. To clean, soak the slides in cleaning solution (p. 570). Wash them overnight in running tap water; one hour in distilled water, changing the solution several times; and in 2 changes of 80% alcohol. Dry. They also can be cleaned by boiling in Alconox. Treat with sub solution (p. 550) and dry. Store the subbed slides in dustfree boxes and use within a month. In the darkroom, frosted ends on slides help to determine the side holding the sections. Add identifying code with Labink.[3] Other marks may wash off in solutions. Slides with frosted ends can be used and identification applied with pencil.

Mount the sections, drain off the water, dry for 10–15 minutes on a slide warmer, and store in dustfree boxes until application of the emulsion. Subbed slides do not require as long a drying period as do albumenized slides, and the sections and emulsion adhere better to the subbed surface than to albumen. (*Suggestion:* Mount subbed slides from a water bath or place a puddle of water on slide before adding sections. Do not try to place sections on dry subbed slides and then add water.)

Sometimes staining is used before emulsion application, but many stains are affected by photographic processing and some stains may remove the tracer. If staining is done first, a protective coating (celloidin, Formvar, or nylon) usually has to be applied to prevent decolorization during development. This layer will reduce the amount of radiation that reaches the emulsion from weak emitters, e.g., tritium and ^{14}C. Feulgen stain is commonly used before emulsion is applied, but most staining is done after development.

---

[3] Black no. 6249, Arthur H. Thomas.

Emulsion Application

The two methods of application are the "stripping film" and the "liquid emulsion" or "dipping" methods. Stripping film is a little more difficult to master than dipping, but it produces a more uniform emulsion. The dipping technique is given in detail here, since it is simpler and less likely to go wrong.

Stripping Film Method

Stripping film emulsion is removed from its glass plate. First, cut the edges of the emulsion with a razor blade, peel off the emulsion, and float it on distilled water with 1% Dupanol (wetting agent) added. Then slip the slide with the section uppermost under the emulsion and lift it out, with the emulsion. To hold the emulsion in place without slipping, fold it under three sides of the slide, dry it in place, and make the exposure (Bogoroch 1951; Pelc 1956; Simmel 1957).

Liquid Emulsion Method

1. Store liquid emulsion at 5°C; 4 months is maximum time. Remove to room temperature, but do not open until in darkroom. The emulsion is a gel at room temperature and has to be melted in a water bath at 42–45°C. This must be done in the darkroom under Wratten series no. 1 safelight, red filter, 15 watt bulb. Work no closer than 3 feet from light.
2. When emulsion is melted, pour the amount to be used into the dipping vessel. Never reuse emulsion. For a few slides a coplin jar is convenient; if it is filled to the top of the slots, the depth will be sufficient for most of the slides. Leave at 42–45°C for 30–45 minutes to attain proper fluidity and to allow bubbles to escape.
3. Test each new bottle of emulsion. Dip a blank slide slowly into the emulsion and withdraw it with a slow uniform motion (4–5 seconds total). Check it for lumpiness, uneven coating, and bubbles. Drain against tissue or gauze. Wipe the back dry and air dry in a vertical position in a light-tight box: 1–2 hours. Develop and fix (see below) and check for background. If the background is objectionable, see Caro (1964) and Brenner (1962) for ways of reducing it.
4. Squashes, spreads, etc. can be dipped as is, but paraffin must be removed from sections. Use 2 changes of xylene, remove the xylene

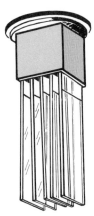

Figure 25-1
Slide holder for autoradiograph processing.

with 2 changes of absolute alcohol, and hydrate the sections to water. If Carbowax has been used, remove it in water. The slides with
mounted tissue are dipped as described above in step 3. Two slides
can be dipped back to back. Coleman (1965) dips 5 at a time by using
a molded plastic holder[4] and completes all processing in the holder
(Fig. 25-1). To reduce cooling of the emulsion, arrange the slides
back to back in an empty coplin jar in the water bath to warm them
before dipping.

5. Drain and separate slides if they have been dipped back to back.
   Slides in plastic holder can be stood on end, holder uppermost, to
   drain. Prescott (1964) arranges them in a neoprene-coated test tube
   rack, where they fit diagonally across the square openings. Place in a
   light-tight box for drying overnight. Allow them to dry slowly. Do
   not use hot air; fast drying increases the background (Sawicki and
   Pawinska 1965), and 40–50% humidity is desirable. An 80% relative
   humidity incubator for slow drying is recommended by Kopriwa and
   Leblond (1962).

   For supporting slides while drying, I use lengths (12–15'') of wood
   or plastic ($1\frac{1}{2}$–2'' square) lying on large blotters (Fig. 25-2). The slides
   lean in a vertical position, backside against the wood, and drain on
   the blotter. For a small number of slides, the side of a closed wooden
   or plastic slide box is a satisfactory back rest. A cool darkroom with
   water running in the sink usually provides an adequate drying atmosphere, but must be completely dark.

6. The slides must be dry before they are boxed for exposure, and they
   must be kept dry, because a wet emulsion is practically insensitive.
   When dry, transfer them to bakelite slide boxes (25 slide capacity)

---

[4] Peel-A-Way plastic slide grip, Peel-A-Way Scientific.

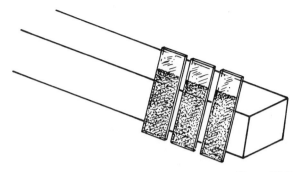

Figure 25-2
Slides drying while resting against a plastic or wood block
or slide box.

no more than 8–9 to a box, allowing 2–3 slots of spacing between
slides. Place 15–25 g of Drierite in a tissue or gauze bag in the
bottom of the box. Do not allow Drierite to get on the emulsion.
Sikov (1965) recommends an encapsulated desiccant, Humi-caps,[5]
that can be rejuvenated in a 60–70°C oven or discarded after use.
Seal the four sides of the box with two lengths of black photographic
tape and wrap in aluminum foil. Store in refrigerator, 4°C during
exposure. Keep the boxes standing on end with the slides in horizontal position, emulsion side up.

Slides in a molded plastic holder (Fig. 25–1) can be wrapped in
several layers of heavy foil for exposure.

7. At intervals of 1, 2, 3, or more weeks, test slides can be removed
   from the exposure box and be developed. If the slides are cold,
   allow them to reach darkroom temperature before development. All
   solutions should be of uniform temperature, 17–18°C. At temperature above 20°C, reticulation occurs in the emulsion.

   a. Develop in D11, D19, or Dektol: 5 minutes.
   b. Rinse in water or treat in a stop-bath of 1% acetic acid: 10
      seconds.
   c. Transfer to Kodak Rapid Fixer: 5 minutes. (Some stains react
      better with the tissue if a hardener is not included in the fixer.)
   d. Wash in running water: 10–15 minutes.
   e. Rinse in distilled water.
   f. Dry, store in dustfree box, or proceed to stain.

General references for autoradiography are Boyd (1955); Caro (1964);
Controls for Radiation, Inc., instruction manual; Fitzgerald et al. (1953);

---

[5] Dri-Aire.

Joftes (1959); Joftes and Warren (1955); Pelc (1958); Perry (1964); Prescott (1964); Sacks (1965).

'For dry mounting of water-soluble materials, see Branton and Jacobson (1961); Fitzgerald (1961); Gallimore et al. (1954); Miller et al. (1964); Sterling and Chichester (1956); Stumpf and Roth (1964).

For double emulsion techniques, see Baserga (1961); Baserga and Nemeroff (1962); Pickworth et al. (1963).

See Durie and Salmon (1975) for high-speed scintillation autoradiography.

Staining

For most experimental work, I find that 5–10 minutes staining with Mayer hematoxylin or Kernechtrot provides satisfactory backgrounds for ARGs. The red of Kernechtrot is a good contrast color for the black of the developed silver grains. Sams and Davies (1967) add Alcian blue and tartrazine to their staining sequence for additional colors. Gude stain (below) is excellent. Keep all solutions cool; never use warm ones.

After mounting ARGs in synthetic resin, dark opaque areas occasionally develop. The resin, instead of penetrating the emulsion, is drawing the clearing agent out of the tissue and leaving it opaque. If this happens, return the slides to the clearing agent (xylene or toluene) long enough to remove all the mounting resin, and the tissues will look clear again. Then transfer the slides through the following solutions:

1. Xylene/resin (2 : 1) for 1–2 hours.
2. Xylene/resin (1 : 1) for 1–2 hours.
3. Xylene/resin (1 : 2) for 1–2 hours.

Mount.

The slides can be left overnight in any of the solutions.

Giemsa Stain (Gude et al. Modification 1955)

PROCEDURE

1. Air dry slides of smears, sections, chromosomes, etc.
2. Stain 1–2 hours in freshly mixed:

Giemsa stock solution	2.5 ml
absolute methyl alcohol	3.0 ml

distilled water	100.0 ml
$M/10$ citric acid	11.0 ml
$M/5$ disodium phosphate	6.0 ml

Use only once.

3. Rinse off stain in distilled water.
4. Air dry and cover with synthetic medium and cover glass.

RESULT

The light blue staining of nuclear material furnishes a good contrast for the black-silver grains of the chromatin tag.

COMMENTS

If this stain is used on tissue sections, follow step 3 with 3 changes isopropanol, 2–3 minutes each; xylene-isopropanol, 2–3 minutes; xylene, 2 changes, 2–3 minutes each (xylene-resin mixes if necessary); and mount. I recommend Gurr's stain as the best Giemsa for ARGs.

Excess emulsion can be removed from the back of slides with a razor blade.

Other stains may be used: Darrow red–light green (Wolberg 1965); methylene blue–basic fuchsin (Bélanger 1961); nuclear fast red–indigo carmine (Mortreuil-Langlois 1962); peroxidase (Popp et al. 1962); toluidine blue–methyl green–pyronin (Stone and Cameron 1964); metanil yellow–hematoxylin (Simmel et al. 1951). Celestin blue and gallocyanin have been recommended, but Deuchar (1962) reports a dense overall graining of film with celestine blue, and Stenram (1962) found that gallocyanin causes a loss of silver grains. See Thurston and Joftes (1963) for a review of staining.

The Use of an Isotope

The isotope in popular use for small cytological structures is tritium (^{3}H). It emits only β radiation, and in water the maximum range of its emitted electrons is 3μ. If an emulsion lies over a point-like source containing tritium, only an area averaging 3μ in diameter will be exposed. Tritium, therefore, has become the isotope of choice for tagging chromosomes.

Mitotic chromosomes cannot be labeled, since DNA is not replicated during mitosis. But mitosis is preceded by the synthesis of DNA, and if radioactive DNA precursor is administered during synthesis the nucleus becomes radioactive and retains the tag during mitosis. It also passes the tag on to its daughter cells. Thymidine (deoxyriboside of thymine, one of the bases of DNA) can be effectively incorporated into DNA, and when it is labeled with tritium it enters the synthetic chain to label the DNA. To obtain efficiently labeled metaphases, therefore, the radioactive thymidine must be given to the cells several hours before they divide, that is, during the DNA-synthetic or S-period. This period varies but is species-specific. Different types of cultures may influence the length of periods—the S-period and the lag period that follows. The amount of thymidine influences the rate of uptake, and incubation temperature is critical. It is necessary to explore all these influences when first using thymidine on cultured cells.

Chromosome Cultures

PROCEDURE

1. Add tritiated thymidine[6] to a final concentration of 1 μc/ml (specific activity 1.9 curies/mmole): 5–6 hours before termination of the culture. Maintain culture at 37°C.
2. Harvest cells at 3, 4, 5, and 6 hours (also 7 and 8 if desired) after introduction of thymidine, but add colchicine or colcemid 2 hours before each harvesting.
3. The thymidine is removed with the medium. Treat with hypotonic solution: 10 minutes.
4. Fix, squash, or spread as usual (p. 484). Stain some preparations if desired.
5. Dried spreads or squashes can be stored at 4°C for about 4 weeks. At room temperature background develops.
6. Coating or stripping techniques may be done with NTB2 (Kodak) or Ilford K5 emulsions.
7. Exposure 2 days.
8. Develop: 5 minutes. Rinse, fix, and wash.
9. Dry and stain in Gude's Giemsa.

COMMENTS

Caution: Handle the tritium thymidine with care. Since it does not emit γ radiation, contamination cannot be easily detected with scintillation coun-

---

[6] Volk Radiochemical Co.; Nuclear Chicago Corp.; New England Nuclear Corp.; Schwarz BioResearch Inc.

ters. Work with the labeled thymidine only on protected areas, covered with "hot mats"[7] or a large sheet of heavy aluminum foil or plastic under a large blotter.

For details about the use of tritiated thymidine, see Frøland (1965); Leblond et al. (1959); Schmid (1965); and Sisken (1964).

---

[7] 19 × 23 paper towels with neoprene backing (300 to package), B. H. Jordan Co., or protective absorbent paper (polyethylene backing), rolls or sheets, New England Nuclear.

Part IV
SOLUTION PREPARATION AND
GENERAL LABORATORY AIDS

Chapter 26
Solution Preparation

ABBREVIATIONS AND TERMS

Laboratory solutions, defined in terms of the methods of measurement used in their preparation, can be grouped as molecular solutions, normal solutions, and percentage solutions. In discussing these the following abbreviations are used:

ml = milliliter	mg = milligram
cc = cubic centimeter	M = molecular solution
g = gram	N = normal solution

Molecular Solutions

A molecular solution (M) contains the molecular weight in grams of the substance made up to 1 liter with distilled water (Cowdry 1952). For example: M oxalic acid $(COOH)_2 \cdot 2H_2O$ is 126 (molecular weight) grams in 1 liter of water.

Molecular weight expressed in grams is called the gram-molecular weight or mole. A millimole is 1/1000 of a mole.

Normal Solutions

A normal solution (N) contains 1 gram-molecular weight of dissolved substance divided by the hydrogen equivalent of the substance (that is, 1 gram equivalent) per liter of solution. N oxalic acid is half the concentration of the M solution above.

Percentage Solutions

In the preparation of percentage solutions, it should be made clear whether the percentage is determined by weight or volume. The percentage should be written out in grams and milliliters, or it should be expressed as follows:

w/v = weight in grams in a 100 ml volume of solution
v/v = volume in milliliters in a 100 ml volume of solution

Although it is erroneous, a long-established practice of technicians is to dilute liquids as though the reagent solution was 100% concentration. In most procedures the correct dilution is indicated in parentheses following the percentage desired. That is, a 1% solution of acetic acid is 1 ml of glacial acetic acid in 99 ml of distilled water.

The table below gives dilutions for 1 N solutions which are sufficiently accurate for histological preparations. To make a normal solution, add

Dilutions for 1 N Solutions*

Reagent	Molecular weight	Percentage assay	Specific gravity	g per liter	ml per liter
acetic acid	60.05	99.7– *100*†	1.0498	1050	57.2
(CH₃COOH)		99.0	1.0524	1042	57.6
		98.0	1.0549	1034	58.0
ammonium	17.03	26	0.904	235	72.4
hydroxide		28†	0.8980	251.4	67.7
(NH₄OH)		30	0.8920	267.6	63.63
formic acid	46.03	96	1.217	1180	39.3
(HCOOH)		98†	1.2183	1194	38.4
		99	1.2202	1208	38.0
		100	1.2212	1221	37.6
hydrochloric acid	36.46	36	1.1789	424.4	85.9
(HCl)		37– *38*†	1.188– 1.192	451.6	80.4
		40	1.1980	479.2	76.0
nitric acid	63.02	69	1.4091	972.3	64.8
(HNO₃)		70†	1.4134	989.4	63.6
		71	1.4176	1006	62.6
		72	1.4218	1024	61.5
sulfuric acid	98.075	95	1.8337	1742	28.1
(H₂SO₄)		96†	1.8355	1762	27.8
		97	1.8364	1781	27.5

  * Calculated from Norbert Adolph Lange, *Handbook of Chemistry* (Sandusky, Ohio: Handbook Publishers, Inc., 1956). *Italicized* figures were used in calculations.
  † Indicates most common form of the reagent. In addition, a calculation has been made for the nearest lower and higher percentages.

enough distilled water to the number of ml shown in the right-hand column to make a combined total of 1 liter.

STOCK SOLUTIONS

Solutions normally found on histology laboratory shelves are referred to as stock solutions. Included here are solutions for general use, physiological solutions, environmental solutions, a few adhesive and buffer solutions, and some stain solubilities.

General-Purpose Solutions

Acid Alcohol

70% ethyl alcohol	100.0 ml
hydrochloric acid, concentrated	1.0 ml

Alkaline Alcohol

70% ethyl alcohol	100.0 ml
ammonia, concentrated	1.0 ml

Or use saturated aqueous sodium or lithium bicarbonate.

Carbol-Xylol

phenol (carbolic acid) melted	1 part
xylene	3 parts

During use, keep covered to reduce evaporation of the xylene.

For creosote-xylol use beechwood creosote in place of phenol. For aniline-xylol use aniline.

Gold Chloride Stock Solution

gold chloride (15 grains).....................	1.0 g
distilled water.............................	100.0 ml

Use only 3–4 times; gold content is reduced at each use.

Lugol Solution

There are various formulas to which this name has been applied; the following are frequently used.

The strongest concentration

iodine crystals	1.0 g
potassium iodide	2.0 g
distilled water..........................	12.0 ml

Weigert variation

iodine crystals	1.0 g
potassium iodide	2.0 g
distilled water..........................	100.0 ml

Gram variation

iodine crystals	1.0 g
potassium iodide	2.0 g
distilled water..........................	300.0 ml

For all solutions, first dissolve the potassium iodide, then the iodine will go into solution readily.

Chatton Agar for Blocking Small Organisms (Gray 1954)

agar	1.3 g

Add agar to:

boiling distilled water	100.0 ml

Stir until dissolved, then add:

formalin, concentrated	2.5 ml

Store in refrigerator.

Scott Solution

sodium bicarbonate	2.0 g
magnesium sulfate	20.0 g
distilled water	1000.0 ml

Add a pinch of thymol to retard molds.

Sodium Thiosulfate Solution (Hypo), 5%

sodium thiosulfate ($Na_2S_2O_3$)	5.0 g
distilled water	100.0 ml

Physiological Solutions (Balanced Salt Solutions, BSS)

Several physiological solutions should be available to satisfy the needs of different types of research. These solutions are useful for land vertebrates and can be imperative at times. A normal or *isotonic* solution is one which contains the right amount of salts to maintain tissues in a normal condition. For example, red blood cells in isotonic fluid will remain unaltered in form and will not lose their hemoglobin, because the osmotic pressure and salt content of both the solution and the blood fluid are the same. If the solution is *hypotonic,* the blood cells will swell, because the osmotic pressure and salt content are less than that of the blood fluid. If the solution is *hypertonic,* the blood cells will shrink, because the osmotic pressure and salt content are greater than that of the blood fluid. For example, distilled water is hypotonic without the addition of the correct quantity of salt.

Physiological Saline

mammals, 0.9% (9 g NaCl/1000 ml water)
birds, 0.75% (7.5 g NaCl/1000 ml water)
salamanders, 0.8% (8 g NaCl/1000 ml water)
frogs, 0.64% (6.4 g NaCl/1000 ml water)
insects, 0.6–0.8% (6–8 g NaCl/1000 ml water)
invertebrates, 0.75% (7.5 g NaCl/1000 ml water)

Earle Solution

sodium chloride	0.68 g
calcium chloride	0.02 g
magnesium sulfate	0.01 g
potassium chloride	0.04 g
sodium bicarbonate	0.014 g
sodium phosphate, monobasic	0.22 g
glucose	0.1 g
distilled water	100.0 ml

Hank Solution

	(1)		(2)	
sodium chloride	0.8	g	0.8	g
calcium chloride	0.02	g	0.014	g
magnesium sulfate	0.02	g	0.02	g
potassium chloride	0.04	g	0.04	g
potassium phosphate, monobasic	0.01	g	0.006	g
sodium bicarbonate	0.127	g	0.035	g
sodium phosphate, dibasic	0.01	g	0.006	g
glucose	0.2	g	0.1	g
distilled water	100.0	ml	100.0	ml

If formula 1 gives too strong a solution, try formula 2.

Locke Solution

	(1)		(2)	
sodium chloride	0.9	g	0.95	g
potassium chloride	0.042	g	.02	g
sodium bicarbonate	0.03	g	.02	g
calcium chloride	0.024	g	.02	g
glucose	0.1	g	0.1	g
distilled water	100.0	ml	100.0	ml

Which of the two formulas is used is a matter of personal preference. In both, decrease sodium chloride content to 0.65 g for cold-bloods; to 0.85 g for birds.

Ringer Solution

sodium chloride	0.9 g
potassium chloride	0.042 g
calcium chloride	0.025 g
distilled water............................	100.0 ml

Best prepared fresh. For cold-bloods, use 0.65 g of sodium chloride; for birds, 0.85 g.

Environmental Solutions

Just as physiological solutions are useful for land vertebrates, environmental solutions are convenient for some invertebrates. In the case of marine invertebrates, such solutions can become necessary for successful preparation and sometimes are recommended as the basic fluid for fixatives. Among the physical properties of sea water, the salt content must be considered; it varies and is complex, but an arbitrary definition of salt content (salinity) has been calculated. Salinity (S 0/00) equals the weight in grams (in vacuo) of solids in 1 kilogram of sea water. The major constituent, which is easily determined, is a silver-precipitating halide. Chlorinity (Cl 0/00) is defined as the weight in grams (in vacuo) of chlorides in 1 kilogram of sea water. Standard sea water is about 34.3243 0/00 salinity and 19.4 0/00 chlorinity (Barnes 1954).

Artificial Sea Water (Hale 1958)

Chlorinity, 19.0/00; salinity, 34.33 0/00. The weights in the right-hand column are for the anhydrous form of the salt; those in parentheses include crystallization water as indicated on the reagent bottle.

$NaCl$......................................		23.991 g
KCl.......................................		0.742 g
$CaCl_2$		1.135 g
($CaCl_2 \cdot 6H_2O$	2.240 g)	
$MgCl_2$		5.102 g
($MgCl_2 \cdot 6H_2O$	10.893 g)	
Na_2SO_4		4.012 g
($Na_2SO_4 \cdot 10H_2O$	9.1 g)	
$NaHCO_3$		0.197 g

NaBr .... 0.085 g
    ($NaBr \cdot 2H_2O$ .... 0.115 g)
$SrCl_2$ .... 0.011 g
    ($SrCl_2 \cdot 6H_2O$ .... 0.018 g)
H_3BO_3 .... 0.027 g

Dissolve in enough distilled water to make 1 liter.

This solution is to be used only for technical purposes, not for aquaria, for which a proper pH condition is critical and must be maintained (Kelley 1965).

Synthetic Spring Water (Kirby 1947)

	(1)	(2)
Na_2SiO_3	15.0 mg	100.0 mg
NaCl	12.0 mg	12.0 mg
Na_2SO_4	6.0 mg	6.0 mg
$CaCl_2$	6.5 mg	6.5 mg
$MgCl_2$	3.5 mg	3.5 mg
$FeCl_3$	4.0 mg	4.0 mg
distilled water	1000.0 ml	1000.0 ml

Adjust with HCl to pH 6.8 to 7.0.

Use formula 1 for first transfer from nature; for succeeding transfers use 2.

Adhesive Solutions

Not all of the adhesive or buffer solutions will be found or required in every laboratory; however, it is practical to keep formulas of the more widely used solutions at hand, so that the information is available if it is needed.

Mayer Albumen Fixative

Beat white of an egg until well broken up, but not stiff, with egg beater and pour into tall cylinder. Let stand until the air brings suspended material to the top (overnight). Pour off liquid from bottom and to it add an equal volume of glycerin. A bit of sodium salicylate, thymol, merthiolate, or formalin (1:100) will prevent growth of molds. Many technicians filter their solution through glass wool, but I have found this of no advantage.

Faulkner and Lillie (1945*b*) substitute dried egg white. A 5% solution of dried egg white in 0.5% NaCl is shaken at intervals for one day. Do not allow it to froth. Filter in a small Büchner funnel with vacuum. Add an equal amount of glycerin and 0.5 ml of 1 : 10,000 merthiolate to each 100 ml of solution.

Masson Gelatin Fixative

gelatin	50.0 mg
distilled water	25.0 ml

This is recommended by many for alkaline silver techniques, when sections tend to float off during or after impregnation. Float sections on solution on slide, and place on warm plate. When sections have spread, drain excess gelatin and blot dry with filter paper. Place in formalin vapor, 40–50°C: overnight.

Haupt Gelatin Fixative (1930)

gelatin	1.0 g
distilled water	100.0 ml

Dissolve at 30°C (not above) in water bath or oven. Add:

phenol (carbolic acid) crystals	2.0 g
glycerol	15.0 ml

Stir well and filter.

Use 2% formalin when mounting sections. This hardens the gelatin; water is not adequate. Some may find the formalin fumes irritating to the eyes and nostrils.

Haupt suggests that if sections tend to loosen, a small uncovered dish of concentrated formalin be placed in an oven with the slides while they are being dried. The formalin tends to make the gelatin insoluble and helps to hold the sections in place.

Bissing's (1974) method reduces the formalin fumes. Combine 1 part of above adhesive with 140 parts of 3% formalin (prepare in distilled water). Flood clean slides with the mixture and float the sections on it. Spread on a warm plate, drain off fluid, and dry for 1 hour.

Weaver Gelatin Fixative (1955)

Solution A

gelatin	1.0 g
calcium propionate[1]	1.0 g
Roccal (1% benzalkonium chloride)[2]	1.0 ml
distilled water	100.0 ml

Solution B

chrome alum [$Cr_2K_2(SO_4)_4 \cdot 24H_2O$]	1.0 g
distilled water	90.0 ml
formalin	10.0 ml

Mix 1 part of A with 9 parts of B. Flood slide with adhesive mixture, add paraffin ribbons, and allow to stretch as usual. Drain off excess adhesive and blot. Wipe edges close to paraffin. Deposits of adhesive should be removed because they pick up stain. Good for sections that are difficult to affix.

Blood Serum Fixative (Priman 1954)

fresh human blood serum (only from negative Wassermann's)	15.0 ml
distilled water	10.0 ml
formalin, 5%	6.0 ml

Filter through filter paper and use same as albumen fixative. Does not stain and sections do not loosen.

"Subbed" Slides (Boyd 1955)

Dissolve 1 g of gelatin in 1 liter of hot distilled water. Cool and add 0.1 g of chromium potassium sulfate. Store in refrigerator. Dip slides several times in the solution. Drain and dry in a vertical position. Store in dustfree box.

Papas (1971) uses 5 g gelatin and 0.5 g chrome alum, but I have not found the added strength is necessary.

---

[1] Mycoba, E. I. du Pont de Nemours and Co.
[2] Rohm & Haas.

Land Adhesive (Gray 1954)

Solution A

| gum arabic | 0.5 g |
| distilled water | 50.0 ml |

Solution B

| potassium dichromate | 0.5 g |
| distilled water | 100.0 ml |

Smear a small amount of A on a slide. Flood slide with solution B, add
sections, and warm to flatten. Drain off fluid against cleansing tissue and
dry. Bilstad[3] dries slides in bright sunlight for several hours; she found it
more effective than artificial light.

Buffer Solutions[4]

The following are a few of the chemicals most commonly used for buffer
solutions. For others consult a chemical handbook.

Molecular Weights of Buffer Ingredients

acetic acid, CH_3COOH	60.05
borax, $Na_2B_4O_7 \cdot 10H_2O$ (sodium tetraborate)	381.43
boric acid, $B(OH)_3$	61.84
citric acid (anhydrous), $C_3H_4(OH)(COOH)_3$	192.12
citric acid crystals, $C_3H_4(OH)(COOH)_3 \cdot H_2O$	210.14
formic acid, $HCOOH$	46.03
hydrochloric acid, HCl	36.465
maleic acid, $HOOCCH\text{-}CHOOH$	116.07
potassium acid phosphate, KH_2PO_4	136.09
sodium acetate, CH_3COONa	82.04
sodium acetate, crystals, $CH_3COONa \cdot 3H_2O$	136.09
sodium barbital, $C_8H_{11}O_3N_2Na$	206.18
sodium bicarbonate, $NaHCO_3$	84.02
sodium carbonate, $NaCO_3$	106.00

---

[3] Personal communication.

[4] The 8-color Panpeha pH indicator strips purchased from The Presto Co. are extremely
useful when a laboratory does not possess a pH meter.

Molecular Weights of Buffer Ingredients

sodium citrate, crystals, $C_3H_4OH(COONa)_3$ $\begin{cases} \cdot 5H_2O \\ 5\tfrac{1}{2}H_2O \end{cases}$		348.17 357.18
sodium citrate, granular, $C_3H_4OH(COONa)_3 \cdot 2H_2O$		294.12
sodium hydroxide, NaOH		40.005
sodium phosphate, monobasic, $NaH_2PO_4 \cdot H_2O$		138.01
sodium phosphate, dibasic, Na_2HPO_4 $\begin{cases} 7H_2O \\ \text{anhydrous} \end{cases}$		268.14 141.98
sulfuric acid, H_2SO_4		98.082
tris (hydroxymethyl) aminomethane $C_4H_{11}NO_3$		121.14

0.2 *M* Acetate Buffer (Gomori)

STOCK SOLUTIONS

Acetic acid

12.0 ml made up to 1000 ml with distilled water.

Sodium acetate

27.2 g made up to 1000 ml with distilled water.

Add a few crystals of camphor to both solutions. For desired pH mix correct amounts as indicated below:

pH	Acetic acid (ml)	Sodium acetate (ml)
3.8	87.0	13.0
4.0	80.0	20.0
4.2	73.0	27.0
4.4	62.0	38.0
4.6	51.0	49.0
4.8	40.0	60.0
5.0	30.0	70.0
5.2	21.0	79.0
5.4	14.5	85.5
5.6	11.0	89.0

Acetate–Acetic Acid Buffer (Walpole)

STOCK SOLUTIONS

M/5 acetic acid

12.0 ml (99% assay) made up to 1000 ml with distilled water.

M/5 sodium acetate

27.2 g made up to 1000 ml with distilled water.

For desired pH mix correct amounts as indicated below.

pH	M/5 acetic acid (ml)	M/5 sodium acetate (ml)
2.70	200.0	0.0
2.80	199.0	1.0
2.91	198.0	2.0
3.08	196.0	4.0
3.14	195.0	5.0
3.20	194.0	6.0
3.31	192.0	8.0
3.41	190.0	10.0
3.59	185.0	15.0
3.72	180.0	20.0
3.90	170.0	30.0
4.04	160.0	40.0
4.16	150.0	50.0
4.27	140.0	60.0
4.36	130.0	70.0
4.45	120.0	80.0
4.53	110.0	90.0
4.62	100.0	100.0
4.71	90.0	110.0
4.80	80.0	120.0
4.90	70.0	130.0

pH	M/5 acetic acid (ml)	M/5 sodium acetate (ml)
4.99	60.0	140.0
5.11	50.0	150.0
5.22	40.0	160.0
5.38	30.0	170.0
5.57	20.0	180.0
5.89	10.0	190.0
6.21	5.0	195.0
6.50	0.0	200.0

0.05 *M* Barbital Buffer (Gomori)

STOCK SOLUTION

Sodium barbital

1.03 g in 50 ml distilled water.

To 50 ml sodium barbital add 0.1 *N* HCl according to the table below to obtain desired pH; dilute to total of 100 ml.

pH	0.1 N HCl (ml)
8.7	5.0
8.5	7.5
8.3	11.0
8.1	15.0
7.9	19.0
7.65	26.0
7.45	31.0
7.3	36.0
7.15	41.0
6.9	43.5

Boric Acid–Borax Buffer (Holmes)

STOCK SOLUTIONS

M/5 boric acid

12.368 g made up to 1000 ml with distilled water.

M/20 borax

19.071 g made up to 1000 ml with distilled water.

For desired pH mix correct amounts as indicated below.

pH	M/5 boric acid (ml)	M/20 borax (ml)
7.4	90.0	10.0
7.6	85.0	15.0
7.8	80.0	20.0
8.0	70.0	30.0
8.2	65.0	35.0
8.4	55.0	45.0
8.7	40.0	60.0
9.0	20.0	80.0

0.2 *M* Phosphate Buffer (Gomori)

STOCK SOLUTIONS

Monobasic sodium phosphate

27.6 g made up to 1000 ml with distilled water.

Dibasic sodium phosphate

53.6 g made up to 1000 ml with distilled water.

For desired pH mix correct amounts as indicated below.

pH	Monobasic sodium phosphate (ml)	Dibasic sodium phosphate (ml)
5.9	90.0	10.0
6.1	85.0	15.0
6.3	77.0	23.0
6.5	68.0	32.0
6.7	57.0	43.0
6.9	45.0	55.0
7.1	33.0	67.0
7.3	23.0	77.0
7.4	19.0	81.0
7.5	16.0	84.0
7.7	10.0	90.0

Phosphate Buffer (Sörensen)

STOCK SOLUTIONS

M/15 dibasic sodium phosphate

9.465 g made up to 1000 ml with distilled water.

M/15 potassium acid phosphate

9.07 g made up to 1000 ml with distilled water.

For desired pH mix correct amounts as indicated on the facing page: some techniques will require dilution with up to 1000 ml of water.

pH	M/15 dibasic sodium phosphate (ml)	M/15 potassium acid phosphate (ml)
5.29	2.5	97.5
5.59	5.0	95.0
5.91	10.0	90.0
6.24	20.0	80.0
6.47	30.0	70.0
6.64	40.0	60.0
6.81	50.0	50.0
6.98	60.0	40.0
7.17	70.0	30.0
7.38	80.0	20.0
7.73	90.0	10.0
8.04	95.0	5.0

Standard Buffer (McIlvaine)

STOCK SOLUTIONS

0.1 M citric acid (anhydrous)

19.212 g made up to 1000 ml with distilled water.

0.2 M disodium phosphate (anhydrous)

28.396 g ($\cdot 7H_2O$, 53.628 g) made up to 1000 ml with distilled water.

For desired pH mix correct amounts as indicated below.

pH	Citric acid (ml)	Disodium phosphate (ml)
2.2	19.6	0.4
2.4	17.76	1.24
2.6	17.82	2.18
2.8	16.83	3.17
3.0	15.89	4.11
3.2	15.06	4.94
3.4	14.3	5.7
3.6	13.56	6.44
3.8	12.9	7.1
4.0	12.29	7.71
4.2	11.72	8.28
4.4	11.18	8.82
4.6	10.65	9.35
4.8	10.14	9.86
5.0	9.7	10.3
5.2	9.28	10.72
5.4	8.85	11.15
5.6	8.4	11.6
5.8	7.91	12.09
6.0	7.37	12.63
6.2	6.78	13.22
6.4	6.15	13.85
6.6	5.45	14.55
6.8	4.55	15.45
7.0	3.53	16.47
7.2	2.61	17.39
7.4	1.83	18.17
7.6	1.27	18.73
7.8	0.85	19.15
8.0	0.55	19.45

0.2 *M* Tris Buffer (Hale)

STOCK SOLUTIONS

0.2 M tris (hydroxymethyl) aminomethane

24.228 g made up to 1000 ml with distilled water.

0.1 N HCl (38% assay)

8.08 ml made up to 1000 ml with distilled water.

To 25 ml 0.2 *M* tris add 0.1 *N* HCl as indicated in the table below and dilute to 100 ml.

pH	0.1 N HCl (ml)
7.19	45.0
7.36	42.5
7.54	40.0
7.66	37.5
7.77	35.0
7.87	32.5
7.96	30.0
8.05	27.5
8.14	25.0
8.23	22.5
8.32	20.0
8.41	17.5
8.51	15.0
8.62	12.5
8.74	10.0
8.92	7.5
9.1	5.0

Tris Maleate Buffer (Gomori)

maleic acid	29.0 g
tris (hydroxymethyl) aminomethane	30.3 g
distilled water.............................	500.0 ml

Add 2 g charcoal, shake, let stand 10 minutes, and filter.

To 40 ml of stock solution add N NaOH (4%) as indicated below and dilute to 100 ml.

pH	Sodium hydroxide (ml)
5.8	9.0
6.0	10.5
6.2	13.0
6.4	15.0
6.6	16.5
6.8	18.0
7.0	19.0
7.2	20.0
7.6	22.5
7.8	24.2
8.0	26.0
8.2	29.0

STAIN SOLUBILITIES

Stain solubilities are useful only occasionally, but they are definitely needed when a saturated solution must be made. Solubilities of different batches of dyes can vary, but the table on the next three pages can be consulted when preparing saturated solutions. Certain dyes whose solubilities were not available are not included, such as pinacyanole (b), Darrow red (b), Sirius supra blue (a), garnet GB (b), and the azures (b). Gallocyanin is considered an amphoteric dye. For complete details concerning these dyes, see Conn, *Biological Stains* (1969).

Stain Solubilities

Stain	In water (percent)		In absolute ethyl alcohol (percent)		Basic or acidic	C.I. number
	26°C*	15°C†	26°C*	15°C†		
acid alizarine blue	—	1.0	—	0.75	a	63015
acid fuchsin	—	45.0	—	3.0	a	42685
acridine orange	—	5.0	—	0.75	b	46005
acridine yellow	—	0.5	—	0.75	b	46025
acriflavine	—	15.0	—	1.0	b	46000
Alcian blue 8GX	—	9.5	—	6.0	b	74240
alizarine red S	7.69	6.5	0.15	0.15	a	58005
amido black 10B	—	3.0	—	3.15	a	20470
aniline blue WS	—	50.0	—	0.0	a	—
auramin O	0.74	1.0	4.49	4.0	b	41000
aurantia	0.0	0.1	0.33	0.55	a	10360
azocarmine G	—	1.0	—	0.1	a	50085
basic fuchsin	0.26–0.39	1.0	5.93–8.16	8.0	b	42500
Biebrich scarlet	—	5.0	0.5	0.25	a	26905
Bismarck brown Y	1.36	1.5	1.08	3.0	b	21000
brilliant cresyl blue	—	3.0	—	2.0	b	51010
celestin blue B	—	2.0	—	1.5	b	51050
chlorantine fast red	—	1.0	—	0.45	a	28160
chlorazol black E	—	6.0	—	0.1	a	30235
chromotrope 2R	19.3	19.0	0.17	0.15	a	16570
chrysoidin Y	0.86	5.5	2.21	4.75	b	11270
Congo red	—	5.0	0.19	0.75	a	22120
coriphosphine O	—	4.75	—	0.6	b	46020
cresyl violet (cresylecht violet)	0.38	9.5	0.25	6.0	b	—
crystal violet (methyl violet 10B)	1.68	9.0	13.87	8.75	b	42555
eosin B	39.11	10.0	0.75	3.0	a	45400
eosin (S) alc. sol. (ethyl eosin)	0.03	0.0	1.13	1.0	a	45386
eosin Y, WS	44.20	44.0	2.18	2.0	a	45380
erythrosin B	11.10	10.0	1.87	5.0	a	45430
erythrosin Y	—	8.5	—	4.5	a	45425
fast green FCF	16.04	4.0	0.35	9.0	a	42053
fluorescein (uranin)	50.02	50.0	7.19	7.0	a	45350
gallocyanin (bisulfite)	—	3.0	—	0.5	b	51030

* From Conn (1969).   † From Gurr (1960).

(continued)

Stain Solubilities (cont.)

Stain	In water (percent)		In absolute ethyl alcohol (percent)		Basic or acidic	C.I. number
	26°C*	15°C†	26°C*	15°C†		
gallocyanin (hypochlorite)	—	0.5	—	1.25	b	51030
hematein	—	1.5	—	7.5	b	75290
hematoxylin	1.75	10.0	60.0	10.0	b	75290
Janus green B	5.18	5.0	1.12	1.0	b	11050
light green SF yellowish	20.35	20.0	0.82	4.0	a	42095
Luxol fast blue MBS	—	0.0	—	3.0	a	—
malachite green	—	10.0	—	8.5	b	42000
martius yellow	4.57	1.0	0.16	0.0	a	10315
metanil yellow	5.36	5.0	1.45	1.5	a	13065
methyl blue	—	50.0	—	0.0	a	42780
methyl violet 2B	2.93	—	15.21	—	b	42535
methyl violet 6B	—	4.7	—	9.5	b	42535
methylene blue	3.55	9.5	1.48	6.0	b	52015
naphthol yellow S	—	12.5	—	0.65	a	10316
neutral red	5.64	4.0	2.45	1.8	b	50040
new methylene blue	13.32	—	1.65	—	b	52030
new fuchsin	1.13	1.0	3.2	8.0	b	42520
nigrosin, alc. sol.	—	0.0	—	2.5	b	50415
nigrosin WS	—	10.0	—	0.0	a	50420
Nile blue sulfate	—	6.0	—	5.0	b	51180
oil red O	—	0.0	—	0.5	a	26125
orange G	10.86	8.0	0.22	0.22	a	16230
orcein	—	2.0	—	4.2	b	—
phloxine B	—	10.5	—	5.0	a	45410
picric acid	1.18	1.2	8.96	9.0	a	10305
Poirrier blue C4B (cotton blue)	soluble	—	soluble	—	a	42780
ponceau 2R	—	5.0	—	0.1	a	16150
ponceau S	—	1.35	—	1.2	a	27195
ponceau de xylidine	—	5.0	—	0.1	a	16150
primilin	—	0.25	—	0.03	a	49000
pyronin B	—	10.0	—	0.5	b	45010
pyronin Y	8.96	9.0	0.6	0.5	b	45005
quinoline yellow	—	0.5	—	0.0	a	47005
rhodamin B	0.78	2.0	1.47	1.75	b	45170

*From Conn (1969).   † From Gurr (1960).

(continued)

Stain Solubilities (cont.)

Stain	In water (percent)		In absolute ethyl alcohol (percent)		Basic or acidic	C.I. number
	26°C*	15°C†	26°C*	15°C†		
rhodamin 6G	—	1.5	—	6.5	b	45160
safranin O	5.45	4.5	3.41	3.5	b	50240
Sudan black B	—	0.0	—	0.25	a	26150
Sudan III	0.0	0.0	0.15	0.25	a	26100
Sudan IV	0.0	0.0	0.09	0.5	a	26105
thioflavine S	—	1.0	—	0.4	a	49010
thioflavine T	—	2.0	—	1.0	b	49005
thionin	0.25	1.0	0.25	1.0	b	52000
Titian yellow	—	1.0	—	0.02	a	19540
toluidine blue O	3.82	3.25	0.57	1.75	b	52040
trypan blue	—	1.0	—	0.02	a	23850
Victoria blue B	—	4.3	—	8.25	b	44045
Victoria blue 4R	3.23	3.0	20.49	20.0	b	42563
water blue I	—	50.0	—	0.0	a	42755
wool green S	—	4.0	—	—	a	44090

* From Conn (1969).   † From Gurr (1960).

Rosin Stock Solution (Colophonium)

white wood rosin .... 10.0 g
absolute alcohol .... 100.0 ml

Rosin Working Solution

rosin stock solution .... 5.0 ml
95% ethyl alcohol .... 40.0 ml

Chapter 27
General Laboratory Aids

LABELING AND CLEANING SLIDES

All slide labels should contain complete information: the name or number of the tissue, the section thickness, the stain and date, and the fixative used.

Before ordinary slide labels are glued on, processed slides must be cleaned to prevent loosening of the labels. Dip the slides (when mountant is thoroughly hardened) in water to which has been added a small amount of ammonia and household cleanser (or glass cleaner). Wipe dry and paste on label.

Labels applied by pressure are easier to use and do not require slide cleaning. They are called Time Microscopic Slide Labels[1] and are made in a standard thickness, $\frac{7}{8}$ by $\frac{7}{8}$ inch, and a "tissue-high" thickness, $\frac{3}{8}$ by $\frac{7}{8}$ inch. The latter is used as an end label to protect the cover glass from sticking to an object laid on it.

Mounting can be untidy, and some of the resin may ooze over the cover glass. If this happens, scrape off excess resin with a razor blade when the slide is completely dry, taking care not to chip or break the cover. Do not clean away the resin too close to the glass; in fact, leave a small band of it to protect the cover from chipping or catching against an object.

---

[1] Professional Tape Co., Inc.

RESTAINING FADED SLIDES

Slides that have been standing for a long time or that have been exposed to bright light frequently fade. If they must be recovered, McCormick offers the following method of bleaching them for restaining.

McCormick Method (1959a)

1. Remove cover glass by soaking slides in xylene until cover glass slips off. Do not force it off; the sections might get torn.
2. Soak longer in xylene to make certain that all of the resin is removed.
3. Hydrate to water.
4. Treat with 0.5% potassium permanganate (0.5 g/100 ml water): 5 minutes.
5. Wash in running water: 5 minutes.
6. Bleach with 0.5% oxalic acid (0.5 g/100 ml water) until colorless: 1–2 minutes. If old stain still remains, repeat steps 4, 5, and 6.
7. Wash thoroughly in running water: 5 or more minutes.
8. Restain, using more dilute stains or staining for a shorter time. Potassium permanganate and oxalic acid make tissues especially sensitive to hematoxylin and aniline nuclear stains.

RECOVERING BROKEN SLIDES

Usually it is practical to remove the bits of broken cover glass by soaking the slide in xylene overnight or longer, until the cover glass pieces slide off unaided. Then mount the broken pieces of slide on another slide (as thin a slide as obtainable) with mounting medium, just as if mounting a cover glass. Do not allow the tissue on the slide to dry out during this process. Cover with a new cover glass and mountant. Allow to dry on warm plate overnight or several days. With a razor blade carefully clean away any mountant exuding from between the slides.

See Geil (1961) for salvaging histologic sections from broken slides.

RESTORING BASOPHILIC PROPERTIES

If tissues have been improperly fixed (unbuffered formalin, overtime in Zenker fixative), stored for long periods in a fixative, or excessively decalcified or dried, the nuclei will lose some of their basophilic properties and stain poorly. After unbuffered formalin and over decalcification, extended time in hematoxylin sometimes is sufficient. But more satisfactory results will follow treatment of the deparaffinized sections in 10% aqueous sodium bicarbonate: 6–8 hours. This treatment definitely improves the staining of Zenker-fixed tissue. Small tissue blocks can be treated before embedding in a 5–10% aqueous solution of sodium bicarbonate overnight. Wash thoroughly in running water before staining slides or processing blocks. Luna (1968) also suggests using 5% aqueous ammonium sulfide or 5% aqueous periodic acid in the same manner. I prefer the bicarbonate treatment.

TWO DIFFERENT STAINS ON ONE SLIDE

Feder and Sidman Method (1957)

This method is good for a quick checking of a tissue against two stains, side by side.

1. Hydrate sections to water.
2. Blot carefully, and while they are still moist, coat alternate sections with silicone grease. A soft brush may be used, or more efficiently use a 1 ml syringe and no. 15 needle that has been clipped off and flattened.
3. Apply first staining procedure.
4. Dehydrate and leave in xylene 15–30 minutes to remove coating. (After sections have been in xylene 5 minutes, blotting with filter paper helps to remove grease.)
5. Rehydrate; apply coating to stained sections.
6. Apply second staining procedure.
7. Repeat step 4.
8. Rinse briefly in absolute alcohol and clear in fresh xylene. Mount.

RECLAIMING AND STORING SPECIMENS

Reclaiming Biopsy Material

Graves Method (1943)

If biopsy material has dried, do not try to process it before first softening and rehydrating it. Place it in physiological saline solution for 1 hour, then fix, dehydrate, clear, and embed as usual.

See Haviland (1963) for recovery of rust-stained formalin-preserved gross tissues.

Reclaiming Dried Gross Specimens[2]

Tissues that have been stored in alcohol or formalin often become completely or partially desiccated and shrunken. If a major catastrophe arises and tissue must be reclaimed, partial recovery can be made with fair returns for microscopic identification. Van Cleave and Ross (1947) restored desiccated helminths and invertebrates to normal size by soaking them in a 0.5% aqueous solution of trisodium phosphate. Trial runs of this nature were made on dried formalin-fixed specimens for 4 hours to 30 days, depending on the hardness and size of tissue. Occasional changing of the fluid seemed to help if the pieces were exceptionally dry; also, warming in a ±40°C oven was helpful. Increasing the concentration of the phosphate did not speed recovery. Finally the tissues were washed for 30 minutes to 2–3 hours, again depending on size.

If the specimen was fixed in a mercuric chloride fixative, the trisodium phosphate was found to be ineffective until the tissue had been pretreated as follows: (1) soaked in water with enough Lugol's solution added to color the solution a deep brown: 1–2 days (if the solution became colorless, it was renewed); (2) washed in running water: 2 hours; (3) treated with 5% sodium thiosulfate: $\frac{1}{2}$–1 hour; and (4) washed: 2 hours.

Only fair results were obtained by the above methods. Considerable shrinkage remained in the cells, and the nuclei stained only lightly. The latter condition improved somewhat after mordanting in mercuric chloride or potassium dichromate.

Since recovery was based on the detergent action of trisodium phosphate, the inevitable question arose: Why not try one of the modern household detergents? A 1% solution of Trend in water was added to dried tissues and kept overnight in a paraffin oven at approximately 58°C. This was followed by 6–8 hours washing and then processing overnight in the Technicon for embedding the next morning. The results were creditable—good enough for tissue identification and some pathological reading. The tissues in this run averaged 10–25 cm in size and were fixed in either formalin or a mercuric chloride fixative. A longer stay in detergent might be required for larger pieces. No pretense is made that results are exceptional: considerable shrinkage remains in the cells, and staining is not brilliant, but it is better than after trisodium phosphate. Therefore, if

[2] Special acknowledgment is made to Robert Ingersoll Howes, Jr., for hours of effort in developing a recovery process for the Los Alamos Medical Center in 1959.

it is essential to check back on a tissue concerning its identity, a malignancy, or some other "matter of life or death," this method of recovery may save the day.

Sandison Method (1955)

Sandison treated mummified material with the following solution and prepared histological sections with good results.

95% ethyl alcohol	30 volumes
1% formalin	50 volumes
5% aqueous sodium carbonate	20 volumes

Leave tissues in solution overnight or until they become soft. Replace one-third of the fluid with 95% alcohol. Repeat until the tissues become firm. Dehydrate, clear, infiltrate, and embed.

Lhotka and Ferreira (1950) deformalizing technique will work on some tissues.

Preserving Gross Specimens in a Pliable Condition

Palkowski Method (1960)

It is economical to preserve some gross specimens, pathological ones in particular, for future teaching or demonstration purposes. Fix as soon as possible in Kaiserling solution: 3–4 hours.

Kaiserling solution

chloral hydrate (a drug)	300.0 g
potassium sulfate	6.0 g
potassium nitrate	114.0 g
sodium sulfate	54.0 g
sodium chloride	66.0 g
sodium bicarbonate	60.0 g
formalin	300.0 ml
distilled water	10,000.0 ml

Drain off excess Kaiserling and place specimens in four times their volume of 80% alcohol: 18–24 hours. Deep freeze at about $-29°C$, sealed in airtight polyethylene bags not much larger than the specimen.

Because of contained air, lungs will float on the fixative; submerge them under cotton soaked with fixative.

The specimens will keep indefinitely this way, retain their color, and become pliable after thawing. After use, they can be returned to deep-freeze with little deterioration.

Color Preservation in Gross Specimens

McCormick Fixation (1961)

formaldehyde, concentrated	50.0 ml
Sörenson's buffer, pH 6.4	950.0 ml
Prague powder[3]	8.5 g
ascorbic acid	1.7 g

Storing Gross Specimens

Sealing tissues for storage has been a serious problem. Bottled storage has risks; any seal can spring a leak, and the resulting evaporation culminates in desiccated tissue. Storing in plastic bags is a more reliable system than the use of bottles. A heavy quality of polyethylene plastic is recommended and is widely available. Lighter grades can split or unseal, whereas the heavy grade, once well sealed, almost supports itself without collapsing. The plastic can be cut to any size and sealed on all four edges, if necessary, with the Pack-Rite Poly Jaw Sealer.[4] A small amount of formalin included in the bag will keep the tissue moist as long as the bag remains sealed. A quantity of small containers can be sealed together in a large bag, affording additional protection against drying. Plastic tags with data can be enclosed or attached to the outside. Storing in bags also saves storage space, preservative, and containers. Medical Associates International, Topeka, Kansas, sells lock-type bags, which can be opened and reclosed, called Bitran liquid-type bags. See also Broadway and Koelle (1960); Gordon (1953); and Lieb (1959).

---

[3] Griffith Laboratories.

[4] Pack-Rite Machines. (The CRC "Plastiseal" Bar Sealer is smaller and not as sturdy as the Pack-Rite machine.)

REMOVING LABORATORY STAINS FROM HANDS AND GLASSWARE

The most frequently used laboratory stains, with suggested treatment for removal, are listed here.

basic fuchsin—Difficult to remove; try strong acetic acid in 95% alcohol, or dilute HCl.

carmine—Strong ammonia water or weak HCl; if stain resists, use them alternately.

chromic acid—Dilute sulfurous acid in water, or concentrated sodium thiosulfate and a few drops of sulfuric acid added.

fast green and similar acid stains—Ammonia water.

hematoxylin—Weak acid or lemon juice.

hemoglobin—Fresh stains: lukewarm to cool water (never hot). Older stains: soften with borax solution, dilute ammonia, or tincture of green soap; finally, treat with 2% aqueous oxalic acid.

iodine—Sodium thiosulfate.

iron alum stains on glassware—Strong NaOH (sticks) in water, followed by strong HCl.

methylene blue—Acid alcohol or tincture of green soap.

most dyes—Tincture of green soap.

osmic acid on glassware—3% H_2O_2 (Carr and Bacsich 1958).

PAS—To remove from clothing, soak in weak solution of potassium permanganate, approximately 0.5%; rinse off and bleach in 1% oxalic acid. Warning: some fabric dyes also bleach.

picric acid—Lithium iodide or carbonate, aqueous.

potassium permanganate—Dilute sulfurous acid, HCl, oxalic acid, or hyposulfite.

safranin and gentian violet—Difficult to remove; try acid alcohol.

silver—Lugol or tincture of iodine, followed by sodium thiosulfate.

Verhoeff—4% aqueous citric acid (Hull and Wegner 1952).

Cleaning Solution for Glassware

Strong

potassium dichromate......................	20.0 g
water	200.0 ml

Dissolve dichromate in water; when cool add very slowly:

sulfuric acid, concentrated	20.0 ml

Weak

potassium dichromate, 2% aqueous		9 parts
sulfuric acid		1 part

This is strong enough for most purposes.

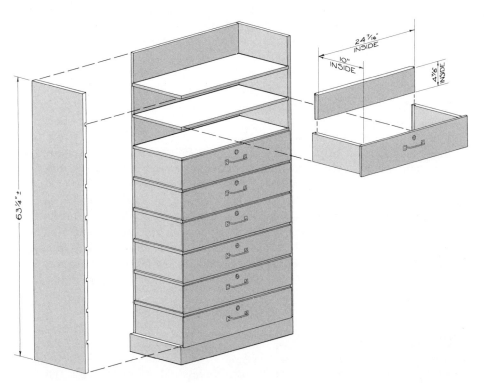

Figure 27-1

A cabinet of drawers designed for use in microtechnique classes at the University of California at Los Angeles. The Cabinets are not a part of the working table but are arranged along a wall of the classroom. The drawers are designed to hold coplin jars, mountant jars, and similar items. The student can carry the drawer to the work table, turn it around with its backside facing forward, remove the back panel, and do the staining right in the drawer. This relieves the work table of staining equipment, and the student can carry all the staining jars at once in one "basket." If locking the drawers is unnecessary, the front panel may be made removable; this eliminates having to turn the drawer around. Drawers may be left out of the upper compartment, or upper two compartments, to make convenient storage nooks. These cabinets are custom made. Standard cabinets can be adapted to this purpose, but their drawers do not have a removable panel. A good cabinet (Unit no. P3) can be obtained from the Kewanee Manufacturing Company, Adrian, Michigan. The interior dimensions of its drawers are $20\frac{15}{16}$ inches long, $5\frac{5}{8}$ inches high, and $12\frac{1}{2}$ inches wide.

LABORATORY SAFETY

Hazardous conditions can occur in microtechnique laboratories. The physical safety of the personnel must be safeguarded at all times and good lab habits established to prevent accidents. Think before you use any chemical, especially dangerous ones.

Have the proper kinds of fire extinguishers on hand; dry chemical and CO_2 extinguishers are recommended, also a sand bucket. All labs should be well ventilated, definitely with an exhaust or hood when noxious chemicals are used. Make a habit of keeping all containers, including staining set-ups, closed when not in use.

Commercial containers of chemicals now have labeled warnings for their safe use and antidotes for misuse. Read them. Make certain the labels are not lost. If some of the contents are transferred to smaller shelf stock containers, permanently place a similar label on that can or bottle. Poisons and carcinogens must be clearly labeled as such. Only a few of these materials are used in a microtechnique lab; but if they are required, handle them with extreme care. Many chemicals are toxic either by inhaling or swallowing: hydrocarbons, methacrylates, epoxys, acids, alkalis, formaldehyde, iodine. Some can be fatal: iodine, methanol, phenol, mercuric chloride. Never allow any chemical to remain on the skin; wash off immediately and with soap. Neutralize acids with mild alkalis and alkalis with weak acids. Whenever eyes are involved, consider medical attention. Liquid spills can be absorbed on paper toweling and allowed to evaporate in a hood or open area outdoors. Solid spills can be picked up on paper to be disposed of safely.

Some Specifics

Acids are highly reactive with water. Never add water to acids (HCl, H_2SO_4, HNO_3) when measuring them for dilution. Add the acid *slowly* to the water and avoid the fumes. Keep concentrated acids in small bottles and not full. Store large stock bottles in a safe place, preferably a cool one (never warm). If acids are involved in a fire, never use water; use a dry chemical or CO_2. Acids have a corrosive action on the skin; wash off immediately with long washing and soap.

The caustic *alkalis* are corrosive and must be washed off thoroughly, preferably first with a weak acid or vinegar and then by repeated washing with water. Get medical attention for any in the eyes.

Alums can cause burns. Wash thoroughly.

The *aromatic amines* are dangerous; benzidine is a carcinogen, as is benzene after prolonged use. Systemic poisoning can occur through the skin and by ingestion and inhalation. Wash off immediately and thoroughly with soap. Rewash several times. Many amines are highly flammable; keep away from flames. They should be used only in well-ventilated labs. Prevent prolonged breathing of them.

Chromic acid is a strong oxidizer and is toxic. It can cause inflammation and ulcers on the skin. Wash off immediately and thoroughly. Avoid inhalation; it can damage the respiratory tract.

Ethers are highly flammable; a can of ether can be ignited by a spark or open flame. Use a fume hood if available. A charge can even take place when pouring ether; ground it by keeping the two containers in contact when pouring. Quickly and tightly cover burning vapors to cut off oxygen. Ether, if long standing in contact with air and exposed to light and especially if stored in clear glass, can contain peroxides and then be explosive. Keep tightly closed in a metal can. Ethers are not highly toxic but will cause dizziness and possible collapse. Dioxane and ethylene oxide are more toxic than the ethyl and isopropyl ethers. They may produce nausea and vomiting, and even liver and kidney damage. They can be absorbed through the skin.

Ethylene glycol: Avoid breathing the vapor and wash well after contact on skin.

Iodine is poisonous; avoid contact. Medical attention is required if it is swallowed.

Nitrocellulose can burn. To dispose of it, allow the solvent to evaporate in a hood or open outside area. At that time avoid breathing it, as oxides of nitrogen are being released. It is highly flammable.

Osmic acid is toxic; keep it from skin and eyes.

Oxalic acid is absorbed in the blood to form calcium oxalate, a dangerous condition. Wash it off immediately. The small amounts used in the lab should be no problem.

Periodic acid is an oxidizer and can cause skin burns. Wash well.

Peroxides are irritating to the skin, eyes, and respiratory tract. The H_2O_2 as used in the lab rarely presents a problem, but wash it off immediately if it drips off the bottle onto the skin. Keep bottle from heat. It can be stored in the refrigerator; in fact, this prolongs its life.

Picric acid can be explosive. Handle with respect.

Silver compounds become dangerous when combined with ammonia. Aging of these solutions or exposure to air or light can form explosive silver compounds. Silver nitride and silver azide may be formed if formalin or alcohol is present in the solution. Do not store silver solutions. Prepare them only just before use and in clean glassware (none with silvered sides). Unused solutions can be inactivated by adding NaCl or dilute HCl.

THF produces eye and nose irritation and dermatitis and, with long exposure, kidney and liver damage. Store in safety cans; if it is spilled, leave the area to allow it to evaporate. Do not spread it by pouring water on it.

Drugs: Chloral hydrate, sodium barbitol (barbiturate), and cocaine, if stocked, must be kept under lock and key in a place provided for drugs.

All shelves holding bottled chemicals and stains in solution should have a barrier across the front to prevent shifting and falling bottles.

SUPPLIERS OF EQUIPMENT, GLASSWARE, AND CHEMICALS

Many companies will forward a catalog upon request. Specialized items have been noted in text footnotes.

Abbott Laboratories, Chicago, Ill.
Air Reduction Chemical and Carbide Co., New York, N.Y.
Aldrich Chemical Co., Fairfield, N.J.
Allied Chemical Corporation, Morristown, N.J.
Amend Drug and Chemical Co., New York, N.Y.
American Optical Corporation, Buffalo, N.Y.
Ames Co., Elkhart, Ind.
Ansul Chemical Co., Marinette, Wisc.
Antara Chemicals, General Aniline and Film Corp., New York, N.Y.
Arnel Products Co. (Sylvania Chemicals), New York, N.Y.

Bailey Instruments Co., Saddle Brook, N.J.
Baltimore Biological Laboratory, Baltimore, Md.
Barnstead Still and Sterilizer Co., Boston, Mass.
Behr-Manning, a Division of Norton Co., Troy, N.Y.
Biochemical and Nuclear Co., Burbank, Calif.
Biomedical Specialties, Los Angeles, Calif.
British Drug Houses, London.

Calbiochem, Ja Jolla, Calif.
Calibrated Instruments, Ardsley, N.J.
Carbide and Chemical Corporation, New York, N.Y.
Cargille Laboratories, Cedar Grove, N.J.
Carolina Biological Supply Co., Burlington, N.C.
Century Laboratories, Chicago, Ill.
Chemical Sales Co., Denver, Colo.
Ciba, Ltd., Montreal, Canada

Clay-Adams Inc., New York, N.Y.
Cole-Parmer Instrument Co., Chicago, Ill.
Corning Glass Works, Corning, N.Y.
Curtin Matheson Scientific Co., Atlanta, Boston, Dallas, and many other
 cities.

Dade Reagents, Miami, Fla.
Dajac Laboratories, Philadelphia, Penn.
Delkote, Wilmington, Del.
Delmar Scientific Glass Products, Maywood, Ill.
Difco Laboratory, Detroit, Mich.
Dow Chemical Co., Indianapolis, Ind.
Dri-Aire, East Norwalk, Conn.
DuPage Kinetic Laboratories, Downers Grove, Ill.
E.I. duPont de Nemours and Co., Wilmington, Del.
DuPont Instruments/Sorvall, Norwalk, Conn.

Eastman Kodak Co. (Eastman Chemicals), Rochester, N.Y.
ESBE Laboratory Supplies, Toronto, Canada

Fisher Scientific Co., Boston, Chicago, Ft. Worth, and many other cities.

Gelman Instrument Co., Chelsea, Mich.
General Biological Supply House, Chicago, Ill.
Glyco Chemical, Williamsport, Penn.
Goldschmidt Chemical Corp., New York, N.Y.
Griffith Laboratories, Chicago, Ill.
Edward Gurr, High Wycombe, Bucks, England (products can be obtained
 from Roboz Surgical Instruments).
G. T. Gurr, London.

Hacker Instruments, Fairfield, N.J.
Harleco and Hemal Stain Co., Danbury, Conn.
Harris Refrigeration Co., Cambridge, Mass.
Hartman-Leddon Co., Philadelphia, Penn.
Hyland Laboratory, Los Angeles, Calif.

Imperial Chemical Industries, Manchester, England.
Industrial Chemicals, Inc. Richmond, Va.
International Equipment Co., Needham Heights, Mass.

B. H. Jordan Co., Great Neck, N.Y.

K and K Laboratories, Inc., Plainview, N.Y.
Kerr Company, Detroit, Mich.
Kessler Chemical Co., Philadelphia, Penn.
O. Kindler, Freiburg, Germany.
Krylon, Norristown, Penn.

Lab-Line Instruments, Melrose Park, Ill.
Lab-Tek Instruments Co., Westmont, Ill.
La Pine Scientific, Berkeley, Calif.; Chicago; Irvington-on-Hudson, N.Y.
Lederle Laboratories, New York, N.Y.
Lerner Laboratories, Stamford, Conn.
Lipshaw Manufacturing Co., Detroit, Mich.

M. E. Matchlett and Son, New York, N.Y.
Matheson Coleman and Bell, products obtained through Curtin Matheson
    Scientific.
Medical Associates International, Topeka, Kansas.
Microbiological Associates, Albany, Calif.; Bethesda, Md.
Monsanto Chemical Corp., Dayton, Ohio.

New England Nuclear Corp., Boston, Mass.
Nuclear Chicago Corp., Des Plaines, Ill.
Nutritional Biochemicals Corp., Cleveland, Ohio.

Ortho Pharmaceutical Corp., Raritan, N.J.
Oxford Labs, San Mateo, Calif.

Pack-Rite Machines, Milwaukee, Wisc.
Peel-A-Way Scientific Co., El Monte, Calif.
Pfatz and Bauer, Inc., New York, N.Y.
Polysciences, Inc., Warrington, Penn., will prepare special compounds.
Preiser Scientific Inc., Charleston, W.Va.; Cincinnati, Ohio; Louisville,
    Ky.
The Presto Co., Glendale, Calif.
Professional Tape Co., Inc., Riverside, Ill.

Radiation, Inc., Cambridge, Mass.
Randolph Products Co., Carlstadt, N.J.
Reichert Optical, Hacker Instruments, West Caldwell, N.J.
Roboz Surgical Instruments, Washington, D.C.
Rohm & Haas Co., Philadelphia, Penn.
Ruger Chemical Co., Irvington-on-Hudson, N.Y.

Sargent Welch Scientific Co., Birmingham, Ala.
Schwarz BioResearch, Inc., Orangeburg, N.Y.
Scientific Products, Evanston, Ill.
Searle Diagnostic, G. O. Searle, High Wycombe, Bucks, England.
Shandon Southern Instruments, Sewickley, Penn.
Shawnigan Resins Corp., Springfield, Mass.
Sigma Chemical Co., St. Louis, Mo.
Eric Sobokta Co., Farmingdale, N.Y.
Strem Chemical, Danvers, Mass.

Technicon Co., Chauncey, N.Y.
Arthur H. Thomas Co., Philadelphia, Penn.

Ultra Violet Products, Inc., San Gabriel, Calif.
Union Carbide Co., New York, N.Y.

Verona Dyestuffs, Union City, N.Y.
Volk Radiochemical Co., Chicago, Los Angeles, New York, Washington, D.C.
VWR Scientific, Atlanta, Los Angeles, Rochester, N.Y., and many other cities.

Ward's Natural Science Establishment, Rochester, N.Y.
Watson and Sons, London.
Edward Weck and Co., Long Island City, N.Y.
Will Corporation, Atlanta; Baltimore; Buffalo, N.Y.; Charleston, W.Va.; New York; Rochester, N.Y.
Worthington Biochemical Corp., Freehold, N.J.

Zeiss-Winkel, New York, N.Y.

References

d'Ablaing, Gerrit; Rogers, Eugene R.; Parker, John W.; and Lukes, Robert J.
  (1970) "A simplified and modified methyl green pyronin stain." *American Journal of Clinical Pathology* 54:667–669.
Abrams, Harvey A., and Elias, Hans.
  (1962) "Acridine orange fluorescence in paraffin sections." *Stain Technology* 37:50–52.
Ackerman, Adolph G.
  (1958) "A combined alkaline phosphatase–PAS staining method." *Stain Technology* 33:269–271.
Adams, C. W. M.
  (1956) "A stricter interpretation of the ferric ferricyanide reaction with particular reference to the demonstration of protein-bound sulphydryl and disulphide groups." *Journal of Histochemistry and Cytochemistry* 4:32–35.
  (1957) "A p-dimethylaminobenzaldehydenitrite method for the histochemical demonstration of tryptophane and related compounds." *Journal of Clinical Pathology* 10:56–62.
Adams, C. W. M., and Sloper, J. C.
  (1955) "Technique for demonstrating neurosecretory material in the human hypothalamus." *Lancet* 268:651–652.
  (1956) "The hypothalamic elaboration of posterior pituitary principles in man, the rat and dog; histochemical evidence derived from a performic acid–Alcian blue reaction for cystine." *Journal of Endocrinology* 13:221–228.
Adams, C. W. M., and Tuqan, N. A.
  (1961) "The histochemical demonstration of protease by a gelatin-silver film substrate." *Journal of Histochemistry and Cytochemistry* 9:469–472.
Adamstone, F. B., and Taylor, A. B.
  (1948) "The rapid preparation of frozen tissue sections." *Stain Technology* 23:109.
Adey, W. R.; Rudolph, Alice F.; Hine, I. F.; and Harritt, Nancy J.
  (1958) "Glees staining of monkey hypothalamus: a critical appraisal of normal and experimental material." *Journal of Anatomy* 92:219–235.
Agrell, Ivar P. S.
  (1958) "Whole mounts of small embryos attached directly to glass slides." *Stain Technology* 33:265–267.
Ahlqvist, John, and Andersson, Leif.
  (1972) "Methyl green–pyronin staining: effects of fixation; use in routine pathology." *Stain Technology* 47:17–22.

Albrecht, Mildred.
(1954) "Mounting frozen sections with gelatin. *Stain Technology* 29:89–90.
(1956) "A simple mechanical aid in staining frozen sections in quantity." *Stain Technology* 31:231–233.
Aldridge, William G., and Watson, Michael L.
(1963) "Perchloric acid extraction as a histochemical technique." *Journal of Histochemistry and Cytochemistry* 11:773–781.
Alfert, Max, and Geschwind, I. I.
(1953) "A selective staining method for basic proteins of cell nuclei." *Proceedings of National Academy of Sciences* 39:991–999.
Alpert, Morton; Jacobowitz, David; and Marks, Bernard H.
(1960) "A simple method for the demonstration of lipofuscin pigment." *Journal of Histochemistry and Cytochemistry* 8:153–158.
Alsop, David W.
(1974) "Rapid single-solution polychrome staining of semithin epoxy sections using polyethylene glycol 200 (PEG 200) as a stain solvent." *Stain Technology* 49:265–272.
Altman, Philip L., and Dittman, Dorothy S., eds.
(1964) *Biology Data Book.* Washington, D.C.: Federation of American Societies for Experimental Biology.
Amano, M.
(1962) "Improved techniques for the enzymatic extraction of nucleic acids from tissue sections." *Journal of Histochemistry and Cytochemistry* 10:204–212.
American Public Health Association.
(1956) *Diagnostic Procedures for Virus and Rickettsial Diseases.* New York: The Association.
Ammerman, Frank.
(1950) "A chrome-alum preparation for delicate and difficult fixations." *Stain Technology* 25:197–199.
Anderson, D. L.
(1961) "Selective staining of ribonucleic acid." *Journal of Histochemistry and Cytochemistry* 9:619–620.
Anderson, Doris, and Grieff, Donald.
(1964) "Direct fluorochroming of Rickettsiae." *Journal of Histochemistry and Cytochemistry* 12:194–196.
Anderson, F. D.
(1959) "Dichromate-chlorate perfusion prior to staining degeneration in brain and spinal cord." *Stain Technology* 34:65–67.
Anderson, Paul J.
(1967) "Purification and quantitation of glutaraldehyde and its effect on several enzyme activities in skeletal muscle." *Journal of Histochemistry and Cytochemistry* 15:652–661.
Anderson, Ronald A.
(1971) "Decalcification of snail shells to facilitate sectioning of the whole organism." *Stain Technology* 46:267–268.
Anthony, Adam, and Clater, Merlin.
(1959) "Alcohol-xylene versus Dioxane in the shrinkage of tissue." *Stain Technology* 34:9–13.
Aoyama, Fumic
(1930) "Eine Modifikation der Cajakschen Methode zur Darstellung des Golgischen Binnenapparates." *Zeitschrift für Wissenschaftliche Mikroskopie und Mikroskopische Technik* 46:489–491.
Aparicio, S. R., and Marsden, P.
(1969) "Application of standard micro-anatomical staining methods to epoxy resin-embedded sections." *Journal of Clinical Pathology* 22:589–592.
Arensburger, Konstantin E., and Markell, Edward K.
(1960) "A simple combination direct smear and fecal concentrate for permanent stained preparations." *American Journal of Clinical Pathology* 34:50–51.

Armstrong, J. A.
   (1956) "Histochemical differentiation of nucleic acids by means of induced fluorescence." *Experimental Cell Research* 11:640–643.
Armstrong, J. A.; Niven, Janet; and Anderson, S. S.
   (1957) "Fluorescence microscopy in the study of nucleic acids." *Nature* 80:1335–1336.
Arn, Von Doris, and Landolt, Ruth.
   (1975) "Direkte Fixierungsmöglichkeit von Snitten auf Objektträgern durch Verwendung des Einbettungsmediums Paraplast-Plus." *Mikroskopie* 31:99–106.
Arnold, James S., and Jee, Webster
   (1954a) "Preparing bone sections for radioautography." *Stain Technology* 29:49–54.
   (1954b) "Embedding and sectioning undercalcified bone and its application to radioautography." *Stain Technology* 29:225–239.
Arnold, Zach M.
   (1952) "A rapid method for concentrating small organisms for sectioning." *Stain Technology* 27:199–200.
Aronson, Willard, and Pharmakis, Thomas.
   (1962) "Enhancement of neotetrazolium staining for succinic dehydrogenase activity with cyanide." *Stain Technology* 37:321.
Arzac, J. P.
   (1950) "A simple histochemical reaction for aldehydes." *Stain Technology* 25:187–194.
Ashley, C. A., and Feder, N.
   (1966) "Glycol methacrylate in Histopathology." *Archives of Pathology* 81:391–397.
Atkinson, William B.
   (1952) "Studies on the preparation and recoloration of fuchsin sulfurous acid." *Stain Technology* 27:153–160.
Augulis, Vaclovas, and Sepinwall, Jerry.
   (1971) "Use of gallocyanine as a myelin stain for brain and spinal cord." *Stain Technology* 46:137–143.
Austin, A. P.
   (1959) "Iron alum aceto-carmine staining for chromosomes and other anatomical features of Rhodophyceae." *Stain Technology* 34:69–75.
Avers, Charlotte J.
   (1963) "An evaluation of various fixings and staining procedures for mitochondria in plant root tissues." *Stain Technology* 38:29–35.

Baker, J. R.
   (1932) "A new method for mitochondria." *Nature* 30:134.
   (1944) "The structure and chemical composition of the Golgi element." *Quarterly Journal of Microscopical Science* 85:1–72.
   (1945) *Cytological Technique.* London: Methuen.
   (1946) "The histochemical recognition of lipine." *Quarterly Journal of Microscopical Science* 87:441–470.
   (1947) "The histochemical recognition of certain guanidine derivatives." *Quarterly Journal of Microscopical Science* 88:115–121.
   (1958) *Principles of Biological Technique.* London: Methuen; New York: John Wiley.
Bancroft, J. D.
   (1963) "Methyl green as a differentiator and counterstain in the methyl violet technique for demonstration of amyloid in fresh cryostat sections." *Stain Technology* 38:336–337.
Bangle, Raymond.
   (1954) "Gomori's paraldehyde fuchsin stain." *Journal of Histochemistry and Cytochemistry* 2:291–299.
Banny, Theresa M., and Clark, George.
   (1950) "New domestic cresyl echt violet." *Stain Technology* 25:195–196.
Barer, R.
   (1956) "Phase contrast and interference microscopy in cytology." *Physical Techniques in Biological Research,* Vol. III. Ed. by Gerald Oster and Arthur W. Pollister. New York: Academic Press.

(1959) "Interference and Polarizing Microscopy." *Analytical Cytology*. Ed. by Robert C. Mellors. New York: Blakiston.

Barka, T., and Anderson, P. J.
(1963) *Histochemistry*. New York: Harper & Row.

Barley, David A.
(1964) "Salivary gland chromosomes of black flies." *Turtox News* 44:298–300.

Barlow, R. M.
(1957) "A tribasic stain for the granules of mast cells." *Journal of Pathology and Bacteriology* 73:272–274.

Barnard, E. A.
(1961) "Acylation and diazonium coupling in protein cytochemistry with special reference to the benzoylation-tetrazonium method." *General Cytochemical Methods*, II. Ed. by J. F. Danielli. New York: Academic Press.

Barnard, John W.; Roberts, J. O.; and Brown, Joseph G.
(1949) "A simple macroscopic staining and mounting procedure for wet sections from cadaver brains." *Anatomical Record* 105:11–17.

Barnes, H.
(1954) "Some tables for the ionic composition of sea water." *Journal of Experimental Biology* 31:582–588.

Barnett, R. J., and Seligman, A. M.
(1952) "Demonstration of protein-bound sulfhydryl and disulfide groups by two new histochemical methods." *Journal of National Cancer Institute* 13:215–216.

Barr, M. L., and Bertram, E. G.
(1949) "A morphological distinction between neurones of the male and female, and the behavior of the nucleolar satellite during accelerated nucleoprotein synthesis." *Nature* 163:676.

Barr, M. L.; Bertram, Fraser; and Lindsay, H. A.
(1950) "The morphology of the nerve cell nucleus, according to sex." *Anatomical Record* 107:283–292.

Barrett, A. M.
(1944) "On the removal of formaldehyde-produced precipitate from sections." *Journal of Pathology and Bacteriology* 56:135–136.

Barrós-Pita, José Carlos.
(1971) "Protective paraffin infiltration of soft tissues in insects to facilitate softening of hard exoskeleton." *Stain Technology* 46:171–175.

Bartholomew, James W., and Mittwer, Tod.
(1950) "The mechanism of the Gram reaction. 1. The specificity of the primary dye." *Stain Technology* 25:103–110.
(1951) "The mechanism of the Gram reaction. 3. Solubilities of dye-iodine precipitates, and further studies of primary dye substitutes. *Stain Technology* 26:231–240.

Bartholomew, James W.; Mittwer, Tod; and Finklestein, Harold.
(1959) "The phenomenon of Gram-positivity; its definition and some negative evidence on the causative role of sulfhydryl groups." *Stain Technology* 34:147–154.

Barton, A. A.
(1959) "The examination of ultrathin sections with a light microscope after electron microscopy." *Stain Technology* 34:348–349.
(1960) "Carbon coated grids for electron microscopy." *Stain Technology* 35:287–289.

Baserga, Renato.
(1961) "Two-emulsion autoradiography for the simultaneous demonstration of precursors of deoxyribonucleic and ribonuclei acids." *Journal of Histochemistry and Cytochemistry* 9:586.

Baserga, Renato, and Nemeroff, Ken.
(1962) "Two-emulsion radioautography." *Journal of Histochemistry and Cytochemistry* 10:628–635.

Batty, I., and Walker, P. D.
(1963) "Differentiation of *Clostridium septicum* and *Clostridium chauvoei* by the use of fluorescent labelled antibodies." *Journal of Pathology and Bacteriology* 85:517–520.

Bauer, H.
    (1933) "Mikroskopisch-chemischer Nachweis von Glykogen un einigen anderen Polysacchariden." *Zeitschrift für Mikroskopisch-Anatomische Forschung* 33:143–160.
Bayley, John.
    (1955) " 'Cracking' and 'chatter' in paraffin sections." *Journal of Pathology and Bacteriology* 70:548–549.
Beamer, Parker R., and Firminger, Harlan I.
    (1955) "Improved methods for demonstrating acid-fast and spirochaetal organisms in histological sections." *Laboratory Investigation* 4:9–17.
Beçak, M. L.; and Nazareth, H. R. S.
    (1962) "Karyotypic studies of two species of South American snakes." *Cytogenetics* 1:305–313.
    (1964) "Chromosomes of cold blood animals from whole blood short-term cultures." *Mammalian Chromosomes Newsletter* 14:55–56.
Beck, Conrad.
    (1938) *The Microscope. Theory and Practice.* London: R. J. Beck.
Beckel, W. E.
    (1959) "Sectioning large heavily sclerotized whole insects." *Nature* 184:1584–1585.
Becker, E. R., and Roudabush, R. L.
    (1935) *Brief Directions in Histological Technique.* Ames, Iowa: Collegiate Press.
Beckert, William H., and Garner, James G.
    (1966) "Staining sex chromatin; Biebrich scarlet and Fast green FCF as a mixture versus their use in sequence." *Stain Technology* 41:141–148.
Bedi, K. S., and Horobin, Richard W.
    (1976) "An alcohol-soluble Schiff's reagent: a histochemical application of the complex between Schiff's reagent and phosphotungstic acid." *Histochemistry* 48:153–159.
Beech, R. H., and Davenport, H. A.
    (1933) "The Bielschowsky staining technic. A study of the factors influencing its specificity for nerve fibers." *Stain Technology* 8:11–30.
Beek, R. M.
    (1955) "Improvements in the squash technic for plant chromosomes." *Aliso* 3:131–133.
Behnke, O., and Rostgaard, J.
    (1963) "A device for mounting glass knives in the ordinary rotary microtome for sectioning plastic-embedded material." *Stain Technology* 38:299–300.
Bélanger, Leonard F.
    (1950) "A method for routine detection of radio-phosphates and other radioactive compounds in tissues. The inverted autograph." *Anatomical Record* 107:149–156.
    (1952) "Improvements in the melted emulsion technique of autoradiography." *Nature* 170:165.
    (1961) "Staining processed radioautographs." *Stain Technology* 36:313–317.
Bélanger, Leonard F., and Bois, Pierre.
    (1964) "A histochemical and histophysiological survey of the effects of different fixatives on the thyroid gland of rats." *Anatomical Record* 148:573–580.
Belling, John.
    (1926) "The iron-aceto-carmine method of fixing and staining chromosomes." *Biological Bulletin, Woods Hole* 50:160–162.
    (1930) *The Use of the Microscope.* New York: McGraw-Hill.
Bencosme, Sergie; Stone, Robert S.; Latta, Harrison; and Madden, Sidney C.
    (1959) "A rapid method for localization of tissue structures or lesions for electron microscopy." *Journal of Biophysical and Biochemical Cytology* 5:508–509.
Bender, M. A., and Eide, P. E.
    (1962) "Separation of monkey (*Ateles*) leukocytes for culture." *Mammalian Chromosomes Newsletter* 8.
Benés, Karel.
    (1960) "The chemical specificity of staining mitochondria in frozen dried and frozen substituted tissue with buffered Amidoblack 10B." *Acta Histochemica* 10:255–264.

Benge, W. P. J.
    (1960) "Staining autoradiographs at low temperatures." *Stain Technology* 35:106–108.
Bennett, H. Stanley.
    (1950) "The microscopical investigation of biological materials with polarized light."
        Handbook of Microscopical Technique, 3d ed. Ed. by Ruth McClung Jones. New
        York: Hoeber.
    (1951) "The demonstration of thiol groups in certain tissues by means of a new colored
        sulfhydryl reagent." *Anatomical Record* 110:231–247.
Bennett, H. Stanley, and Watts, Ruth M.
    (1958) "The cytochemical demonstration and measurement of sulfhydryl groups by
        azo-aryl mercaptide coupling with special reference to mercury orange." *General
        Cytochemical Methods*, I. Ed. by J. F. Danielli. New York: Academic Press.
Bennet, H. Stanley; Wyrick, A. Dean; Lee, Sarah W.; and McNeill, John H.
    (1976) "Science and art in preparing tissues embedded in plastic for light microscopy,
        with special reference to glycol methacrylate, glass knives and simple stains."
        Stain Technology 51:71–97.
Bensley, R. R., and Bensley, S. H.
    (1938) *Handbook of Histological and Cytological Technique*. Chicago: University of
        Chicago Press.
Bensley, Sylvia H.
    (1952) "Pinacyanol erythrosinate as a stain for mast cells." *Stain Technology* 27:269–
        273.
    (1959) "The scope and limitations of histochemistry." *American Journal of Medical
        Technology* 25:15–32.
Bergeron, John A., and Singer, Marcus.
    (1958) "Metachromasy: an experimental and theoretical re-evaluation." *Journal of
        Biophysical and Biochemical Cytology* 4:433–457.
Berkowitz, Lillian R.; Fiorello, Olga; Kruger, Lawrence; and Maxwell, David S.
    (1968) "Selective staining of nervous tissue for light microscopy following preparation
        for electron microscopy." *Journal of Histochemistry and Cytochemistry* 16:808–
        814.
Berman, Irwin, and Kaplan, Henry S.
    (1959) "The cultivation of mouse bone marrow in vivo." *Blood* 14:1040–1046.
Berman, Irwin, and Newby, Earlene J.
    (1963) "Autoradiography and staining of hematopoietic cells grown on Millipore mem-
        branes in vivo." *Stain Technology* 38:62–65.
Bernhardt, H.; Gourley, R. D.; Young, J. M.; Shepherd, M. C.; and Killian, J. J.
    (1961) "A modified membrane-filter technic for detection of cancer cells in body fluids."
        American Journal of Clinical Pathology 36:462–464.
Berthrong, Morgan, and Barhite, Melvin.
    (1964) "A technic for preparation of tissue sections of needle-aspirated material from
        bone marrow." *American Journal of Clinical Pathology* 42:207–211.
Berton, William M., and Phillips, W. R.
    (1961) "A film leader technic for the study of cells in culture." *Laboratory Investigation*
        10:373–380.
Bertram, E. G.
    (1958) "A non-sticking coverglass weight." *Stain Technology* 33:143–144.
Berube, G. R.; Powers, Margaret M.; Kerkay, Julius; and Clark, George.
    (1966) "The gallocyanin–chrome alum stain: influence of methods of preparation on its
        activity and separation of active staining compound." *Stain Technology* 41:73–
        81.
Betts, Anthony.
    (1961) "The substitution of acridine orange in the periodic acid–Schiff stain." *American
        Journal of Clinical Pathology* 36:240–243.
Bharadwaj, T. P., and Love, Robert.
    (1959) "Staining mitochondria with hematoxylin after formalin-sublimate fixation; a
        rapid method." *Stain Technology* 34:331–334.

Bielschowsky, M.
   (1902) "Die Silber impragnation der Axencylinder." *Neurologisches Centralblatt* 21:579–584.
Birge, Wesley J., and Tibbits, Donald F.
   (1961) "The use of sodium chloride containing fixatives in minimizing cellular distortion in histological and cytochemical preparations." *Journal of Histochemistry and Cytochemistry* 9:409–414.
Bissing, Donald R.
   (1974) "Haupt's gelatine adhesive mixed with formalin for affixing paraffin sections to slides." *Stain Technology* 49:116–117.
Bitensky, Lucille.
   (1963) "Modification to the Gomori phosphatase technique for controlled temperature frozen sections." *Quarterly Journal of Microscopical Science* 104:193–196.
Black, M. M., and Ansley, H. R.
   (1964) "Histone staining with ammoniacal silver." *Science* 43:693.
Black, M. M.; Ansley, H. R.; and Mandl, R.
   (1964) "On cell specificity of histones." *Archives of Pathology* 78:350–368.
Black, M. M., and Jones, Edward Wilson.
   (1971) "Macular Amyloidosis: a study of 21 cases with special reference to the role of the epidermis in its histogenesis." *British Journal of Dermatology* 84:199.
Blackburn, D. T., and Christophel, D. C.
   (1976) "A method of permanently mounting biological tissue cleared in Herr's four-and-a-half clearing fluid." *Stain Technology* 51:125–130.
Blandau, R. J.
   (1938) "A method of eliminating the electrification of paraffin ribbons." *Stain Technology* 13:139–141.
Blank, Harvey, and McCarthy, Philip L.
   (1950) "A general method for preparing histologic sections with a water-soluble wax." *Journal of Laboratory and Clinical Medicine* 36:776.
Blank, Harvey; McCarthy, Philip L.; and Delamater, Edward D.
   (1951) "A non-vacuum freezing-dehydrating technic for histology, autoradiography, and microbial cytology." *Stain Technology* 26:193–197.
Block, David P., and Godman, Gabriel C.
   (1955) "A microphotometric study of the synthesis of desoxyribonucleic acid and nuclear histone." *Journal of Biophysical and Biochemical Cytology* 1:17–28.
Block, M.; Smaller, Victoria; and Brown, Jessie.
   (1953) "An adaptation of the Maximow technique for preparation of sections of hematopoietic tissue." *Journal of Laboratory and Clinical Medicine* 42:145–151.
Bogen, Emil.
   (1941) "Detection of tubercle bacilli by fluorescence microscopy." *American Review of Tuberculosis* 44:267–271.
Bogoroch, Rita.
   (1951) "Detection of radio-elements in histological slides by coating with stripping emulsion—the strip-coating technic." *Stain Technology* 26:43–50.
Bohorfoush, Joseph G.
   (1963) "A thiosulfate diluent for Wright's stain." *Stain Technology* 38:292–293.
Bokdawala, F. D., and George, J. C.
   (1964) "Histochemical demonstration of muscle lipase." *Journal of Histochemistry and Cytochemistry* 12:768–771.
Borror, Arthur C.
   (1968) "Nigrosin-HgCl$_2$-formalin; a stain-fixative for ciliates (Protozoa, Ciliophora)." *Stain Technology* 43:293–295.
Borysko, Emil.
   (1956) "Recent developments in methacrylate embedding. 1. A study of the polymerization damage phenomenon by phase contrast microscopy." *Journal of Biophysical and Biochemical Cytology* 2, Part 2, Supplement: 3–14.

Borysko, Emil, and Sapranauskas, P.
    (1954) "A technique for comparison phase-contrast and electron microscope studies of cells grown in tissue culture with evaluation of time-lapse cinemicrographs." *Johns Hopkins Hospital Bulletin* 95:68–80.

Bouchard, Frances M.
    (1963) "Acceleration of Giemsa staining by heat." *Stain Technology* 38:288.

Boyd, G. A.
    (1955) *Autoradiography in Biology and Medicine.* New York: Academic Press.

Boyd, I. A.
    (1962) "Uniform staining of nerve endings in skeletal muscle with gold chloride." *Stain Technology* 37:225–230.

Bradley, Muriel V.
    (1948*a*) "An aceto-carmine squash technic for mature embryo sacs." *Stain Technology* 23:29–40.
    (1948*b*) "A method for making aceto-carmine squashes permanent without removal of the cover slip." *Stain Technology* 23:41–44.
    (1957) "Sudan black B and aceto-carmine as a combination stain." *Stain Technology* 32:85–86.

Brandi, A. James; Malferrari, Rosella; and Massignani, Adrian.
    (1962) "Collodion fixation and protection of cytological smears for transportation and staining by Papanicolaou's method." *Stain Technology* 37:183–185.

Branton, Daniel, and Jacobson, Louis J.
    (1962) "Dry, high resolution autoradiography." *Stain Technology* 37:239–242.

Braunstein, H., and Adriano, S. M.
    (1961) "Fluorescent stain for tubercle bacilli in histologic sections." *American Journal of Clinical Pathology* 36:37–40.

Brecher, George.
    (1949) "New methylene blue as a reticulocyte stain." *American Journal of Clinical Pathology* 19:895–896.

Brenner, R. M.
    (1962) "Controlled oxidation of background grains in radioautographs." *Journal of Histochemistry and Cytochemistry* 10:678.

Bridges, C. B.
    (1937) "The vapour method of changing reagents and of dehydration." *Stain Technology* 12:51–52.

Bridges, Charles H., and Luna, Lee.
    (1957) "Kerr's improved Warthin-Starry technic. Study of permissible variations." *Laboratory Investigation* 6:357–367.

Briggs, Donald K.
    (1958) "Recent studies in chromatin sex determination." *Transactions New York Academy of Sciences* Ser. II, 20:500–504.

Broadway, Charles B., and Koelle, Donald G.
    (1960) "Use of polyethylene plastic bags for storage of pathologic tissue." *Technical Bulletin of the Registry of Medical Technologists* 30:17–21.

Brooke, M. M., and Donaldson, A. W.
    (1950) "Use of a surface active agent to prevent transfer of malarial parasites between blood films during mass staining procedures." *Journal of Parasitology* 36:84.

Brown, Charles D., and Fleming, Loraine.
    (1965) "A modified micromethod for the culturing of lymphocytes from peripheral blood for the assessment of chromosome morphology." *Mammalian Chromosomes Newsletter* 15:110–113.

Brown, James O.
    (1948) "A simplified and convenient method for the double embedding of tissues." *Stain Technology* 23:83–89.

Bruemmer, Nancy C.; Carver, Michael J.; and Thomas, Lloyd E.
    (1957) "A tryptophan histochemical method." *Journal of Histochemistry and Cytochemistry* 5:140–144.

Brunk, Ulf T., and Ericsson, Jan L. E.
  (1972) "The demonstration of acid phosphatase in *in vitro* cultured tissue cells. Studies on the significance of fixation, tonicity and permeability." *Histochemical Journal* 4:349–363.
Bryan, John H. D., and Hughes, Rebecca L.
  (1976) "A simple and inexpensive static eliminator for paraffin sectioning." *Stain Technology* 50:397–398.
Buck, R. C., and Jarvic, C. E.
  (1959) "A simple automatic ultramicrotome." *Stain Technology* 34:109–111.
Buckley, Sonja M.; Whitney, Elinor; and Rapp, Fred.
  (1955) "Identification by fluorescent antibody of development forms of psittacosis virus in tissue culture." *Proceedings of the Society of Experimental Biology and Medicine* 90:226–230.
Bullivant, S.
  (1965) "Freeze substitution and supporting techniques." *Laboratory Investigation* 14:1178–1195.
Bullivant, S., and Hotchin, J.
  (1960) "Chromyl chloride, a new stain for electron microscopy." *Experimental Cell Research* 21:211–214.
Bulmer, D.
  (1962) "Observations on histological methods involving the use of phosphotungstic and phosphomolybdic acids, with particular reference to staining with phosphotungstic acid hematoxylin." *Quarterly Journal of Microscopical Science* 103:311–324.
Burdette, W. J.
  (1962) *Methodology in Human Genetics.* San Francisco: Holden-Day.
Burkholder, Peter M.; Littell, Andrew H.; and Klein, Paul G.
  (1961) "Sectioning at room temperature of unfixed tissues, frozen in a gelatin matrix, for immunohistologic procedures." *Stain Technology* 36:89–91.
Burns, J., and Neame, P. B.
  (1966) "Staining of blood cells with periodic acid/salicylol hydrazide (PA–SH). A fluorescent method for demonstrating glycogen." *Blood* 28:674–682.
Burrows, Robert B.
  (1967) "Improved preparation of polyvinyl alcohol–HgCl$_2$ fixative used for fecal smears." *Stain Technology* 42:93.
Burrows, William.
  (1954) *Textbook of Microbiology.* Philadelphia: W. B. Saunders.
Burstone, M. S.
  (1957a) "Polyvinyl acetate as a mounting medium for azo-dye procedures." *Journal of Histochemistry and Cytochemistry* 5:196.
  (1957b) "Polyvinyl pyrrolidone as a mounting medium for stains for fat and for azo-dye procedures." *American Journal of Clinical Pathology* 28:429–430.
  (1958) "The relationship between fixation and techniques for the histochemical localization of hydrolytic enzymes." *Journal of Histochemistry and Cytochemistry* 6:322–339.
  (1959a) "New histochemical techniques for the demonstration of tissue oxidase (cytochrome oxidase)." *Journal of Histochemistry and Cytochemistry* 7:112–121.
  (1959b) "Acid phosphatase activity of calcifying bone and dentin matrices." *Journal of Histochemistry and Cytochemistry* 7:147–148.
  (1960a) "Histochemical demonstration of cytochrome oxidase with new amine reagents." *Journal of Histochemistry and Cytochemistry* 8:63–70.
  (1960b) "Postcoupling, noncoupling, and fluorescence techniques for the demonstration of alkaline phosphatase." *Journal of the National Cancer Institute* 24:1199–1218.
  (1962) *Enzyme Histochemistry and its Application in the Study of Neoplasms.* New York: Academic Press.
Burstone, M. S., and Flemming, T. J.
  (1959) "A new technique for the histochemical study of smears." *Journal of Histochemistry and Cytochemistry* 7:203.

Burstone, M. S., and Folk, J. E.
   (1956) "Histochemical demonstration of aminopeptidase." *Journal of Histochemistry and Cytochemistry* 4:217–226.
Burton, George J.
   (1958) "Preparation of thick malarial blood smears by the tapping method." *Mosquito News* 18:228–229.
Bush, Vannevar, and Hewitt, Richard E.
   (1952) "Frozen sectioning: A new and rapid method." *American Journal of Pathology* 28:863–867.
Bussolati, G., and Bassa, Tatiana.
   (1974) "Thiosulfation aldehyde fuchsin (TAF) procedure for staining of pancreatic B cells." *Stain Technology* 49:313–315.
Buzzell, Gerald R.
   (1975) "Double embedding technique for light microscope histology." *Stain Technology* 50:285–287.

Cameron, D. A.
   (1956) "A note on breaking glass knives." *Journal of Histochemistry and Cytochemistry* 2:57–59.
Cameron, M. L., and Steele, J. E.
   (1959) "Simplified aldehyde-fuchsin staining of neurosecretory cells." *Stain Technology* 34:265–266.
Caratzali, Alexandre; Phelps, Annchesi; and Turpin, Raymond.
   (1957) "Variations quantitatives des corpuscules sexuels des granulocytes neutrophiles au sours du cycle menstruel." *Bulletin de l'Académie de Medicine* 141:496–501.
Cares, A.
   (1945) "A note on stored formaldehyde and its easy reconditioning." *Journal of Technical Methods* 25:67–70.
Carey, Eben J.
   (1941) "Experimental pleomorphism of motor nerve plates as a mode of functional protoplasmic movement." *Anatomical Record* 81:393–413.
Carleton, H. M., and Leach, E. H.
   (1947) *Histological Technique*. New York: Oxford University Press.
Carlo, Ravetto.
   (1964) "Alcian blue–Alcian yellow: a new method for the identification of different acidic groups." *Journal of Histochemistry and Cytochemistry* 12:44–45.
Carmichael, G. G.
   (1963) "A tetrazolium salt reduction method for demonstrating lipo-proteins in tissue sections." *Journal of Histochemistry and Cytochemistry* 11:738–740.
Caro, Lucian G.
   (1964) "High resolution autoradiography." *Methods in Cell Physiology*, I. Ed. by David M. Prescott. New York: Academic Press.
Carr, D. H., and Walker, J. E.
   (1961) "Carbol fuchsin as a stain for human chromosomes." *Stain Technology* 36:233–236.
Carr, L. A., and Bacsich, P.
   (1958) "Removal of osmic acid stains." *Nature* 182:1108.
Carr, Lawrence B.; Rambo, Oscar N.; and Feichtmeier, Thomas V.
   (1961) "A method of demonstrating calcium in tissue sections using chloranilic acid." *Journal of Histochemistry and Cytochemistry* 4:415–417.
Carter, C. H., and Leise, J. M.
   (1958) "Specific staining of various bacteria with a single fluorescent antiglobulin." *Journal of Bacteriology* 76:152–154.
Carver, M. Joseph; Brown, F. Christine; and Thomas, Lloyd E.
   (1953) "An arginine histochemical method using Sakaguchi's new reagent." *Stain Technology* 28:89–91.

Case, Norman M.
(1953) "The use of a cation exchange in decalcification." *Stain Technology* 28:155–158.
Cason, Jane E.
(1950) "A rapid one-step Mallory–Heidenhain stain for connective tissue." *Stain Technology* 25:225–226.
Caspersson, T.; Zech, L.; and Johansson, C.
(1970) "Differential binding of alkylating fluorochromes in human chromosomes." *Experimental Cell Research* 60:315–319.
Caspersson, T.; Zech, L.; Johansson, C.; and Modest, E. J.
(1970) "Identification of human chromosomes by DNA-binding fluorescent agents." *Chromosoma* 30:215–227.
Casselman, W. B. Bruce.
(1959) *Histochemical Technique*. London: Methuen; New York: John Wiley.
Castañeda, M. Ruiz.
(1939) "Experimental pneumonia produced by typhus Rickettsia." *American Journal of Pathology* 15:467–475.
Caulfield, James B.
(1957) "Effects of varying the vehicle for osmic acid in tissue fixation." *Journal of Biophysical and Biochemical Cytology* 3:827–829.
Cavanagh, J. B.; Passingham, R. J.; and Vogt, J. A.
(1964) "Staining of sensory and motor nerves in muscles with Sudan black B." *Journal of Pathology and Bacteriology* 88:89–92.
Cejková, J.; Bolková, A.; and Lojda, Z.
(1973) "A study of acid mucopolysaccharides in cold microtome sections of normal and experimentally hydrated bovine corneas." *Histochemie* 36:167–172.
Celarier, Robert P.
(1956) "Tertiary butyl alcohol dehydration of chromosome smears." *Stain Technology* 31:155–157.
Chang, J. P.
(1956) "Staining mitochondria in frozen dried tissues." *Experimental Cell Research* 11:643–646.
Chang, J. P., and Hori, Samuel H.
(1961) "The section-freeze-substitution technique. I. Method." *Journal of Histochemistry and Cytochemistry* 9:292–300.
(1961) "The section-freeze-substitution technique. II. Application to localization of enzymes and other chemicals." *Annales d'Histochemie* 6:419–432.
Chang, J. P.; Hori, Samuel H ; and Yokoyama, Masao.
(1970) "A modified section freeze-substitution technique." *Journal of Histochemistry and Cytochemistry* 18:683–684.
Chang, Sing Chen.
(1972) "Hematoxylin-eosin staining of plastic embedded tissue sections." *Archives of Pathology* 93:344–351.
Chaplin, A. J., and Grace, S. R.
(1976) "An evaluation of some complexing methods for the histochemistry of calcium." *Histochemistry* 47:263–269.
Chapman, D. M.
(1975) "Dichromatism of bromphenol blue, with an improvement in the mercuric bromphenol blue technic for protein." *Stain Technology* 50:25–30.
(1977) "Eriochrome cyanin as a substitute for haematoxylin and eoxin." *Canadian Journal of Medical Technology* 39:65–66.
Chatton, E., and Lwoff, A.
(1930) "Imprégnation, par diffusion argentique, de l'infraciliature des ciliés marins et d'eau douce, après fixation cytologique et sans dessiccation." *Comptes Rendus Hebdomadaires des Séances et Mémoires de la Société de Biologie* 104:834–836.
(1935) "Les ciliés apostomes—première partie. Aperçu historique et général étude monographique des genres et des espèces." *Archives de Zoologie Expérimentale et Générale* 77:1–453.

(1936) "Technique pour l'étude des protozaires, spécialement de leurs structures superficielles (cenétome et argyrome)." *Bulletin de la Société Française de Microscopie* 5:25–39.

Chen, Tze-Tuan.
  (1942) "A staining rack for handling cover-glass preparations." *Stain Technology* 17:129–130.
  (1944a) "Staining nuclei and chromosomes in protozoa." *Stain Technology* 19:83–90.
  (1944b) "The nuclei in avian parasites. I. The structure of nuclei in *Plasmodium elongatum* with some considerations on technique." *American Journal of Hygiene* 40:26–34.

Chessick, Richard D.
  (1953) "Histochemical study of the distribution of esterases." *Journal of Histochemistry and Cytochemistry* 1:471–485.

Chesterman, W., and Leach, E. H.
  (1958) "A bleaching method for melanin and two staining methods." *Quarterly Journal Microscopical Science* 99:65–66.

Chèvremont, M., and Fréderic, J.
  (1943) "Une nouvelle méthode histochimique de mise en évidence des substances à fonction sulfhydrile. Application à l'épiderme, au poil et à la levure." *Archives de Biologie* 54:589–605.

Chiffelle, Thomas L., and Putt, Frederick A.
  (1951) "Propylene and ethylene glycol as solvents for Sudan IV and Sudan black B." *Stain Technology* 26:51–56.

Chipps, A. D., and Duff, G. L.
  (1942) "Glycogen infiltration of the liver cell nuclei." *American Journal of Pathology* 18:645–660.

Chresman, C. L.
  (1976) "Modified procedures for G-banding of mouse embryo chromosomes." *Stain Technology* 51:307–309.

Christenson, Leroy P.
  (1965) *Human Chromosome Methodology*. Ed. by Jorge J. Yunis. New York: Academic Press.

Christophers, A. J.
  (1956) "The differential leukocyte count: observations on the error due to method of spreading." *Medical Journal of Australia* 1:533–536.

Chubb, James C.
  (1963) "Acetic acid as a diluent and dehydrant in the preparation of whole, stained helminths." *Stain Technology* 37:179–182.

Chung, C. F., and Chen, Christian, M. C.
  (1970) "Restoring exhausted Schiff's reagent." *Stain Technology* 45:91–92.

Churg, Jacob, and Prado, Artie.
  (1956) "A rapid Mallory trichrome stain (chromotrope–aniline blue)." *Archives of Pathology* 62:505–506.

Churukian, Charles J., and Schenk, Eric A.
  (1976) "Iron Gallein elastic method—a substitute for Verhoeff's elastic tissue stain." *Stain Technology* 51:213–217.

Clark, George.
  (1947) "A simplified method for embedding cellular contents of body fluids in paraffin." *American Journal of Clinical Pathology* 17:256.
  (1945) "A simplified Nissl stain with thionin." *Stain Technology* 20:23–24.

Clark, George; Reed, C. S.; and Brown, F. M.
  (1973) "An evaluation and modification of Cole's hematoxylin." *Stain Technology* 48:189–191.

Clark, R. F., and Hench, M. E.
  (1962) "Practical application of acridine orange stain in the demonstration of fungi." *American Journal of Clinical Pathology* 37:237–238.

Clayden, E. C.
    (1952) "A discussion on the preparation of bone sections by the paraffin wax method with special reference to the control of decalcification." *Journal of Medical Laboratory Technologists* 10:103.

Clendenin, Thomas M.
    (1969) "Intraperitoneal colchicine and hypotonic KCL for enhancement of abundance and quality of meiotic chromosome spreads from hamster testes." *Stain Technology* 44:63–69.

Cocke, E. C.
    (1938) "A method for fixing and staining earthworms." *Science* 87:443–444.

Cohen, Isadore.
    (1949) "Sudan black B—a new stain for chromosome smear preparation." *Stain Technology* 24:117–184.

Cohen, Sophia M.; Gordon, Irving; Rapp, Fred; Macaulay, John C.; and Buckley, Sonia M.
    (1955) "Fluorescent antibody and complement-fixation tests of agents isolated in tissue culture from measles patients." *Proceedings of the Society for Experimental Biology and Medicine* 90:118–122.

Cole, E. C.
    (1933) "Ferric chloride as a mordant for phosphate ripened hematoxylin." Mimeo from author.
    (1943) "Studies on hematoxylin stains." *Stain Technology* 18:125–142.

Cole, Madison B., and Sykes, Stephen M.
    (1974) "Glycol methacrylate in light microscopy: a routine method for embedding and sectioning animal tissues." *Stain Technology* 49:387–400.

Cole, Wilbur V.
    (1946) "A gold chloride method for motor-end plates." *Stain Technology* 21:23–24.

Cole, Wilbur V., and Mielcarek, J. E.
    (1962) "Fluorochroming nuclei of gold chloride–stained motor endings." *Stain Technology* 37:35–39.

Coleman, Edward J.
    (1965) "A simplified autoradiographic dipping procedure—slides handled in groups of five." *Stain Technology* 40:240–241.

Collins, Doris N.; Katz, Sidney S.; and Harris, Albert H.
    (1961) "Preserving sputum for examination in the cytology laboratory." *American Journal of Clinical Pathology* 36:92–93.

Collins, E. M.
    (1969) "Improved paraffin sectioning with etched microtome knife edges." *Stain Technology* 44:33–37.

Comings, David E.
    (1975) "Chromosome banding." *Journal of Histochemistry and Cytochemistry* 23:461–462.

Conger, Alan D.
    (1960) "Dentist's Sticky wax: a cover sealing compound for temporary slides." *Stain Technology* 35:225.

Conger, Alan D., and Fairchild, Lucile M.
    (1953) "A quick-freeze method for making smear slides permanent." *Stain Technology* 28:281–283.

Conklin, James S.
    (1963) "Staining reactions of mucopolysaccharides after formalin containing fixatives." *Stain Technology* 38:56–59.

Conn, H. J.
    (1946) "The development of histological staining." *Ciba Symposia* 7.
    (1948) *History of Staining.* Geneva, N.Y.: Biotech Publications.
    (1969) *Biological Stains.* 8th ed. rev. by R. D. Lillie. Baltimore: Williams and Wilkins.

Conn, H. J., and Emmel, Victor M.
    (1960) *Staining Procedures.* Baltimore: Williams and Wilkins.

Conrad, M. E., Jr., and Crosby, W. H.
  (1961) "Bone marrow biopsy: modification of the Vim-Silverman needle." *Journal of Laboratory and Clinical Medicine* 57:642–645.
Conroy, J. D., and Toledo, A. B.
  (1976) "Metachromasia and improved histologic detail with toluidine blue–hematoxylin and eosin." *Veterinary Pathology* 13:78–80.
Controls for Radiation, Inc.
  n.d. *Con-Rad/Joftes Fluid Emulsion Radioautography Instruction Manual.* Cambridge, Mass.
Coolidge, Barbara J.
  (1972) "Commercial glycol-base antifreeze with added detergent as a frozen section clarifier and adherence aid." *Stain Technology* 47:170–171.
Coons, Albert H.
  (1956) "Histochemistry with labeled antibody." *International Review of Cytology.* Ed. by G. H. Bourne and J. F. Danielli. New York: Academic Press, V:1–23.
  (1958) "Fluorescent antibody methods." *General Cytochemical Methods,* I. Ed. by J. F. Danielli. New York: Academic Press, pp. 399–422.
Coons, Albert H.; Creech, H. J.; Jones, Norman; and Berliner, Ernest.
  (1942) "The demonstration of pneumococcal antigen in tissues by the use of fluorescent antibody." *Journal of Immunology* 45:159–170.
Coons, Albert H., and Kaplan, Melvin H.
  (1950) "Localization of antigen in tissue cells. II. Improvements in a method for the detection of antigen by means of fluorescent antibody." *Journal of Experimental Medicine* 91:1–13.
Coons, Albert H.; Leduc, Elizabeth H.; and Connolly, Jeanne M.
  (1955) "Studies on antibody production I. A method for the histochemical demonstration of specific antibody and its application to a study of the hyperimmune rabbit." *Journal of Experimental Medicine* 102:49–60.
Coons, Albert H.; Leduc, Elizabeth H.; and Kaplan, Melvin H.
  (1951) "Localization of antigen in tissue cells. VI. The fate of injected foreign proteins in the mouse." *Journal of Experimental Medicine* 93:173–188.
Coons, Albert H.; Snyder, John C.; Cheever, F. Sargent; and Murry, Edward S.
  (1950) "Localization of antigen in tissue cells. IV. Antigens of rickettsiae and mumps virus." *Journal of Experimental Medicine* 91:31–37.
Cordova, Margarito R., and Ploanco, E. R.
  (1962) "The use of plain agar in attaching paraffin sections to slides." *Technical Bulletin of the Registry of Medical Technologists* 32:129.
Corliss, John O.
  (1953) "Silver impregnation of ciliated protozoa by the Chatton–Lwoff technic." *Stain Technology* 28:97–100.
Cosslett, V. E.
  (1947) *The Electron Microscope.* New York: Interscience Publishers.
Cosslett, V. E., and Nixon, W. C.
  (1960) *X-ray Microscopy.* Cambridge, England: Cambridge University Press.
Courtright, R. C.
  (1966) "Use of polyester resins as a mounting medium for parasites." *Transactions American Microscopical Society* 85:319–320.
Cowdry, E. V.
  (1952) *Laboratory Technique in Biology and Medicine.* 3d ed. Baltimore: Williams and Wilkins.
Crabb, Edward D.
  (1949) "A rapid rosin-celloidin-paraffin method for embedding tissues." *Stain Technology* 24:87–91.
Craig, Elson L.; Frajola, Walter J.; and Greider, Marie H.
  (1962) "An embedding technique for electron microscopy using Epon 812." *Journal of Cell Biology* 12:190–194.

Cramer, Adelbert D.; Rogers, Eugene R.; Parker, John W.; and Lukes, Robert J.
(1973) "The Giemsa stain for tissue sections: an improved method." *American Journal of Clinical Pathology* 60:148–156.

Crandall, Frank B.
(1961) "An improved diamond knife holder for ultra-microtomy." *Stain Technology* 36:34–36.

Crary, D. D.
(1962) "Modified benzyl alcohol clearing on alizarin-stained specimens without loss of flexibility." *Stain Technology* 37:124–125.

Crozier, R. H.
(1968) "An acetic acid dissociation air-drying technique for insect chromosomes, with aceto-lactic orcein staining." *Stain Technology* 43:171–173.

Culling, C. E. A.
(1957) *Handbook of Histopathological Technique.* London: Butterworth.

Culling, Charles, and Vassar, Philip.
(1961) "Desoxyribose nucleic acid. A fluorescent histochemical technique." *Archives of Pathology* 71:88/76–92/80.

Cumley, R. W.; Crow, J. F.; and Griffin, A. B.
(1939) "Clearing specimens for demonstration of bone." *Stain Technology* 14:7–11.

Cunningham, G. J.; Bitensky, L.; Chayen, J.; and Silcox, A. A.
(1961) "The preservation of cytological and histochemical detail by a controlled temperature freezing and sectioning technique." *Annales d'Histochimie* 6:433–436.

Dalton, A. J.
(1955) "A chrome-osmium fixative for electron microscopy." *Anatomical Record* 121:281.

Danielli, J. F.
(1953) *Cytochemistry.* New York: John Wiley.

Daoust, R.
(1957) "Localization of deoxyribonuclease in tissue sections. A new approach to the histochemistry of enzymes." *Experimental Cell Research* 12:203–211.
(1961) "Localization of Deoxyribonuclease activity by the substrate film method." *General Cytochemical Methods,* II. Ed. by J. F. Danielli. New York: Academic Press.
(1964) "In vitro binding of nucleic acids to tissue sections after removal of tissue nucleic acids." *Journal of Histochemistry and Cytochemistry* 12:640–645.
(1968) "The localization of enzyme activities by substrate film methods—Evaluation and perspectives." *Journal of Histochemistry and Cytochemistry* 16:540–545.

Daoust, R., and Amano, Hamko.
(1960) "The localization of ribonuclease activity in tissue secretions." *Journal of Histochemistry and Cytochemistry* 8:131–134.

Darrow, Mary A.
(1952) "Synthetic orcein as an elastic tissue stain." *Stain Technology* 27:329–332.

Dart, Leroy H., and Turner, Thomas R.
(1959) "Fluorescence microscopy in exfoliative cytology." *Laboratory Investigation* 8, Part II:1513–1522.

Davenport, H. A.
(1948) "Protargol: old and new." *Stain Technology* 23:219–220.
(1960) *Histological and Histochemical Technics.* Philadelphia: W. B. Saunders.

Davenport, H. A., and Combs, C. M.
(1954) "Golgi's dichromate silver method. 3. Chromating fluids." *Stain Technology* 29:165–173.

Davenport, H. A.; Windle, W. F.; and Buch, R. H.
(1934) "Block staining of nervous tissue. IV. Embryos." *Stain Technology* 9:5–10.

Davidson, William M., and Smith, D. Robertson.
(1954) "A morphological sex difference in the polymorphonuclear neutrophil leucocytes." *British Medical Journal,* 2 July: 6–7.

Davies, Howard G.
(1958) "The determination of mass and concentration by microscopic interferometry." *General Cytochemical Methods,* I. Ed by J. F. Danielli. New York, Academic Press.

Davies, Helen, and Harmon, Pinkney J.
(1949) "A suggestion for prevention of loose sections in the Bodian Protargol method." *Stain Technology* 24:249.

Dawar, Bhagirath L.
(1973) "A combined relaxing agent and fixative for Triclads (Planarians)." *Stain Technology* 48:93.

Debruyn, P. P. H.; Farr, R. S.; Banks, Hilda; and Morthland, F. W.
(1953) "In vivo and in vitro affinity of diaminoacridine for nucleoproteins." *Experimental Cell Research* 4:174–180.

Debruyn, P. P. H.; Robertson, C.; and Farr, Richard S.
(1950) "In vivo affinity of diamino-acridines for nuclei." *Anatomical Record* 108:279–307.

Deck, J. David, and Desouza, G.
(1959) "A disrupting factor in silver staining techniques." *Stain Technology* 34:287.

Decosse, J. J., and Aiello, N.
(1966) "Feulgen hydrolysis: effect of acid and temperature." *Journal of Histochemistry and Cytochemistry* 14:601–604.

DeDuve, C.
(1959) "Lysosomes, a new group of cytoplasmic particles." *Subcellular Particles.* Ed. by T. Hayashi. New York: Ronald Press.
(1963*a*) "The lysosome." *Scientific American* 208 (May):64.
(1963*b*) "General properties of lysosomes: the lysosome concept." *Lysosomes.* Ed. by A. V. S. deReuck and M. P. Cameron. Boston: Little, Brown.

De Giusti, Dominic, and Ezman, Leon.
(1955) "Two methods for serial sectioning of arthropods and insects." *Transactions of American Microscopical Society* 74:197–201.

De Harven, Etienne.
(1958) "A new technique for carbon films." *Journal of Biophysical and Biochemical Cytology* 4:133–134.

Deitch, Arline Douglas.
(1955) "Microspectrophotometric study of the binding of the anionic dye, naphthol yellow S, by tissue sections and by purified protein." *Laboratory Investigation* 4:324–351.
(1961) "An improved Sakaguchi reaction for microspectrophotometric use." *Journal of Histochemistry and Cytochemistry* 9:477–483.
(1964) "A method for the cytophotometric estimation of nucleic acids using methylene blue." *Journal of Histochemistry and Cytochemistry* 12:451–461.

Deitch, Arline Douglas; Wagner, Dieter; and Richart, Ralph M.
(1968) "Conditions influencing the intensity of the Feulgen reaction." *Journal of Histochemistry and Cytochemistry* 16:371–379.

Delameter, Edward D.
(1948) "Basic fuchsin as a nuclear stain." *Stain Technology* 23:161–176.
(1951) "A staining and dehydrating procedure for handling of micro-organisms." *Stain Technology* 26:199–204.

De La Torre, Luis, and Salisbury, G. W.
(1962) "Fading of Feulgen-stained bovine spermatozoa." *Journal of Histochemistry and Cytochemistry* 10:39–41.

Delez, Arthur L., and Davis, Olive Stull.
(1950) "The use of oxalic acid in staining with phloxine and hematoxylin." *Stain Technology* 25:111–112.

Del Vecchio, P. R.; Dewitt, S. H.; Borelli, J. I.; Ward, J. B.; Wood, T. A., Jr.; and Malmgren, R. A.

(1959) "Application of millipore fixation technique to cytologic material." *Journal of the National Cancer Institute* 22:427–432.

de Martino, C.; Capanna, E.; Civitelli, M. V.; and Procicchiani, G.
(1965) "A silver staining reaction for chromosomes and nuclei." *Histochemie* 5:78–85.

Demke, Donald D.
(1952) "Staining and mounting helminths." *Stain Technology* 27:135–139.

Dempsey, Edward W., and Lansing, Albert I.
(1954) "Elastic Tissue." *International Review of Cytology*, III. Ed. by G. H. Bourne and J. B. Danielli. New York: Academic Press.

Dempster, Wilfred Taylor.
(1944a) "Principles of microscopic illumination and the problem of glare." *Journal of the Optical Society of America* 34:695–710.
(1944b) "Visual factors in microscopy." *Journal of the Optical Society of America* 34:711–717.
(1944c) "Properties of paraffin relating to microtechnique." *Michigan Academy of Science, Arts, and Letters* 29:251–264.

Denver Report
(1960) "A proposed standard system of nomenclature of human mitotic chromosomes." *Lancet* 1:1063–1065.

DePalma, P. A., and Young, G. G.
(1963) "Rapid staining of *Candida albicans* in tissue by periodic acid oxidation, basic fuchsin, and light green." *Stain Technology* 38:257–259.

DeRenzis, Frank A., and Schechtman, A.
(1973) "Staining by neutral red and trypan blue in sequence for assaying vital and nonvital cultured cells." *Stain Technology* 48:135–136.

DeSouza, E. J., and Kothare, S. N.
(1959) "A method for the cytochemical demonstration of succinic dehydrogenase in human leukocytes." *Journal of Histochemistry and Cytochemistry* 7:77–79.

Deuchar, E. M.
(1962) "Staining sections before autoradiographic exposure: excessive background graining caused by celestine blue." *Stain Technology* 37:324.

Dewitt, S. H.; Del Vecchio, P. R.; Borelli, J. I.; and Hilberg, A. W.
(1957) "A method for preparing wound washings and bloody fluids for cytologic evaluation." *Journal of the National Cancer Institute* 19:115–122.

Diacumakos, E. G.; Day, Emerson; and Kopac, M. J.
(1960) "A new plastic mounting medium for cytologic and histologic preparations." *Laboratory Investigation* 9:499–502.

Diamond, L. S.
(1945) "A new rapid stain technic for intestinal protozoa, using Tergitol-hematoxylin." *American Journal of Clinical Pathology* 15:68–69.

Diegenbach, P. C.
(1970) "Commercial fabric softeners for facilitating sectioning of refractory paraffin-embedded tissues." *Stain Technology* 45:303–304.

Donaldson, Patricia T.; Lillie, R. D.; and Pizzolato, Philip.
(1973) "Staining mast cells in sublimate-fixed guinea pig tissue." *Stain Technology* 48:47–48.

Dowding, Grace L.
(1959) "Plastic embedding of undecalcified bone." *American Journal of Clinical Pathology* 32:245–249.

Drets, Maximo E., and Shaw, Margery W.
(1971) "Specific banding patterns of human chromosomes." *Proceedings of the National Academy of Sciences* 68:2073–2077.

Duddy, James A., and Curran, Charles S.
(1962) "Mechanical counting of serial sections cut by a sliding microtome." *Stain Technology* 37:113–114.

Dunn, R. C.
(1946) "A hemoglobin stain for histologic use based on the cyanol-hemoglobin reaction." *Archives of Pathology* 41:676–677.
DuPont de Nemours and Company
(n.d.) *"Elvanol" polyvinyl alcohol for adhesives and binders.* Dupont Vinyl Products Bulletin V 2–254.
Durie, B., and Salmon, S.
(1975) "High speed scintillation autoradiography." *Science* 190:1093–1095.
Dutt, Mihir K.
(1974) "Cytochemical localization of nucleic acids with gallocyanin." *Acta Histochemica* 48:149–151.
Dyer, A. F.
(1963) "The use of lacto-propionic orcein in rapid squash methods for chromosome preparations." *Stain Technology* 38:85–90.
Dziabis, Marvin Dean.
(1958) "Luxol fast blue MBS, a stain for gross brain sections." *Stain Technology* 33:96–97.

Eapen, J.
(1960) "The effect of alcohol–acetic formalin, Zenker's fluid, and gelatin on the activity of lipase." *Stain Technology* 35:227–228.
Earle, W. R.
(1939) "Iron hematoxylin stain containing high concentration of ferrous iron." *Science* 89:323–324.
Eayrs, J. T.
(1950) "An apparatus for fixation and supravital staining of tissues by perfusion method." *Stain Technology* 25:137–142.
Edwards, J. H., and Young, R. B.
(1961) "Chromosome analysis from small volumes of blood." *Lancet* 2:48–49.
Egozcue, J., and Vilarasau de Egozcue, M.
(1966) "Simplified culture and chromosome preparations of primate leukocytes." *Stain Technology* 41:173–178.
Ehrenrich, Theodore, and Kerpe, Stase.
(1959) "A new rapid method of obtaining dry fixed cytological smears." *Journal of American Medical Association* 170:94–95.
Einarson, L.
(1951) "On the theory of gallocyanin-chromalum staining and its application for quantitative estimation of basophilia. A selective staining of exquisite progressivity." *Acta Pathologica et Microbiologica Scandinavica* 28:82–102.
Electron Microscopy Society of America
(1953) "Proceedings." *Journal of Applied Physiology* 24.
Elftman, Herbert.
(1952) "A direct silver method for the Golgi apparatus." *Stain Technology* 27:47–52.
(1954) "Controlled chromation." *Journal Histochemistry and Cytochemistry* 2:1–8.
(1956) "Response of the anterior pituitary to dichromate oxidation." *Journal of Histochemistry and Cytochemistry* 4:410.
(1957a) "A chrome-alum fixative for the pituitary." *Stain Technology* 32:25–28.
(1957b) "Phospholipid fixation by dichromate-sublimate." Stain Technology 32:29–31.
(1958) "Effects of fixation in lipoid histochemistry." *Journal of Histochemistry and Cytochemistry* 6:317–321.
(1959a) "Combined aldehyde-fuchsin and PAS staining of the pituitary." *Stain Technology* 34:77–80.
(1959b) "Aldehyde-fuchsin for pituitary cytochemistry," *Journal of Histochemistry and Cytochemistry* 7:98–100.
(1959c) "A Schiff reagent of calibrated sensitivity," *Journal of Histochemistry and Cytochemistry* 7:93–97.

(1960) "Hematoxylin as a pituitary stain." *Stain Technology* 35:97–101.

(1963) "Combined Schiff procedures." *Stain Technology* 38:127–130.

Elftman, Herbert, and Elftman, Alice G.

(1945) "Histological methods for the demonstration of gold in tissues." *Stain Technology* 20:59–62.

Elias, Julius M.

(1969) "Effects of temperature, poststaining rinses and ethanol-butanol dehydrating mixtures on methyl green–pyronin staining." *Stain Technology* 44:201–204.

Elleder, M., and Lojda, Z.

(1973) "Studies in lipid histochemistry. XI. New, rapid, simple and selective method for the demonstration of phospholipids." *Histochemie* 36:149–166.

Ellinger, P.

(1940) "Fluorescence microscopy in biology." *Biological Reviews of the Cambridge Philosophical Society* 15:323–350.

Elston, Robert N., and Sheehan, John F.

(1967) "Freon-Aerosol freezing of squashes or smears for removal of cover glass." *Stain Technology* 42:317.

Emig, William H.

(1959) *Microtechnique; Test and Laboratory Exercises.* Colorado Springs: The author.

Emmel, Victor M., and Cowdry, E. V.

(1964) *Laboratory Technique in Biology and Medicine.* Baltimore: Williams and Wilkins.

Endicott, K. M.

(1945) "Plasma or serum as a diluting fluid for thin smears of bone marrow." *Stain Technology* 20:25–26.

Enerbäck, Lennart.

(1969) "Detection of histamine in mast cells by O-phthalaldehyde reaction after liquid fixation." *Journal of Histochemistry and Cytochemistry* 17:757–759.

Engen, Paul C.

(1974) "Double embedding again." *Stain Technology* 49:375–380.

Engström, Arne.

(1956) "Historadiography." *Physical Techniques in Biological Research,* Vol. III. Ed. by Gerald Oster and Arthur W. Pollister. New York: Academic Press.

(1959) "X-ray microscopy." *Analytical Cytology.* Ed. by Robert C. Mellors. New York: Blakiston.

Enlow, Donald H.

(1954) "A plastic seal method for mounting sections of ground bones." *Stain Technology* 29:21–22.

(1961) "Decalcification and staining of ground thin sections of bone." *Stain Technology* 36:250–251.

Epple, August.

(1967) "A staining sequence for A, B, and D cells of pancreatic islets." *Stain Technology* 42:53–61.

Ericsson, Jan L. E.

(1965) "Transport and digestion of hemoglobin in the proximal tubule. I. Light microscopy and cytochemistry of acid phosphatase." *Laboratory Investigation* 14:1–15.

Ericsson, Jan L. E., and Biberfeld, P.

(1967) "Studies on aldehyde fixation. Fixation rates and their relation to fine structure and some histochemical reactions in liver." *Laboratory Investigation* 17:281–298.

Essner, Edward.

(1970) "Observations on hepatic and renal peroximes (microbodies) in the developing chick." *Journal of Histochemistry and Cytochemistry* 18:80–92.

Evans, T. C.

(1947) "Radioautographs in which tissue is mounted directly on photographic plate." *Proceedings of the Society for Experimental Biology and Medicine* 64:313–315.

Everett, Mona M., and Miller, William A.
 (1974) "The role of phosphotungstic and phosphomolybdic acids in connective tissue staining I. Histochemical studies." *Histochemical Journal* 6:25–34.
Ewen, A. B.
 (1962) "An improved aldehyde-fuchsin staining technique for neurosecretory products in insects." *Transactions of the American Microscopical Society* 81:94–96.

Fahimi, H. Daruish.
 (1967) "Perfusion and immersion fixation of rat liver with glutaraldehyde." *Laboratory Investigation* 16:737–750.
Falck, Bengt, and Hillarp, Nils-ake.
 (1959) "A note on the chromaffin reaction." *Journal of Histochemistry and Cytochemistry* 7:149.
Farber, Emmanuel, and Bueling, Ernest.
 (1956) "Histochemical localization of specific oxidative enzymes. V. The dissociation of succinic dehydrogenase from carriers by lipase and the specific histochemical localization of the dehydrogenase with phenazine methosulfate and tetrazolium salts." *Journal of Histochemistry and Cytochemistry* 4:357–362.
Farber, Emmanuel, and Louvriere, Connie D.
 (1956) "Histochemical localization of specific oxidative enzymes. IV. Soluble oxidation-reduction dyes as aids in histochemical localization of oxidative enzymes with tetrazolium salts." *Journal of Histochemistry and Cytochemistry* 4:347–356.
Farnsworth, Marjorie.
 (1956) "Rapid embedding of minute objects in paraffin." *Stain Technology* 31:295–296.
 (1963) "Handling insect eggs during fixation." *Stain Technology* 38:300–301.
Farquhar, Marilyn G.
 (1956) "Preparation of ultrathin tissue sections for electron microscopy." *Laboratory Investigation* 5:317–337.
Farquhar, Marilyn G., and Rinehart, J. F.
 (1954) "Electron microscopic studies on the anterior pituitary glands of castrate rats." *Endocrinology* 54:516–541.
Faulkner, R. R., and Lillie, R. D.
 (1945a) "A buffer modification of the Warthin-Starry silver method for spirochetes in single paraffin sections." *Stain Technology* 20:81–82.
 (1945b) "Dried egg white for Mayer's albumin fixative." *Stain Technology* 20:99–100.
Faust, E. C.; Russell, P. F.; and Jung, R. C.
 (1970) *Clinical Parasitology.* 4th ed. Philadelphia: Lea and Febiger.
Favorsky, B. A.
 (1930) "Eine Modifikation des silber Impregnations-verfahrens Rámon y Cajal für das periphere Nervensystem." *Anatomischer Anzeiger* 70:376–378.
Feder, Ned.
 (1962) "Polyvinyl alcohol as an embedding medium for lipid and enzyme histochemistry." *Journal of Histochemistry and Cytochemistry* 10:341–347.
Feder, Ned, and Sidman, Richard.
 (1957) "A method for applying different stains to alternate serial sections on a single microscope slide." *Stain Technology* 32:271–273.
 (1958) "Methods and principles of fixation by freeze-substitution." *Journal of Biophysical and Biochemical Cytology* 4:593–600.
Fenton, J. C. B., and Innes, James.
 (1945) "A staining method for malaria parasites in thick blood films." *Transactions of the Royal Society of Tropical Medicine and Hygiene* 39:87–90.
Ferguson-Smith, Malcolm A.
 (1961) "Chromosomes and Human Disease." *Progress in Medical Genetics.* Ed. by Arthur G. Steinberg. New York: Grune and Stratton.
 (1962) "The identification of human chromosomes." *Proceedings of the Royal Society of Medicine* 55:471–475.

Férnandez-Moran, H.
(1953) "A diamond knife for ultrathin sectioning." *Experimental Cell Research* 5:255.
(1956) "Application of a diamond knife for ultrathin sectioning to the study of the fine structure of biological tissues and metals." *Journal of Biophysical and Biochemical Cytology, Supplement* 2:29–31.
(1960) "Low temperature preparation techniques for electron microscopy of biological specimens based on rapid freezing with liquid helium II. Freezing and drying of biological materials. Part III." *Annals of New York Academy of Sciences* 85:689–713.
Ferreira, Alberto Vaz, and Combs, C. Murphy.
(1951) "Deterioration of nitrocellulose solutions caused by light." *Stain Technology* 26:81–84.
Feulgen, R.
(1914) "Über die Kohlenwassenstoffgruppe der echten Nukleinsaüre." *Zeitschrift für physiologische chemie* 92:154–158.
Feulgen, R., and Rossenbeck, H.
(1924) "Mikroskopisch-Chemischer Nachweis einer Nukleinsaüre von Typus Thymusnucleinsaüre und die darauf beruhende elektive Färbung von Zellkernen in mikroskopischen Präparaten." *Zeitschrift für physiologische chemie* 135:203–248.
Field, J. W.
(1941) "Further notes on a method of staining malarial parasites in thick blood films." *Transactions of the Royal Society of Tropical Medicine and Hygiene* 35:35–42.
Firminger, Harlan I.
(1950) "Carbowax embedding for obtaining thin tissue sections and study of intracellular lipids." *Stain Technology* 25:121–123.
Fischer, H. W.
(1957) "A technic for radiography of lymph nodes and vessels." *Laboratory Investigation* 6:522–527.
Fisher, E. R., and Haskell, A. E.
(1954) "Combined Gomori methods for demonstration of pancreatic alpha and beta cells." *American Journal of Clinical Pathology* 24:1433–1434.
Fitzgerald, Patrick J.
(1961) "Dry-mounting autoradiographic technic for intracellular localization of water-soluble compounds in tissue sections." *Laboratory Investigation* 10:846–856.
Fitzgerald, Patrick J., and Pohlmann, M.
(1966) "Use of the silver-hydroquinone sequence for the display of reticular fibers." *Stain Technology* 41:267–272.
Fitzgerald, Patrick J.; Simmel, Eva; Winstein, Jerry; and Martin, Cynthia.
(1953) "Radioautography: theory, technic and applications." *Laboratory Investigation* 2:181–222.
Fitz-William, William G.; Jones, Georgeanna Seegar; and Goldberg, Benjamin.
(1960) "Cryostat techniques: methods for improving conservation and sectioning of tissues." *Stain Technology* 35:195–204.
Flax, Martin, and Caulfield, James.
(1962) "Use of methacrylate embedding in light microscopy." *Archives of Pathology* 74:387–395.
Flax, Martin, and Pollister, Arthur W.
(1949) "Staining of nucleic acids by Azure A." *Anatomical Record* 105:536–537.
Foley, James O.
(1943) "A Protargol method for staining nerve fibers in frozen or celloidin sections." *Stain Technology* 18:27–33.
Foot, Nathan Chandler.
(1929) "Comments on the impregnation of neuroglia with ammoniacal silver salts." *American Journal of Pathology* 51:223–238.
Ford, C. E.
(1962) "Methodology of chromosomal analysis in man." *National Cancer Institute Monograph* 7:105–113.

Ford, C. E., and Hamerton, J. E.
  (1956) "The chromosomes of man." *Nature* 178:1020–1023.
Ford, E. H. R., and Woollam, D. H. M.
  (1963) "A colchicine, hypotonic citrate, air drying sequence for foetal mammalian chromosomes." *Stain Technology* 38:271–274.
Ford, Lee.
  (1965) "Leukocyte culture and chromosome preparations from adult dog blood." *Stain Technology* 40:317–320.
Fox, M., and Zeiss, J. M.
  (1961) "Chromosome preparation from fresh and cultural tissues using a modification of the drying technique." *Nature* 192:1213–1214.
Frandsen, John C.
  (1964) "Serial sections: simplified handling of paraffin ribbons on a floating-out bath. *Stain Technology* 39:279–282.
Frankel, Howard H., and Peters, Robert L.
  (1964) "A modified calcium-cobalt method for the demonstration of alkaline phosphatase." *American Journal of Clinical Pathology* 42:324–327.
Frankel, J., and Heckmann, K.
  (1968) "A simplified Chatton-Lwoff silver impregnation procedure for use in experimental studies with ciliates." *Transactions American Microscopical Society* 87:317–321.
Freed, Jerome J.
  (1955) "Freeze-drying technics in cytology and cytochemistry." *Laboratory Investigation* 4:106–121.
Freeman, Barbara L.; Moyer, Elizabeth K.; and Lassek, Arthur M.
  (1955) "The pH of fixing fluids during fixation of tissues. *Anatomical Record* 121:593–600.
Freeman, James A.
  (1964) *Cellular Fine Structure.* New York: McGraw-Hill.
Friedland, Lester M.
  (1951) "A note on frozen section technic." *American Journal of Clinical Pathology* 21:797.
Friend, William G.
  (1963) "A microstrainer for handling small ova." *Stain Technology* 38:205–206.
Frigerio, Norman A., and Shaw, Michael J.
  (1969) "A simple method for determination of glutaraldehyde," *Journal of Histochemistry and Cytochemistry* 17:176.
Frøland, Anders.
  (1965) "Photographic recording and dye staining of chromosomes for autoradiography and morphology." *Stain Technology* 40:41–43.
Frost, H. M.
  (1959) "Staining of fresh, undecalcified thin bone sections." *Stain Technology* 34:135–146.
Fullmer, Harold M., and Lillie, R. D.
  (1958) "The peracetic acid–aldehyde fuchsin stain." *Journal of Histochemistry and Cytochemistry* 6:391.

Gabor, D.
  (1948) *The Electron Microscope.* New York: Chemical Publishing Co.
Gage, Simon Henry.
  (1943) *The Microscope.* Ithaca, N.Y.: Comstock.
Gairdner, B. M.
  (1969) "Deteriorated paraformaldehyde: an insidious cause of failure in aldehyde-fuchsin staining." *Stain Technology* 44:52–53.
Galigher, A. E.
  (1934) *The Essentials of Practical Microtechnique.* Privately published.

Galigher, A. E., and Kozloff, Eugene N.
  (1964) *Essentials of Practical Microtechnique*. Philadelphia: Lea and Febiger.
Gallimore, John C.; Bauer, E. C.; and Boyd, George A.
  (1954) "A non-leaching technic for autoradiography." *Stain Technology* 29:95–98.
Galtsoff, Paul S.
  (1956) "Simple method of making frozen sections." *Stain Technology* 31:231.
Garcia Poblete, Eduardo; Herraez, Reyes Flores; and Lopez de Rego Martinez, Jacobo.
  (1976) "Modification of Heindenhain's Azocarmine–aniline blue for formaldehyde-fixed tissues." *Mikroskopie* 32:114–116.
Gardner, D. L.
  (1958) "Preparation of bone marrow sections." *Stain Technology* 33:295–297.
Gardner, Harold H., and Punnett, Hope H.
  (1964) "An improved squash technique for human male meiotic chromosomes: softening and concentration of cells; mounting in Hoyer's medium." *Stain Technology* 39:245–248.
Gatenby, J. Brontë, and Beams, H. W.
  (1950) *The Microtomists's Vade-Mecum*. London: J. and A. Churchill.
Gavin, Mary Ann, and Lloyd, Bolivar, J., Jr.
  (1959) "Knives of high silica content glass for thin sectioning." *Journal of Biophysical and Biochemical Cytology* 5:507.
Gavin, Thelma.
  (1938) "Spirochetal stain on paraffin sections." *American Journal of Clinical Pathology, Technical Supplement* 2: 144–145.
Gay, H., and Anderson, T. F.
  (1954) "Serial sections for electron microscopy." *Science* 120:1071–1073.
Geil, Robert G.
  (1961) "A simple technic for salvage of histologic sections from broken microslides." *Technical Bulletin of the Registry of Medical Technologists* 31:195–196.
Gelei, J. V.
  (1932) "Eine neue Goldmethode zur Ciliatenforschung und eine neue Ciliate: *Colpidium pannonicum.*" *Archiv für Protistenkunde* 77:219–230.
  (1935) "Eine neue Abänderung der Klein'schen trockenen Silbermethode und das Silberliniensystem von *Glaucoma scintillans.*" *Archiv für Protistenkunde* 84:446–455.
Gengozian, N.; Batson, J. S.; and Dide, P.
  (1964) "Hematologic and cytogenetic evidence for hematopoietic chimerism in the marmoset, *Tamarinus nigricollis.*" *Cytogenetics* 3:384–393.
George, J. C.; and Ambadkar, P. M.
  (1963) "Histochemical demonstration of lipids and lipase activity in rat testis." *Journal of Histochemistry and Cytochemistry* 11:420–425.
George, J. C., and Iype, P. Thomas.
  (1960) "Improved histochemical demonstration of lipase activity." *Stain Technology* 35:151–152.
George, K. P.
  (1971) "Quinacrine mustard—a selective fluorescent stain for the Y chromosome in human tissues for routine cytogenetic screening." *Stain Technology* 46:34–36.
Geren, B. B., and McCullock, D.
  (1951) "Development and use of the Minot Rotary microtome for thin sectioning." *Experimental Cell Research* 2:97–102.
German, W. M.
  (1939) "Hortega's silver impregnation methods. Technique and applications." *American Journal of Clinical Pathology Technical Supplement* 3:13–19.
Gersh, Isidore.
  (1932) "The Altman technique for fixation by drying while freezing." *Anatomical Record* 53:309–337.
  (1956) "The preparation of frozen-dried tissue for electron microscopy." *Journal of Biophysical and Biochemical Cytology, Supplement* 2:37–43.

(1959) "Fixation and staining." *The Cell,* I. Ed. by Jean Brachet and Alfred E. Mirsky. New York: Academic Press.

Gettner, I. R., and Ornstein, L.
(1956) "Microtomy." *Physical Techniques in Biological Research,* III. Ed. by Gerald Oster and Arthur W. Pollister. New York: Academic Press.

Geyer, Günther.
(1962) "Histochemical-methylation with methanol and thionylchloride." *Acta Histochemica* 14:284–296.

Gill, G. W.; Frost, J. K., and Miller, K. A.
(1974) "A new formula for a half-oxidized hematoxylin solution that neither overstains nor requires differentiation." *Acta cytologica* 18:300–311.

Giménez, D. F.
(1964) "Staining rickettsiae in yolk sac cultures." *Stain Technology* 39:135–140.

Giolli, Roland A.
(1965) "A note on the chemical mechanism of the Nauta-Gygax technique." *Journal of Histochemistry and Cytochemistry* 13:206–210.

Giovacchini, Rubert P.
(1958) "Affixing Carbowax sections to slides for routine staining." *Stain Technology* 33:274–278.

Gladden, Margaret H.
(1970) "A modified pyridine-silver stain for teased preparation of motor and sensory nerve endings in skeletal muscle." *Stain Technology* 45:161–164.

Glauert, Audrey M.
(1965) "The fixation and embedding of biological specimens." *Techniques for Electron Microscopy.* Ed. by Desmond H. Kay. Philadelphia: F. A. Davis.

Glauert, Audrey M., ed.
(1974) *Practical Methods in Electron Microscopy,* II and III. New York: Elsevier.

Glauert, Audrey M., and Glauert, R. H.
(1958) "Araldite as an embedding medium for electron microscopy." *Journal of Biophysical and Biochemical Cytology* 4:191–194.

Glegg, R. E.; Clermont, Y.; and Leblond, C. P.
(1952) "The use of lead tetra-acetate, benzidine, O-dianisidine, and a 'Film Test' in investigating the periodic-acid–Schiff technic." *Stain Technology* 27:277–305.

Glenner, George G.
(1957) "Simultaneous demonstration of bilirubin, hemosiderin and lipofuscin pigments in tissue sections." *American Journal of Clinical Pathology* 27:1–5.
(1963) "A re-evaluation of the ninhydrin-Schiff reaction." *Journal of Histochemistry and Cytochemistry* 11:285–286.

Glenner, George G., and Lillie, R. D.
(1957a) "A rhodocyan technic for staining anterior pituitary." *Stain Technology* 32:187–190.
(1957b) "The histochemical demonstration of indole derivatives by the post-coupled p-dimethylaminobenzylidene reaction." *Journal of Histochemistry and Cytochemistry* 5:279–296.
(1959) "Observations on the diazonium coupling reaction for the histochemical demonstration of tyrosine: metal chelation and formazon variants." *Journal of Histochemistry and Cytochemistry* 7:416–421.

Glick. David.
(1949) *Techniques of Histo- and Cytochemistry.* New York: Interscience Publishers.
(1962) *Quantitative Chemical Techniques of Histo- and Cytochemistry.* 2 vols. New York: Interscience Publishers.

Glick, David, and Malstrom, B. G.
(1952) "Studies in histochemistry. XXIII. A simple and efficient freezing-drying apparatus for the preparation of embedded tissue." *Experimental Cell Research* 3:125–235.

Glynn, J. H.
  (1935) "The application of the Gram stain to paraffin sections." *Archives of Pathology* 20:896–899.
Goh, Kong-OO.
  (1965) "Human cytogenetics." *DM, Disease-a-Month*, April.
Goland, Philip P.; Jason, Robert S.; and Berry, Kathryn P.
  (1954) "Combined Carbowax-paraffin technic for microsectioning fixed tissues." *Stain Technology* 29:5–8.
Goldberg, Arthur F.
  (1964) "Acid phosphatase activity in Auer bodies." *Blood* 24:305–308.
Goldfischer, Sidney; Essner, Edward; and Novikoff, Alex B.
  (1964) "The localization of phosphatase activities at the level of ultrastructure." *Journal of Histochemistry and Cytochemistry* 12:72–95.
Goldman, Morris.
  (1949) "A single solution ironhematoxylin stain for intestinal protozoa." *Stain Technology* 24:57–60.
  (1968) *Fluorescent Antibody Methods*. New York: Academic Press.
Goldstein, D. J.
  (1962) "Ionic and non-ionic bonds in staining, with special reference to the action of urea and sodium chloride on the staining of elastic fibers and glycogen." *Quarterly Journal of Microscopical Science* 103:477–492.
Gomori, George.
  (1936) "Microtechnical demonstration of iron." *American Journal of Pathology* 12:655–663.
  (1941a) "The distribution of phosphatase in normal organs and tissues." *Journal of Cellular and Comparative Physiology* 17:71–84.
  (1941b) "Observations with differential stains on human islets of Langerhans." *American Journal of Pathology* 17:395–406.
  (1946) "The study of enzymes in tissue sections." *American Journal of Clinical Pathology* 16:347–352.
  (1948) "Chemical character of enterochromaffin cells." *Archives of Pathology* 45:48–55.
  (1950a) "An improved histochemical technic for acid phosphatase." *Stain Technology* 25:81–85.
  (1950b). "A rapid one-step trichrome stain." *American Journal of Clinical Pathology* 20:662–664.
  (1950c) "Aldehyde-fuchsin; a new stain for elastic tissue." *American Journal of Clinical Pathology* 20:665–666.
  (1950d) "Sources of error in enzymatic histochemistry." *Journal of Laboratory and Clinical Medicine* 35:802–809.
  (1951) "Alkaline phosphatase of cell nuclei." *Journal of Laboratory and Clinical Medicine* 37:526–531.
  (1952) *Microscopic Histochemistry*. Chicago: University of Chicago Press.
  (1953) "Human esterases." *Journal of Laboratory and Clinical Medicine* 42:445–453.
  (1954a) "The histochemistry of mucopolysaccharides." *British Journal of Experimental Pathology* 35:377–380.
  (1954b) "Histochemistry of the enterochromaffin substance." *Journal of Histochemistry and Cytochemistry* 2:50–53.
  (1955) "Histochemistry of human esterases." *Journal of Histochemistry and Cytochemistry* 3: 3:479–484.
  (1956) "Histochemical methods for acid phosphatase." *Journal of Histochemistry and Cytochemistry* 4:453–461.
Gonzalez, Romeo.
  (1959a) "The removal of mercury after fixation in sublimate-containing mixtures." *Stain Technology* 34:111–112.
  (1959b) "Differentiation of mastocyte granules by tartrazine counterstaining after PAS procedure." *Stain Technology* 34:173–174.

Gordon, Harold.
    (1953) "Method for storing wet histologic accessions and disposing of autopsy material." *Laboratory Investigation* 2:152–153.
Gough, J., and Fulton, J. D.
    (1929) "A new fixative for mitochondria." *Journal of Pathology and Bacteriology* 32:765–769.
Gower, W. Carl.
    (1939) "A modified stain and procedure for trematods." *Stain Technology* 14:31–32.
Gradwohl, R. B. H.
    (1963) *Clinical Laboratory Methods and Diagnosis,* II. St. Louis: C. V. Mosby.
Graupner, Heinz, and Weissberger, Arnold.
    (1931) "Uber der Verwendung des Dioxanes beim Einbetten mikroskopischer Objekte. Mitteilungen zur mikroskopischen Technik I." *Zoologischer Anzeiger* 96:204–206.
    (1933) "Die Verwendung von Lösungen in Dioxan als Fiexierungsmittel für Gefrierschnitte." *Zoologischer Anzeiger* 102:39–44.
Graves, K. D.
    (1943) "Restoration of dried biopsy tissue." *American Journal of Clinical Pathology,* Technical Section 7:111.
Gray, Peter.
    (1954) *The Microtomists's Formulary and Guide.* New York: Blakiston.
    (1964*a*) "Making dry mounts of Foraminifera and Radiolaria." *Ward's Bulletin* 6.
    (1964*b*) *Handbook of Basic Microtechnique.* 3d ed. New York: Blakiston.
Green, James A.
    (1956) "Luxol Fast Blue MBS: a stain for phase contrast microscopy." *Stain Technology* 31:219–221.
Greenblatt, Robert A., and Manautou, Jorge Martinez.
    (1957) "A simplified staining technique for the study of chromosomal sex in oral mucosal and peripheral blood smears." *American Journal of Obstetrics and Gynecology* 74:629–534.
Greenstein, J. S.
    (1957) "A rapid phloxine–methylene blue oversight stain for formalin-fixed material." *Stain Technology* 32:75–77.
    (1961) "A simplified five-dye stain for sections and smears." *Stain Technology* 36:87–88.
Gregg, V. R., and Puckett, W. O.
    (1943) "A corrosive sublimate fixing solution for yolk-laden amphibian eggs." *Stain Technology* 18:179–180.
Gridley, Mary Francis.
    (1951) "A modification of the silver impregnation method of staining reticular fibers." *American Journal of Clinical Pathology* 21:897–899.
    (1953) "A stain for fungi in tissue sections." *American Journal of Clinical Pathology* 23:303–307.
    (1957) *Manual of Histologic and Special Staining Technics.* Washington, D.C.: Armed Forces Institute of Pathology.
Griffin, A. B.
    (1960) "Mammalian pachytene chromosome mapping and somatic chromosome identification." *Journal of Cellular and Comparative Physiology* 56:113–121.
Griffin, Lawrence E., and McQuarrie, Agnes M.
    (1942) "Iron hematoxylin staining of salivary gland chromosomes in Drosophila." *Stain Technology* 17:41–42.
Grimley, Philip M.
    (1964) "A tribasic stain for thin sections of plastic-embedded, osmic acid fixed tissues." *Stain Technology* 39:229–233.
Grimley, Philip M.; Albrecht, Joseph M.; and Michelitch, Herman J.
    (1965) "Preparation of large Epoxy sections for light microscopy as an adjunct to fine-structure studies." *Stain Technology* 40:357–366.

Groat, Richard A.
   (1949) "Initial and persisting staining power of solutions of iron hematoxylin lake."
      Stain Technology 24:157–163.
Grocott, R. G.
   (1955) "Stain for fungi in tissue sections and smears, using Gomori's methenamine–
      silver nitrate method." *American Journal of Clinical Pathology* 25:975–979.
Grunbaum, Benjamin; Geary, John R., Jr.; and Glick, David.
   (1956) "Studies in Histochemistry: the design and use of improved apparatus for the
      preparation and freezing-drying of fresh-frozen sections of tissue." *Journal of
      Histochemistry and Cytochemistry* 4:555–560.
Guard, Hormez, R.
   (1959) "A new technic for differential staining of the sex chromatin and the determina-
      tion of its incidence in exfoliated vaginal epithelial cells." *American Journal of
      Clinical Pathology* 32:145–151.
Gude, W. D.
   (1968) *Autoradiographic Techniques.* Englewood Cliffs: Prentice-Hall.
Gude, W. D., and Odell, T. T.
   (1955) "Vinisil as a diluent in making bone marrow smears." *Stain Technology* 30:27–28.
Gude, W. D.; Upton, Arthur C.; and Odell, T. T.
   (1955) "Giemsa staining of autoradiograms prepared with stripping film." *Stain
      Technology* 30:161–162.
Guillery, R. W.; Shirra, B.; and Webster, K. E.
   (1961) "Differential impregnation of degenerating nerve fibers in paraffin-embedded ma-
      terial." *Stain Technology* 36:9–13.
Gurr, Edward.
   (1956) *A Practical Manual of Medical and Biological Staining Techniques.* New York:
      Interscience.
   (1958) *Methods of Analytical Histology and Histochemistry.* London: Leonard Hill.
   (1969) *Encyclopaedia of Microscopic Stains.* London: Edward Gurr; Baltimore:
      Williams and Wilkins.
Gurr, G. T.
   (1953) *Biological Staining Methods.* London: G. T. Gurr.
Guyer, M. F.
   (1953) *Animal Micrology.* Chicago: University of Chicago Press.

Hack, M. H.
   (1952) "A new histochemical technique for lipids applied to plasmal cells." *Anatomical
      Record* 112:275–301.
Hajian, Ahmad.
   (1961) "Note on trichrome stain." *Technical Bulletin of the Registry of Medical
      Technologists* 31:92.
Hale, Arthur J.
   (1952) "The effect of temperature and of relative humidity on sectioning of tissues em-
      bedded in polyethylene glycol wax." *Stain Technology* 27:189–192.
   (1955) "The effect of formalin on the periodic acid Schiff staining of certain types of
      mucus." *Journal of Histochemistry and Cytochemistry* 3:421–429.
   (1957) "The histochemistry of polysaccharides." *International Review of Cytology,* VI.
      Ed. by G. H. Bourne and J. F. Danielli. New York: Academic Press.
Hale, C. W.
   (1946) "Histochemical demonstration of acid polysaccharides in animal tissue." *Nature*
      157:802.
Hale, Dean M.; Cromartie, William J.; and Dobson, Richard L.
   (1960) "Luxol fast blue as a selective stain for dermal collagen." *Journal of Investiga-
      tive Dermatology* 35:293–294.
Hale, L. J.
   (1958) *Biological Laboratory Data.* London: Methuen; New York: John Wiley.

Hall, C. E.
  (1953) *Introduction to Electron Microscopy*. New York: McGraw-Hill.
Halmi, Nicholas S.
  (1950) "Two types of basophils in the anterior pituitary of the rat and their respective cytophysiological significance." *Endocrinology* 47:289–299.
Ham, Arthur Worth.
  (1957) *Histology*. 3d ed. Philadelphia: Lippincott.
Hamerton, John L.
  (1961) "Sex chromatin and human chromosomes. *International Review of Cytology*, XII. Ed by G. H. Bourne and J. F. Danielli. New York: Academic Press.
Hamiyn, J. H.
  (1957) "Application of the Nauta-Gygax technic for degenerating axons to mounted sections." *Stain Technology* 32:123–126.
Hancox, N. M.
  (1957) "Experiments on the fundamental effects of freeze substitution." *Experimental Cell Research* 13:263–275.
Hanker, J. S.; Yates, P. E.; Clapp, D. H.; and Anderson, W. A.
  (1972) "New methods for the demonstration of lysosomal hydrolases by the formation of osmium blacks." Histochemie 30:201.
Hansen, D. W.; Hunter, D. T.; Richards, D. F.; and Allred, L.
  (1970) "Acridine orange in the staining of blood parasites." *Journal of Parasitology* 56:386–387.
Hanson, A. A., and Oldemeyer, D. L.
  (1951) "Staining root tip smears with aceto-carmine." *Stain Technology* 26:241–242.
Hanson, V., and Hermodsson, L. H.
  (1960) "Freeze-drying of tissues for light and electron microscopy." *Journal of Ultrastructure Research* 4:332–348.
Harada, Kiyoshi.
  (1957) "Selective staining of mast cell granules with chrysoidin stain." *Stain Technology* 32:183–186.
  (1973) "Effect of prior oxidation on the acid-fastness of mycobacteria." *Stain Technology* 48:269–273.
  (1976) "The nature of mycobacterial acid-fastness." *Stain Technology* 51:255–260.
  (1976) "Periodic acid–methenamine silver stain for mycobacteria in tissue sections." *Stain Technology* 51:278–280.
Hardonk, M. J., and vanDuijan, J.
  (1964a) "The mechanism of the Schiff reaction as studied with histochemical model systems. *Journal of Histochemistry and Cytochemistry* 12:748–751.
  (1964b) "A quantitative study of the Feulgen reaction with the aid of histochemical model systems." *Journal of Histochemistry and Cytochemistry* 12:752–757.
  (1964c) "Studies on the Feulgen reaction with histochemical model systems." *Journal of Histochemistry and Cytochemistry* 12:758–767.
Harmon, John W.
  (1950) "The selective staining of mitochondria." *Stain Technology* 25:69–72.
Harnden, D. G.
  (1959) "A human skin culture technique used for cytological examinations." *British Journal of Experimental Pathology* 41:31–37.
Harnden, D. G., and Brunton, Sheila.
  (1965) "The skin culture technique." *Human Chromosome Methodology*. Ed. by Jorge J. Yunis. New York: Academic Press.
Harris, G. W., and Donovan, B. T.
  (1966) *The Pituitary Gland*. 3 vols. Berkeley: University of California Press.
Hartman-Leddon Company.
  (1952) *Informative Bulletin 306-2*. Philadelphia.
Hartroft, W. S.
  (1951) "Fluorchromy as an aid in the resolution of the specific granules of the islets of Langerhans." *Nature* 168:1000.

Hartz, P. H.
   (1945) "Frozen sections from Bouin-fixed material in histopathology." *Stain Technology* 20:113–114.
   (1947) "Simultaneous histologic fixation and gross demonstration of calcification." *American Journal of Clinical Pathology* 17:750.
Haupt, A. W.
   (1930) "A gelatin fixative for paraffin sections." *Stain Technology* 5:97.
Hause, Welland A.
   (1959) "Saw for preparation of blocks of bone." *Technical Bulletin of the Registry of Medical Technicians* 29:101.
Haust, M. Daria.
   (1958) "Tetrahydrofuran (THF) for dehydration and infiltration." *Laboratory Investigation* 7:58–67.
   (1959) "Tetrahydrofuran (THF) for routine dehydration clearing and infiltration." *Technical Bulletin of the Registry of Medical Technicians* 29:33–37.
Haviland, Thomas N.
   (1963) "Restoration of rust-stained formalin-preserved gross specimens." *American Journal of Clinical Pathology* 39:364.
Hayat, M. A., ed.
   (1970) *Principles and Techniques of Electron Microscopy,* Vol. I. New York: Van Nostrand.
   (1972) Vol. II.
Heady, Judith, and Rogers, T. Edwin.
   (1962) "Turtle blood cell morphology." *Proceedings of the Iowa Academy of Sciences* 69:587–590.
Hegner, Robert W.; Cort, William W.; and Root, Francis M.
   (1927) *Outlines of Medical Zoology.* New York: Macmillan.
Hendrickson, Anita; Kunz, Sandra; and Kelly, Douglas E.
   (1968) "NaOH–HIO$_4$ treatment of osmium–collidine fixed Epoxy sections to facilitate staining after autoradiography." *Stain Technology* 43:175–176.
Herlant, Marc.
   (1960) "Étude critique de deux techniques nouvelles destinées a mettre en évidence les différentes catégories cellulaires présentes dans la glande pituitaire." *Bulletin de Microscopie Appliquée* 10:37–44.
Herr, Barbara Evelyn; Coleman, Paul D.; and Griggs, Robert C.
   (1976) "A Bodian method for mounted frozen sections. *Stain Technology* 51:261–265.
Hetherington, Duncan C.
   (1936) "Pinacyanol as a supravital stain for blood." *Stain Technology* 11:153–154.
Hicks, J. D., and Matthaei, E.
   (1955) "Fluorescence in Histology." *Journal of Pathology and Bacteriology* 70:1–12.
   (1958) "A selective fluorescence stain for mucin." *Journal of Pathology and Bacteriology* 75:473–476.
Hillary, B. B.
   (1938) "Permanent preparations from rapid cytological technics." *Stain Technology* 13:161–167.
Hilleman, Howard H., and Lee, C. H.
   (1953) "Organic chelating agents for decalcification of bones and teeth." *Stain Technology* 28:285–286.
Himes, Marion, and Moriber, Louis.
   (1956) "A triple stain for desoxyribonucleic acid, polysaccharides and proteins." *Stain Technology* 31:67–70.
Histochemical Society
   (1953) "Proceedings." *Journal of Histochemistry and Cytochemistry* 1:387–388.
Hodgman, Charles D.
   (1957) *Handbook of Chemistry and Physics.* Cleveland: Chemical Rubber Publishing Co.

Hoefert, Lynn L.
(1968) "Polychromatic stains for thin sections of *Beta* embedded in Epoxy resin." *Stain Technology* 43:145–151.
Hoffmann, E. O., and Miller, M. J.
· (1975) "Immunofluorescent staining of amoebae in routine paraffin-embedded tissues." *Journal of Parasitology* 61:1104.
Holczinger, L., and Bálint, Zsuzsa.
(1961) "The staining properties of 'masked' lipids." *Acta Hitochemica* 11:284–288.
Hollander, David H.
(1963) "An oil-soluble anti-oxidant in resinous mounting media to inhibit fading of Romanowsky stains." *Stain Technology* 38:288–289.
Hollister, Gloria.
(1934) "Clearing and dyeing fish for bone study." *Zoologica* 12:89–101.
Holmes, W. C.
(1929) "The mechanism of staining. The case for the physical theories." *Stain Technology* 4:75–80.
Holt, Margaret; Cowing, R. F.; and Warren, S.
(1949) "Preparation of radioautographs of tissues without loss of water soluble P^{32}." *Science* 110:328–329.
Holt, Margaret; Sommers, S. C.; and Warren, S.
(1952) "Preparation of tissue sections for quantitative histochemical studies." *Anatomical Record* 112:177–186.
Holt, Margaret, and Warren, Shields.
(1950) "A radioautograph method for detailed localization of radioactive isotopes in tissues without isotope loss." *Proceeding of the Society for Experimental Biology and Medicine* 73:545.
(1953) "Freeze-drying tissues for autoradiography." *Laboratory Investigation* 2:1–14.
Holt, S. J.
(1956) "The value of fundamental studies of staining reactions in enzyme histochemistry, with reference to indoxyl methods for esterases." *Journal of Histochemistry and Cytochemistry* 4:541–554.
Holt, S. J.; Hobbiger, Eluned E.; and Pawan, G. L. S.
(1960) "Preservation of integrity of rat tissues for cytochemical staining purposes." *Journal of Biophysical and Biochemical Cytology* 7:383–386.
Hood, R. C. W. S., and Neill, W. M.
(1948) "A modification of alizarine red S technic for demonstrating bone formation." *Stain Technology* 23:209–218.
Hori, Samuel H.
(1963) "A simplified acid hematein test for phospholipids." *Stain Technology* 38:221–225.
Hori, Samuel H., and Chang, Jeffrey P.
(1963) "Demonstration of lipids in mitochondria and other cellular elements." *Journal of Histochemistry and Cytochemistry* 11:115–116.
Horikawa, Masakatsu, and Kuroda, Yukiaki.
(1959) "In vitro cultivation of blood cells of *Drosophila melanogaster* in a synthetic medium." *Nature* 184:2017–2018.
Hörmann, H.; Grassman, W.; and Fries, G.
(1958) "Über den Mechanismus der Schiffschen Reaktion." *Justus Liebigs Annalen der Chemie* 616:125–147.
Horn, Robert G., and Spicer, S. S.
(1964*a*) "Sulfated mucopolysaccharide and basic protein in certain granules of rabbit leukocytes." *Laboratory Investigation* 13:1–15.
(1964*b*) "Sulfated mucopolysaccharides in azurophile granules of immature granulocytes of the rabbit." *Journal of Histochemistry and Cytochemistry* 12:33.
Horobin, R. W.; Flemming, L.; and Kevill-Davies, I. M.
(1974) "Basic Fuchsin–ferric chloride: a simplification of Weigert's Resorcin-fuchsin stain for elastic fibers." *Stain Technology* 49:207–210.

Horobin, R. W., and Kevill-Davies, I. M.
   (1971) "Basic Fuchsin in acid alcohol: a simplified alternative to Schiff reagent." *Stain Technology* 46:53–58.
Hotchkiss, R. D.
   (1948) "A microchemical reaction resulting in the staining of polysaccharide structures in fixed tissues." *Archives of Biochemistry* 16:131–141.
Houck, C. E., and Dempsey, E. W.
   (1954) "Cytological staining procedures applicable to methacrylate embedded tissues." *Stain Technology* 29:207–212.
Hrushovetz, S. B., and Harder, C. Elizabeth.
   (1962) "Permanent mounting of unstained and aceto-orcein stained cells in the water soluble medium, Abopon." *Stain Technology* 37:307–311.
Hsu, T. C., and Kellogg, Douglas.
   (1960) "Primary cultivation and continuous propagation in vitro of tissues from small biopsy specimens." *Journal of the National Cancer Institute* 25:221–235.
Huber, William M., and Caplin, Samuel M.
   (1947) "Simple plastic mount for preservation of fungi and small arthropods." *Archives of Dermatology and Syphilology* 56:763–765.
Hukill, Peter, and Putt, Frederick A.
   (1962) "A specific stain for iron using 4,7-diphenyl-1,10-phenanthroline (Bathophenanthroline)." *Journal of Histochemistry and Cytochemistry* 10:490–494.
Hull, Susan, and Wegner, Sally.
   (1952) "Removal of stains." *Stain Technology* 27:224.
Hulton, William E.
   (1958) "A survey of the application of the 'molecular' membrane filter to the study of cerebrospinal fluid cytology." *American Journal of Clinical Pathology* 30:407–410.
Humason, Gretchen L., and Lushbaugh, C. C.
   (1960) "Selective demonstration of elastin, reticulum, and collagen by silver, orcein, and aniline blue." *Stain Technology* 35:209–214.
   (1961) "A quick pinacyanole stain for frozen sections." *Stain Technology* 36:257–258.
   (1969) "Sirius supra blue FGL-CF; superior to aniline blue in the combined elastin, reticulum and collagen stain." *Stain Technology* 44:105–106.
Humphrey, A. A.
   (1935) "Di-nitrosoresorcinol—a new specific for iron in tissue." *Archives of Pathology* 20:256–258.
   (1936) "A new rapid method for frozen section diagnosis." *Journal of Laboratory and Clinical Medicine* 22:198–199.
Hungerford, David A.
   (1965) "Leukocytes cultures from small inocula of whole blood and the preparation of metaphase chromosomes by treatment with hypotonic KCI." *Stain Technology* 40:333–338.
Hungerford, David A., and Nowell, Peter C.
   (1962) "Chromosome studies in human leukemia." *Journal of the National Cancer Institute* 29:545–565.
Hutchinson, H. E.
   (1953) "The significance of stainable iron in sternal marrow sections." *Blood* 8:236–248.
Hutner, S. H.
   (1934) "Destaining agents for iron alum hematoxylin." *Stain Technology* 9:57–59.
Hutton, W. E.
   (1953) "Ninhydrin staining of tissue sections." *Stain Technology* 28:173–175.

Ibanez, Michael L.; Russell, William O.; Chang, Jeffrey P.; Speece, Arthur J.
   (1960) "Cold chamber frozen sections for operating room diagnosis routine surgical stains." *Laboratory Investigation* 9:275–278.

Idelman, Simon.
  (1964) "Modification de la technique de Luft en vue de la conservation des lipides en microscopie électronique." *Journal de Microscopie* 3:715–718.
Ingram, R. L.; Otken, L. B.; and Jumper, J. R.
  (1961) "Staining of malarial parasites by the fluorescent antibody technic." *Proceedings of the Society for Experimental Biology and Medicine* 106:52–54.
Irugalbandara, Z. E.
  (1960) "Simplified differentiation of Nissl granules stained by toluidine blue in paraffin sections." *Stain Technology* 35:47–48.

Jackson, John F.
  (1961) "Supravital blood studies, using acridine orange fluorescence." *Blood* 17:643–649.
Jacobson, Stanley.
  (1963) "Handling sections in bulk, with special reference to the Nauta technic." *Stain Technology* 38:262–263.
Jagatic, Juraj, and Weiskopf, Robert.
  (1966) "A fluorescent method for staining mast cells." *Archives of Pathology* 82:430–433.
Janigan, David T.
  (1965) "The effects of aldehyde fixation on acid phosphatase activity in tissue block." *Journal of Histochemistry and Cytochemistry* 13:476–483.
Jennings, Robert B.
  (1951) "A simple apparatus for dehydration of frozen tissues." *Archives of Pathology* 52:195–197.
Jha, Raj K.
  (1976) "An improved polychrome staining method for thick epoxy sections." *Stain Technology* 51:159–162.
Joftes, David L.
  (1959) "Liquid emulsion autoradiography with tritium." *Laboratory Investigation* 8:131–148.
Joftes, David L., and Warren, S.
  (1955) "Simplified liquid emulsion radioautography." *Journal of the Biological Photographic Association* 23:145.
Johanson, Donald.
  (1940) *Plant Microtechnique.* New York: McGraw-Hill.
Johnston, Muriel E.
  (1960) "Film as a histological tissue support." *Journal of Histochemistry and Cytochemistry* 8:139.
Jona, Roberto.
  (1963) "Squashing under Scotch Tape 665 for autoradiographic and permanent histologic preparations." *Stain Technology* 38:91–95.
Jones, Rose M.; Thomas, Wilbur A.; and O'Neal, Robert M.
  (1959) "Embedding of tissues in Carbowax." *Technical Bulletin of the Registry of Medical Technologists* 29:49–52.

Kaback, M. M.; Saksela, E.; and Mellman, W. J.
  (1964) "The effect of 5-bromide-oxyuridine on human chromosomes." *Experimental Cell Research* 34:182–186.
Kabat, Elvin A., and Furth, Jacob.
  (1941) "A histochemical study of the distribution of alkaline phosphatase in various normal and neoplastic tissues." *American Journal of Pathology* 17:303–318.
Kallman, J.
  (1971) "Aldehyde-fuchsin followed by toluidine blue O for pancreatic islet cells." *Stain Technology* 46:210–211.

Kaniwar, K. C.
(1960) "Note on the specificity of mercuric bromophenol blue for the cytochemical detection of proteins." *Experimentia* 16:355.

Kaplan, Melvin H.; Coons, Albert H.; and Deane, Helen Wendler.
(1950) "Localization of antigen in tissue cells. III. Cellular distribution of pneumococcal polysaccharides, types II and III, in the mouse." *Journal of Experimental Medicine* 91:15–29.

Kaplow, Leonard S.
(1963) "Cytochemistry of leukocyte alkaline phosphatase." *American Journal of Clinical Pathology* 39:439–449.

Kaplow, Leonard S., and Burstone, M. S.
(1963) "Acid-buffered acetone as a fixative for enzyme cytochemistry." *Nature* 200:690–691.
(1964) "Cytochemical demonstration of acid phosphatase in hematopoietic cells in health and in various hematological disorders using azo dye techniques." *Journal of Histochemistry and Cytochemistry* 12:805–811.

Kassel, Robert, and Melnitsky, Ida.
(1951) "Embedding and staining ameboid forms." *Stain Technology* 26:167–171.

Kasten, Frederick H.
(1960) "The chemistry of Schiff's reagent." *International Review of Cytology,* X. Ed. by G. H. Bourne and J. F. Danielli. New York: Academic Press.
(1962) "Some comments on a recent criticism of the ninhydrin-Schiff reaction." *Journal of Histochemistry and Cytochemistry* 10:769–770.
(1965) "Loss of DNA and protein, and changes in DNA during a 30-hour cold perchloric acid extraction of cultured cells." *Stain Technology* 40:127–135.

Kasten, Frederick H., and Burton, Vivian.
(1959) "A modified Schiff's solution." *Stain Technology* 34:289.

Kasten, Frederick H.; Burton, Vivian; and Glover, P.
(1959) "Fluorescent Schiff-type reagents for cytochemical detection of polyaldehyde moieties in sections and smears." *Nature* 184:1797–1798.

Kasten, Frederick H.; Burton, Vivian; and Lofland, Sue.
(1962) "Schiff-type reagents in cytochemistry. Detection of primary amine dye impurities in pyronin B and pyronin Y (G)." *Stain Technology* 37:277–291.

Kasten, Frederick H., and Lala, Raymond.
(1975) "The Feulgen reaction after glutaraldehyde fixation." *Stain Technology* 50:197–201.

Kasten, Frederick H., and Sandritter, W.
(1962) "Crystal violet contamination of methyl green and purification of methyl green—a historical note." *Stain Technology* 37:253–255.

Katline, Vicki Castrogiovanni.
(1962) "Retention of nuclear staining by phosphomolybdic-phosphotungstic mordanting." *Stain Technology* 37:193–195.

Kay, Desmond.
(1965) *Techniques for Electron Microscopy.* Philadelphia: F. A. Davis.

Keeble, S. A., and Jay, R. F.
(1962) "Fluorescent staining for the differentiation of intracellular ribonucleic acid and deoxyribonucleic acid." *Nature* 193:695–696.

Keller, G. J.
(1945) "A reliable Nissl stain." *Journal of Technical Methods* 25:77–78.

Kelley, William E.
(1965) "New aquaria and synthetic sea salts provide a notable advance in marine biology." *Ward's Bulletin* 5.

Kellogg, D. S., and Deacon, W. E.
(1964) "A new rapid immunofluorescent staining technic for identification of *Treponema pallidum* and *Neisseria gonorrhoeae.*" *Proceedings of the Society for Experimental Biology and Medicine* 115:963–965.

Kelly, John W.; Morgan, Paul N.; and Saini, Nirmal.
    (1962) "Detection of tissue fungi by sulfation and metachromatic staining." *Archives of Pathology* 73:70–73.
Kemali, Milena.
    (1976) "A modification of the rapid Golgi method." *Stain Technology* 51:169–172.
Kenny, Michael; Dyckman, Jacob; and Aronson, Stanley M.
    (1971) "Acid fast staining for hooklets of Echinococcus." *Stain Technology* 46:160–161.
Kent, Sidney P.
    (1961) "A study of mucins in tissue sections using the fluorescent antibody technique. I. The preparation and specificity of bovine submaxillary gland mucin antibody." *Journal of Histochemistry and Cytochemistry* 9:491–497.
Kerbaugh, Mildred A.
    (1960) "Identification of beta hemolytic streptococci, group A, by the fluorescent method." *Official Journal of American Medical Technologists* 22:13–16.
Kerenyi, N.
    (1959) "Congo red as a simple stain for the beta cells of the hypophysis." *Stain Technology* 34:343–346.
Kerenyi, N., and Taylor, W. A.
    (1961) "Niagara blue 4B as a fast simple stain for the adenohypophysis." *Stain Technology* 36:169–172.
Kerr, Donald A.
    (1938) "Improved Warthin-Starry method of staining spirochetes in tissue sections." *American Journal of Clinical Pathology, Technical Supplement* 2:63–67.
Kessel, J. F.
    (1925) "The distinguishing characteristics of the intestinal protozoa of man." *China Medical Journal*, Feb.: 1–57.
Khudr, Gabriel, and Kint, Benirschke.
    (1973) "Quinacrine fluorescence microscopy of formalin-fixed tissues." *Stain Technology* 48:193–194.
Kidder, George W.
    (1933) "Studies on *Conchophthirius mytili* De Morgan. I. Morphology and division." *Archiv für Protistenkunde* 79:1–24.
Kimball, R. F., and Perdue, Stella W.
    (1962) "Quantitative cytochemical studies on *Paramecium*." *Experimental Cell Research* 27:405–415.
Kimmel, D., and Jee, W. S. S.
    (1975) "A rapid plastic embedding technique for preparation of three-micron thick sections of decalcified hard tissue." *Stain Technology* 50:83–86.
Kiossoglou, Kosmos A.; Wolman, Irving J.; and Garrison, Mortimer, Jr.
    (1963) "Fetal hemoglobin-containing erythrocytes. I. Counts of cells stained by the acid elution method compared with alkali denaturation methods." *Blood* 21:553–560.
Kirby, Harold.
    (1947) *Methods in the Study of Protozoa*. University of California Syllabus Series. Berkeley: University of California Press.
Klásterská, I., and Natarajan, A. T.
    (1976) "A modified C-bond technique for staining of diffuse diplotene chromosomes." *Stain Technology* 51:209–211.
Klein, B. M.
    (1926) "Über eine neue Eugentumlichkeit der Pellicula von Chilodon uncinatus Ehrbg." *Zoologischer Anzeiger* 67:160–162.
Klessen, Chr.
    (1972) "Histochemical staining of zymogen granules of pancreatic acinar cells using a permanganate-HID-technique." *Histochemie* 30:366.
Klinger, H. B.
    (1958) "The fine structure of the sex chromatin body." *Experimental Cell Research* 14:207–211.

Klinger, H. B., and Hammond, Daniel O.
    (1971) "Rapid chromosome and sex-chromatin staining with pinacyanol." *Stain Technology* 46:43–47.
Klinger, H. B., and Ludwig, Kurt S.
    (1957) "A universal stain for the sex chromatin body." *Stain Technology* 32:235–244.
Klionsky, B., and Marcoux, L.
    (1960) "Frozen storage of incubation media for enzyme histochemistry." *Journal of Histochemistry and Cytochemistry* 8:329.
Kloeck, John M., and Sweaney, Henry C.
    (1943) "Binocular fluorescent microscopy." *American Journal of Clinical Pathology, Technical Section* 7:96–98.
Klüver, H., and Barrera, E.
    (1953) "A method for the combined staining cells and fibers in the nervous system." *Journal of Neuropathology and Experimental Neurology* 12:400–403.
Koenig, Harold.
    (1963) "Intravital staining of lysosomes by basic dyes and metallic ions." *Journal of Histochemistry and Cytochemistry* 11:120–121.
Koenig, Harold; Groat, Richard A.; and Windle, William F.
    (1945) "A physical approach to perfusion-fixation of tissues with formalin." *Stain Technology* 20:13–22.
Koneff, Alexei A.
    (1936) "An iron hematoxylin–aniline blue staining method for routine laboratory use." *Anatomical Record* 66:173.
    (1938) "Adaptation of the Mallory-Azan staining method to the anterior pituitary of the rat." *Stain Technology* 13:49–52.
Koneff, Alexei A., and Lyons, W. R.
    (1937) "Rapid embedding with hot low-viscosity nitrocellulose." *Stain Technology* 12:57–59.
Kopriwa, Beatrix Markus, and Leblond, C. P.
    (1962) "Improvements in the coating technique of radioautography." *Journal of Histochemistry and Cytochemistry* 10:269–284.
Kornhauser, S. I.
    (1930) "Hematein. Its advantages for general laboratory usage." *Stain Technology* 5:13–15.
    (1943) "A quadruple stain for strong color contrasts." *Stain Technology* 18:95–97.
    (1945) "A revised method for the 'Quad' stain." *Stain Technology* 20:33–35.
    (1952) "Orcein and elastic fibers." *Stain Technology* 27:131–134.
Korson, Roy.
    (1964) "A silver stain for deoxyribonucleic acid." *Journal of Histochemistry and Cytochemistry* 12:875–879.
Kosenow, W.
    (1961) "Nuclear appendages in leucocytes and the sex pattern of chromosomes." *The Physiology and Pathology of Leukocytes*. Ed. by Dorothea Zucker-Franklin and Herbert Braunsteiner. New York: Grune and Stratton.
Koulischer, L., and Mulnard, J.
    (1962) "Staining of Chromosomes." *Lancet* 1:917.
Kraijian, Aram A., and Gradwohl, R. B. H.
    (1952) *Histopathological Technic*. St Louis: C. V. Mosby.
Krichesky, Boris.
    (1931) "A modification of Mallory's triple stain." *Stain Technology* 6:97–98.
Krishan, Awtar.
    (1962) "Avian microchromosomes: as shown by prefixation treatment with colchicine, squashing, and hematoxylin staining." *Stain Technology* 37:335–337.
Kristensen, Harold K.
    (1948) "An improved method for decalcification." *Stain Technology* 23:151–154.
Kropp, B.
    (1954) "Grinding thin sections of plastic embedded bone." *Stain Technology* 29:77–80.

Krus, Stefan; Andrade, Zilton A.; and Barka, Tibor.
   (1961) "Histochemical demonstration of specific phosphatases of the liver preserved in
      hypertonic sucrose solution." *Journal of Histochemistry and Cytochemistry*
      9:487–490.
Kruszynski, J.
   (1954) "Selective demonstration of the Golgi structure and of mitochondria." *Stain
      Technology* 29:151–155.
Krutsay, M.
   (1960) "A versatile Resorcin-fuchsin formula: I: Combined with formaldehyde. II: Used
      after periodic acid. III: Used after HCL hydrolysis." *Stain Technology* 35:283–
      284.
   (1962*a*) "Permanganate–resorcin–fuchsin: a selective stain for elastic tissue." *Stain
      Technology* 37:250–251.
   (1962*b*) "The preparation of iron-hematoxylin from alum hematoxylin." *Stain Technol-
      ogy* 37:249.
Kubie, Lawrence, S., and Davidson, David.
   (1928) "The ammoniacal silver solution as used in neuropathology." *Archives of
      Neurology and Psychiatry* 19:888–903.
Kuhn, Geraldine D., and Lutz, Earnest L.
   (1958) "A modified polyester embedding medium for sectioning." *Stain Technology*
      33:1–7.
Kurnick, N. B.
   (1952) "Histological staining with methyl green–pyronin." *Stain Technology* 27:233–
      242.
   (1955) "Histochemistry of nucleic acids." *International Review of Cytology*, IV. Ed. by
      G. H. Bourne and J. F. Danielli. New York: Academic Press.
Kutlík, Von Igor E.
   (1968) "Über die Argentaffinität des geformten Bilirubins in den Geweben." *Acta His-
      tochemica* 5:213–224.
   (1970) "Nachweis von Eisen in den Geweben mittels Chlorathämatoxylinfärbung."
      Acta Histochemica 37:259–267.
Kuyper, Ch. M. A.
   (1957) "Identification of mucopolysaccharides by means of fluorescent basic dyes."
      Experimental Cell Research 13:198–200.
Kwan, S. K.
   (1970) "Sticky wax infiltration in the preparation of sawed undecalcified bone sections."
      Stain Technology 45:177–181.

LaBauve, P. M.; LaBauve, R. J.; and Petersen, D. F.
   (1965) "A digitized comparator for karyotype analysis." *Journal of Heredity* 56:47–52.
Lacey, Paule E., and Davies, J.
   (1959) "Demonstration of insulin in mammalian pancreas by the fluorescent antibody
      method." *Stain Technology* 34:85–89.
LaCour, L.
   (1935) "Technic for studying chromosome structure." *Stain Technology* 10:57–60.
   (1941) "Acetic-orcein: a new stain-fixative for chromosomes." *Stain Technology*
      16:169–174.
LaCroix, J. Donald, and Preiss, Sister Rose Frederick.
   (1960) "Lamination of thermoplastic sheets as a means of mounting histological mate-
      rial." *Stain Technology* 35:331–337.
Lagunoff, David; Phillips, Michael; and Benditt, Earl P.
   (1961) "The histochemical demonstration of histamine in mast cells." *Journal of His-
      tochemistry and Cytochemistry* 9:534–541.
Lamkie, N. Joan, and Burstone, M. S.
   (1962) "Vinylpyrrolidonevinyl acetate copolymers as mounting media for azo and other
      dyes." *Stain Technology* 37:109–110.

Lampros, Steve J.
  (1962) "A rapid method for preparing paraffin sections." *American Journal of Clinical Pathology* 37:77.
Landing, Benjamin H.
  (1954) "Histological study of the anterior pituitary." *Laboratory Investigation* 3:348–368.
Landing, Benjamin H., and Hall, Hazel E.
  (1955) "Differentiation of human anterior pituitary cells by combined metalmordant and mucoprotein stains." *Laboratory Investigation* 4:275–278.
  (1956) "Selective demonstration of histidine." *Stain Technology* 31:197–200.
Lane, Bernard P., and Europa, Dominic L.
  (1965) "Differential staining of ultrathin sections of Epon-embedded tissues for light microscopy." *Journal of Histochemistry and Cytochemistry* 13:579–582.
Lange, Norbert Adolph.
  (1956) *Handbook of Chemistry.* Sandusky, Ohio: Handbook Publishers.
Lartique, Donald J., and Fite, George L.
  (1962) "The chemistry of the acid-fast reaction." *Journal of Histochemistry and Cytochemistry* 10:611–618.
Lascano, Eduardo F.
  (1959) "A new silver method for the Golgi apparatus." *Archives of Pathology* 68:499–500.
Lasky, A., and Greco, J.
  (1948) "Argentaffin cells of the human appendix." *Archives of Pathology* 46:83–84.
Lawless, D. K.
  (1953) "A rapid permanent mount stain technic for the diagnosis of the intestinal protozoa." *American Journal of Tropical Medicine and Hygiene* 2:1137.
Lazarus, Sidney S.
  (1958) "A combined periodic acid–Schiff trichrome stain." *Archives of Pathology* 66:767–772.
Leach, E. H.
  (1946) "Curtis' substitute for van Gieson stain." *Stain Technology* 21:107–109.
Leach, W. B.
  (1960) "A method for the histological examination of bone marrow granules." *Canadian Medical Association Journal* 83:717–719.
Leaver, R. E.; Evans, B. J.; and Corrin, B.
  (1977) "Identification of Gram-negative bacteria in histological sections using Sandiford's counterstain." *Journal of Clinical Pathology* 30:290–291.
Leblond, C. P.; Messier, B.; and Kopriwa, Beatrix.
  (1959) "Thymidine-H^3 as a tool for the investigation of the renewal of cell populations." *Laboratory Investigation* 8:296–308.
Leduc, Elizabeth H.; Coons, Albert H.; and Connolly, Jeanne M.
  (1955) "Studies on anti-body production. II. The primary and secondary responses in the popliteal lymph node of the rabbit." *Journal of Experimental Medicine* 102:61–71.
Lee, M. Raymond.
  (1969) "A widely applicable technic for direct processing of bone marrow for chromosomes of vertebrates." *Stain Technology* 44:155–158.
Lehmann, F. E., and Mancuso, V.
  (1957) "Improved fixative for astral rays and nuclear membrane of Tubifex embryos." *Experimental Cell Research* 13:161–164.
Lendrum, Alan C.
  (1944) "On the cutting of tough and hard tissues embedded in paraffin." *Stain Technology* 19:143–144.
Lendrum, Alan C., and McFarlane, David.
  (1940) "A controllable modification of Mallory's trichrome staining method." *Journal of Pathology and Bacteriology* 50:38–40.

Lennox, Bernard.
  (1956) "A ribonuclease-gallocyanin stain for sexing skin biopsies" *Stain Technology* 31:167–172.
Leske, R., and Von Mayersbach, H.
  (1969) "The role of histochemical and biochemical preparation methods for the detection of glycogen." *Journal of Histochemistry and Cytochemistry* 17:527–538.
Lev, Maurice, and Thomas, John T.
  (1955) "An improved method for fixing and staining frozen sections." *American Journal of Clinical Pathology* 25:465.
Lev, Robert, and Stoward, Peter J.
  (1969) "On the use of eosin as a fluorescent dye to demonstrate mucous cells and other structures in tissue sections." *Histochemie* 20:363–377.
Levine, N. D., and Morrill, C. C.
  (1941) "Chlorazol black E, a simple connective tissue stain." *Stain Technology* 16:121–122.
Lewis, Louis W.
  (1945) "Method for affixing celloidin sections." *Stain Technology* 20:138.
Lewis, P. R., and Shute, C. C. D.
  (1963) "Alginate gel; an embedding medium for facilitating the cutting and handling of frozen sections." *Stain Technology* 38:307–310.
Lhotka, J. F., and Davenport, H. A.
  (1947) "Differential staining of tissue in the block with picric acid and the Feulgen reaction." *Stain Technology* 22:139–144.
  (1949) "Deterioration of Schiff's reagent." *Stain Technology* 24:237–239.
  (1951) "Aldehyde reactions in tissues in relation to the Feulgen technic." *Stain Technology* 26:35–41.
Lhotka, J. F., and Ferreira, Alberto Vaz.
  (1950) "A comparison of deformalinizing technics." *Stain Technology* 25:27–32.
Lhotka, J. F., and Myhre, Byron A.
  (1953) "Periodic acid–Foot stain for connective tissue." *Stain Technology* 28:129–133.
Lieb, Ethel.
  (1959) "The plastic (mylar) sack as an aid in the teaching of pathology." *American Journal of Clinical Pathology* 32:385–392.
Liebman, Emil.
  (1951) "Permanent preparations with the Thomas arginine histochemical test." *Stain Technology* 26:261–263.
Lillie, R. D.
  (1929) "A brief method for the demonstration of mucin." *Journal of Technical Methods and Bulletin of International Association of Medical Museums* 12:120–121.
  (1940) "Further experiments with the Masson trichrome modifications of the Mallory connective tissue stain." *Stain Technology* 15:17–22.
  (1944) "Acetic methylene blue counterstain in staining tissues for acid fast bacilli." *Stain Technology* 18:45.
  (1945) "Studies on selective staining of collagen with acid anilin dyes." *Journal of Technical Methods and Bulletin of International Association of Medical Museums* 25:1.
  (1946) "A simplified method of preparation of di-ammine-silver hydroxide for reticulum impregnation; comments on the nature of the so-called sensitization before impregnation." *Stain Technology* 21:69–72.
  (1949) "Studies on the histochemistry of normal and pathologic mucin in man and laboratory animals." *Bulletin, International Association of Medical Museums* 29:1–53.
  (1951a) "Histochemical comparison of the Casella, Bauer, and periodic acid oxidation Schiff leucofuchsin technics." *Stain Technology* 26:123–126.
  (1951b) "Simplification of the manufacture of Schiff reagent for use in histochemical procedures." *Stain Technology* 26:163–165.

(1951c) "The allochrome procedure. A differential method segregating the connective tissues, collagen, reticulum, and basement membranes into two groups." *American Journal of Clinical Pathology* 21:484–488.

(1952) "Staining of connective tissue." *Archives of Pathology* 54:220–233.

(1953) "Factors influencing periodic acid–Schiff reaction of collagen fibers." *Journal of Histochemistry and Cytochemistry* 1:353–361.

(1954a) "Argentaffin and Schiff reactions after periodic and oxidation and aldehyde blocking reactions." *Journal of Histochemistry and Cytochemistry* 2:127–136.

(1954b) *Histopathologic Technic and Practical Histochemistry.* 2d ed. New York: Blakiston.

(1955) "The basophilia of melanins." *Journal of Histochemistry and Cytochemistry* 3:453–454.

(1956a) "Nile blue staining technic of the differentiation of melanin and lipofuscins." *Stain Technology* 31:151–153.

(1956b) "The mechanism of Nile blue staining of lipofuscins." *Journal of Histochemistry and Cytochemistry.* 4:377–381.

(1956c) "The *p*-dimethylaminobenzaldehyde reaction for pyrroles in histochemistry: melanins, enterochromaffin, zymogen granules, lens." *Journal of Histochemistry and Cytochemistry* 4:118–129.

(1957a) "Ferrous ion uptake." *Archives of Pathology* 64:100–103.

(1957b) "The xanthydrol reaction for pyrroles and indoles in histochemistry: zymogen granules, lens, enterochromaffin, and melanins." *Journal of Histochemistry and Cytochemistry* 5:188–195.

(1957c) "Metal reduction reactions in melanins." *Journal of Histochemistry and Cytochemistry* 5:325–333.

(1957d) "Adaptation of the Morel Sisley protein diazotization procedure to the histochemical demonstration of protein bound tyrosine." *Journal of Histochemistry and Cytochemistry* 5:528–532.

(1958) "Acetylation and nitrosation of tissue amines in histochemistry." *Journal of Histochemistry and Cytochemistry* 6:352–362.

(1960) "Metal chelate reaction of enterochromaffin." *Journal of Histochemistry and Cytochemistry* 9:44–48.

(1961) "Investigation on the structure of the enterochromaffin substance." *Journal of Histochemistry and Cytochemistry* 9:184–189.

(1962) "Glycogen in decalcified tissue." *Journal of Histochemistry and Cytochemistry* 10:763–765.

(1964a) "Histochemical acylation of hydroxyl and aminogroups. Effect on the periodic acid–Schiff reaction, anionic and cationic dye and van Gieson collagen stains." *Journal of Histochemistry and Cytochemistry* 12:821–841.

(1964b) "Studies on histochemical acylation procedures. I. Phenols." *Journal of Histochemistry and Cytochemistry* 12:522–529.

(1965) *Histopathologic Technic and Practical Histochemistry.* 3d ed. New York: McGraw-Hill.

Lillie, R. D., and Burtner, H. J.
(1953a) "The ferric ferricyanide reduction test in histochemistry." *Journal of Histochemistry and Cytochemistry* 1:87–92.

(1953b) "Stable sudanophilia of human neutrophil leucocytes in relation to peroxidase and oxidase." *Journal of Histochemistry and Cytochemistry* 1:8–26.

Lillie, R. D.; Burtner, H. J.; and Henson, J. P.
(1953) "Diazosafranin for staining enterochromaffin." *Journal of Histochemistry and Cytochemistry* 1:154–159.

Lillie, R. D., and Earle, W. R.
(1939) "Iron hematoxylins containing ferric and ferrous iron." *American Journal of Pathology* 15:765–770.

Lillie, R. D., Gilmer, P. R., Jr.; and Welsh, R. A.
(1961) "Black periodic and black Bauer methods for tissue polysaccharides." *Stain Technology* 36:361–363.

Lillie, R. D., and Glenner, G. G.
(1957) "Histochemical aldehyde blockade by aniline in glacial acetic acid." *Journal of Histochemistry and Cytochemistry* 5:167–169.
Lillie, R. D.; Gutiérrez, Anselmo; Madden, Dolores; and Henderson, Raljean.
(1968) "Acid orcein-iron and acid orcein-copper stains for elastin." *Stain Technology* 43:203–206.
Lillie, R. D., and Henderson, R.
(1968) "A short chromic acid–hematoxylin stain for frozen sections of formol fixed brain and spinal cord." *Stain Technology* 43:121–122.
Lillie, R. D.; Henson, Jacquelin, P.; Greco, J.; and Burtner, Helen C. J.
(1957) "Metal reduction reactions of melanins: silver and ferric ferricyanide reduction by various reagents in vitro." *Journal of Histochemistry and Cytochemistry* 5:311–324.
Lillie, R. D.; Henson, Jacquelin, P.; Greco, J.; and Carson, Joseph C.
(1960) "Azo-coupling rate of enterochromaffin with various diazonium salts." *Journal of Histochemistry and Cytochemistry* 9:11–21.
Lillie, R. D.; Lasky, A.; Greco, J.; Burtner, H.; Jacquier; and Jones, P.
(1951) "Decalcification of bone in relation to staining and phosphatase technics." *American Journal of Clinial Pathology* 21:711–722.
Lillie, R. D., and Pizzolato, P.
(1968) "Histochemical studies of oxidation and reduction reactions of the bile pigments in obstructive Icterus, with some notes on hematoidin." *Journal of Histochemistry and Cytochemistry* 16:17–28.
Lillie, R. D.; Pizzolato, P.; and Donaldson, P. T.
(1973) "Iron Gallein in van Gieson technics, replacing iron hematoxylin." *Stain Technology* 48:348–349.
Lillie, R. D.; Windle, W. F.; and Zirkle, Conway.
(1950) "Interim report of use committee on histologic mounting media: Resinous media." *Stain Technology* 25:1–9.
Lillie, R. D.; Zirkle, Conway; Dempsey, Edward; and Greco, Jacqueline F.
(1953) "Final report of the committee on histological mounting media." *Stain Technology* 28:57–80.
Lin, C. W., and Corlett, Michael.
(1969) "The use of a xylene spray for restoring compressed paraffin serial sections." *Stain Technology* 44:159–160.
Lison, Lucien.
(1954) "Alcian blue 8G with chlorantine fast red 5B. A technic for selective staining of mucopolysaccharides." *Stain Technology* 29:131–138.
Litwin, Jan A., and Kasprzyk, Józef M.
(1975) "PAS reaction performed on semithin Epon sections following removal of the resin by NaOH in absolute ethanol." *Acta Histochemica* 55:98–103.
Litwin, Jan A.; Kasprzyk, Józef M.; and Cichocki, Tadeusz.
(1974) "Light microscopic differential staining of epoxy-embedded adenohypophysis." *Acta Histochemica* 52:17–22.
Liu, Winifred.
(1960) *An Introduction to Gynecological Exfoliative Cytology. A Working Manual for Cytotechnicians.* Springfield, Ill.: Charles C Thomas.
Ljungberg, Otto.
(1970) "Cresyl fast violet—a selective stain for human C-cells." *Acta Pathologica et Microbiologica Scandinavica* 78:618–620.
Lockard, Isabel, and Reers, Bernard L.
(1962) "Staining tissue of the central nervous system with luxol fast blue and neutral red." *Stain Technology* 37:13–16.
Lodin, Z.; Faltin J.; and Sharma, K.
(1967) "Attempts at standardization of a highly sensitive Schiff reagent." *Acta Histochemica* 26:244–254.

Lodin, Z.; Marés, V.; Karásek, J.; and Skřivanová, P.
    (1967) "Studies on nervous tissue. II. Changes of sizes of nuclei of nervous cells after fixation and after further histological treatment of the nervous tissue." *Acta Histochemica* 28:297–312.
Lojda, Z., and Havránková, E.
    (1975) "The histochemical demonstration of aminopeptidase with bromoindolyl leucinamide." *Histochemistry* 43:355–366.
London Conference on the Normal Karotype.
    (1964) [Report] *Annals of Human Genetics* 27:295–296.
Long, Margaret E.
    (1948) "Differentiation of myofibrillae, reticular and collagenous fibrillae in vertebrates." *Stain Technology* 23:69–75.
Loots, J. M.; Loots, G. P.; and Joubert, W. S.
    (1977) "A silver impregnation method for nverous tissue suitable for routine use with mounted sections." *Stain Technology* 52:85–87.
Love, Aldyth M., and Vickers, T. H.
    (1972) "Durable staining of cartilage in foetal rat skeleton by methylene blue." *Stain Technology* 47:7–11.
Love, Robert.
    (1957) "Distribution of ribonucleic acid in tumor cells during mitosis." *Nature* 180:1338–1339.
    (1962) "Improved staining of nucleoproteins of the nucleolus." *Journal of Histochemistry and Cytochemistry* 10:227.
Love, Robert, and Liles, Roscoe H.
    (1959) "Differentiation of nucleoproteins by inactivation of protein-bound amino groups and staining with toluidine blue and ammonium molybdate." *Journal of Histochemistry and Cytochemistry* 7:164–181.
Love, Robert, and Rabotti, Giancarlo.
    (1963) "Studies of the cytochemistry of nucleo-proteins. III. Demonstration of deoxyribonucleic-ribonucleic acid complexes in mammalian cells." *Journal of Histochemistry and Cytochemistry* 11:603–612.
Low, Frank N.
    (1955) "A fixation method for the electron microscopy of endoplasmic reticulum." *Anatomical Record* 121:332–333.
Low, Frank N., and Freeman, James A.
    (1956) "Some experiments with chromium compounds as fixers for electron microscopy." *Journal of Biophysical and Biochemical Cytology* 2:629–631.
Lowry, Robert J.
    (1963) "Aceto-iron-hematoxylin for mushroom chromosomes." *Stain Technology* 38:149–155.
Lucas, Alfred M., and Jamroz, Casimir.
    (1961) *Atlas of Avian Hematology.* Agriculture Monograph 25. Washington, D.C.: U.S. Department of Agriculture.
Lucas, R. B.
    (1952) "Observations on the electrolysis method of decalcification." *Journal of Pathology and Bacteriology* 64:654–657.
Luft, John H.
    (1956) "Permanganate—a new fixative for electron microscopy." *Journal of Biophysical and Biochemical Cytology* 2:799–801.
    (1959) "The use of acrolein as a fixative for light and electron microscopy." *Anatomical Record* 133:305.
    (1961) "Improvements in epoxy resin embedding method." *Journal of Biophysical and Biochemical Cytology* 9:409–414.
Luna, Lee G.
    (1968) *Manual of Histologic Staining Methods of the Armed Forces Institute of Pathology.* 3d ed. New York: McGraw-Hill.

Luna, Lee G., and Ballou, E. F.
  (1959) "Better paraffin sections with the aid of vacuum." *American Journal of Medical Technology* 25:411–413.
Lycette, R. M.; Danforth, W. F.; Koppel, J. L.; and Olwin, J. H.
  (1970) "The binding of Luxol fast blue ARN by various biological lipids." *Stain Technology* 45:155–160.
Lynch, Mathew J., and Inwood, Martin J. H.
  (1963) "Gold as a permanent stain for amyloid." *Stain Technology* 38:259–260.
Lynch, Mathew J.; Raphael, Stanley S.; Mellor, Leslie D.; Spare, Peter D.; and Inwood, Martin J. H.
  (1969) *Medical Laboratory Technology and Clinical Pathology.* Philadelphia: W.B. Saunders.
Lynn, Joseph A.; Martin, James H.; and Race, George J.
  (1966) "Recent improvements of histologic technics for the combined light and electron microscopic examination of surgical specimens." *American Journal of Clinical Pathology* 45:704–713.

McCann, Jo Ann.
  (1977) "Methyl green as a cartilage stain; human embryos." *Stain Technology* 46:263–265.
McClintock, Barbara.
  (1929) "A method for making aceto-carmine smears permanent." *Stain Technology* 4:53.
McClung, C. E.
  (1939) *Handbook of Microscopical Technique.* New York: Hoeber.
McCormick, James B.
  (1959*a*) "Technic for restaining faded histopathologic slides." *Technical Bulletin of the Registry of Medical Technologists* 29:13–14.
  (1959*b*) "One hour paraffin processing technic for biopsy of bone marrow and other tissues, and specimens of fluid sediment." *American Journal of Clinical Pathology* 31:278–279.
  (1961) "Color preservation of gross museum specimens." *Archives of Pathology* 72:82–85.
McCormick, William F., and Coleman, Sidney A.
  (1962) "A membrane filter technic for cytology of spinal fluid." *Technical Bulletin of the Registry of Medical Technologists* 32:113–119.
McCurdy, L. E., and Burstone, M. S.
  (1966) "Water-soluble plastic mounting media." *Journal of Histochemistry and Cytochemistry* 14:427–428.
McDonald, Donald M.
  (1964) "Silver impregnation of Golgi apparatus with subsequent nitrocellulose embedding." *Stain Technology* 39:345–349.
McGee-Russell, S. M.
  (1958) "Histochemical methods for calcium." *Journal of Histochemistry and Cytochemistry* 6:22–42.
McGuire, S. R., and Opel, H.
  (1969) "Resorcin-fuchsin staining of neurosecretory cells." *Stain Technology* 44:235–237.
McManus, J. F. A.
  (1946) "The histological demonstration of mucin after periodic acid." *Nature* 158:202.
  (1948) "Histological and histochemical uses of periodic acid." *Stain Technology* 23:99–108.
  (1961) "Periodate oxidation techniques." *General Cytochemical Methods,* II. Ed. by J. F. Danielli. New York: Academic Press.
McManus, J. F. A., and Cason, J. E.
  (1950) "Carbohydrate histochemistry studied by acetylation techniques. I. Periodic acid methods." *Journal of Experimental Medicine* 91:651.

McManus, J. F. A., and Mowry, Robert W.
(1958) "Effects of fixation on carbohydrate histochemistry." *Journal of Histochemistry and Cytochemistry* 6:309–316.
(1960) *Staining Methods: Histologic and Histochemical.* New York: Hoeber.

Maggi, V., and Riddle, P. N.
(1965) "Histochemistry of tissue culture cells: a study of the effects of freezing and of some fixatives." *Journal of Histochemistry and Cytochemistry* 13:310–317.

Mahoney, Roy.
(1968) *Laboratory Techniques in Zoology.* London: Butterworth.

Makowski, E. L.; Prem, K. A.; and Kaiser, I. H.
(1956) "Detection of sex of fetuses by the incidence of sex chromatin body in nuclei of cells in amniotic fluid." *Science* 123:542–543.

Malhotra, S. K.
(1961) "Coloration of the Golgi-Nissl network in a vertebrate neuron by Sudan black." *Quarterly Journal of Microscopical Science* 102:387–389.

Malinin, Theodora I.
(1961) "Feulgen-oxidized tannin-azo (FOTA) technique for concomitant staining of DNA and protein." *Stain Technology* 36:198–200.

Mallory, Frank B.
(1944) *Pathological Technique.* Philadelphia: W. B. Saunders.

Manheimer, Leon H., and Seligman, Arnold.
(1949) "Improvement in the method for histochemical demonstration of alkaline phosphatase and its use in a study of normal and neoplastic tissues." *Journal of the National Cancer Institute* 9:181–199.

Manns, E.
(1960) "Combined myelin-Nissl stain." *Stain Technology* 35:349–351.

Manufacturing Chemists' Association.
(1957) *Technical Data on Plastics.* Washington, D.C.

Manwell, Reginald D.
(1945) "The JSB stain for blood parasites." *Journal of Laboratory and Clinical Medicine* 30:1078–1082.

Marberger, Eve; Boccabella, Rita A.; and Nelson, Warren O.
(1955) "Oral smear as a method of chromosomal sex detection." *Proceedings of the Society for Experimental Biology and Medicine* 89:488–489.

Marengo, P.
(1967) "Thin sectioning microtome knife edges from a modified automatic sharpener." *Stain Technology* 42:216–217.

Margolis, G., and Pickett, J. P.
(1956) "New applications of the Luxol fast blue myelin stain." *Laboratory Investigation* 5:459–474.

Markert, Clement L., and Hunter, Robert L.
(1959) "The distribution of esterases in mouse tissue." *Journal of Histochemistry and Cytochemistry* 7:42–49.

Marshall, J. D.; Eveland, W. C.; and Smith, C. W.
(1958) "Superiority of fluorescein isothiocyanate (Riggs) for fluorescent-antibody technic with a modification of its application." *Proceedings of the Society for Experimental Biology and Medicine* 98:898.

Marshall, J. D.; Iverson, L.; Eveland, W. C.; and Kase, Alice.
(1961) "Application and limitations of the fluorescent antibody stain in the specific diagnosis of cryptococcosis." *Laboratory Investigation* 10:719–728.

Marti, Walter J., and Johnson, Berkley H.
(1951) "Acid fast staining technic for histological sections." *American Journal of Clinical Pathology* 21:793.

Martin, John H.
(1966) "A different organism for the demonstration of giant salivary chromosomes." *Turtox News* 44:178–180.

Martin, John H.; Lynn, J. A.; and Nickey, W. M.
  (1966) "A rapid polychrome stain for epoxy-embedded tissue." *American Journal of Clinical Pathology* 46:250–251.
Marwah, A. S., and Weinmann, J. P.
  (1955) "A sex difference in epithelial cells of human gingiva." *Journal of Periodontology* 26:11–13
Masek, Bertha, and Birns, Monroe.
  (1961) "Advantages of the polyvinyl alcohol–glycerol embedding method for enzyme histochemistry." *Journal of Histochemistry and Cytochemistry* 9:634–635.
Massignani, Adriana, and Malferrari, Rosella.
  (1961) "Phosphotungstic acid–eosin combined with hematoxylin as a stain for Negri bodies in paraffin sections." *Stain Technology* 36:5–8.
Massignani, Adriana, and Refinetti, Eloisa Misasi.
  (1958) 'The Papanicolaou stain for Negri bodies in paraffin sections." *Stain Technology* 33:197–199.
Masson, P.
  (1928) "Carcinoids (argentaffin cell tumors) and nerve hyperplasia of the appendicular mucosa." *American Journal of Pathology* 4:181–211.
Mayer, D. M.; Hampton, J. C.; and Rosario, B.
  (1961) "A simple method for removing the resin from epoxy-embedded tissue." *Journal of Biophysical and Biochemical Cytology* 9:909–910.
Mayner, Doris A., and Ackerman, G. Adolph.
  (1962) "Histochemical demonstration of tissue ribonuclease activity." *Journal of Histochemistry and Cytochemistry* 10:687.
  (1963) "Tissue localization of ribonuclease activity by the substrate film technique." *Journal of Histochemistry and Cytochemistry* 11:573–577.
Melander, Yngve, and Wingstrand, Karl Georg.
  (1953) "Gomori's hematoxylin as a chromosome stain." *Stain Technology* 28:217–223.
Mellors, Robert C.
  (1959) *Analytical Cytology.* New York: McGraw-Hill.
Mellors, Robert C.; Siegel, Malcolm; and Pressman, David.
  (1955) "Analytic pathology. Histochemical demonstration of antibody localization in tissues with special reference to the antigenic components of kidney and lung." *Laboratory Investigation* 4:69–89.
Meloan, Susan N., and Puchtler, Holde.
  (1974) "Iron alizarin blue S stain for nuclei." *Stain Technology* 49:301–304.
Melvin, Dorothy M., and Brooke, M. M.
  (1955) "Triton X-100 in Giemsa staining of blood parasites." *Stain Technology* 30:269–275.
Mendelow, Harvey, and Hamilton, James B.
  (1950) "A new technique for rapid freezing and dehydration of tissues for histology and histochemistry." *Anatomical Record* 107:443–451.
Menzies. D. W.
  (1959) "Picro-Gomori method." *Stain Technology* 34:294–295.
  (1962*a*) "A methylene blue–basic fuchsin stain for mast cells in paraffin sections." *Stain Technology* 37:43–44.
  (1962*b*) "Paraffin-beeswax-stearic acid: an embedding mass for thin sections." *Stain Technology* 37:235–238.
  (1963*a*) "Red–blue staining of hydrolysed nucleic acids in paraffin sections." *Stain Technology* 38:157–160.
  (1963) "Romanowsky-type staining: enzymatic analysis of nuclear metachromasia in formalin-fixed paraffin sections." *Stain Technology* 38:161–163.
Mercer, E. H., and Birbeck, C.
  (1961) *Electron Microscopy. An Handbook for Biologists.* Springfield, Ill.: Charles C Thomas.

Merton, H.
(1932) "Gestalterhaltende Fixierungsversuche an besonders kontractilen Infusorien."
Archiv für Protistenkunde 77:449–521.
Meryman, H. T.
(1959) "Sublimation freze-drying without vacuum. *Science* 130:628–629.
(1960) "Freezing and drying of biological materials. Part III. Principles of freeze-drying." *Annals of the New York Academy of Sciences* 85:501–734.
Metcalf, R. L., and Paton, R. L.
(1944) "Fluorescence microscopy applied to Entomology and allied fields." *Stain Technology* 19:11–27.
Mettler, Fred A.
(1932) "The Marchi method for demonstrating degenerated fiber connections within the central nervous system." *Stain Technology* 7:95–106.
Mettler, Fred A., and Hanada, Ruth E.
(1942) "The Marchi method." *Stain Technology* 17:111–116.
Mettler, Fred A.; Mettler, Cecelia C.; and Strong F. C.
(1936) "The cellosolve-nitrocellulose technic." *Stain Technology* 11:165–166.
Mettler, Sidney, and Bartha, Alexander S.
(1948) "Brilliant cresyl blue as a stain for chromosome smear preparations." *Stain Technology* 23:27–28.
Metz, G.
(1976) "Mahon's myelin stain for celloidin-embedded nervous tissue." *Stain Technology* 51:59–61.
Meyer, James R.
(1945) "Prefixing with paradichlorobenzene to facilitate chromosome study." *Stain Technology* 20:121–125.
Meyer-Arendt, Jurgen R.
(1959) "Mylar film as carrier material for microradiography." *Journal of Histochemistry and Cytochemistry* 7:351–352.
Mikat, K., and Mikat, Dorothy M.
(1973) "Fixation of tissue by formaldehyde vapor during centrifugation; a means of enhancing morphological visibility by cellular flattening." *Stain Technology* 48:33–37.
Miller, Bernard Joseph.
(1937) "The use of dioxane in the preparation of histological sections." *The Mendel Bulletin.* December:5–9.
Miller, O. L.; Stone, G. E.; and Prescott, D. M.
(1964) "Autoradiography of water-soluble materials." *Methods in Cell Physiology,* I. Ed. by D. M. Prescott. New York: Academic Press.
Miller, P. J.
(1971) "An elastin stain." *Medical Laboratory Technology* 28:148–149.
Milligan, Mildred.
(1946) "Trichrome stain for formalin-fixed tissue." *American Journal of Clinical Pathology, Technical Section.* 10:184–185.
Millonig, G.
(1961) "Advantages of a phosphate buffer for OsO_4 solutions in fixation." *Journal of Applied Physiology* 32:1637.
Minick, O. T.
(1963) "Low temperature storage of epoxy embedding resins." *Stain Technology* 38:131–133.
Mitchell, B. S.
(1975) "A rapid, reliable modification of Mallory's phosphotungstic acid hematoxylin (PTAH) method for astrocytes using Susa fixative as a mordant." *Medical Laboratory Technology* 32:331–333.
Mittwer, Tod; Bartholomew, T. W.; and Kallman, Burton J.
(1950) "The mechanism of the Gram reaction. 2. The function of iodine in the Gram stain." *Stain Technology* 25:169–179.

Moffat, D. B.
    (1958) "Demonstration of alkaline phosphate and periodic acid–Schiff positive material in the same section." *Stain Technology* 33:225–228.
Mohr, John L.
    (1950) "On the natural coloring matter, brazilin, and its use in microscopical technique." *Micro-notes* 5:4–16.
Mohr, John L., and Wehrle, William.
    (1942) "Notes on mounting media." *Stain Technology* 17:157–160.
Moldovanu, G.
    (1961) "L'Identification de la chromatine du sexe chez les chimères hématologiques canines." *Revue Francaise d'Etudes Cliniques et Biologiques* 6:165–167.
Moline, S. W., and Glenner, G. G.
    (1964) "Ultra-rapid tissue freezing in liquid nitrogen." *Journal of Histochemistry and Cytochemistry* 12:777–783.
Moliner, Enrique Ramon.
    (1957) "A chlorate-formaldehyde modification of the Golgi method." *Stain Technology* 32:105–116.
Mollenhauer, Hilton H.
    (1964) "Plastic embedding mixtures for use in electron microscopy." *Stain Technology* 39:111–114.
Molnar, Livia M.
    (1974) "Double embedding with nitrocellulose and paraffin." *Stain Technology* 49:311.
Moloney, William C.; McPherson, Kenneth; and Fliegelman, Lila.
    (1960) "Esterase activity in leukocytes demonstrated by use of napthol AS-D chloracetate substrate." *Journal of Histochemistry and Cytochemistry* 8:200–207.
Monroe, C. W., and Spector, B.
    (1963) "Tannic acid, iron hematoxylin, Alcian blue and basic fuchsin for staining islets and reticular fibers of the pancreas." *Stain Technology* 38:187–192.
Moody, M. D.; Ellis, E. C.; and Updyke, L.
    (1958) "Staining bacterial smears with fluorescent antibody. IV. Grouping streptococci with antibody." *Journal of Bacteriology* 75:553–560.
Moore, A. R.
    (1962) "Collecting and preserving the developmental stages of the Pacific coast Bat Starfish, *Patiria miniata*." *Ward's Bulletin* 1.
Moore, Keith L., and Barr, Murry L.
    (1955) "Smears from the oral mucosa in the detection of chromosomal sex." *Lancet* 269:57–58.
Moore, Keith L., Graham, Margaret A.; and Barr, Murray L.
    (1953) "The detection of chromosomal sex in hermaphrodites from a skin biopsy." *Surgery, Gynecology and Obstetrics* 96:641–648.
Moorman, Albert E.
    (1971) "A modification of Heidenhain's iron-haematoxylin technique, as used to stain smears of Drosophila larval polytene chromosomes and neuroblast cells." *Canadian Journal of Zoology* 49:132.
Moran, Thomas J.; Radcliffe, Merle L.; and Tevault, Isabelle H.
    (1947) "Rapid method of staining tubercle bacilli with Tergitol." *American Journal of Clinical Pathology* 17:75–77.
Moree, R.
    (1944) "Control of the ferric ion concentration in iron-acetocarmine staining." *Stain Technology* 19:103–108.
Morris, Russell E., and Benton, Robert S.
    (1956) "Studies on demineralization of bone. IV. Evaluation of morphology and staining characteristics of tissues after demineralization." *American Journal of Clinical Pathology* 26:882–898.
Morrison, John H., and Kronheim, Suzanne.
    (1962) "The cytochemical demonstration of succinic dehydrogenase in mouse leukocytes." *Journal of Histochemistry and Cytochemistry* 10:402–411.

Morrison, Maurice, and Samwick, A. A.
  (1940) "Restoration of overstained Wright films and a new method of staining blood smears." *American Journal of Clinical Pathology, Technical Supplement* 4:92–93.
Mortreuil-Langlois, Micheline.
  (1962) "Staining sections coated with radiographic emulsion; a nuclear fast red, indigo-carmine sequence." *Stain Technology* 37:175–177.
Mote, R. F.; Muhm, R. L.; and Gigstad, D. C.
  (1975) "A staining method using Acridine orange and Auramine O for fungi and mycobacteria in bovine tissue." *Stain Technology* 50:5–9.
Mossman, H. W.
  (1937) "The Dioxan technic." *Stain Technology* 12:147–156.
Movat, Henry Z.
  (1955) "Demonstration of all connective tissue elements in a single section." *Archives of Pathology* 60:289–295.
Mowry, Robert W.
  (1956) "Alcian blue technique for histochemical study of acidic carbohydrates." *Journal of Histochemistry and Cytochemistry* 4:407.
  (1959) "Effect of periodic acid used prior to chromic acid on the staining of polysaccharides by Gomori's methenamine silver." *Journal of Histochemistry and Cytochemistry* 7:288.
  (1960) "Revised method producing improved coloration of acidic mucopolysaccharides with Alcian blue 8GX supplied currently." *Journal of Histochemistry and Cytochemistry* 8:323.
  (1963) "The special value of methods that color both acidic and vicinal hydroxyl groups in the histochemical study of mucins, with revised directions for the colloidal iron stain, the use of Alcian blue 8GX and their combinations with the periodic acid–Schiff reaction." *Annals of the New York Academy of Sciences* 106:402–423.
Mowry, Robert W., and Emmel, Victor M.
  (1966) "The coloration of carbohydrate polyanions by National fast blue compared with that obtained with Alcian blue 8GX." *Journal of Histochemistry and Cytochemistry* 14:799–800.
Mowry, Robert W., and Winkler, C. H.
  (1956) "The coloration acidic carbohydrates of bacteria and fungi in tissue sections with special reference to capsules of *Crytococcus neoformans, Pneumococcus* and *Staphylococcus.*" *American Journal of Pathology* 32:628–629.
Mullen, John P., and McCarter, John C.
  (1941) "A mordant preparing formaldehyde-fixed neuraxis tissue for phosphotungstic acid hematoxylin staining." *American Journal of Pathology* 17:289–291.
Mundkur, B., and Brauer, B.
  (1966) "Selective localization of nucleolar protein with Amidoblack 10B." *Journal of Histochemistry and Cytochemistry* 14:94–103.
Munoz, Frank, and Charipper, Harry A.
  (1943) *The Microscope and Its Use.* New York: Chemical Publishing Co.
Murgatroyd, L. B.
  (1971) "Chemical and spectrometric evaluation of glycogen after routine histological fixatives." *Stain Technology* 46:111–119.
Murgatroyd, L. B., and Horobin, R. W.
  (1969) "Specific staining of glycogen and hematoxylin and certain anthraquinone dyes." *Stain Technology* 44:59–62.
Murín, A.
  (1960) "Substitution of cellophane for glass covers to facilitate preparation of permanent squashes and smeas." *Stain Technology* 35:351–353.

Nachlas, Marvin; Tsou, Kwan-Chung; Desouza, Eustace; Cheng, Chao-Shing; and Seligman, Arnold M.
  (1957) "Cytochemical demonstration of succinic dehydrogenase by the use of a new

p-nitrophenyl substituted ditetrazole." *Journal of Histochemistry and Cytochemistry* 5:420–436.

Nairn, R. C.
(1962) *Fluorescent Protein Tracing.* Edinburgh and London: E. and S. Livingstone.

Nakane, Paul K., and Pierce, G. Barry.
(1967) "Enzyme labeled antibodies: Preparation and application for the localization of antigens." *Journal of Histochemistry and Cytochemistry* 14:929–931.

Naoumenko, Julia, and Feigin, Irwin.
(1974) "A simple silver solution for staining reticulin." *Stain Technology* 49:153–155.

Nash, David, and Plaut, W.
(1964) "On the denaturation of chromosomal DNA *in situ.*" *Proceedings of the National Academy of Sciences* 51:731–735.

Nassar, Tamir K., and Shanklin, William M.
(1961) "Simplified procedure for staining reticulum." *Archives of Pathology* 71:611/21–614/24.

Nassonov, D. N.
(1923) "Das Golgische Binnennetz und seine Beziehungen zu der Sekretion. Untersuchungen über einige Amphibiendrüsen." *Archiv für Mikroskopische Anatomie und Entwicklungsmechanik* 97:136–186.
(1924) "Das Golgische Binnennetz und seine Beziehungen zu der Sekretion. Morphologische und experimentelle Untersuchugen an einigen Säugetierdrüsen." *Archiv für Mikroskopische Anatomie und Entwicklungsmechanik* 100:433–472.

Nauman, R. V.; West, P. W.; Trou, F.; and Geake, G. C.
(1960) "A spectroscopic study of the Schiff reaction as applied to the quantitative determination of sulfur dioxide." *Annals of Chemistry* 32:1307–1311.

Nauta, W. J. H., and Gygax, P. A.
(1951) "Silver impregnation of degenerating axon terminals in the central nervous system. I. Technic. Chemical notes." *Stain Technology* 26:5–11.

Nauta, W. J. H., and Ryan, Lloyd F.
(1952) "Selective silver impregnation of degenerating axons in the central nervous system." *Stain Technology* 27:175–179.

Navagiri, S. S., and Dubey, P. N.
(1976) "Simple method of staining amyloid deposits." *Journal of Anatomical Society of India* 25:45.

Nayebi, M.
(1971) "Immunofluorescent technique for diagnosis of *Entamoeba histolytica* strains." *Medical Laboratory Technology* 28:413.

Nedelkoff, B., and Christopherson, William M.
(1962) "A millipore filter technic for cytologic examination of body fluids." *American Journal of Clinical Pathology* 37:97–103.

Needham, George Herbert.
(1958) *The Practical Use of the Microscope.* Springfield, Ill.: Charles C Thomas.

Nelson, Eric V.
(1974) "A simple technique for sectioning honey bee abdomens." *Stain Technology* 49:117–118.

Neurath, Peter W.
(1968) "Computer-aided chromosome analysis system." *Laboratory Management,* March: 14–16.
(1970) "Computer-aided chromosome analysis system [II]." *Laboratory Management,* September: 20–21.

Newcomer, Earl A.
(1940) "An osmic impregnation method for mitochondria in plant cells." *Stain Technology* 15:89–90.
(1959) "Feulgen staining of tissues prior to embedding and sectioning." *Stain Technology* 34:349–350.

Newcomer, Earl A., and Donnelly, Grace M.
  (1963) "Leukocyte culture for chromosome studies in the domestic fowl." *Stain Technology* 38:54–56.
Nickerson, Mark.
  (1944) "A dry ice freezing unit for rotary microtomes." *Science* 100:177–178.
Noble, Glenn A.
  (1944) "A five-minute method for staining fecal smears." *Science* 100:37–38.
Nolte, D. J.
  (1948) "A modified technic for salivary gland chromosomes." *Stain Technology* 23:21–25.
Norris, William P., and Jenkins, Phyllis.
  (1960) "Epoxy resin embedding in contrast radioautography of bones and teeth." *Stain Technology* 35:253–260.
Norton, William T.; Korey, Saul R.; and Brotz, Miriam.
  (1962) "Histochemical demonstration of unsaturated lipids by a bromine-silver method." *Journal of Histochemistry and Cytochemistry* 10:83–88.
Notenboom, C. D.; van de Veerdonk, F.; and de Kramer, J. C.
  (1967) "A fluorescent modification of the Sakaguchi reaction on arginine." *Histochemie* 9:117–121.
Novelli, Amato.
  (1962) "A short method for chondriome." *Journal of Histochemistry and Cytochemistry* 10:102–103.
Novikoff, Alex B.
  (1960) "Biochemical and staining reactions of cytoplasmic constituents." *Developing Cell Systems and their Control.* Ed. by Dorothea Rudnick. New York: Ronald Press.
  (1961a) *"Mitochondria* (Chondriosomes)." *The Cell,* II. Ed by Jean Brachet and Alfred E. Mirsky. New York: Academic Press.
  (1961b) "Lysosomes and Related Particles." The Cell, II. Ed. by Jean Brachet and Alfred E. Mirsky. New York: Academic Press.
Novikoff, Alex B.; Goldfischer, Sidney; and Essner, Edward.
  (1961) "The importance of fixation in a cytochemical method for the Golgi apparatus." *Journal of Histochemistry and Cytochemistry* 9:459–460.
Novikoff, Alex B., and Podber, Estelle.
  (1957) "The contribution of differential centrifugation to the intracellular localization of enzymes." *Journal of Histochemistry and Cytochemistry* 5:552–564.
Novikoff, Alex B.; Shin, Woo-Young; and Drucker, Joan.
  (1960) "Cold acetone fixation for enzyme localization in frozen sections." *Journal of Histochemistry and Cytochemistry* 8:37–40.
Novotney, G. E. K., and Novotny, E.
  (1974) "Glees silver for degenerating nerve tissue." *Stain Technology* 49:273.
  (1977) "Triple staining of normal and degenerating nervous tissue." *Stain Technology* 52:97–99.
Nowell, Peter C.
  (1960) "Phytohemagglutin: an initiator of mitosis in cultures of normal human leucocytes." *Cancer Research* 20:462–466.
Noyes, Wilbur F.
  (1955) "Visualization of Egypt 101 virus in the mouse's brain and in cultured human carcinoma cells by means of fluorescent antibody." *Journal of Experimental Medicine* 102:243–247.
Noyes, Wilbur F., and Watson, Barbara K.
  (1955) "Studies on the increase of vaccine virus in cultured human cells by means of fluorescent antibody technique." *Journal of Experimental Medicine* 102:237–242.
Nurnberger, John J.
  (1955) "Ultraviolet microscopy and microspectroscopy." *Analytical Cytology.* Ed. by Robert C. Mellors. New York: Blakiston.

Ogawa, K.; Mizuno, N.; and Okamoto, M.
   (1961) "Cytochemistry of cultured neural tissue. III. Heterogeneity of lysosomes."
      Journal of Histochemistry and Cytochemistry 9:635.
Ohno, Susumo.
   (1965) "Direct handling of germ cells." *Human Chromosome Methodology.* Ed. by
      Jorge J. Yunis. New York: Academic Press.
Ojeda, J. L.; Barbosa, E.; and Bosque, P. Gomez.
   (1970) "Selective skeletal staining in whole chicken embryos; a rapid Alcian blue tech-
      nique." *Stain Technology* 45:137–138.
Ores, Richard O.
   (1971) "Advantages of epoxy resin as a mounting medium for light microscopy." *Stain
      Technology* 46:315–317.
Oster, Gerald.
   (1955) "X-ray diffraction techniques and their application to the study of biomolecular
      structures." *Analytical Cytology.* Ed. by Robert C. Mellors. New York: Blakis-
      ton.
Owen, G.
   (1955) "Use of propylene phenoxetol as a relaxing agent." *Nature* 175:434.
Owen, G.; and Steedman, H. F.
   (1956) "Preservation of animal tissues, with a note on staining solutions." *Quarterly
      Journal of Microscopical Science* 97:319–321.
   (1958) "Preservation of molluscs." *Proceedings of the Malacological Society of London*
      33:101–103.

Paddy, J. F.
   (1970) "Metachromasy of dyes in solution." NATO Advanced Study Institute. St. Mar-
      gherita [Report]. Ed. by E. A. Balasz. London: Academic Press.
Paget, G. E., and Eccleston, Enid.
   (1959) "Aldehyde-thionin: a stain having similar properties to aldehyde-fuchsin." *Stain
      Technology* 34:223–226.
   (1960) "Simultaneous specific demonstration of thyrotroph, gonadotroph and acidophil
      cells in the anterior hypophysis." *Stain Technology* 35:119–122.
Palade, G. E.
   (1952) "A study of fixation for electron microscopy." *Journal of Experimental Medicine*
      95:285–297.
Palkowsky, Walter.
   (1960) "A new method of deep freezing pathologic specimens in a pliable state." *Techni-
      cal Bulletin of the Registry of Medical Technologists* 30:187.
Palmer, M. Winnogene, and McDonald, L. W.
   (1963) "A device for the continuous chilling of microtome knife and paraffin blocks."
      American Journal of Clinical Pathology 40:633.
Palmgren, Axel.
   (1954) "Tape for microsectioning of very large, hard, or brittle specimens." *Nature*
      174:46.
Pantin, C. F. A.
   (1946) *Notes on Microscopical Technique for Zoologists.* Cambridge, England: Cam-
      bridge University Press.
Papanicolaou, George N.
   (1942) "A new procedure for staining vaginal smears." *Science* 95:438–439.
   (1947) "The cytology of the gastric fluid of carcinoma of the stomach." *Journal of the
      National Cancer Institute* 7:357–360.
   (1954) *Atlas of Exfoliative Cyology.* Cambridge, Mass.: Harvard University Press.
   (1957) "The cancer diagnostic potential of uterine exfoliative cytology." *CA, Bulletin of
      Cancer Progress* 7:125–135.
Pappas, Peter W.
   (1971) "The use of a chrome alum–gelatin (subbing) solution as a general adhesive for
      paraffin sections." *Stain Technology* 46:121–124.

Parsons, D. F.
  (1970) *Some Biological Techniques in Electron Microscopy.* New York: Academic Press.
Past, Wallace L.
  (1961) "The histologic demonstration of iron in osseous tissue." *American Journal of Pathology* 39:443–449.
Patau, K.
  (1960) "The identification of individual chromosomes, especially in man." *American Journal of Human Genetics* 12:250–276.
  (1961) "Chromosome identification and the Denver Report." *Lancet* 1:933–934.
  (1965) "Identification of chromosomes." *Human Chromosome Methodology.* Ed. by Jorge J. Yunis. New York: Academic Press.
Patau, K.; Therman, Eva; Smith, David W.; Inhorn, Stanley L.; and Picken, Bruce F.
  (1961) "Partial-trisomy syndromes. I. Sturge-Weber's disease." *American Journal of Human Genetics* 13:287–298.
Patten, Bradley M.
  (1952) *Early Embryology of the Chick.* 4th ed. New York: Blakiston.
Patten, Stanley F., and Brown, Kenneth A.
  (1958) "Freeze-solvent substitution technic. A review with application to fluorescence microscopy." *Laboratory Investigation* 7:209–223.
Pauly, H.
  (1964) "Über die Konstitution des Histidins." *Hoppe-Seylers Zeitschrift für Physiologische Chemie* 42:508–518.
Pearse, A. G. Everson.
  (1949) "The cytochemical demonstration of gonadotropic hormone in the human anterior hypophysis." *Journal of Pathology and Bacteriology* 61:195–202.
  (1950) "Differential stain for human and animal anterior hypophysis." *Stain Technology* 25:95–102.
  (1955) "Copper phthalocyanins as phospholipid stains." *Journal of Pathology and Bacteriology* 70:554–557.
  (1968) *Histochemistry, Theoretical and Applied,* I. 3d ed. Baltimore, Williams and Wilkins.
  (1972) *Histochemistry, Theoretical and Applied,* II. 3d ed. London: Churchill Livingston.
Pearson, Bjarne.
  (1958) "Improvement in the histochemical localization of succinic dehydrogenase by the use of nitroneotetrazolium chloride." *Journals of Histochemistry and Cytochemistry* 6:112–121.'
Pearson, Bjarne; Wolf, Paul; and Andrews, Marian.
  (1963) "The histochemical demonstration of leucine aminopeptidase by means of a new indoyl compound." *Laboratory Investigation* 12:712–720.
Peary, Joseph Igor.
  (1955) "Freeze dehydration for permanent mounts after aceto-orcein stain." *Stain Technology* 30:213–230.
Pease, Daniel C.
  (1964) *Histological Techniques for Electron Microscopy.* New York: Academic Press.
Pease, Daniel C., and Baker, R. F.
  (1948) "Sectioning techniques for electron microscopy using conventional microtome." *Proceedings of the Society for Experimental Biology and Medicine* 67:470–474.
  (1950) "Electron microscopy of the kidney." *American Journal of Anatomy* 87:349–390.
Pelc, S. R.
  (1956) "A stripping film technique for autoradiography." *International Journal of Applied Radiation and Isotopes* 1:172–177.
  (1958) "Autoradiography as a cytochemical method, with special reference to C^{14} and S^{35}." *General Cytochemical Methods,* I. Ed. by J. F. Danielli. New York: Academic Press.

Peltier, Leonard F.
   (1954) "The demonstration of fat emboli in tissue sections using phosphin 3R, a water-
      soluble fluorochrome." *Journal of Laboratory and Clinical Medicine* 43:321–323.
Pequeno, Rubens Alves.
   (1960) "A simple pen method for preparing linear blood films for accurate differential
      leukocyte counts." *Technical Bulletin of the Registry of Medical Technologists*
      30:193–196.
Perry, Linda J.; Harbison, Robert M.; and Lumb, Roger H.
   (1975) "The use of Herr four-and-a-half clearing fluid for the rapid microscopic examina-
      tion of thick sections of normal and neoplastic tissues." *Stain Technology* 50:47–
      50.
Perry, Robert P.
   (1964) "Quantitative Autoradiography." *Methods in Cell Physiology,* I. Ed. by David
      M. Prescott, New York: Academic Press.
Persidsky, Maxim D.
   (1954) "Restoration of deteriorated temporary aceto-carmine preparations." *Stain
      Technology* 29:278.
Peters, H.
   (1961*a*) "A glass sieve for carrying loose frozen and celloidin sections through all stages
      of processing." *Stain Technology* 36:201.
   (1961*b*) "An albumen air drying sequence for attaching celloidin sections to slides."
      Stain Technology 36:248–250.
Petko, M.
   (1974) "The use of 1:9 dimethyl–methylene blue for the demonstration of thyroid
      cells." *Stain Technology* 49:65–67.
Picciano, Dante J., and McKinnell, Robert G.
   (1977) "A short term culture method for the preparation of chromosome spreads from
      Rana pipiens." *Stain Technology* 52:101–103.
Pickett, John Phillip; Bishop, Carl M.; Chick, Ernest W.; and Baker, Roger D.
   (1960) "A simple fluorescent stain for fungi." *American Journal of Clinical Pathology*
      34:197–202.
Pickett, John Phillip, and Klavins, J. V.
   (1961) "Demonstration of elastic tissue and iron or elastic tissue and calcium simulta-
      neously." *Stain Technology* 36:371–374.
Pickett, John Phillip, and Sommer, Joachim R.
   (1960) "Thirty-five mm film as mounting base, and plastic spray as cover glass, for
      histologic sections." *Archives of Pathology* 69:13–21, 239–247.
Pickworth, J. W.; Cotton, K.; and Skyring, A. P.
   (1963) "Double emulsion autoradiography for identifying tritium-labelled cells in
      sections." *Stain Technology* 38:237–244.
Pienaar, Uys de Villiers.
   (1962) *Hematology of Some South African Reptiles.* Johannesburg: Witwatersrand Uni-
      versity Press.
Pitelka, Dorothy Riggs.
   (1945) "Morphology and taxonomy of flagellates of the genus *Peranema dujardin.*"
      Journal of Morphology 76:179–190.
Pizzolato, Philip, and McCrory, Phil
   (1962) "Light influence on von Kóssa's silver calcium reaction in the myocardium."
      Journal of Histochemistry and Cytochemistry 10:102.
Poirier, Louis J.; Ayotte, Robert A.; and Gauthier, Claude.
   (1954) "Modification of the Marchi technic." *Stain Technology* 29:71–75.
Poley, R. W., and Fobes, C. D.
   (1964) "Fuchsinophilia in early myocardial infarction." *Archives of Pathology* 77:325–
      329.
Pollak, O. J.
   (1944) "A rapid trichrome stain." *Archives of Pathology* 37:294.

Pollister, Arthur W., and Pollister, Priscilla F.
(1957) "The structure of the Golgi apparatus." *International Review of Cytology*, VI. Ed. by G. H. Bourne and J. F. Danielli. New York: Academic Press.

Pool, Charlotte R.
(1969) "Hematoxylin-eosin staining of OsO_4-fixed epon-embedded tissue; prestaining oxidation by acidified H_2O_2." *Stain Technology* 44:75–79.
(1973) "Prestaining oxidation by acidified H_2O_2 for revealing Schiff-positive sites in epon-embedded sections." *Stain Technology* 48:123–126.

Popp, Raymond A.; Gude, William D.; and Popp, Diana M.
(1962) "Peroxidase staining combined with autoradiography for study of eosinophilic granules." *Stain Technology* 37:243–247.

Popper, Hans, and Szanto, Paul B.
(1950) "Fluorescence Microscopy." *Handbook of Microscopical Technique*. Ed. by Ruth McClung Jones. New York: Hoeber.

Porter, Keith R., and Blum, J.
(1953) "A study in microtomy for electron microscopy." *Anatomical Record* 117:685–710.

Pottz, Glenn; Rampey, James H.; and Furmandean, Benjamin.
(1964) "A method for rapid staining of acid-fast bacteria in smears and sections of tissue." *American Journal of Clinical Pathology* 42:552–554.

Powell, Ervin W., and Brown, Geraldine.
(1975) "A critique of silver impregnation methods." *Mikroskopie* 31:77–84.

Powers, Margaret, and Clark, George
(1955) "An evaluation of cresyl echt violet as a Nissl stain." *Stain Technology* 30:83–92.
(1963) "A note on Darrow red." *Stain Technology* 38:289–290.

Powers, Margaret; Clark, George; Darrow, Mary; and Emmel, Victor M.
(1960) "Darrow red, a new basic dye." *Stain Technology* 35:19–22.

Preece, Ann.
(1959) *A Manual for Histologic Technicians*. Boston: Little Brown.

Prescott, D. M.
(1964) "Autoradiography with liquid emulsion." *Methods in Cell Physiology*, I. Ed. by D. M. Prescott. New York: Academic Press.

Prescott, D. M., and Bender, M. A.
(1964) "Preparation of mammalian metaphase chromosomes for autoradiography." *Methods in Cell Physiology*, I. Ed. by D. M. Prescott. New York: Academic Press.

Prescott, D. M., and Carrier, R. F.
(1964) "Experimental procedures and cultural methods for *Euplotes eurystomus* and *Amoeba proteus*." *Methods in Cell Physiology*, I. Ed. by D. M. Prescott. New York: Academic Press.

Priman, Jacob.
(1954) "Blood serum as an adhesive for paraffin sections." *Stain Technology* 29:105–107.

Procknow, John J.; Connelly, A. Philip, Jr.; and Ray, C. George.
(1962) "Fluorescent antibody technique in histoplasmosis." *Archives of Pathology* 73:313–324.

Proescher, Frederick.
(1933) "Pinacyanol as a histological stain." *Proceedings of the Society for Experimental Biology and Medicine* 31:79–81.
(1934) "Contribution to the staining of neuroglia." *Stain Technology* 9:33–38.

Proescher, Frederick, and Arkush, A. S.
(1928) "Metallic lakes of the oxazines (gallamin blue, gallocyanin, and celestin blue) as nuclear stain substitutes for hematoxylin." *Stain Technology* 2:28–38.

Puchtler, Holde.
(1958) "Significance of the iron hematoxylin method of Heidenhain." *Journal of Histochemistry and Cytochemistry* 6:401–402.

Puchtler, Holde; Chandler, A. B.; and Sweat, Faye.
   (1961) "Demonstration of fibrin in tissue sections by the Rosindole method." *Journal of Histochemistry and Cytochemistry* 9:340.
Puchtler, Holde, and Leblond, C. P.
   (1958) "Histochemical analysis of cell membranes and associated structures as seen in the intestinal epithelium." *American Journal of Anatomy* 102:1–31.
Puchtler, Holde; Meloan, Susan N.; and Terry, Mary S.
   (1969) "On the history and mechanism of alizarin and alizarin red S stains for calcium." *Journal of Histochemistry and Cytochemistry* 17:110–124.
Puchtler, Holde; Rosenthal, Sanford I.; and Sweat, Faye.
   (1964) "Revision of the amidoblack stain for hemoglobin." *Archives of Pathology* 78:76–78.
Puchtler, Holde, and Sweat, Faye.
   (1960) "Commercial resorcin-fuchsin as a stain for elastic fibers." *Stain Technology* 35:347–348.
   (1962) "Some comments on the ninhydrin-Schiff reaction." *Journal of Histochemistry and Cytochemistry* 10:365.
   (1963a) "A combined hemoglobin-hemosiderin stain." *Archives of Pathology* 75:588–590.
   (1963b) "Influence of various pretreatments on the staining properties of connective tissue fibers." *Annales d'Histochemie* 8:189–198.
   (1964a) "On the mechanism of sequence iron-hematein stains." *Histochemie* 4:197–208.
   (1964b) "Histochemical specificity of stain-methods for connective tissue fibers: resorcin-fuchsin and van Gieson's picro-fuchsin." *Histochemie* 4:24–34.
   (1965) "Congo red as a stain for fluorescence microscopy of amyloid." *Journal of Histochemistry and Cytochemistry* 13:693–694.
   (1966) "A review of early concepts of amyloid in context with contemporary chemical literature from 1839 to 1859." *Journal of Histochemistry and Cytochemistry* 14:123–134.
Puchtler, Holde; Sweat, Faye; and Doss, Nancy O.
   (1963) "A one-hour phosphotungstic acid-hematoxylin stain." *American Journal of Clinical Pathology* 40:334.
Puchtler, Holde; Sweat, Faye; and Levine M.
   (1962) "On the binding of Congo red by amyloid." *Journal of Histochemistry and Cytochemistry* 10:355–364.
Puchtler, Holde; Waldrop, Faye Sweat; Meloan, Susan N.; Terry, Mary S.; and Conner, H. M.
   (1970) "Methacarn (Methanol-Carnoy) fixation. Practical and theoretical considerations." *Histochemie* 21:97–116.
Puchtler, Holde; Waldrop, Faye Sweat; Terry, Mary S.; and Conner, H. M.
   (1969) "A combined PAS-myofibril stain for demonstration of early lesions of striated muscle." *Journal of Microscopy* 89:329–338.
Puchtler, H.; Waldrop, F. S.; and Valentine, L. S.
   (1973) "Fluorescence microscopic distinction between elastin and collagen." *Histochemie* 35:17–30.
Puck, Theodore T.; Cieciura, Steven J.; and Robinson, Arthur.
   (1958) "Genetics of somatic mammalian cells. III. Long-term cultivation of euploid cells from human and animal subjects." *Journal of Experimental Medicine* 108:945–956.
Pugh, M. H., and Savchuck, W. B.
   (1958) "Suggestions on the preparation of undecalcified bone for microradiography." *Stain Technology* 33:287–293.
Putt, F. A.
   (1948) "Modified eosin counterstain for formaldehyde-fixed tissues." *Archives of Pathology* 45:72.
   (1951) "A benzidine-thionine method for the demonstration of hemoglobin in formaldehyde-fixed, paraffin-embedded tissue." *Archives of Pathology* 52:293.

(1956) "Flaming red as a dye for the demonstration of lipids." *Laboratory Investigation* 5:377–379.

(1971) "Alcian dyes in calcium chloride: a routine selective method to demonstrate mucins." *Yale Journal of Biology and Medicine* 43:279–282.

Putt, F. A., and Huskill, Peter B.

(1962) "Alcian green, a routine stain for mucins." *Archives of Pathology* 74:169–170.

Quary, W. B.

(1957) "Experimental cyanine red, a new stain for nucleic acid and mucopolysaccharides." *Stain Technology* 32:175–182.

Quintarelli, G.; Cifonelli, J. A.; and Zito, R.

(1971) "On phosphotungstic acid staining, II." *Journal of Histochemistry and Cytochemistry* 19:648–653.

Rae, C. A.

(1955) "Masson's trichrome stain after Petrunkewitsch or Susa fixation." *Stain Technology* 30:147–148.

Rafalko, Stanley.

(1946) "A modified Feulgen technic for small and diffuse chromatin elements." *Stain Technology* 21:91–94.

Ralph, P. H.

(1941) "The histochemical demonstration of hemoglobin in blood cells and tissue smears." *Stain Technology* 16:105–106.

Raman, K.

(1955) "A method of sectioning aspirated bone-marrow." *Journal of Clinical Pathology* 8:265–266.

Ramon y Cajal, Santiago.

(1903) "Un sencillo metodo de coloracion selectiva del reticilo proto-plasmico y sus effectos en los diversos organos nerviosis." *Trabajos del Laboratorio Investigaciones Biologica* 2:129–221.

(1910) "Las formulas del proceder del nitrato de plata reducido." *Trabajos del Laboratorio Investigaciones Biologica* 8:1–26.

Ramon y Cajal, Santiago, and De Castro, F.

(1933) *Elementos de tecnica micrografica del sistema nerviosa* Madrid: Tipografia Artistica.

Randolph, L. F.

(1935) "A new fixing fluid and a revised schedule for the paraffin method in plant cytology." *Stain Technology* 10:95–96.

Ray, H. N., and Hajra, B.

(1962) "Hydrosulfite-Schiff; limitations in use as shown by periodic-Schiff reactions in protozoa." *Stain Technology* 37:75–77.

Reaven, Eve P., and Cox, Alvin J.

(1963) "The histochemical localization of histidine in the human epidermis and its relationship to zinc binding." *Journal of Histochemistry and Cytochemistry* 11:782–790.

Reid, J. D., and Sarantakos, G.

(1966) "Infiltrating and embedding with mixtures of polyethylene glycols and polyvinyl-acetate resins." *Stain Technology* 41:207–210.

Reid, J. D., and Taylor, D.

(1964) "An improved method for embedding tissues, using polyethylene glycols, with incorporation of low-viscosity nitrocellulose." *American Journal of Clinical Pathology* 41:513–516.

Renaud, Serge

(1959) "Superiority of alcoholic over aqueous fixation in the histochemical detection of calcium." *Stain Technology* 34:267–271.

Repak, Arthur J., and Levine, Alan B.
    (1967) "A new use for an old stain." *Turtox News* 45:227.
Richards, Oscar W.
    (1941) "An efficient method for the identification of M. tuberculosis with a simple
    fluorescence microscope." *American Journal of Clinical Pathology, Technical
    Section* 5:1–8.
    (1949) *The Effective Use and Proper Care of the Microtome.* American Optical Com-
    pany.
    (1954) *The Effective Use and Proper Care of the Microscope.* American Optical
    Company.
Richards, Oscar W.; Kline, Edmund K.; and Leach, Raymond E.
    (1941) "Demonstration of tubercle bacilli by fluorescence microscopy." *American Re-
    view of Tuberculosis* 44:255–266.
Richardson, Howard L.
    (1960) "The advantages of a glycol fixative for the preparation of Papanicolaou smears."
    Technical Bulletin of the Registry of Medical Technologists 30:15.
Richardson, K. C.; Jarrett, L.; and Finke, E. H.
    (1960) "Embedding in epoxy resins for ultrathin sectioning in electron microscopy."
    Stain Technology 35:313–323.
Richmond, Gordon W., and Bennett, Leslie.
    (1938) "Clearing and staining of embryos for demonstration of ossification." *Stain
    Technology* 13:77–79.
Riis, Povl.
    (1957) "The chromatin structure in lymphocyte cells from peripheral blood." *Acta
    Haematologica* 18:168–174.
Rinderknecht, H.
    (1960) "A new technique for the fluorescent labelling of proteins." *Experientia* 16:430–
    431.
Rinehart, J. F., and Abul-Haj, S. K.
    (1951a) "An improved method for histologic demonstration of acid mucopolysac-
    charides in tissues." *Archives of Pathology* 52:189–194.
    (1951b) "Histological demonstration of lipids in tissue after dehydration and embedding
    in polyethylene glycol." *Archives of Pathology* 51:666–669.
Riopel, James L.
    (1962) "Carbowax for embedding and serial sectioning of botanical material." *Stain
    Technology* 37:357–362.
Ritter, Carl; Di Stephano, H. S.; and Farah, A.
    (1961) "A method for the cytophotometric estimation of ribonucleic acid." *Journal of
    Histochemistry and Cytochemistry* 9:97–102.
Roden, D. B.
    (1975) "Nitrocellulose sectioning of heads of larval Cerambycidae (Celopter)." *Stain
    Technology* 50:207–208.
Rodriquez, Jose, and Deinhardt, Friedrich.
    (1960) "Preparation of a semipermanent medium for fluorescent antibody studies." *Vir-
    ology* 12:316–317.
Roels, F.; Schiller, B.; and Goldfischer, S.
    (1970) "Peroxisomes and lysosomes in the toad, *Bufo marinus.*" *Journal of His-
    tochemistry and Cytochemistry* 18:681–682.
Roman, Nickolas; Perkins, Stephen F.; Perkins, Edwin M., Jr.; and Dolnick, Ethel H.
    (1967) "Orcein-hematoxylin in iodized ferric chloride as a stain for elastic fibers, with
    methanil yellow counterstaining." *Stain Technology* 42:199–202.
Romeis, B.
    (1948) *Mikroskopische Technik.* München: Leibniz Verlag.
Roozenmond, R. C.
    (1969) "The effect of calcium chloride and formaldehyde on the release and composition
    of phospholipids from cryostat sections of rat hypothalamus." *Journal of His-
    tochemistry and Cytochemistry* 17:273–279.

Roque, Augustine L.; Jafarey, Naeem A.; and Coulter, Priscilla.
(1965) "A stain for the histochemical demonstration of nucleic acids." *Experimental and Molecular Pathology* 4:266–274.

Rosa, Charles G.
(1953) "Preparation and use of aldehyde fuchsin stain in the dry form." *Stain Technology* 28:299–302.

Rosenthal, Sanford I.; Puchtler, Holde; and Sweat, Faye.
(1965) "Paper chromatography of dyes." *Archives of Pathology* 80:190–196.

Rosewater, Joseph.
(1963) "An effective anesthetic for giant clams and other molluscs." *Turtox News* 41:300–302.

Rothenbacher, Hans J., and Hitchcock, Dorothy J.
(1962) "Heat fixation and Giemsa staining for flagella and other cellular structures of trichomonads." *Stain Technology* 37:111–113.

Rothfels, K. H., and Siminovitch, L.
(1958) "An air drying technique for flattening chromosomes in mammalian cells grown in vitro." *Stain Technology* 33:73–77.

Rothstein, N.
(1958) "Vital staining of blood parasites with acridine orange." *Journal of Parasitology* 44:588–595.

Rothstein, N., and Brown, M. L.
(1960) "Vital staining and differentiation of microfilaria." *American Journal of Veterinary Research* 21:1090–1094.

Rouiller, C. H.
(1960) "Physiological-pathological changes in mitochondrial morphology." *International Review of Cytology,* XI. Ed. by G. H. Bourne and J. F. Danielli. New York: Academic Press.

Rubin, Robert.
(1951) "A rapid method for making permanent mounts of nematodes." *Stain Technology* 26:257–260.

Ruch, Fritz.
(1955) "Birefringence and dichroism of cells and tissues." *Physical Techniques in Biological Research,* I. Ed. by Gerald Oster and Arthur W. Pollister. New York: Academic Press.

Ruddell, Craig L.
(1967a) "Hydroxyethyl methacrylate combined with polyethylene glycol 400 and water; an embedding medium for routine 1-2 micron sectioning." *Stain Technology* 42:119–123.
(1967b) "Embedding media for 1-2 micron sectioning. 2. Hydroxyethyl methacrylate combined with 2-butoxyethanol." *Stain Technology* 42:253–255.

Rumpf, P.
(1935) "Recherches physico-chimiques sur la reaction colorie des aldehydes, dite 'Reaction de Schiff'." *Annales de Chimie* 3:327.

Russell, Robert.
(1973) "The Bodian stain for nerve fibers and nerve endings." *Laboratory Medicine* 4:40.

Rutenberg, Alexander M.; Gofstein, Ralph; and Seligman, Arnold M.
(1950) "Preparation of a new tetrazolium salt which yields a blue pigment on reduction and its use in the demonstration of enzymes in normal and neoplastic tissues." *Cancer Research* 10:113–121.

Rutenberg, Alexander M.; Wolman, Moshi; and Seligman, Arnold M.
(1953) "Comparative distribution of succinic dehydrogenase in six mammals and modification in the histochemical technic." *Journal of Histochemistry and Cytochemistry* 1:66–81.

Ruth, E. B.
(1946) "Demonstration of the ground substance of cartilage, bone, and teeth." *Stain Technology* 27–30.

Sabatini, D. D.; Bensch, K. G.; and Barrnett, R. J.
    (1963) "Cytochemistry and electron microscopy. The preservation of cellular ultrastructure and enzymatic activity by aldehyde fixation." *Journal of Cell Biology* 17:19–58.
Sabatini, D. D.; Miller, Fritz; and Barrnett, J.
    (1964) "Aldehyde fixation for morphological and enzyme histochemical studies with the electron microscope." *Journal of Histochemistry and Cytochemistry* 12:57–71.
Sachs, Leo.
    (1953) "Simple methods for mammalian chromosomes." *Stain Technology* 28:169–172.
Sacks, Jacob.
    (1965) "Tracer techniques: stable and radioactive isotopes." *Physical Techniques in Biological Research*, II. Ed. by Gerald Oster and Arthur W. Pollister. New York: Academic Press.
Sage, Martin.
    (1972) "Polyethylene glycol distearate 600 with 10% 1-Hexadecanol; a superior embedding wax for warm climates." *Stain Technology* 47:313–315.
St. Amand, George Sims, and St. Amand, W.
    (1951) "Shortening maceration time for alizarine red S preparations." *Stain Technology* 26:271.
Sainte-Marie, Guy.
    (1962) "A paraffin embedding technique for studies employing immunofluorescence." *Journal of Histochemistry and Cytochemistry* 10:250–256.
Sakesela, E., and Moorhead, P. S.
    (1962) "Enhancement of secondary constrictions and the heterochromatic X in human cells." *Cytogenetics* 1:225–244.
Sallmen, Henry, and Sherman, Mary.
    (1961) "A method for the rapid processing of celloidin sections." *Technical Bulletin of the Registry of Medical Technologists* 31:189–191.
Salthouse, T. N.
    (1958) "Tetrahydrofuran and its use in insect histology." *Canadian Entomologist* 90:555–557.
    (1962) "Luxol-fast blue ARN: a new solvent dye with improved staining qualities for myelin and phospholipids." *Stain Technology* 37:313–316.
    (1964) "Luxol-fast blue G as a myelin stain." *Stain Technology* 39:123.
Sams, Alice.
    (1963) "A celloidin infiltration–frozen section sequence for enhanced preservation of phosphatases in bone." *Stain Technology* 38:1–8.
Sams, Alice, and Davies, Freda M. R.
    (1967) "Commercial varieties of nuclear fast red; their behaviour in staining after autoradiography." *Stain Technology* 42:269–276.
Sanders, Phyllis C., and Humason, Gretchen L.
    (1964) "Culture and slide preparation of leukocytes from blood of *Macaca*. *Stain Technology* 39:209–213.
Sandison, A. T.
    (1955) "The histological examination of mummified material." *Stain Technology* 30:277–283.
Sandritter, W.
    (1962) "Is gallocyanin chromalum a quantitative stain for nucleic acids?" *Journal of Histochemistry and Cytochemistry* 10:680–681.
Sani, Guelfo; Citti, Ugo; Caramazza, Giuliano; and Quinto, Pietro.
    (1964) *Fluorescence Microscopy in the Cytodiagnosis of Cancer*. Springfield, Ill.: Charles C Thomas.
Santamarina, Enrique.
    (1964) "A formalin-Wright staining technique for avian blood cells." *Stain Technology* 39:267–274.
Sasaki, Masao, Makeno, Sajiro.
    (1963) "The demonstration of secondary constrictions in human chromosomes by means of a new technique." *American Journal of Human Genetics* 15:24–33.

Sasaki, Motomichi.
(1961) "Observations on the modification in size and shape of chrmosomes due to technical procedures." *Chromosoma* 11:514–522.
Sato, Taizan, and Shamoto, Mikihiro.
(1973) "A simple rapid polychrome stain for epoxy-embedded tissue." *Stain Technology* 48:223–227.
Saunders, A. M.
(1962) "Acridine orange staining for the identification of acid mucopolysaccharides." *Journal of Histochemistry and Cytochemistry* 10:683.
(1964) "Histochemical identification of acid mucopolysaccharides with acridine orange." *Journal of Histochemistry and Cytochemistry* 12:164–170.
Sawicki, Wojciech, and Pawinska, Maria.
(1965) "Effect of drying on unexposed autoradiographic emulsion in relation to background." *Stain Technology* 40:67–68.
Saxena, P. N.
(1957) "Formalin-chloride fixation to improve silver impregnation of the Golgi apparatus." *Stain Technology* 32:203–207.
Schajowicz, F., and Cabrini, R. L.
(1955) "The effect of acids (decalcifying solutions) and enzymes on the histochemical behavior of bone and cartilage." *Journal of Histochemistry and Cytochemistry* 3:122–129.
(1956) "Chelating agents as histological and histochemical decalcifiers." *Stain Technology* 31:129–133.
(1959) "Histochemical demonstration of acid phosphatase in hard tissues." *Stain Technology* 34:59–63.
Schantz, A., and Schecter, A.
(1965) "Iron-hematoxylin and safranin O as a polychrome stain for epon sections." *Stain Technology* 40:479–482.
Scheres, J. M. J. C.
(1977) "R- and CT-banding of human chromosomes with basic fuchsin." *Histochemistry* 52:349–353.
Schiff, Robert; Quinn, Loyd Y.; and Bryan, J. H. D.
(1967) "A safranin fast green stain for the differentiation of the nuclei of rumen protozoa." *Stain Technology* 42:75–80.
Schiffer, Lewis M., and Vaharu, Tiju.
(1962) "Acridine orange as a useful chromosomal stain." *Technical Bulletin of the Registry of Medical Technologists* 32:91–92.
Schlegel, Jorgen Ulrik.
(1949) "Demonstration of blood vessels and lymphatics with a fluorescent dye in ultraviolet light." *Anatomical Record* 105:433–443.
Schleicher, Emil Maro.
(1951) "Floating solutions for mounting paraffin sections." *American Journal of Clinical Pathology* 21:900.
Schleifstein, J.
(1937) "A rapid method for demonstrating Negri bodies in tissue sections." *American Journal of Public Health* 27:1283–1285.
Schmid, Werner.
(1965) "Autoradiography of human chromosomes." *Human Chromosome Methodology*. Ed. by Jorge J. Yunis. New York: Academic Press.
Schmidt, Robert W.
(1956) "Simultaneous fixation and decalcification of tissue." *Laboratory Investigation* 5:306–307.
Schneider, John D.
(1963) "A simplified Gram stain for demonstrating fungi in tissues." *Technical Bulletin of the Registry of Medical Technologists* 33:195–197.
Schubert, Maxwell, and Hamerman, David.
(1956) "Metachromasia: chemical theory and histochemical use." *Journal of Histochemistry and Cytochemistry* 4:159–189.

Schultz, J., and St. Lawrence, P.
  (1949) "A cytological basis for a map of the nucleolar chromosome in man." *Journal of Heredity* 40:31–38.
Schultze, Brigitte.
  (1969) "Autoradiography at the cellular level." *Physical Techniques in Biological Research,* III. 2d ed. Ed. by Arthur W. Pollister. New York: Academic Press.
Schwind, Joseph L.
  (1950) "The supravital method in the study of the cytology of the blood and marrow cells." *Blood* 5:597–622.
Scott, Gordon H.
  (1928) "A method for making permanent preparations of supravitally stained blood cells." *Anatomical Record* 38:233–237.
Scott, H. R., and Clayton, B. P.
  (1953) "A comparison of the staining affinities of aldehyde-fuchsin and the Schiff reagent." *Journal of Histochemistry and Cytochemistry* 1:336–352.
Scott, Jesse F.
  (1955) "Ultraviolet absorption spectrophotometry." *Physical Techniques in Biological Research,* I. Ed. by Gerald Oster and Arthur W. Pollister. New York: Academic Press.
Scott, J. E.
  (1976) "Phosphotungstic acid 'Schiff-reactive' but not a 'glycol reagent.'" *Histochemistry* 48:1084–1085.
Scott, John E., and Mowry, Robert W.
  (1970) "Alcian blue—a consumer's guide." *Journal of Histochemistry and Cytochemistry* 18:842.
Seabright, M.
  (1971) "A rapid banding technique for human chromosomes." *Lancet* 2:91–92.
Seal, S. H.
  (1956) "A method for concentrating cancer cells suspended in large quantities of fluid." *Cancer* 9:866–868.
Seligman, Arnold M.; Wasserkrug, Hannah L.; Deb, Chandicharan; and Hanker, Jacob S.
  (1968) "Osmium-containing compounds with multiple basic or acidic groups as stains for ultrastructure." *Journal of Histochemistry and Cytochemistry* 16:87–101.
Seman, G.
  (1961) "Technique de coloration des lobules hétérochromatiques des leucocytes (spécialement des mononucléés) valeur de cette technique pour la determination du sexe." *Revue Française d'Études Cliniques et Biologiques* 6:161–165.
Serra, J. A.
  (1946) "Histochemical tests for proteins and amino acids; the characterization of basic proteins." *Stain Technology* 21:5–18.
  (1958) "Cytochemical demonstration of masked lipids." *Science* 128:28–29.
Shanklin, D. R., and Laite, Melville B.
  (1963) "Pickett–Sommer film strip technique." *Archives of Pathology* 75:91–93.
Shanklin, W. M., and Nassar, T. K.
  (1959) "Luxol fast blue combined with periodic acid Schiff procedure for cytological staining of kidney." *Stain Technology* 34:257–260.
Shanklin, W. M.; Nasser, T. K.; and Issidorides, M.
  (1959) "Luxol fast blue as a selective stain for alpha cells in the human pituitary." *Stain Technology* 34:55–58.
Shaver, Evelyn L.
  (1962) "The chromosomes of the opossum, *Didelphis virginiana.*" *Canadian Journal of Genetics and Cytology* 4:62–68.
Sheehan, Dezna.
  (1960) "A comparative study of the histologic techniques for demonstrating chromaffin cells." *American Journal of Medical Technology* 26:237–240.
Shelley, Walter B.
  (1963) "Rapid technic for obtaining leukocytes from small samples of blood." *American Journal of Clinical Pathology* 39:433–435.

(1969) "Fluorescent staining of elastic tissue with rhodamine B and related xanthene dyes." *Histochemie* 20:244–249.

Shillaber, Charles Patten.
(1944) *Photomicrography, in Theory and Practice.* New York: John Wiley.

Shires, T. K.; Johnson, M.; and Richter, K. M.
(1969) "Hematoxylin staining of tissues embedded in epoxy resins." *Stain Technology* 44:21–25.

Shute, P. G., and Maryon, M. E.
(1966) *Laboratory Technique for the Study of Malaria.* 2d ed. London: J. and A. Churchill.

Shyamasundari, Koka, and Rao, Kona Hanumantha.
(1975) "A procedure for the simultaneous demonstration of neurosecretory and mucosubstances in tissue sections." *Acta Histochemica* 54:272–274.

Sidman, R. L.; Mottla, P. A.; and Feder, N.
(1961) "Improved polyester wax embedding for histology." *Stain Technology* 36:279–284.

Sieracki, Joseph; Michael, James E.; and Clark, Daisy A.
(1960) "The demonstration of beta cells in pancreatic islets and their tumors." *Stain Technology* 35:67–69.

Sikov, Melvin R.
(1965) "A commercial encapsulated desiccant for use in slide containers during autoradiographic exposure." *Stain Technology* 40:239.

Sills, Bernard, and Marsh, Walton H.
(1959) "A simple technic for staining fat with oil-soluble azo dyes." *Laboratory Investigation* 8:1006–1009.

Silver, Maurice L.
(1942) "Colloidal factors controlling silver staining." *Anatomical Record* 82:507–529.

Silverstein, Arthur M.
(1957) "Contrasting fluorescent labels for two antibodies." *Journal of Histochemistry and Cytochemistry* 5:94–95.

Simmel, Eva B.
(1957) "The use of a fast, coarse grain stripping film for radioautography." *Stain Technology* 32:299–300.

Simmel, Eva B.; Fitzgerald, P. J.; and Godwin, J. T.
(1951) "Staining of radioautography with metanil yellow and iron hematoxylin." *Stain Technology* 26:25–28.

Simonds, Herbert R.; Weith, Archie J.; and Bigelow, M. H.
(1949) *Handbook of Plastics.* New York: Van Nostrand.

Simpson, Charles F.; Carlisle, J. W.; and Mallard, L.
(1970) "Rhodanile blue: a rapid and selective stain for Heinz bodies." *Stain Technology* 45:221–223.

Singer, Marcus.
(1952) "Factors which control the staining of tissues with acid and basic dyes." *International Review of Cytology,* I. Ed. by G. H. Bourne and J. F. Danielli. New York: Academic Press.

Sinha, Ranendra Nath.
(1953) "Sectioning insects with sclerotized cutile." *Stain Technology* 28:249–253.

Sisken, Jesse E.
(1964) "Methods for measuring the length of the mitotic cycle and the timing of DNA synthesis for mammalian cells in culture." *Methods in Cell Physiology,* I. Ed. by D. M. Prescott. New York: Academic Press.

Slidders, W.; Fraser, D. S.; Smith, R.; and Lendrum, A. C.
(1958) "On staining the nucleus red." *Journal of Pathology and Bacteriology* 75:466–468.
(1969) "A stable iron-haematoxylin solution for staining chromatin of cell nuclei." *Journal of Microscopy* 90:61–65.

Sloper, J. C., and Adams, C. W. M.
(1956) "The hypothalamic elaboration of posterior pituitary principles in man. Evidence

derived from hypophysectomy." *Journal of Pathology and Bacteriology* 72:587–602.

Smith, Arthur.
(1962) "Tissue shrinkage caused by attachment of paraffin sections to slides: its effects on staining." *Stain Technology* 37:339–345.

Smith, C. W.; Marshall, J. D.; and Eveland, W. C.
(1959) "Use of contrasting fluorescent dyes as counterstain in fixed tissue preparations." *Proceedings of the Society of Experimental Biology and Medicine* 102:179–181.

Smith, E. J.; Puchtler, Holde; and Sweat, Faye.
(1966) "Investigation of the chemical mechanism of trichrome stain." *Laboratory Investigation* 15:1141–1142.

Smith, G. Stewart.
(1943) "A danger attending the use of ammoniacal solutions of silver." *Journal of Pathology and Bacteriology* 55:227–228.

Smith, Luther.
(1947) "The acetocarmine technique." *Stain Technology* 22:17–31.

Smith, Robert E.
(1970) "Comparative evaluation of two instruments and procedures to cut non-frozen sections." *Journal of Histochemistry and Cytochemistry* 18:590–591.

Smyth, J. D.
(1944) "A technic for mounting free-living protozoa." *Science* 100:62.

Snodgrass, Ann B., and Dorsey, Charles H.
(1963) "Egg albumin embedding: a procedure compatible with neurological staining techniques." *Stain Technology* 38:149–155.

Snodgrass, Ann B., and Lacey, Leonard B.
(1961) "Luxol fast blue staining of degenerating myelinated fibers." *Anatomical Record* 140:83–89.

Snodgrass, M. A.; Dorsey, Charles H.; Bailey, George W. H.; and Dickson, Larry G.
(1972) "Conventional histopathologic staining methods compatible with Epon-embedded osmicated tissue." *Laboratory Investigation* 26:329–337.

Snow, Richard.
(1963) "Alcoholic hydrochloric acid–carmine as a stain for chromosomes in squash preparations." *Stain Technology* 38:9–13.

Sodeman, W. A., Jr., and Jeffery, G. M.
(1966) "Indirect fluorescent antibody test for malaria antibody." *Public Health Reports* 81:1037–1041.

Solcia, E.; Capella, C.; and Vassallo, G.
(1969) "Lead hematoxylin as a stain for endocrine cells." *Histochemie* 20:116–126.

Solcia, E.; Vassallo, G.; and Capella, C.
(1968) "Selective staining of endocrine cells by basic dyes after acid hydrolysis." *Stain Technology* 43:257–263.

Sommer, Joachim R., and Pickett, John Phillip.
(1961) "Thirty-five mm picture film for cytologic smears." *Archives of Pathology* 71:79–81.669–671.

Spatz, Maria.
(1960) "Bismarck brown as a selective stain for mast cells." *American Journal of Clinical Pathology* 34:285–287.

Spicer, S. S.
(1960) "A correlative study of the histochemical properties of rodent acid mucopolysaccharides." *Journal of Histochemistry and Cytochemistry* 8:18–33.
(1961) "Differentiation of nucleic acids by staining at controlled pH and by a Schiff-methylene blue sequence." *Stain Technology* 36:337–340.
(1962) "Histochemically selective acidophilia of basic nucleoproteins in chromatin and nucleoli at alkaline pH." *Journal of Histochemistry and Cytochemistry* 10:691–703.

Spicer, Samuel S.; Horn, Robert G.; and Leppi, T. John.
(1967) "The connective tissue." *Histochemistry of Connective Tissue Mucopolysaccharides*. International Academy of Pathology Monograph. Ed. by Bernard M. Wagner and David E. Smith. Baltimore: Williams and Wilkins.

Spicer, S. S., and Lillie, R. D.
(1961) "Histochemical identification of basic proteins with Biebrich scarlet at alkaline pH." *Stain Technology* 36:365–370.

Spicer, S. S., and Meyer, D. B.
(1960) "Histochemical differentiation of acid mucopolysaccharides by means of combined aldehyde fuchsin–Alcian blue staining." *American Journal of Clinical Pathology, Technical Section,* 33:453–456.

Spurlock, B. O.; Skinner, M. S.; and Kattine, A. A.
(1966) "A simple rapid method for staining epoxy-embedded specimens for light microscopy with the polychromatic stain, Paragon-1301." *American Journal of Clinical Pathology* 46:252–258.

Srivastava, Probodh K., and Lasley, J. F.
(1968) "Leucocyte culture and chromosome preparations from pig blood." *Stain Technology* 43:187–190.

Stafford, E. S.
(1934) "An eosin–methylene blue technique for rapid diagnosis." *Johns Hopkins Hospital Bulletin* 55:229.

Stafford, Marvin W.
(1960) "An improved method for labelling frosted-end microslides used in exfoliative cytology." *Technical Bulletin of the Registry of Medical Technologists* 30:26–27.

Stagg, Franklin B., and Tappen, N. C.
(1963) "A device for reducing condensation within the rotary microtome adapted for cutting frozen sections and for reducing tissue distortion." *Stain Technology* 38:352–354.

Stapp, Paul, and Cumley, Russell W.
(1936) "A technic for clearing large insects." *Stain Technology* 11:105–106.

Stearn, Allen E., and Stearn, Esther.
(1929) "The mechanism of staining explained on a chemical basis. I. The reaction between dyes, proteins, and nucleic acid." *Stain Technology* 4:111–119.
(1930) "The mechanism of staining explained on a chemical basis. II. General considerations." *Stain Technology* 5:17–24.

Steedman, H. F.
(1947) "Ester wax: a new embedding medium." *Quarterly Journal of Microscopical Science* 88:123–133.
(1950) "Alcian blue 8G: a new stain for mucin." *Quarterly Journal of Microscopical Science* 91:477–479.
(1960) *Section Cutting in Microscopy*. Springfield, Ill.: Charles C Thomas.

Steil, W.
(1936) "Modified Wright's method for staining blood smears." *Stain Technology* 11:99–100.

Steinman, Irvin D.
(1955) "A vinyl plastic coating for stained thin smears." *Stain Technology* 30:49–50.

Stenram, Unne.
(1962) "Loss of silver grains from radioautographs stained by gallocyanin–chrome alum." *Stain Technology* 37:231–234.

Sterling, Clarence, and Chichester, C. O.
(1956) "Autoradiography of water-soluble materials in plant tissues." *Stain Technology* 31:227–230.

Sternberg, S. S.; Cronin, A. P.; and Philips, F. S.
(1964) "Histochemical demonstration of zinc by the formation of a fluorescent chelate of 8-hydroxyquinoline (oxine): comparison with dithizone reactions." *Proceedings of the Federation of American Societies for Experimental Biology* 23:575.

Stone, G. E., and Cameron, I. L.
    (1964) "Methods for using *Tetrahymena* in studies of the normal cell cycle." *Methods in Cell Physiology,* I. Ed. by David M. Prescott. New York: Academic Press.
Stowell, Robert E.
    (1941) "Effect on tissue volume of various methods of fixation, dehydration, and embedding." *Stain Technology* 16:67–83.
    (1945) "Feulgen reaction for thymonucleic acid." *Stain Technology* 20:45–58.
    (1946) "The specificity of the Feulgen reaction for thymonucleic acid." *Stain Technology* 21:137–148.
Straus, Werner.
    (1964) "Factors affecting the cytochemical reaction of peroxidase with benzidine and the stability of the blue reaction product." *Journal of Histochemistry and Cytochemistry* 12:462–469.
    (1967) "Methods for the study of small phagosomes and their relationship to lysosomes with horseradish peroxidase as a 'marker' protein." *Journal of Histochemistry and Cytochemistry* 15:375–380.
Strike, Thomas A.
    (1962) "A device for adapting the rotary microtome to frozen sectioning." *Stain Technology* 37:187–189.
Structure and Biochemistry of Mitochondria Symposium 1953 [Proceedings]. *Journal of Histochemistry and Cytochemistry* 1:179–276.
Stumpf, Walter, and Roth, Loyd J.
    (1964) "Vacuum freeze drying of frozen sections for dry-mounting high-resolution autoradiography." *Stain Technology* 39:219–223.
    (1966) "Freeze-drying of small tissue samples and thin frozen sections below $-60°C$." *Journal of Histochemistry and Cytochemistry* 15:243–251.
Swank, R. L., and Davenport, H. A.
    (1934a) "Marchi's staining method. I. Studies of some of underlying mechanisms involved." *Stain Technology* 9:11–19.
    (1934b) "Marchi's staining method. II. Fixation." *Stain Technology* 9:129–135.
    (1935a) "Marchi's staining method. III. Artifacts and effects of perfusion." *Stain Technology* 10:45–52.
    (1935b) "Chlorate-osmic-formalin method for staining degenerating myelin." *Stain Technology* 10:87–90.
Swanson, David W., and McKee, Mary E.
    (1964) "Direct cover-glass culture and staining for chromosome analysis." *Stain Technology* 39:117–119.
Swartz, Frank J., and Nagy, Ernest R.
    (1963) "Feulgen stain stability in relation to three mounting media and exposure to light." *Stain Technology* 38:179–185.
Sweat, Faye; Meloan, Susan N.; and Puchtler, Holde.
    (1968) "A modified one-step trichrome stain for demonstration of fine connective tissue fibers." *Stain Technology* 43:227–231.
Sweat, Faye; Puchtler, Holde; and Rosenthal, Sanford I.
    (1964) "Sirius red F3BA as a stain for connective tissue." *Archives of Pathology* 78:69–72.
Sweat, Faye; Puchtler, Holde; and Sesta, J. J.
    (1968) "PAS-phosphomolybdic acid–Sirius supra blue FGL-CF." *Archives of Pathology* 86:33–39.
Sweat, Faye; Puchtler, Holde; and Woo, Pauline.
    (1964) "A light-fast modification of Lillie's allochrome stain." *Archives of Pathology,* 78:73–75.
Swigart, Richard H.; Wagner, Charles E.; and Atkinson, William B.
    (1960) "The preservation of glycogen in fixed tissues and tissue sections." *Journal of Histochemistry and Cytochemistry* 8:74–75.

Symeonidis, A.
  (1935) "Neue Anwendungsmöglichkeiten der Gefrierschneidemethode mit Messer tief-
  kühlung bei fixierten Gewebe." *Centralblatt für allgemeine Pathologie und
  Pathologische Anatomie* 63:245–246.

Taft, E. B.
  (1951) "The problem of standardized technique for the methyl-green-pyronin stain."
  Stain Technology 26:205–212.
Taft, P. D., and Arizaga-Cruz, J. M.
  (1960) "A comparison of the cell block, Papanicolaou, and millipore filter technics for
  the cytological examination of serous fluids." *Technical Bulletin of the Registry
  of Medical Technologists* 30:189–192.
Taft, P. D., and Lojananond, Preecha.
  (1962) "An evaluation of fluorescence microscopy in gynecologic exfoliative cytology."
  American Journal of Clinical Pathology 37:334–336.
Takaya, Kinichi.
  (1967) "Luxol fast blue MBS and phloxine; a stain for mitochondria." *Stain Technology*
  42: 207–211.
Takeuchi, Jun.
  (1962) "Staining sulfated mucopolysaccharides in sections by means of acriflavine."
  Stain Technology 37:105–107.
Tan, K. H.
  (1973) "Peracetic acid as an oxidizer to replace permanganate in the staining of
  neurosecretory cells." *Stain Technology* 48:140–141.
Tanaka, Ryuso.
  (1961) "Aceto-basic fuchsin as a stain for nucleoli and chromosomes of plants." *Stain
  Technology* 36:325–327.
Tandler, C. J.
  (1955) "The reaction of nucleoli with ammoniacal silver in darkness; additional data."
  Journal of Histochemistry and Cytochemistry 3:196–202.
  (1974) "A method for the selective removal of deoxyribonucleic acid from tissue
  sections." *Stain Technology* 49:147–152.
Tarkham, A. A.
  (1931) "The effect of fixatives and other reagents on cell size and tissue bulk." *Journal
  of the Royal Microscopical Society* 51:387–399.
Tauxe, W. Newlon; Moser, Arthur H.; and Boyd, George A.
  (1954) "Etymology of autoradiography." *Science* 120:149–150.
Taylor, A. I.
  (1963) "Sex chromatin in the newborn. *Lancet* 1:912–914.
Taylor, John D.
  (1965) "Gelatin embedding on the tissue carrier for thin serial sections in the cryostat."
  Stain Technology 40:29–31.
Taylor, William Ralph.
  (1967) "An enzyme method of clearing and staining small vertebrates." *Proceedings of
  U.S. National Museum* 122:1–17.
Tepper, H. B., and Gifford, E. M.
  (1962) "Detection of ribonucleic acid with pyronin." *Stain Technology* 37:52–53.
Terner, Jacob Y., and Clark, Georger.
  (1960a) "Gallocyanin–chrome alum: I. Technique and specificity." *Stain Technology*
  35:167–168.
  (1960b) "Gallocyanin–chrome alum: II. Histochemistry and specificity." *Stain
  Technology* 35:305–311.
Thieme, George.
  (1965) "Small tissue dryers with high capacity for rapid freeze-drying." *Journal of Histochemistry and Cytochemistry* 13:386–389.

Thomas, John T.
  (1953) "Phloxine–methylene blue staining of formalin fixed tissue." *Stain Technology* 28:311–312.
Thomas, Lloyd E.
  (1950) "An improved arginine histochemical method." *Stain Technology* 25:143–148.
Thomison, John B.
  (1961) "Combination of millipore filtration and fluorescence microscopy in cytologic examination." *American Journal of Clinical Pathology* 35:407–410.
Thompson, E. C.
  (1957) "A tray for staining frozen sections in quantity." *Stain Technology* 32:255–257.
  (1961) "Simultaneous staining of reticulocytes and Heinz bodies with new methylene blue N in dogs given iproniazid." *Stain Technology* 36:38–39.
Thompson, Samuel Wesley.
  (1966) *Selected Histochemical and Histological Methods.* Springfield, Ill.: Charles C Thomas.
Thurston, Jean M., and Joftes, David L.
  (1963) "Stains compatible with dipping radioautography." *Stain Technology* 38:231–234.
Tijo, J. H., and Levan, A.
  (1956) "The chromosome number of man." *Hereditas-Genetiskt Arkiv* 42:1–6.
Tijo, J. H., and Puck, T. T.
  (1958a) "The somatic chromosomes of man." *Proceedings of the National Academy of Sciences* 44:1229–1237.
  (1958b) "Genetics of somatic mammalian cells. II. Chromosomal constitution of cells in tissue culture." *Journal of Experimental Medicine* 108:259–269.
Tijo, J. H., and Wang, J.
  (1962) "Chromosome preparations of bone marrow cells without prior in vitro culture or in vivo colchicine administration." *Stain Technology* 37:17–20.
Tilden, I. L., and Tanak, Masso.
  (1945) "Fite's fuchsin-formaldehyde method for acid fast bacilli applied to frozen sections." *American Journal of Clinical Pathology* 9:95–97.
Tobie, John E.
  (1958) "Certain technical asects of fluorescence microscopy and the Coons fluorescent antibody technique." *Journal of Histochemistry and Cytochemistry* 6:271–277.
Tomlinson, W. J., and Grocott, R. G.
  (1944) "A simple method of staining malaria protozoa and other parasites in paraffin sections." *American Journal of Clinical Pathology* 14:316–326.
Tonna, Edgar A.
  (1958) "Factors which influence the latent image in autoradiography." *Stain Technology* 33:255–260.
  (1959) "Decalcification of bone by means of an automatic continuous flow apparatus." *American Journal of Clinical Pathology* 31:160–164.
Tonna, Edgar A., and Love, Robert A.
  (1961) "High pressure plastic lamination for flexible mounts of cover-slip preparations." *Stain Technology* 36:213–218.
Toren, David A.
  (1963) "A Giemsa-trichrome stain for mast cells in paraffin sections." *Stain Technology* 38:249–250.
Trott, J. R.
  (1961a) "The presence of glycogen in the rat liver following in vitro processing in decalcifying agents." *Journal of Histochemistry and Cytochemistry* 9:699–702.
  (1961b) "An evaluation of methods commonly used for the fixation and staining of glycogen." *Journal of Histochemistry and Cytochemistry* 9:703–710.
Trott, J. R.; Gorenstein, S. L.; and Peikoff, M. D.
  (1962) "A chemical and histochemical investigation of glycogen in rat liver and palate following treatment with various fixatives and ethylenediamine tetraacetic acid." *Journal of Histochemistry and Cytochemistry* 10:245–249.

Trump, Benjamin F.; Smuckler, Edward A.; and Benditt, Earl P.
  (1961) "A method for staining epoxy sections for light microscopy." *Journal of Ultrastructure Research* 5:343–348.
Tuan, Hsu-Chuan.
  (1930) "Picric acid as a destaining agent for iron alum hematoxylin." *Stain Technology* 5:135–138.
Tunturi, Archie R.
  (1973) "A method for impregnating myelinated axons in adult central nervous system." *Stain Technology* 48:297–304.
Turchini, Jean-Pascal, and Malet, Paul.
  (1965) "Long conservation of histoenzymatic activities of fresh tissues in glycerol." *Journal of Histochemistry and Cytochemistry* 13:405–406.

Uber, Fred M.
  (1936) "Mcirotome knife sharpeners operating on the abrasive ground glass principle." *Stain Technology* 11:93–98.
Ueckert, Edwin.
  (1960) "Polyester embedding for sectioning hard and soft tissues." *Stain Technology* 35:261–265.
Umiker, William, and Pickle, Larry.
  (1960) "The cytologic diagnosis of lung cancer by fluorescence microscopy: acridine orange fluorescence technic in routine screening and diagnosis." *Laboratory Investigation* 9:613–624.
Upadhya, M. D.
  (1963) "The use of a α-bromo-naphthalene, rapid hot fixation, and distributed pressure squashing for chromosomes of Triticinae." *Stain Technology* 38:293–295.
Uzman, L. Lahut.
  (1956) "Histochemical localization of copper with rubeanic acid." *Laboratory Investigation* 5:299–305.

Vacek, Z., and Plackova, A.
  (1959) "Silver impregnation of nerve fibers in teeth after decalcification with ethylenediaminetetraacetic acid." *Stain Technology* 34:1–3.
Van Cleave, Harley J., and Ross, Jean A.
  (1947) "A method of reclaiming dried zoological specimens." *Science* 105:318.
Van Duijn, P.
  (1961) "Acrolein-Schiff, a new staining method for proteins." *Journal of Histochemistry and Cytochemistry* 9:234–241.
Vasquez, J. J., and Dixon, F. J.
  (1956) "Immunohistochemical analysis of amyloid by the fluorescent technique." *Journal of Experimental Medicine* 104:727.
Vassar, Philip S., and Culling, Charles F. A.
  (1959) "Fluorescent stains with special reference to amyloid and connective tissue." *Archives of Pathology* 68:487–498.
  (1962) "Fluorescent amyloid staining of casts in myeloma nephrosis." *Archives of Pathology* 73:59–63.
Vernino, David M., and Laskin, David M.
  (1960) "Sex chromatin in memmalian bone." *Science* 132:675–676.
Verrling, John M., and Thompson, Donald E.
  (1972) "A polychrome stain for use in Parasitology." *Stain Technology* 47:164–165.
Vickers, A. E. J.
  (1956) *Modern Methods of Microscopy.* New York: nterscience Publishers.
Vidal, O. R.; Aya, T.; and Sandberg, A. A.
  (1971) "Glutaraldehyde–ammoniacal silver carbonate-formaldehyde staining of histones of mitotic chromosomes." *Stain Technology* 46:89–92.

Villyaneuva, A. R.; Hattner, R. S.; and Frost, H. M.
   (1964) "A tetrachrome stain for fresh, mineralized bone sections useful in the diagnosis of bone diseases." *Stain Technology* 39:87–94.
Veincent, W. S.
   (1955) "Structure and chemistry of nucleoli." *International Review of Cytology,* IV. Ed. by G. H. Bourne and J. F. Danielli. New York: Academic Press.
Vinegar, Ralph.
   (1961) "Metachromatic differential fluorochroming of living and dead ascites tumor cells with acridine orange." *Cancer Resarch* 16:900–906.
Vlachos, John.
   (1959) "Desaturation of staining solutions for lipids, a means of avoiding precipitation on stained sections. *Stain Technology* 34:292.
Von Bertalanffy, Ludwig, and Bickis, Ivar.
   (1956) "Identification of cytoplasmic basophilia (ribonucleic acid) by fluorescence microscopy." *Journal of Histochemistry and Cytochemistry* 4:481–493.
Von Bertalanffy, Ludwig; Masin, Marianna; and Masin, Francis.
   (1958) "A new and rapid method for diagnosis of vaginal and cervical cancer by fluorescence microscopy." *Cancer* 11:873–887.
Von Bertalanffy, Ludwig, and Von Bertalanffy, Felix D.
   (1960) "A new method for cytological diagnosis of pulmonary cancer." *Annals of the New York Academy of Sciences* 84:225–238.
Vyas, A. B.
   (1972) "Taking the sting out of preserving scorpions." *Turtox News* 49:25.

Wachstein, Max, and Meisel, Elizabeth.
   (1964) "Demonstration of peroxidase activity in tissue sections." *Journal of Histochemistry and Cytochemistry* 12:538–544.
Wade, W. H.
   (1952) "Notes on the carbowax method of making tissue sections." *Stain Technology* 27:71–79.
   (1957) "A modification of the Fite formaldehyde (Fite I) method for staining acid fast bacilli in paraffin sections." *Stain Technology* 32:287–292.
Wagner, Bernard M., and Shapiro, Sylvia H.
   (1957) "Application of Alcian blue as a histochemical method" *Laboratory Investigation* 6:472–477.
Waldrop, Faye Sweat, and Puchtler, Holde.
   (1975) "Luxol fast blue MBSN–Levafix red violet E-2BL." *Archives of Pathology* 99:529–532.
Walker, P. M. B.
   (1958) "Ultraviolet microspectrophotometry." *General Cytochemical Methods,* I. Ed. by J. F. Danielli. New York: Academic Press.
Wall, Patrick.
   (1950) "Staining and recognition of fine degenerating nerve fibers." *Stain Technology* 25:125–126.
Walls, G. L.
   (1932) "The hot celloidin technic for animal tissues." *Stain Technology* 7:135–145.
   (1936) "A rapid celloidin method for the rotary microtome." *Stain Technology* 11:89–92.
   (1938) "The microtechnique of the eye with suggestions as to material." *Stain Technology* 13:69–72.
Walsh, Richard J., and Love, Robert.
   (1963) "Studies of the cytochemistry of nucleoproteins. II. Improved staining methods with toluidine blue and ammonium molybdate." *Journal of Histochemistry and Cytochemistry* 11:188–196.
Warmke, Harry E.
   (1935) "A permanent root tip smear method." *Stain Technology* 10:101–103.
   (1946) "Precooling combined with chromosomo-acetic fixation in studies of somatic chromosomes in plants." *Stain Technology* 21:87–90.

Warren, O.
   (1944) "Some facts about nerve ending preparations." *Turtox News* 22.
Wartman, W. B.
   (1943) "Notes on the Field's method of staining parasites in thick blood films." *Army Medical Bulletin* 68:173–177.
Wassersug, Richard J.
   (1976) "A procedure for differential staining of cartilage and bone in whole formalin-fixed vertebrates." *Stain Technology* 51:131–134.
Watson, Barbara K.
   (1952) "Distribution of mumps virus in tissue cultures as determined by fluorescein-labeled antiserum." *Proceedings of the Society of Experimental Biology and Medicine* 79:222–224.
Watson, M. L.
   (1956) "Carbon films and specimen stability." *Journal of Biophysical and Biochemical Cytology, Supplement* 2:31–37.
Weaver, Harry Lloyd.
   (1955) "An improved gelatine adhesive for paraffin sections." *Stain Technology* 30:63–64.
Weiner, S.
   (1957) "A simplified vacuum embedder." *Stain Technology* 32:195–196.
Weiss, Jules.
   (1954) "The nature of the reaction between orcein and elastin." *Journal of Histochemistry and Cytochemistry* 2:21–28.
Weissmann, Gerald.
   (1964) "Lysosomes." *Blood* 24:594–606.
   (1968) "The many-faceted lysosome." *Hospital Practice* 31–39.
Welshons, W. J.; Gibson, B. H.; and Scandyln, B. J.
   (1962) "Slide processing for the examination of male mammalian meiotic chromosomes." *Stain Technology* 37:1–5.
Werth, Von G.
   (1953) "Fluoreszenz mikroskopische Beobachtungen an menschlichem Knochenmark." *Acta Haematologica* 10:209–222.
West, William T.
   (1962) "The Pearse-Sless cryostat: improvement in the antiroll plate." *Stain Technology* 37:320.
Westfall, Jane A.
   (1961) "Obtaining flat serial sections for electron microscopy." *Stain Technology* 36:36–38.
Wheatley, Walter B.
   (1951) "A rapid procedure for intestinal amoebae and flagellates." *American Journal of Clinical Pathology* 21:990–991.
White, Lowell E., Jr.
   (1960) "Enhanced reliability in silver impregnation of terminal axonal degeneration—original Nauta method." *Stain Technology* 35:5–9.
White, Robert G.; Coons, Albert H.; and Connolly, Jeanne M.
   (1955) "Studies on antibody production. II. The alum granuloma. IV. The role of a wax fraction of myobacterium tuberculosis in adjuvant emulsions on the production of antibody to egg albumin." *Journal of Experimental Medicine* 102:73–103.
Wieland, H., and Scheuing, G.
   (1921) "Die Farbreaktion mit Aldehyden." *Berichte der Deutschen Chemischen Gesellschaft* 54:2527–2555.
Wigglesworth, V. B.
   (1959) "A simple method for cutting sections in the 0.5 to 1μ range, and for sections of chitin." *Quarterly Journal of Microscopical Science* 100:315–320.
Wijffels, C. C. B. M.
   (1971) "Freeze-drying by cryosorption; a simple device having a tissue capacity of 500 milligrams." *Stain Technology* 46:33–34.

Wilcox, Aimee.
    (1943) *Manual for the Microscopical Diagnosis of Malaria in Man*. National Institute of
        Public Health, Bulletin 180. Washington, D.C.: U.S. Government Printing Office.
Wilder, H. C.
    (1935) "An improved technique for silver impregnation of reticular fibers." *American
        Journal of Pathology* 11:817–819.
Wilhelm, Walter E., and Smott, Al.
    (1966) "Permanent demonstration of food vacuoles in *Paramecium* stained in relief."
        Turtox News 44:158.
Williams, Dorothy L.; Lafferty, Diana A.; and Webb, Sondra L.
    (1970) "An air drying method for the preparation of dictyotene chromosomes from
        ovaries of Chinese hamsters." *Stain Technology* 45:133–135.
Williams, G., and Jackson, D. S.
    (1956) "Two organic fixatives for acid mucopolysaccharides." *Stain Technology*
        31:189–191.
Williams, R. C., and Backus, R. C.
    (1949) "The electron micrographic structure of shadow-cast films." *Journal of Applied
        Physics* 20:98–106.
Williams, R. C., and Kallman, F.
    (1954) "Examination of tissue-cultured Hela cells by electron microscopy of serial
        sections." *Journal of Applied Physics* 25:1455.
Williams, R. C., and Wyckoff, R. W. G.
    (1944) "Thickness of electron microscopic objects." *Journal of Applied Physics*
        15:712–715.
Williams, T. D.
    (1957) "Mounting and preserving serial celloidin sections." *Stain Technology* 32:97.
Williams, Terry.
    (1962) "The staining of nervous elements by the Bodian method. I. The influence of
        factors preceding impregnation." *Quarterly Journal of Microscopical Science*
        103:155–161.
Winteringham, F. P. W.; Harrison A.; and Hammond, J. H.
    (1950) "Autoradiography of water-soluble tracers in histological sections." *Nature*
        165:149–152.
Witten, Victor H., and Holmstrom, Vera.
    (1953) "New histologic technics for autoradiography." *Laboratory Investigation*
        2:368–375.
Wittman, W.
    (1962) "Aceto-iron-hematoxylin for staining chromosomes in squashes of plant mate-
        rial." *Stain Technology* 37:27.
    (1963) "Permanent-type mounting of aceto-iron-hematoxylin squashes in corn syrup."
        Stain Technology 40:161–164.
Wolberg, William H.
    (1965) "Darrow red–light green as a stain for autoradiographs." *Stain Technology* 40:
        90.
Wolfe, Hubert J.
    (1964) "Techniques for the histochemical localization of extremely water soluble acid
        mucopolysaccharides." *Journal of Histochemistry and Cytochemistry* 12:217–
        218.
Wolman, M.
    (1955) "Problems of fixation in cytology, histology, and histochemistry." *International
        Review of Cytology*, IV. Ed. by G. H. Bourne and J. F. Danielli. New York:
        Academic Press.
    (1956) "A histochemical method for the differential staining of acidic tissue components
        (particularly ground-substance polysaccharides)." *Bulletin of the Research
        Council of Israel, Section E, Experimental Medicine* 6E:27–35.
    (1961) "Differential staining of acidic tissue components by the improved bi-col
        method." *Stain Technology* 36:21–25.

Woodruff, Lois A., and Norris, William P.
(1955) "Sectioning of undecalcified bone; with special reference to radioautographic applications." *Stain Technology* 30:179–188.
Woods, Philip S., and Pollister, A. W.
(1955) "An ice-solvent method of drying frozen tissue for plant cytology." *Stain Technology* 30:123–131.
Wright, J. H.
(1902) "A rapid method for the differential staining of blood films and malarial parasites." *Journal of Medical Research* 7:138–144.
Wróblewska, Joanna.
(1969) "Chromosome preparations from mouse embryos during early organogenesis: dissociation after fixation, followed by air drying." *Stain Technology* 44:147–150.

Yaeger, James A.
(1958) "Methacrylate embedding and sectioning of calcified bone." *Stain Technology* 33:220–239.
Yamada, K.
(1963) "Staining of sulfated polysaccharides by means of Alcian blue." *Nature* 197:789.
Yamaguchi, Ben T., Jr., and Braunstein, Herbert.
(1965) "Flourescent stain for tubercle bacilli in histologic sections. II. Diagnostic efficiency in granulomatous lesions of the liver." *Technical Bulletin of the Registry of Medical Technologists* 35:184–187.
Yang, J., and Scholten, T. H.
(1976) "Celestine blue B stain for intestinal protozoa." *American Journal of Clinical Pathology* 65:715–718.
Yasuma, Akiyasu, and Ichikawa, Toyoki.
(1953) "Ninhydrin-Schiff and Alloxan-Schiff staining. A new histochemical staining method for protein." *Journal of Laboratory and Clinical Medicine* 41:296–299.
Yenson, Jack.
(1968) "Removal of epoxy resin from histological sections following halogenation." *Stain Technology* 43:344–346.
Yerganian, G.
(1957) "Cytologic maps of some isolated human pachytene chromosomes." *American Journal of Human Genetics* 9:42–54.
(1963) "Cytogenetic analysis." *Methodology in Mammalian Genetics.* Ed. by Walter J. Burdette. San Francisco: Holden-Day.
Yetwin, I. J.
(1944) "A simple permanent mounting medium for *Necator americanus.*" *Journal of Parasitology* 30:201.
Young, Irving.
(1962) "The use of a sponge base as an aid in preparing frozen sections of small objects." *Technical Bulletin of the Registry of Medical Technologists* 32:76.
Youngpeter, John M.
(1964) "Spalteholz preparations." *Ward's Bulletin* 3:7.
(1967) "Spalteholz preparations." *Ward's Bulletin* 6:7.

Zambernard, Joseph; Block, Matthew; Vatter, Albert; and Trenner, Lew.
(1969) "An adaptation of methacrylate embedding for routine histologic use." *Blood* 33:444–450.
Zbar, Marcus J., and Winter, William J.
(1959) "A method of concentrating particles of bone marrow for paraffin sections." *American Journal of Clinical Pathology* 32:41–44.
Zeiger, K.; Harders, H.; and Müller, W.
(1951) "Der Strugger-effekt an der Nervenzelle." *Protoplasma* 40:76–84.
Zinn, Donald J., and Morin, Lorraine P.
(1962) "The use of commercial citric juices in gold chloride staining of nerve endings." *Stain Technology* 37:380–381.

Zirkle, Conway.
  (1937) "Aceto-carmine mounting media." *Science* 85:528.
  (1940) "Combined fixing, staining, and mounting media." *Stain Technology* 15:139–153.
Zlotnik, I.
  (1960) "The initial cooling of tissues in the freezing-drying technique." *Quarterly Journal of Microscopical Science* 101:251–254.
Zuck, Robert K.
  (1947) "Simplified permanent aceto-carmine smears with a water-miscible mountant." *Stain Technology* 22:109–110.
Zugibe, Frederick T.
  (1970) *Diagnostic Histochemistry*. St. Louis: C. V. Mosby.
Zugibe, Frederick T.; Brown, Kenneth D.; and Last, Jules H.
  (1959) "A new technique for the simultaneous demonstration of lipid and acid mucopolysaccharides on the same tissue section." *Journal of Histochemistry and Cytochemistry* 7:101–106.
Zugibe, Frederick T.; Fink, Marilyn L.; and Brown, Kenneth W.
  (1958) "Carobowax flotation method." *Journal of Histochemistry and Cytochemistry* 6:381.
  (1959) "Carbowax 400, a new solvent for oil red O and sudan IV for staining carbowax embedded and frozen sections." *Stain Technology* 34:33–37.
Zugibe, Frederick T.; Kopaczyk, Krystyna C.; Cape, William E.; and Last, Jules H.
  (1958) "A new carbowax method for routinely performing lipid, hematoxylin, and eosin and elastic staining techniques on adjacent freeze dried or formalin fixed tissues." *Journal of Histochemistry and Cytochemistry* 6:133–138.
Zwaan, J., and Van Dam, A. F.
  (1961) "Rapid separation of fluorescent antisera and unconjugated dye." *Acta Histochemica* 11:306–308.

Index